Safety Technology

Trainee Guide

Boston   Columbus   Indianapolis   New York   San Francisco   Amsterdam
Cape Town   Dubai   London   Madrid   Milan   Munich   Paris   Montreal   Toronto   Delhi
Mexico City   Sao Paulo   Sydney   Hong Kong   Seoul   Singapore   Taipei   Tokyo

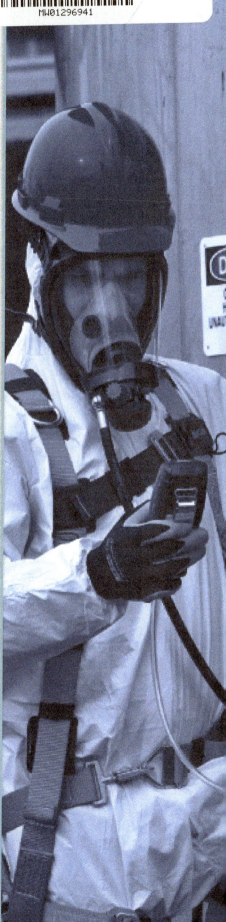

NCCER

President: Don Whyte
Vice President: Steve Greene
Chief Operations Officer: Katrina Kersch
HVAC Project Manager: Chris Wilson
Senior Development Manager: Mark Thomas
Senior Production Manager: Tim Davis

Quality Assurance Coordinator: Karyn Payne
Desktop Publishing Coordinator: James McKay
Permissions Specialists: Kelly Sadler
Production Specialist: Kelly Sadler
Production Assistance: Hannah Payne
Editors: Graham Hack, Debie Hicks

Writing and development services provided by Topaz Publications, Liverpool, NY

Lead Writer/Project Manager: Veronica Westfall
Desktop Publisher: Joanne Hart
Art Director: Alison Richmond

Permissions Editor: Andrea LaBarge
Writers: Veronica Westfall

Pearson

Director of Alliance/Partnership Management: Andrew Taylor
Editorial Assistant: Collin Lamothe
Program Manager: Alexandrina B. Wolf
Assistant Content Producer: Alma Dabral
Digital Content Producer: Jose Carchi
Director of Marketing: Leigh Ann Simms

Senior Marketing Manager: Brian Hoehl
Composition: NCCER
Printer/Binder: LSC Communications
Cover Printer: LSC Communications
Text Fonts: Palatino and Univers

Credits and acknowledgments for content borrowed from other sources and reproduced, with permission, in this textbook appear at the end of each module.

Copyright © 2018, 2003 by NCCER, Alachua, FL 32615, and published by Pearson, New York, NY 10013. All rights reserved. Printed in the United States of America. This publication is protected by Copyright and permission should be obtained from NCCER prior to any prohibited reproduction, storage in a retrieval system, or transmission in any form or by any means, electronic, mechanical, photocopying, recording, or likewise. For information regarding permission(s), write to: NCCER Product Development, 13614 Progress Blvd., Alachua, FL 32615.

ISBN-13: 978-0-13-444636-3
ISBN-10: 0-13-444636-4

25 2023

Preface

To the Trainee

The safety technician has one of the most important roles in construction, and it is a career that is in high demand. Construction safety professionals are responsible for the safety of all employees on a job site. Safety technicians must have strong communication skills and the ability to analyze necessary measures to prevent work-related incidents.

Safety technicians implement company-wide safety programs by analyzing data to create procedures to prevent incidents. Safety technicians plan for task- and site-specific hazards. They arrange safety meetings and training and track all safety-related work permits. They must abide by OSHA standards, respond to OSHA site inspections, and follow required recordkeeping procedures. When incidents occur, safety technicians must assist in investigations, file reports, and analyze the incident to correct any potential hazards that may have caused an incident.

The NCCER Construction Site Safety training program boasts a range of training credentials for workers in both the field and in safety management. You can earn the Construction Site Safety Orientation (CSSO) credential by completing NCCER Module 00101, *Basic Safety*. By completing Field Safety and some additional modules from Safety Technology, you can earn a Construction Site Safety Supervisor (CSSS) credential. In addition, by completing the entirety of Field Safety and Safety Technology, you can earn a Construction Site Safety Technician (CSST) training credential. For details on which modules are needed to earn these credentials, visit the NCCER website at **www.nccer.org**.

In addition to NCCER's safety credentials, the NCCER Safety curriculum is a recognized exam study source for the Board of Certified Safety Professionals (BCSP) Construction Health and Safety Technician (CHST), Safety Trained Supervisor (STS), and Safety Trained Supervisor Construction (STSC) certifications.

We wish you the best as you begin an exciting and promising career.

New with *Safety Technology*

NCCER is proud to release *Safety Technology* in our improved instructional systems design, in which the sections of each module are directly tied to learning objectives. There are additional study questions and updated graphics and photos.

Many important updates have been made to the content. OSHA's Respirable Crystalline Silica Standard for Construction (29 *CFR* 1926.1153) is covered in "Hazard Recognition, Environmental Awareness, and Occupational Health." That module also has new sections on wildlife and land protection, noise and hearing protection, and ergonomics. The module "Positive Safety Communication" has methods to achieve behavior modification on the job site to achieve better safety policy adherence. "Site-Specific Safety Plans" offers new coverage of emergencies, including threats of violence or terrorism, and plans for multi-employer job sites. The module "OSHA Inspections and Recordkeeping" walks through what to expect during and after a job-site safety inspection, including OSHA's latest requirements.

We invite you to visit the NCCER website at **www.nccer.org** for information on the latest product releases and training, as well as online versions of the *Cornerstone* magazine and Pearson's NCCER product catalog.

Your feedback is welcome. You may email your comments to **curriculum@nccer.org** or send general comments and inquiries to **info@nccer.org**.

NCCER Standardized Curricula

NCCER is a not-for-profit 501(c)(3) education foundation established in 1996 by the world's largest and most progressive construction companies and national construction associations. It was founded to address the severe workforce shortage facing the industry and to develop a standardized training process and curricula. Today, NCCER is supported by hundreds of leading construction and maintenance companies, manufacturers, and national associations. The NCCER Standardized Curricula was developed by NCCER in partnership with Pearson, the world's largest educational publisher.

Some features of the NCCER Standardized Curricula are as follows:

- An industry-proven record of success
- Curricula developed by the industry, for the industry
- National standardization providing portability of learned job skills and educational credits
- Compliance with the Office of Apprenticeship requirements for related classroom training (*CFR 29:29*)
- Well-illustrated, up-to-date, and practical information

NCCER also maintains the NCCER Registry, which provides transcripts, certificates, and wallet cards to individuals who have successfully completed a level of training within a craft in NCCER's Curricula. *Training programs must be delivered by an NCCER Accredited Training Sponsor in order to receive these credentials.*

Special Features

In an effort to provide a comprehensive and user-friendly training resource, this curriculum showcases several informative features. Whether you are a visual or hands-on learner, these features are intended to enhance your knowledge of the construction industry as you progress in your training. Some of the features you may find in the curriculum are explained below.

Introduction

This introductory page, found at the beginning of each module, lists the module Objectives, Performance Tasks, and Trade Terms. The Objectives list the knowledge you will acquire after successfully completing the module. The Performance Tasks give you an opportunity to apply your knowledge to real-world tasks. The Trade Terms are industry-specific vocabulary that you will learn as you study this module.

Figures and Tables

Photographs, drawings, diagrams, and tables are used throughout each module to illustrate important concepts and provide clarity for complex instructions. Text references to figures and tables are emphasized with *italic* type.

Notes, Cautions, and Warnings

Safety features are set off from the main text in highlighted boxes and categorized according to the potential danger involved. Notes simply provide additional information. Cautions flag a hazardous issue that could cause damage to materials or equipment. Warnings stress a potentially dangerous situation that could result in injury or death to workers.

Trade Features

Trade features present technical tips and professional practices based on real-life scenarios similar to those you might encounter on the job site.

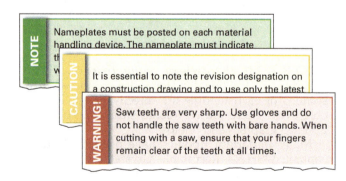

Bowline Trivia

Some people use this saying to help them remember how to tie a bowline: "The rabbit comes out of his hole, around a tree, and back into the hole."

Case History

Case History features emphasize the importance of safety by citing examples of the costly (and often devastating) consequences of ignoring best practices or OSHA regulations.

Going Green

Going Green features present steps being taken within the construction industry to protect the environment and save energy, emphasizing choices that can be made on the job to preserve the health of the planet.

Did You Know

Did You Know features introduce historical tidbits or interesting and sometimes surprising facts about the trade.

Step-by-Step Instructions

Step-by-step instructions are used throughout to guide you through technical procedures and tasks from start to finish. These steps show you how to perform a task safely and efficiently.

Trade Terms

Each module presents a list of Trade Terms that are discussed within the text and defined in the Glossary at the end of the module. These terms are presented in the text with **bold, blue** type upon their first occurrence. To make searches for key information easier, a comprehensive Glossary of Trade Terms from all modules is located at the back of this book.

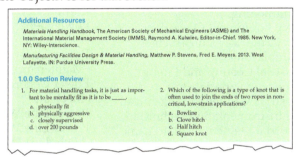

Section Review

Each section of the module wraps up with a list of Additional Resources for further study and Section Review questions designed to test your knowledge of the Objectives for that section.

Review Questions

The end-of-module Review Questions can be used to measure and reinforce your knowledge of the module's content.

iii

NCCER Standardized Curricula

NCCER's training programs comprise more than 80 construction, maintenance, pipeline, and utility areas and include skills assessments, safety training, and management education.

Boilermaking
Cabinetmaking
Carpentry
Concrete Finishing
Construction Craft Laborer
Construction Technology
Core Curriculum: Introductory Craft Skills
Drywall
Electrical
Electronic Systems Technician
Heating, Ventilating, and Air Conditioning
Heavy Equipment Operations
Heavy Highway Construction
Hydroblasting
Industrial Coating and Lining Application Specialist
Industrial Maintenance Electrical and Instrumentation Technician
Industrial Maintenance Mechanic
Instrumentation
Ironworking
Manufactured Construction Technology
Masonry
Mechanical Insulating
Millwright
Mobile Crane Operations
Painting
Painting, Industrial
Pipefitting
Pipelayer
Plumbing
Reinforcing Ironwork
Rigging
Scaffolding
Sheet Metal
Signal Person
Site Layout
Sprinkler Fitting
Tower Crane Operator
Welding

Maritime

Maritime Industry Fundamentals
Maritime Pipefitting
Maritime Structural Fitter

Green/Sustainable Construction

Building Auditor
Fundamentals of Weatherization
Introduction to Weatherization
Sustainable Construction Supervisor
Weatherization Crew Chief
Weatherization Technician
Your Role in the Green Environment

Energy

Alternative Energy
Introduction to the Power Industry
Introduction to Solar Photovoltaics
Power Generation Maintenance Electrician
Power Generation I&C Maintenance Technician
Power Generation Maintenance Mechanic
Power Line Worker
Power Line Worker: Distribution
Power Line Worker: Substation
Power Line Worker: Transmission
Solar Photovoltaic Systems Installer
Wind Energy
Wind Turbine Maintenance Technician

Pipeline

Abnormal Operating Conditions, Control Center
Abnormal Operating Conditions, Field and Gas
Corrosion Control
Electrical and Instrumentation
Field and Control Center Operations
Introduction to the Pipeline Industry
Maintenance
Mechanical

Safety

Field Safety
Safety Orientation
Safety Technology

Supplemental Titles

Applied Construction Math
Tools for Success

Management

Construction Workforce Development Professional
Fundamentals of Crew Leadership
Mentoring for Craft Professionals
Project Management
Project Supervision

Spanish Titles

Acabado de concreto: nivel uno (*Concrete Finishing Level One*)
Aislamiento: nivel uno (*Insulating Level One*)
Albañilería: nivel uno (*Masonry Level One*)
Andamios (*Scaffolding*)
Carpintería: Formas para carpintería, nivel tres (*Carpentry: Carpentry Forms, Level Three*)
Currículo básico: habilidades introductorias del oficio (*Core Curriculum: Introductory Craft Skills*)
Electricidad: nivel uno (*Electrical Level One*)
Herrería: nivel uno (*Ironworking Level One*)
Herrería de refuerzo: nivel uno (*Reinforcing Ironwork Level One*)
Instalación de rociadores: nivel uno (*Sprinkler Fitting Level One*)
Instalación de tuberías: nivel uno (*Pipefitting Level One*)
Instrumentación: nivel uno, nivel dos, nivel tres, nivel cuatro (*Instrumentation Levels One through Four*)
Orientación de seguridad (*Safety Orientation*)
Paneles de yeso: nivel uno (*Drywall Level One*)
Seguridad de campo (*Field Safety*)

Acknowledgments

This curriculum was revised as a result of the farsightedness and leadership of the following sponsors:

ABC National
Alaska Training Center
Carolina Bridge Company
The Haskell Company
LPR Construction Company
MasTec
Northern Industrial Training

Safety Advantage, LLC
Safety Council of Texas City
Southland Safety, LLC
STARCON International, Inc.
TIC – The Industrial Company
Tri-City Electrical Contractors

This curriculum would not exist were it not for the dedication and unselfish energy of those volunteers who served on the Authoring Team. A sincere thanks is extended to the following:

Jo Ballagh
David Burman
Paul Fontenot II
Dell Husted
Earl Hyatt
Dan Nickel
Mike Powers

Jarrett Quoyle
Brett Richardson
Lance Simons
Ronald Sokol
Cameron Strother
Ed Valencia
Chris Williams

NCCER Partners

American Council for Construction Education
American Fire Sprinkler Association
Associated Builders and Contractors, Inc.
Associated General Contractors of America
Association for Career and Technical Education
Association for Skilled and Technical Sciences
Construction Industry Institute
Construction Users Roundtable
Design Build Institute of America
GSSC – Gulf States Shipbuilders Consortium
ISN
Manufacturing Institute
Mason Contractors Association of America
Merit Contractors Association of Canada
NACE International
National Association of Women in Construction
National Insulation Association
National Technical Honor Society
National Utility Contractors Association
NAWIC Education Foundation
North American Crane Bureau
North American Technician Excellence
Pearson

Prov
SkillsUSA®
Steel Erectors Association of America
U.S. Army Corps of Engineers
University of Florida, M. E. Rinker Sr., School of Construction Management
Women Construction Owners & Executives, USA

Contents

Module One
Introduction to Safety Technology
Describes the responsibilities of a safety technician and identifies the basic components of a safety program. It also provides an overview of regulatory requirements. (Module ID 75201; 2.5 Hours)

Module Two
Positive Safety Communication
Explains how to support an effective safety culture on the job site, including communication techniques, motivation, and responding to behavioral issues. (Module ID 75205; 2.5 Hours)

Module Three
Hazard Recognition, Environmental Awareness, and Occupational Health
Covers environmental and safety hazards. It explains how to evaluate risks and identify appropriate methods of hazard control. It also discusses environmental regulations for hazardous materials and describes the elements of a medical surveillance program. (Module ID 75219; 5 Hours)

Module Four
Job Safety Analysis and Pre-Task Planning
Covers provides guidance on safety performance analysis and employee coaching. It also explains how to complete job and task safety planning. (Module ID 75220; 5 Hours)

Module Five
Safety Data Tracking and Trending
Covers how to conduct safety inspections, audits, and employee safety observations. It discusses both traditional and predictive methods of performance measurement, and explains how to analyze safety data in order to prevent future incidents. (Module ID 75221; 5 Hours)

Module Six
Site-Specific Safety Plans
Explains how to use pre-bid checklists to identify hazards and develop a site safety plan. It also describes how to develop an emergency action plan. (Module ID 75222; 5 Hours)

Module Seven
Safety Orientation and Safety Meetings
Covers describes how to prepare and deliver effective training using both formal safety meetings and tailgate talks. (Module ID 75223; 5 Hours)

Module Eight
Permits and Policies
Provides an overview of the various work permits required on a construction site. It also provides detailed procedures for completing a hot work permit, lockout/tagout, and confined-space entry permit. (Module ID 75224; 5 Hours)

Module Nine
Incident Investigations, Policies, and Analysis

Describes how to conduct an incident investigation, including employee interviews and reporting requirements. It also explains how to analyze an incident to determine the root cause and prevent future incidents. (Module ID 75225; 5 Hours)

Module Ten
OSHA Inspections and Recordkeeping

Discusses the OSHA requirements for recordkeeping and explains how to manage the safety and health records for a job site. It also covers the two main types of OSHA inspections. (Module ID 75226; 5 Hours)

Glossary

Index

SAFETY TECHNOLOGY

- Module Nine — OSHA Inspections and Recordkeeping (75226)
- Module Nine — Incident Investigations, Policies, and Analysis (75225)
- Module Eight — Permits and Policies (75224)
- Module Seven — Safety Orientation and Safety Meetings (75223)
- Module Six — Site-Specific Safety Plans (75222)
- Module Five — Safety Data Tracking and Trending (75221)
- Module Four — Job Safety Analysis and Pre-Task Planning (75220)
- Module Three — Hazard Recognition, Environmental Awareness, and Occupational Health (75219)
- Module Two — Positive Safety Communication (75205)
- Module One — Introduction to Safety Technology (75201)

This course map shows all of the modules in *Safety Technology*. The suggested training order begins at the bottom and proceeds up. Skill levels increase as you advance on the course map. The local Training Program Sponsor may adjust the training order.

Introduction to Safety Technology

Overview

This module describes the responsibilities of a safety technician and identifies the basic components of a safety program. It also provides an overview of regulatory requirements.

Module 75201

Trainees with successful module completions may be eligible for credentialing through NCCER's Registry. To learn more, go to **www.nccer.org** or contact us at 1.888.622.3720. Our website has information on the latest product releases and training, as well as online versions of our *Cornerstone* magazine and Pearson's product catalog.

Your feedback is welcome. You may email your comments to **curriculum@nccer.org**, send general comments and inquiries to **info@nccer.org**, or fill in the User Update form at the back of this module.

This information is general in nature and intended for training purposes only. Actual performance of activities described in this manual requires compliance with all applicable operating, service, maintenance, and safety procedures under the direction of qualified personnel. References in this manual to patented or proprietary devices do not constitute a recommendation of their use.

Copyright © 2018 by NCCER, Alachua, FL 32615, and published by Pearson, New York, NY 10013. All rights reserved. Printed in the United States of America. This publication is protected by Copyright, and permission should be obtained from NCCER prior to any prohibited reproduction, storage in a retrieval system, or transmission in any form or by any means, electronic, mechanical, photocopying, recording, or likewise. To obtain permission(s) to use material from this work, please submit a written request to NCCER Product Development, 13614 Progress Blvd., Alachua, FL 32615.

75201 V2

From *Safety Technology,* Trainee Guide. NCCER.
Copyright © 2018 by NCCER. Published by Pearson. All rights reserved.

75201
INTRODUCTION TO SAFETY TECHNOLOGY

Objectives

When you have completed this module, you will be able to do the following:

1. Identify the roles and responsibilities of a safety technician.
    a. Describe the responsibilities of a safety technician.
    b. Define important safety terms.
2. Identify incident causes and costs.
    a. Identify the causes of incidents.
    b. Identify the costs of incidents.
3. Identify the basic components of a safety program.
    a. Identify essential safety program policies and procedures.
    b. Identify effective means and practices for providing safety orientation and training.
4. Identify regulatory requirements with which construction industry safety technicians need to be familiar.
    a. Identify Occupational Safety and Health Administration (OSHA) requirements.
    b. Identify Environmental Protection Agency (EPA) requirements.
    c. Identify Department of Transportation (DOT) requirements.
    d. Identify other relevant health and safety requirements.

Performance Tasks

This is a knowledge-based module; there are no Performance Tasks.

Trade Terms

Consensus standards
Construction User's Round Table (CURT)
Culm
Direct labor costs
Experience modification rate (EMR)
Gross income
Gross vehicle weight rating (GVWR)
Gross unloaded vehicle weight (GUVW)
Job safety analysis (JSA)
Net income
Profit
Release of energy
Surface mines
Unwarrantable failures

Industry Recognized Credentials

If you are training through an NCCER-accredited sponsor, you may be eligible for credentials from NCCER's Registry. The ID number for this module is 75201. Note that this module may have been used in other NCCER curricula and may apply to other level completions. Contact NCCER's Registry at 888.622.3720 or go to www.nccer.org for more information.

Contents

1.0.0 Roles and Responsibilities of a Safety Technician 1
  1.1.0 Safety Technician Responsibilities 1
  1.2.0 Important Safety Terms 2
2.0.0 Incident Causes and Costs 4
  2.1.0 Causes of Incidents 4
    2.1.1 Level I (Direct Causes) 4
    2.1.2 Level II (Indirect Causes) 4
    2.1.3 Level III (Root Causes) 5
  2.2.0 The Costs of Incidents 6
    2.2.1 Calculating Incident Costs 6
    2.2.2 Incident Experience vs. Future Insurance Costs 7
    2.2.3 Cost of Administering an Effective Safety Program 8
3.0.0 Basic Components of a Safety Program 9
  3.1.0 Policies and Procedures 9
    3.1.1 Policy Statement 9
    3.1.2 Alcohol and Drug Abuse Policies and Programs 9
    3.1.3 Safety Policies and Procedures 10
  3.2.0 Safety Orientation and Training 10
    3.2.1 Safety Meetings and Employee Involvement 11
    3.2.2 Emergency Action Plans and Methods for Dealing with the Media 11
    3.2.3 Inspections, Employee Observations, and Audits 11
    3.2.4 Incident Investigation and Analysis 12
    3.2.5 Recordkeeping 12
    3.2.6 Program Evaluation and Follow-Up 12
    3.2.7 Nine Industry Best Practices 13
4.0.0 Section FourRegulatory Requirements 14
  4.1.0 OSHA Requirements 14
  4.2.0 EPA Requirements 15
  4.3.0 DOT Requirements 16
    4.3.1 Commercial Motor Vehicle Safety Act of 1986 16
    4.3.2 Endorsements, Restrictions, and Waivers 17
    4.3.3 Other CDL-Related Requirements 17
  4.4.0 Other Requirements 18
    4.4.1 MSHA Requirements 18
    4.4.2 ANSI and ASME Requirements 19
    4.4.3 NFPA Requirements 20
    4.4.4 DHS and Coast Guard Requirements 20
    4.4.5 USACE and NAVFAC Requirements 21

Figures and Tables

Figure 1   Three levels of causation. 5
Figure 2   Costs associated with incidents. 6
Figure 3   OSHA's online injury costs calculator. 7

Table 1   Table 1 CMV Endorsement Letter Designations 17

Section One

1.0.0 Roles and Responsibilities of a Safety Technician

Objective

Identify the roles and responsibilities of a safety technician.
a. Describe the responsibilities of a safety technician.
b. Define important safety terms.

Trade Terms

Job safety analysis (JSA): A careful study of a job or task to find all of the associated hazards and identify methods of safeguarding workers against each hazard.

Release of energy: Events or conditions that release energy from systems, machines, or pieces of equipment.

The job of a safety technician is both serious and satisfying. In this role, you will have the opportunity to save lives, prevent injuries and disabilities, and help your employer to be profitable and competitive. This is important considering that on a typical working day, four to five construction workers will die from on-the-job injuries and nine hundred will be seriously injured. That means that more than one thousand construction workers are killed on the job every year.

One way to reduce illness and injury on a job site is to be fully aware of the causes and costs of incidents. In addition, it also helps to have a company-wide standardized safety program. This program should provide the basic framework for working safely and efficiently. Safety technicians are an important part of a company's safety program. Some of their main responsibilities include coordinating safety policies and safety procedures, performing audits and inspections, recordkeeping, and risk analysis.

Safety technicians are also responsible for knowing the local, state, and national regulations with which the company must comply. These safety regulations are established to achieve injury-free and incident-free work sites. Keep in mind that all of these regulations are minimum standards; many progressive companies have policies and procedures that are more stringent.

1.1.0 Safety Technician Responsibilities

One of a safety technician's most important responsibilities is to serve as a resource to site management on safety, health, and in some cases, environmental regulations. Carrying out this responsibility includes interacting with workers, subcontractors, and the public. A safety technician is a trainer, motivator, auditor, planner, and advisor—in short, a key player in any organization.

Safety technicians also have the following responsibilities:

- In the absence of a site safety manager or supervisor, represent the company during visits by regulatory agencies.
- Provide safety training for both new and experienced workers.
- Participate in the development, review, or revision of any *Job Safety Analysis (JSA)* or Task Safety Analysis (TSA), as well as work plans, incident reporting forms, and emergency action plans.
- Audit and inspect the job site or work activities.
- Anticipate, identify, and have management correct safety hazards.
- Focus safety programs on the prevention of the Fatal Four (falls, caught-in/between, struck by, and electrocution).
- Audit compliance with regulatory requirements.
- Use proper coaching techniques to correct unsafe behavior and provide recognition for safe behavior.
- Conduct safety meetings.
- Audit compliance with work permits and permit-required work areas.
- Assist site management in conducting incident investigations.
- Analyze data gathered during incident investigations.
- Monitor industrial hygiene and perform employee exposure monitoring.
- Monitor air quality, sound level, etc., and ensure that all workers have the appropriate PPE for each hazard.
- Manage the site safety and health recordkeeping system.
- Serve as a liaison between the job site and insurance company representatives.
- Provide or coordinate first aid and access to follow-up medical care.

1.2.0 Important Safety Terms

Some of the basic terms and concepts that a safety technician needs to know are defined as follows:

- *Safety* – A general term denoting an acceptable level of risk, or the relative freedom from or low probability of harm. It is the control of recognized hazards to attain an acceptable level of risk.
- *Risk* – A measure of both the probability and the consequences of all hazards of an activity or condition. It is a subjective evaluation of relative failure potential. It is the chance of injury, damage, or loss.
- *Hazards* – Conditions, changing sets of circumstances, or behaviors that present the potential for injury, illness, or property damage.
- *Incidents* – Unplanned and sometimes harmful or damaging events that interrupt the normal progress of a task. Incidents are invariably preceded by an unsafe act, condition, or a combination of both.
- *Near miss* – An undesired event that, under slightly different circumstances, could have resulted in personal harm or property damage or any undesired loss of resources. An example of a near miss would be a falling object that narrowly misses a worker below.
- *Unsafe acts* – Sometimes called unsafe behavior or placing yourself at risk; these are behavioral departures from an accepted, normal, or correct procedure or practice. An unsafe act is an unnecessary exposure to a hazard or conduct that reduces the degree of safety normally present.
- *Unsafe conditions* – Any physical states that deviate from that which is acceptable, normal, or correct. An unsafe condition is any physical state that results in a reduction in the degree of safety normally present, such as the uncontrolled release of energy or the presence of a physical or environmental hazard.

It is important to learn these terms and conditions because safety technicians hear, see, and say them often.

Accident or Incident?

The safety industry is moving away from the term *accident* and using the word *incident* instead. This is because calling an event an accident implies that it was unpreventable. The primary purpose of a safety program is to identify hazards and prevent incidents.

Additional Resources

Basic Safety Administration: A Handbook for the New Safety Specialist, Fred Fanning, CSP. 2003. Des Plaines, IL: The American Society of Safety Engineers (ASSE).

Construction Project Safety, John Schaufelberger and Ken-Yu Lin. 2014. Hoboken, NJ: John Wiley & Sons.

Construction Safety Planning, David V. MacCollum, P.E., CSP. 1995. Hoboken, NJ: John Wiley & Sons.

OSHA Safety and Health Regulations for Construction, **www.osha.gov**

1.0.0 Section Review

1. Safety technicians are responsible for participating in the development, review, or revision of work plans, incident reporting forms, and _____.
    a. material safety data forms
    b. OSHA inspection procedures
    c. emergency action plans
    d. CDL documents

2. The safety term for conditions, changing sets of circumstances, or behaviors that present the potential for injury, illness, or property damage is _____.
    a. risks
    b. hazards
    c. incidents
    d. near misses

Section Two

2.0.0 INCIDENT CAUSES AND COSTS

Objective

Identify incident causes and costs.
a. Identify the causes of incidents.
b. Identify the costs of incidents.

Trade Terms

Construction User's Round Table (CURT): CURT describes itself as an autonomous organization that provides a forum for the exchange of information, views, practices and policies of various owners at the national level. Similar groups, called Local User Councils, function at the local level and seek to address problems of cost, quality, safety, and overall cost-effectiveness in their respective areas.

Direct labor costs: Costs that can be directly related to an incident such as medical costs, workers' compensation insurance, benefits, and liability and property damage insurance payments.

Experience modification rate (EMR): A numeric factor used in determining workers' compensation costs. It rises for contractors with poor incident experience and falls for those with good incident experience.

Gross income: Income before deductions (i.e., taxes).

Net income: Income after deductions (i.e., taxes and expenses).

Profit: Net income over a given period of time.

Incidents can result in personal injury or death, environmental impacts, and equipment or property damage. Safety policies and procedures must be followed at all times to minimize the occurrence of incidents. Safety technicians must be aware of the causes and costs of incidents as well as how to prevent them.

2.1.0 Causes of Incidents

The causes of incidents can be classified by three different factors: direct causes, indirect causes, and root causes. These are also called the three levels of incident causation (*Figure 1*). Each level of causation represents one of these factors.

2.1.1 Level I (Direct Causes)

Level I represents direct causes of incidents. These are incidents resulting from the uncontrolled release of energy which may or may not cause injury or property damage, but they are still dangerous. When investigating an incident, look closely at energy sources that may have been released unintentionally. Common examples include the following:

- Contact with the uncontrolled release of energy (chemical, electric, pneumatic, hydraulic, etc.)
- Being struck by an object or equipment
- Uncontrolled body motions, such as would be caused by touching a hot or energized surface
- Being caught in between two objects
- Being struck by debris during the detonation of explosives
- Being cut or scraped by a jagged edge
- Exposure to chemical releases or other environmental hazards

These types of incidents are preventable if equipment is properly maintained and workers are properly trained and follow established safety guidelines. Workers can protect themselves from these situations by remaining out of the line of fire of equipment, traffic, or other sources of energy. Workers also should be aware of sources of potential energy (the energy possessed by an object due to its position or state) and sources of kinetic energy (the energy of an object's motion).

2.1.2 Level II (Indirect Causes)

Level II represents indirect causes of incidents. Indirect causes are factors that contribute to an incident but are not the main cause. Indirect causes are also known as unsafe acts and conditions. In the past, investigators merely tried to identify the indirect causes of incidents, rather than the reasons behind the unsafe act or condition. This proved to be ineffective because indirect causes are often the symptom of a greater problem.

Unsafe Acts – It is important to be able to recognize when a worker's behavior is unsafe. The following is a list of the most common unsafe acts found on a job site:

- Lack of proper and appropriate training
- Failing to use appropriate personal protective equipment
- Failing to communicate potentially hazardous conditions or unsafe behaviors to co-workers

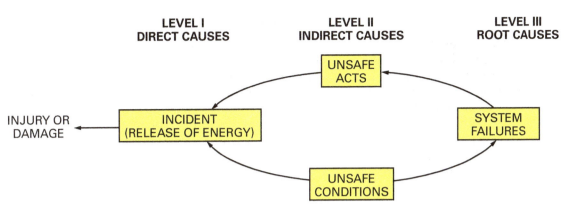

Figure 1 Three levels of causation.

- Failing to follow instructions or procedures
- Using incorrect or defective tools or equipment
- Lifting improperly
- Taking an improper working position
- Making safety devices inoperable
- Operating equipment at improper speeds
- Operating equipment without authority or proper training
- Servicing, repairing, or adjusting equipment while it is in motion or energized
- Loading or placing equipment or supplies improperly or dangerously
- Using equipment improperly
- Working under a suspended load or in an obviously dangerous area
- Working while impaired by alcohol or drugs (either legal or illegal)
- Engaging in horseplay

Unsafe Conditions – Unsafe conditions are physical conditions that are different from acceptable, normal, or correct conditions. The following is a list of the most common unsafe conditions:

- Poor housekeeping
- Poor illumination
- Poor ventilation
- Congested workplaces
- Defective tools, equipment, or supplies
- Extreme temperatures
- Excessive noise/pressure
- Fire and explosive hazards
- Hazardous atmospheric conditions
    - Gases
    - Dusts/fibers
    - Fumes
    - Vapors
- Inadequate supports or guards
- Inadequate warning systems
- Weather hazards

Most incidents are caused by unsafe acts and conditions. Safety technicians must be aware of these causes and use as many preventive measures as necessary. Preventive measures may include design modification, proper training, appropriate equipment maintenance, and good work site housekeeping.

2.1.3 Level III (Root Causes)

Level III represents the root causes of incidents. Root causes are hazardous conditions or unsafe work practices that management can correct, and when corrected, will reduce the likelihood or eliminate the possibility of future incidents. Root causes not only affect single incidents being investigated, they also affect future incidents and other work problems. Root causes should be corrected as soon as they are found as this will help to prevent incidents and save lives.

Lean Construction

The term *lean construction* describes construction methods that are designed to eliminate waste throughout a project's lifecycle. Methods of lean construction include proper storage of equipment, staging only equipment that will be needed that day, and ensuring that nothing hits the floor or is wasted. Lean construction also encourages all parties to share best practices and establishes a partnership between suppliers and end-users to reduce scheduling delays and the over-ordering or under-ordering of supplies and materials.

Source: www.leanconstruction.org

2.2.0 The Costs of Incidents

Incidents are very costly. When they occur, everyone involved loses, including the injured worker, the employer, and the insurance company. Incident costs are often classified by the term *total incurred*. There are direct (insured) and indirect (uninsured) costs associated with incidents.

Direct, or insured, costs include medical costs and workers' compensation (WC) insurance benefits, as well as liability and property damage insurance payments. Of the three direct costs, workers' compensation insurance benefits are the costliest. Direct costs of incidents are not generally fixed. Rather, they vary depending upon the severity of the injuries and damages.

Indirect, or uninsured, costs are the hidden costs involved with an incident. Examples include the costs associated with uninsured property damage, equipment damage, production delays, supervisory time, case management, retraining, company image, and worker morale.

The costs associated with incidents can be compared to an iceberg, as shown in *Figure 2*. The tip of the iceberg represents the direct costs, which are the costs that can be seen. The larger, indirect costs are unseen. Studies indicate that the indirect costs of incidents are usually two to seven times greater than the direct costs. Safety technicians need to be aware of the overall effects of incidents.

2.2.1 Calculating Incident Costs

Real dollars are lost when workplace incidents occur. These dollars have a tremendous effect on the company's profit margin. For instance, consider a company with a gross income of two million dollars. The company operates on a profit margin of three percent of its gross income. This means that its net income is $60,000 after taxes and expenses. If the company suffers a loss of $50,000 due to incidents, it must increase its gross income by $1,716,667 to make up for the incident. The company will have to earn almost two million dollars more just to make up for the $50,000 lost from incidents. This is a good example of why preventing incidents is so important.

OSHA's $afety Pays program provides an online calculator (*Figure 3*) that employers can use to estimate the impact of occupational injuries and illnesses on company profitability. The calculator uses a company's profit margin, the average costs of a specified injury or illness, and an indirect cost multiplier to project the amount of sales a company would need to generate to cover those costs.

> **NOTE**
> For more information on calculating incident costs, refer to OSHA's website at **www.osha.gov**.

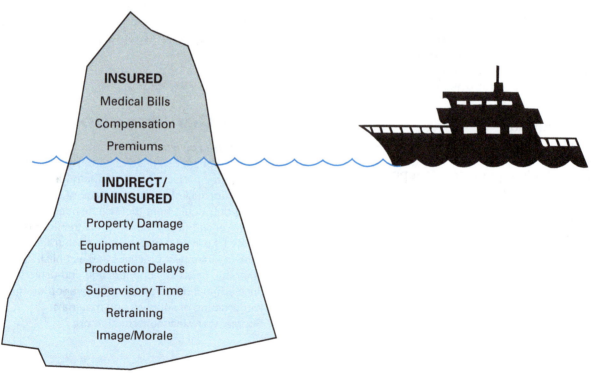

Figure 2 Costs associated with incidents.

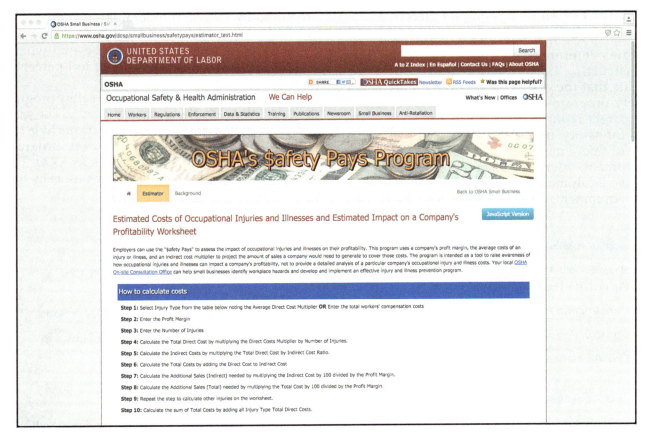

Figure 3 OSHA's online injury costs calculator.

How EMR Affects Costs

OSHA's $afety Pays program shows how implementing safety improvements and reducing EMR can benefit a company's bottom line. OSHA offers a free online $afety Pays estimator program at **www.osha.gov**. After implementing safety improvements, companies can expect to benefit not only from a reduction in injuries and lost time, but also in a lower EMR and reduced workers' compensation premiums.

2.2.2 Incident Experience vs. Future Insurance Costs

A contractor's incident experience has a significant effect on future insurance costs. Contractors with poor incident experience generally pay more for workers' compensation insurance than those with good records. The difference is due to their workers' compensation experience modification rate (EMR).

Experience rating is a method of modifying future workers' compensation insurance premiums by comparing a particular company's actual losses (loss runs) to the losses normally expected for that company's type of work. Loss runs are claims made against insurance (*i.e.*, how much money they spent and how much they set aside for a case) and count against a company when determining their EMR. The average rate for a particular class of work is called the *book rate* or *manual rate*.

A new company starts with an EMR of 1 or 100 percent. After a specified number of years (typically four years) and premiums paid, an EMR is established. Contractors with better-than-average loss experience have a modifier/multiplier of less than 100 percent (EMR of 1), which is a credit factor. Those with experience worse than average will have a modifier of over 100 percent, which is a penalty factor. The EMR is multiplied by the total WC cost to determine what the company will actually pay for WC insurance.

2.2.3 Cost of Administering an Effective Safety Program

Studies referenced in the A-3 report of the Construction User's Round Table (CURT) estimate that the cost of maintaining an effective safety program is approximately 2.5 percent of the direct labor costs. This cost includes salaries for safety, medical, and clerical personnel, cost of safety meetings, inspections of tools and equipment, orientation meetings, personal protective equipment, and miscellaneous supplies and equipment. Most of these items are already required by state and federal laws, and contractors are required to pay for them anyway. The added cost of a safety program beyond what is legally required is probably closer to 1 percent of the direct labor costs.

Implementing a comprehensive safety program can actually save the company money. The same studies referenced in the A-3 report conclude that for every dollar invested in a safety program, the contractor could save as much as $3.20. This means that the contractor actually profits by having an effective safety program.

Additional Resources

Basic Safety Administration: A Handbook for the New Safety Specialist, Fred Fanning, CSP. 2003. Des Plaines, IL: The American Society of Safety Engineers (ASSE).

Construction Project Safety, John Schaufelberger and Ken-Yu Lin. 2014. Hoboken, NJ: John Wiley & Sons.

Construction Safety Planning, David V. MacCollum, P.E., CSP. 1995. Hoboken, NJ: John Wiley & Sons.

OSHA Safety and Health Regulations for Construction, www.osha.gov

2.0.0 Section Review

1. The causes of incidents can be classified as direct causes, indirect causes, and _____.
   a. collateral causes
   b. probable causes
   c. root causes
   d. secondary causes

2. Direct costs associated with incidents include workers' compensation insurance benefits and liability and property damage insurance payments in addition to _____.
   a. uninsured equipment damage
   b. production delays
   c. supervisory time
   d. medical costs

SECTION THREE

3.0.0 BASIC COMPONENTS OF A SAFETY PROGRAM

Objective

Identify the basic components of a safety program.
  a. Identify essential safety program policies and procedures.
  b. Identify effective means and practices for providing safety orientation and training.

There are certain basic elements common to the majority of effective safety programs. The degree and extent to which these elements are included in your program will vary depending on the size and nature of your company's operations. The most basic components of a safety program include the following:

- Management support and a policy statement
- A policy on alcohol and drug abuse
- Safety policies and procedures
- Orientation and training
- Access to first aid and follow-up medical care
- Pre-project and pre-task safety planning
- Safety meetings and employee involvement
- Sub-contractor management
- Emergency action plans and methods for dealing with the media
- Inspections, employee observations, and audits
- Incident investigation and analysis
- Recordkeeping
- Program evaluation and follow-up

Safety programs can be developed using a number of resources, including data from OSHA, ANSI, the National Safety Council, Occupational Health & Safety Advisory Services (OHSAS), and other safety organizations. In order to comply with specific international, federal, state, and local safety regulations, the following additional safety program elements may also be required:

- Hazard communication
- Bloodborne pathogens
- Welding, burning, and cutting
- Respiratory protection
- Personal protective equipment
- Elevated work and fall protection
- Hearing conservation
- Confined-space entry
- Lockout/tagout
- Cranes, hoists, and other lifting devices
- Mobile equipment operations
- Trenching and excavation

3.1.0 Policies and Procedures

The key components of a company's safety program are expressed in policies and procedures that include the company's policy statement, its alcohol and drug abuse policies and programs, and the safety rules that must be followed.

3.1.1 Policy Statement

A policy statement outlines management's philosophies and goals toward safety and loss prevention. It is signed by the president or chief executive officer (CEO) and communicated to all affected employees.

A good policy statement recognizes management's responsibility for safety and loss prevention without diminishing each employee's role. It expresses a strong, positive, and realistic commitment to safety. Management support and commitment are the most important factors contributing to the success of any safety program. Safety technicians need to be familiar with their company's management policy statement and make sure that it is available to all workers.

3.1.2 Alcohol and Drug Abuse Policies and Programs

Over 14 percent of construction workers have an alcohol or drug abuse problem, as reported in the 2015 National Survey on Drug Use and Health. This is an annual survey sponsored by the Substance Abuse and Mental Health Services Administration (SAMHSA), an agency within the US Department of Health and Human Services (DHHS). Even the use of legal substances, such as certain prescription drugs, can affect safety on the job site. It is important to know if a worker is on a prescription drug and to be aware of how the prescribed drug could affect the worker's performance on the job.

The negative impacts of drug and alcohol abuse include the following:

- Increased worker's compensation and healthcare costs
- Decreased productivity
- Increased tools and materials costs
- Increased absenteeism

Many owners and contractors have recognized these concerns and are implementing programs to combat the situation. Make sure you are aware of your company's drug and alcohol abuse programs. Be prepared to advise supervisors and other employees of the key elements of your company's program, including signs and symptoms of drug and alcohol abuse. It is important to stay informed on trends in drug abuse, including synthetic drugs and the abuse of prescription drugs.

Some larger companies sponsor employee assistance programs (EAPs). If your company has an EAP, make sure the information about the program is available to everyone. A good relationship between workers and safety professionals makes it easier to spot changes in normal behavior and offer assistance for emerging issues.

3.1.3 Safety Policies and Procedures

Safety policies and procedures are a necessary part of any program. They must be logical and they should be prepared and presented in terms that are easily understood. Policies and procedures are of no value unless they are clearly communicated and workers buy into the safety process. Everyone on site must also understand the consequences of failing to follow them.
Safety policies and procedures are generally divided into the following categories:

- General/company safety policies and procedures that apply to all employees
- Client-specific, job-specific, or site-specific safety policies and procedures
- Craft or special safety policies and procedures that apply to a specific type of work operation

Safety policies and procedures can be presented as posters on a site, or during orientation, safety meetings, or safety training. Make sure safety policies and procedures are available to all workers at all times, so that workers know the requirements as well as you do.

Progressive contractors plan safety into the job at the bid stage. This requires a systematic management process that involves the estimator, project manager, and the safety department, at a minimum. Likewise, daily pre-task or pre-job safety planning can improve safety, quality, and productivity on any job. Safety technicians help implement the job-specific safety plan and participate in pre-job or pre-task safety planning.

3.2.0 Safety Orientation and Training

Safety training is an important tool for preventing incidents. Training can be broken down into the following major categories: new employee orientation; craft- and/or task-specific training; and site-specific training.

Statistics have shown that employees who have been on the job 30 days or less account for 25 percent of all construction injuries. This clearly shows the need for an effective orientation

Legalized Marijuana

In light of recent shifts in laws that apply to the use of marijuana, it would be wise to review your company's alcohol and drug abuse policy to make sure it clearly conveys your company's stance on marijuana and the workplace. As of 2017, many states have passed medical marijuana laws and some states have approved the recreational use of marijuana by adults. However, the sale and use of marijuana remains illegal under federal law in every state, and workplace drug testing is mandated by federal law for millions of workers including commercial drivers and many other persons in safety-sensitive positions.

Acceptance of federal funding requires compliance with the Drug-Free Workplace Act of 1988, which prohibits contractors from allowing the use or possession of controlled substances in the workplace. To date, marijuana is still considered a controlled substance under federal law. Therefore, even in states that permit medical and/or recreational marijuana use under state law, many employers must test employees and job applicants for marijuana and hold those persons who test positive accountable under federal government mandates.

It is generally recommended that employers continue to use a zero tolerance drug-free workplace standard. It should be clearly communicated to employees that your company's workplace substance abuse and testing polices have not changed, and that these policies still apply to marijuana use. Emphasize that the ultimate goal of workplace drug and alcohol programs is to protect the health and safety of all employees as well as the productivity of the workforce. Just as it is unacceptable for employees to be performing job duties while intoxicated, working while under the influence of marijuana will not be tolerated.

program. Safety training should be conducted for the following situations:

- New employees
- Existing employees assigned to a new job
- New jobs

In order for safety training to be effective, it should cover the following topics:

- Correct work procedures
- Donning and doffing, use, maintenance, and limitations of any required personal protective equipment
- The hazards and safeguards associated with any harmful materials used
- Where to go for help
- Site emergency reporting and response procedures

Informal safety training will likely be done by the first-line supervisor. Your role in these types of training sessions involves guidance and support. In more formal safety sessions, you may be responsible for coordinating or conducting all safety sessions.

3.2.1 Safety Meetings and Employee Involvement

Safety meetings are often used to maintain employee interest and involvement. They are a key element in any safety program. Safety meetings can vary from a formal presentation to short, five-minute toolbox or tailgate talks.

Safety meetings, when properly conducted, can be used to do the following:

- Exchange information regarding specific safety matters
- Provide an outlet for workers to discuss critical issues
- Provide a written record of the actions taken
- Establish an effective communications link between management and employees
- Focus on the hazards and safeguards associated with a planned task

Both federal and state governmental regulations require employers to train their employees on the hazards and safeguards associated with the work they are doing. Safety meetings can be used as training sessions.

Safety technicians should provide safety meeting topics to crew leaders and supervisors. In some cases, they plan and conduct the safety meetings themselves, or they may be asked to provide feedback on meetings conducted by others. This helps to ensure that the message of the meeting, which is always safety, is presented effectively and efficiently. Safety technicians may be required to maintain the records of all site safety meetings.

3.2.2 Emergency Action Plans and Methods for Dealing with the Media

Emergency action plans can reduce the response time to job site emergencies. In many cases, advance planning can also reduce the severity of the incident. For example, if the names, addresses, and phone numbers of the nearest medical, fire, police, and emergency response agencies are posted at each job site phone, it will be easier to get help when it is needed.

As part of the emergency action plan, everyone should be familiar with the site emergency reporting and response procedures. Another part of the emergency action plan is the availability of prompt access to first aid and follow-up medical care. Emergency action plans should also require that at least two people on each job site be trained in basic first aid and cardiopulmonary resuscitation (CPR).

Lives are saved when all of these elements are incorporated into emergency action plans. Safety technicians are responsible for implementing such a plan and making sure that everyone on the job site has been properly trained to follow the plan.

One area often overlooked after an incident is communication with the news media. When serious incidents happen, the press usually responds. How a company reacts can have a dramatic effect on public perception. Safety technicians should be familiar with the organization's policy on discussing incidents with the news media. Only trained company personnel should be authorized to talk to the media. If your company has no media communication policy, one should be developed.

3.2.3 Inspections, Employee Observations, and Audits

Safety technicians have special responsibilities. They are a resource to site personnel for safety, health, and in some cases, environmental matters. They serve as a site consultant and auditor and help to find and correct safety hazards. Safety technicians may work under the direction of a safety manager, but are more likely to work on their own.

Safety technicians coordinate and/or perform the following tasks:

- Safety inspections – Used to detect unsafe conditions and make plans to correct them
- Employee observations – Performed to help detect unsafe work practices and procedures
- Safety audits – Designed to monitor the use and effectiveness of the company's safety policies and procedures

Supervisors are also expected to perform safety inspections and employee observations. Safety technicians may be asked to train the supervisors to do so. They should audit the quality, content, and follow-up of the crew leader's audits and observations and provide feedback as needed.

It is important to know how to do each of these tasks effectively and efficiently so that problems can be found and fixed as soon as possible.

3.2.4 Incident Investigation and Analysis

All incidents, injuries, illnesses, and near misses must be reported and promptly investigated. Incident investigations are important because they accomplish the following:

- Determine the causes of incidents
- Help prevent re-occurrences
- Satisfy insurance company requirements and government regulations
- Document facts and preserve evidence for legal purposes
- Detect trends and identify potential problem areas

The front-line supervisor generally performs the primary incident investigation function because it is generally agreed that front-line supervisors are the most familiar with the work area and employees. Depending on the nature of the incident and other conditions, incidents may also be investigated by a safety technician, a safety committee, or management.

Safety technicians, regardless of their level of participation, should always be available as a resource to the supervisor to provide any guidance or support that is needed. They may be responsible for maintaining site incident investigation records and completing supporting documents required by the company, OSHA, or the insurance company.

3.2.5 Recordkeeping

Recordkeeping is a critical part of a company's safety and health program. The data that is collected helps keep track of work-related injuries and illnesses. Once this data is gathered and analyzed, it can be used to help identify and correct problem areas. This will help to prevent future illnesses and injuries. Recordkeeping not only provides information to management about illness and injury, it also informs workers about incidents that happen in the work area. When workers are aware of injuries, illnesses, and hazards in the workplace, they are more likely to follow safe work practices and report workplace hazards.

Recordkeeping is also required by OSHA. This agency uses specific illness and injury information as part of the agency's site-specific inspection targeting program. The Bureau of Labor Statistics (BLS) also uses injury and illness records as the source data for the Annual Survey of Occupational Injuries and Illnesses. This report shows safety and health trends nationwide and industry-wide.

Safety technicians are responsible for making sure recordkeeping is done correctly. They must know all of the OSHA requirements for recording and classifying workplace illnesses and injuries and must properly report them as needed. They may be required to coordinate all of the site safety and health records required by the firm. To do so, they must be familiar with the firm's recordkeeping and personal privacy policies. Always follow the most stringent requirements for recordkeeping.

3.2.6 Program Evaluation and Follow-Up

Program evaluation and follow-up are some of the most important, but often neglected, elements of an effective safety program. To be assured that your program is meeting the company's goals and objectives, it must be evaluated on a regular basis and modified as needed. The safety technician, along with the project manager, has the responsibility of evaluating the job site safety program on a regular basis and making any needed improvements. OSHA has requirements for the annual evaluation of some policies, such as lockout/tagout and emergency action plans.

3.2.7 Nine Industry Best Practices

Recent follow-ups to the original study conducted in the early 1990s by the Construction Industry Institute (CII) Committee, titled "Making Zero Accidents a Reality", emphasize that the following best practices have the greatest positive impact on safety performance:

- Demonstrated management commitment
- Staffing for safety
- Safety planning for pre-project and pre-task items
- Safety training and education
- Worker involvement and participation
- Recognition and rewards
- Sub-contractor management
- Incident reporting and investigation
- Drug and alcohol testing

Safety technicians should be aware of these industry best practices and, to the extent possible, integrate them into site-specific safety and health activities.

Additional Resources

ANSI/AIHA/ASSE Z10-2012, Occupational Health and Safety Management Systems, 2012. www.ansi.org

Basic Safety Administration: A Handbook for the New Safety Specialist, Fred Fanning, CSP. 2003. Des Plaines, IL: The American Society of Safety Engineers (ASSE).

Construction Project Safety, John Schaufelberger and Ken-Yu Lin. 2014. Hoboken, NJ: John Wiley & Sons.

Construction Safety Planning, David V. MacCollum, P.E., CSP. 1995. Hoboken, NJ: John Wiley & Sons.

OHSAS 18001, Occupational Health and Safety Management, www.nsai.ie

OSHA Safety and Health Regulations for Construction, www.osha.gov

---

3.0.0 Section Review

1. Management's philosophies and goals toward safety and loss prevention for a company are outlined in the company's _____.
   a. employee assistance program
   b. management support and policy statement
   c. worker's compensation policy
   d. executive orders and press releases

2. Safety training should be conducted when an existing employee _____.
   a. applies for worker's compensation benefits
   b. fails a drug screening
   c. is assigned to a new job or task
   d. passes craft certification testing

SECTION FOUR

4.0.0 REGULATORY REQUIREMENTS

Objective

Identify regulatory requirements with which construction industry safety technicians need to be familiar.
   a. Identify Occupational Safety and Health Administration (OSHA) requirements.
   b. Identify Environmental Protection Agency (EPA) requirements.
   c. Identify Department of Transportation (DOT) requirements.
   d. Identify other relevant health and safety requirements.

Trade Terms

Consensus standards: Standards developed through the cooperation of all parties who have an interest in the use of the standard (e.g., the National Electrical Code®). Consensus standards rely on the expertise of manufacturers, inspectors, craft professionals, maintenance personnel, and safety professionals.

Culm: Coal refuse.

Gross vehicle weight rating (GVWR): The maximum allowed gross vehicle weight for a vehicle.

Gross unloaded vehicle weight (GUVW): The weight of a liquid cargo trailer without liquids; also known as dry weight.

Surface mines: Mines in which the mining operation is done near the surface of the ground.

Unwarrantable failures: An unwarrantable failure violation (second within 90 days) refers to a situation in which a mine operator knew or should have known that a violation existed and yet failed to take corrective action.

Safety technicians have a responsibility to know which local, state, and federal requirements apply to the work being done on their site. Agencies that provide regulatory requirements applicable to the construction industry include the following:

- Occupational Safety and Health Administration (OSHA)
- Environmental Protection Agency (EPA)
- Department of Transportation (DOT)
- Mine Safety and Health Administration (MSHA)
- US Department of Homeland Security (DHS)
- US Coast Guard (USCG)
- US Army Corps of Engineers (USACE)
- US Naval Facilities Engineering Command (NAVFAC)
- Consensus standards developed by organizations such as the American National Standards Institute (ANSI), the American Society of Mechanical Engineers (ASME), and the National Fire Protection Association (NFPA)

NOTE: Consensus standards become regulatory when incorporated by reference in a regulatory standard.

Make sure you are familiar with all of these agencies and their regulations. It could cost a company a great deal of money in fines and lawsuits if it is not in compliance.

4.1.0 OSHA Requirements

The passage of the Occupational Safety and Health Act of 1970 led to the establishment of the Occupational Safety and Health Administration, also known as OSHA. This act was passed to ensure safe and healthful working conditions for all workers. The act also provides for the following activities:

- Authorizes enforcement of the standards developed under the act
- Assists and encourages states in their efforts to assure safe and healthful working conditions
- Supplies research, information, education, and training in the field of occupational safety and health

Safety technicians are required to know and follow the rules and regulations established by OSHA. OSHA offers several sources on standards and policies related to the construction industry. OSHA's website at **www.osha.gov** provides detailed information and resources including the following:

- Standards and policies
- Statistics and data
- Reference documents
- Forms
- Training resources

For additional information about OSHA and the services that this agency provides, contact the OSHA or State Plan Office in your area. Take the time to learn which OSHA regulations apply to the work that is being done on your site. Failure to do so could mean serious injuries, loss of lives, or equipment loss or damage.

The Occupational Safety and Health Act also created the National Institute for Occupational Safety and Health (NIOSH). NIOSH is part of the Centers for Disease Control and Prevention (CDC) within the US Department of Health and Human Services (DHHS).

Unlike OSHA, NIOSH does not issue standards that are enforceable under US law. Rather, NIOSH's mandate is to develop recommendations for health and safety standards, in addition to conducting research and providing information, education, and training to help ensure safe and healthful working conditions. NIOSH may also conduct on-site investigations to determine the toxicity of materials used in workplaces, and to test and certify personal protective equipment and hazard measurement instruments.

To accomplish its mission, NIOSH developed a partnership program, the National Occupational Research Agenda (NORA). NORA's goal is to facilitate research collaboration among universities, businesses, professional societies, government agencies, and worker organizations. NORA is organized into several industrial sectors, one of which is the construction sector. Sector councils head the research for the sectors and help implement the national research agenda.

> **NOTE**
> For more information on NIOSH and NORA, visit the CDC's website at **www.cdc.gov**.

4.2.0 EPA Requirements

The US Environmental Protection Agency was established in 1970 to consolidate in one agency a variety of federal research, monitoring, standard-setting and enforcement activities to ensure environmental protection. The EPA's mission is to protect human health and to safeguard the natural environment including the air, water, and land upon which life depends.

It is important to understand how work being done on a construction site affects the environment. This includes knowing the major environmental regulations covering the release, treatment, storage, and disposal of potentially hazardous materials into the air, water, and soil. Safety technicians need to have a working knowledge of the following EPA laws:

- *Comprehensive Environmental Response, Compensation, and Liability Act (CERCLA)* – This law imposes liability on owners or operators of a facility from which there is a release of hazardous substances into the environment.
- *Resource Conservation and Recovery Act (RCRA)* – This law regulates the generation, treatment, storage, and disposal of hazardous waste. RCRA requires the owner/operator of a facility to undertake corrective action to clean up a facility used for the treatment, storage, or disposal of hazardous waste. There is also a complex permit program with which the owner/operator must comply.
- *Underground Storage Tank Regulations (under RCRA and state control)* – This law imposes operating, reporting, financial assurance, and potential cleanup obligations on persons and companies owning and operating underground petroleum storage tanks (USTs). The rules and regulations covering USTs are quite detailed and prescriptive. In addition, many states have their own regulations and enforcement policies.
- *Clean Air Act (CAA) of 1990* – This act is a complex, multi-faceted statute that is designed to regulate air emissions from stationary and mobile sources.
- *Clean Water Act (CWA)* – This act prohibits the discharges of pollutants from point sources and storm water into navigable waters of the United States without a permit. The act imposes liability on the person who is responsible for the operation and/or equipment that results in a discharge.
- *Safe Drinking Water Act (SDWA)* – This act imposes federal drinking water standards on virtually all public water systems. The act requires the establishment of drinking water standards for maximum contaminant levels (MCLs).
- *Toxic Substances Control Act (TSCA)* – This act governs the manufacture and use of chemical products. The importing and exporting of chemicals are regulated under the act. In addition, there are specific regulations regarding the use, management, storage, and disposal of polychlorinated biphenyl (PCB) materials.
- *Oil Pollution Act of 1990* – This act imposes strict liability on responsible parties for removal costs and damages resulting from discharges of oil into navigable waters of the United States. An owner/operator of an onshore facility that results in a discharge is considered a responsible party.

In addition, job-specific EPA certification (*e.g.,* refrigerant use or asbestos abatement) may be required. Safety technicians must also be familiar with EPA permit requirements.

4.3.0 DOT Requirements

DOT requirements cover transportation as well as pipeline operations. According to the US Bureau of Labor Statistics, more fatal work injuries resulted from transportation incidents than from any other event in 2013. Roadway incidents alone accounted for nearly one out of every four fatal work injuries. Outside of roadway incidents, pedestrian vehicular incidents made up the second greatest number of transportation-related fatal injuries. For these reasons, any safety program for construction workers should stress safe driving practices and the importance of maintaining fleet safety. Likewise, workers must be required to follow the standard guidelines for driving safely, such as using seatbelts, not talking on phones while operating vehicles, and yielding for pedestrians.

There are a number of additional driving safety issues that are of special concern for construction workers. Driving certain commercial motor vehicles (CMVs) requires special skills and knowledge. Prior to the implementation of the federal Commercial Driver's License (CDL) Program, in a number of states and the District of Columbia any person licensed to drive an automobile could also legally drive a tractor trailer or bus. Even in many of the states that did have a classified licensing system, a person's skills were not tested on the type of vehicle that he or she would be driving. As a result, many drivers were operating motor vehicles that they were not qualified to drive. This led to many serious incidents. Fortunately, because of CDL programs, this is no longer the case.

Safety technicians may be responsible for knowing CDL rules and regulations that affect their company. They need to make sure that any worker operating commercial vehicles has a current CDL. They must also know the regulatory requirements established by the Department of Transportation's Commercial Driver's License Program and the standards set by the Commercial Motor Vehicle Safety Act of 1986.

4.3.1 Commercial Motor Vehicle Safety Act of 1986

The goal of the Commercial Motor Vehicle Safety Act of 1986 is to improve highway safety by ensuring that drivers of large trucks, heavy equipment, and buses are qualified to operate those vehicles, and to remove unsafe and unqualified drivers from the highways. The Act retained the state's right to issue a driver's license but established minimum national standards that states must meet when licensing CMV drivers.

The Act makes it illegal to hold more than one license by requiring states to adopt testing and licensing standards for truck and bus drivers. This helps to verify a person's ability to operate a specific type of vehicle. It also cuts back on the number of drivers with multiple traffic tickets. Because drivers can only have one license, they cannot register in another state after accumulating several citations in order to have a clean record.

Note that the Act does not require drivers to obtain a separate federal license. It only requires states to upgrade their existing testing and licensing programs, if necessary, and conform to the federal minimum standards.

Since April 1, 1992, drivers have been required to have a CDL in order to drive a CMV. The Federal Highway Administration (FHWA) developed and issued standards for testing and

---

Case History

Flint Water Crisis

The water crisis that occurred in Flint, Michigan illustrates the importance of following the requirements for safe drinking water. In 2014, the city of Flint changed its water source from the treated Detroit Water and Sewerage Department to water sourced from the Flint River. Flint River water is corrosive, and it was not treated with a corrosion inhibitor as required. This water caused lead from aging pipes to leach into the water supply, exposing between 6,000 and 12,000 children to extremely high levels of lead and many potential health problems. On January 5, 2016, the city was declared to be in a state of emergency, and several lawsuits were filed against government officials. On April 20, 2016, criminal charges were filed against three people in regards to the crisis.

The Bottom Line: Failure to follow the appropriate regulatory guidelines can result in personal injury, financial consequences, and criminal negligence.

licensing CMV drivers. Among other things, the standards require states to issue CDLs to their CMV drivers only after the driver passes knowledge and skills tests administered by the state related to the type of vehicle to be operated—for example, a crane, forklift, or personnel lift. Drivers need CDLs if they are in interstate, intrastate, or foreign commerce and drive a vehicle that meets one of the following classifications of a CMV:

- *Class A* – Any combination of vehicles with a gross vehicle weight rating (GVWR) of 26,001 or more pounds, provided the gross unloaded vehicle weight (GUVW) of the vehicle(s) being towed is in excess of 10,000 pounds.
- *Class B* – Any single vehicle with a GVWR of more than 26,001 pounds, or any vehicle towing another vehicle that is not in excess of a 10,000 pounds GUVW.
- *Class C* – Any single vehicle or combination of vehicles that does not meet the definition of Class A or Class B but is either designed to transport 16 or more passengers, including the driver, or is designated for hazardous materials.

4.3.2 Endorsements, Restrictions, and Waivers

Drivers who operate special types of CMVs need to pass additional tests in order to be properly licensed. *Table 1* shows the test subjects and the endorsement letter that is assigned to a CDL after the test is passed.

> **NOTE**
> If a driver either fails the air brake component of the general knowledge test or performs the skills test in a vehicle not equipped with air brakes, an air brake restriction is issued, prohibiting the driver from operating a CMV equipped with air brakes.

Table 1 CMV Endorsement Letter Designations

Test Subject	Endorsement
Double/triple trailers	T
Passenger	P
Tank vehicle	N
Hazardous materials	H
Combination of tank vehicle and hazardous materials	X

All active duty military drivers were waived from the CDL requirements by the Federal Highway Administration. A state, at its discretion, may waive firefighters, emergency response vehicle drivers, farmers, and drivers removing snow and ice in small communities from the CDL requirements, subject to certain conditions.

4.3.3 Other CDL-Related Requirements

There are several other requirements related to the CDL legislation that affect commercial drivers, their employers, and states with CDL programs. They include the following:

- *Penalties* – The federal penalty to a driver who violates the CDL requirements is a civil penalty of up to $2,500 or, in aggravated cases, criminal penalties of up to $5,000 in fines and/or up to 90 days in prison. An employer is also subject to a penalty of up to $10,000 if he or she knowingly uses a driver to operate a CMV without a valid CDL.
- *Blood alcohol content standards* – The Federal Highway Administration (FHWA) has established 0.04 percent as the blood alcohol concentration (BAC) level at or above which a CMV driver is deemed to be driving under the influence of alcohol and is subject to the disqualification sanctions in the Act. This standard applies to a CMV driver regardless of the type of vehicle in use at the time of measurement. States maintain a BAC level between 0.08 percent and 0.10 percent for non-CMV drivers.
- *Employee notifications* – Within 30 days of a conviction for any traffic violation, except parking, a driver must notify his/her employer, regardless of the nature of the violation or the type of vehicle that was driven at the time. If a driver's license is suspended, revoked, or canceled, or if the employee is disqualified from driving, FHWA must notify the employer. The notification must be made by the end of the next business day following receipt of the notice of the suspension, revocation, cancellation, lost privilege, or disqualification. Employers may not knowingly use a driver who has more than one license or whose license is suspended, revoked, canceled, or is disqualified from driving. Violation of this regulation may result in civil or criminal penalties.

- *Disqualifications* – The following is a list of reasons for disqualifications from the CDL program.
    - Conviction while driving a CMV
    - Two or more serious traffic violations within a three-year period
    - One or more violations of an out-of-service order within a 10-year period
    - Driving under the influence of a controlled substance or alcohol
    - Leaving the scene of an incident
    - Using a CMV to commit a felony
    - Using a CMV to commit a felony involving manufacturing, distributing, or dispensing controlled substances.

> **NOTE:** US DOT employers may not accept recreational or medical marijuana use as a valid excuse for a positive test result on a DOT drug test.

> **NOTE:** For more information about CDL requirements, visit the DOT's Federal Motor Carrier Safety Administration (FMCSA) website at www.fmcsa.dot.gov

4.4.0 Other Requirements

Safety technicians need to be familiar with requirements and recommendations from other agencies and organizations that can apply to the construction industry. These include requirements from the following:

- US Mine Safety and Health Administration (MSHA)
- American National Standards Institute (ANSI)
- American Society of Mechanical Engineers (ASME)
- National Fire Protection Association (NFPA)
- Department of Homeland Security (DHS)
- US Coast Guard (USCG)
- US Army Corps of Engineers (USACE)
- US Naval Facilities Engineering Command (NAVFAC)

4.4.1 MSHA Requirements

The US Mine Safety and Health Act of 1977 established the Mine Safety and Health Administration (MSHA). The mission of MSHA is to ensure safe and healthful working conditions for miners and any other workers doing work at a mine, rock quarry, or mine material-processing facility. It is important for safety technicians to understand MSHA requirements because any construction or maintenance done at a mine or facility that handles mine products (such as portland cement, coal, gravel, and limestone) falls under MSHA regulations, not OSHA regulations.

Mine operators must comply with the safety and health standards enforced by MSHA, an agency within the Department of Labor (DOL). Like OSHA, MSHA may issue citations and propose penalties for violations. Unlike employees under OSHA, mine employees are subject to government sanctions for violating safety standards, such as smoking in or near mines or mining machinery. Similarly, employers and other supervisory personnel may be held personally liable for civil penalties or may be prosecuted criminally for violations of Mine Act standards.

> **NOTE:** Mine Act standards are listed in 30 *CFR*, Parts 70, 71, 74, 75, 77, and 90.

Although MSHA is similar to OSHA in many ways, it is significantly different in four key areas:

- Inspections
- Withdrawal orders
- Miner training
- Incident, injury, and illness reporting

A MSHA representative must inspect underground mines in their entirety at least four times a year. Surface mines are to be inspected at least two times a year. The mine operator's representative and a representative authorized by the miners must be given the opportunity to accompany the MSHA inspector during the inspection. These representatives must be given the opportunity to participate in the post-inspection conference as well.

Whenever the inspector observes a condition that appears to be a violation of the standards, he or she must issue a citation. If, in the opinion of the mine inspector, an imminent danger condition exists, then the inspector must issue a withdrawal order. Other situations for which MSHA may issue a withdrawal order include the following:

- During a follow-up inspection, if MSHA finds that a mine operator has failed to abate a cited violation and there is no valid reason to extend the abatement period, MSHA must issue a withdrawal order until the violation is abated.
- If a mine operator fails to abate a breathable dust violation for which a citation has been issued and the abatement period has expired,

MSHA must either extend the abatement period or issue a withdrawal order.
- If two violations make up unwarrantable failures to comply with the standards that are found during the same inspection, or if the second unwarrantable violation is found within 90 days of the first, a withdrawal order must be issued.
- Miners may be ordered withdrawn from a mine if they have not received the safety training required by the Act. Miners who are withdrawn for this reason are protected by the Mine Act from discharge or loss of pay.

> **NOTE:** A withdrawal order does not stop issuance of a citation and proposed penalty.

Mine operators are required to have a safety and health training program approved by MSHA. The program must include the following:

- At least 8 hours of annual refresher training are required for all miners.
- At least 40 hours of instruction for new underground miners, to include the following:
    - Rights of miners and their representatives under the Act
    - Use of self-rescue and respiratory devices
    - Hazard recognition
    - Escapeways
    - Walk-around training
    - Emergency procedures
    - Basic ventilation
    - Basic roof control
    - Electrical hazards
    - First aid
    - Safety and health aspects of the task assignment
- At least 24 hours of instruction are required for new surface miners. The training must include all of the same items for underground miners, with the exception of:
    - Escapeways
    - Basic ventilation
    - Basic roof control

> **NOTE:** The Mine Act requires that training be conducted during normal working hours and that the miners must be paid at their normal rate during the training period. Regulations concerning the training and retraining of miners are stated in 30 *CFR*, Part 48.

All mine operators are required to report incidents immediately to the nearest MSHA district or sub-district office. Also, operators must investigate and submit to MSHA, upon request, an investigation report on incidents and occupational injuries.

MSHA defines the following as incidents for the purpose of reporting incidents, injuries, and illnesses under the Mine Act:

- An unplanned flood of a mine by a liquid or gas
- A fatality at a mine
- An injury to an individual at a mine that has a reasonable potential to result in the worker's death
- An injury that may result in death
- Entrapment for more than 30 minutes
- An unplanned ignition or explosion of gas or dust
- An unplanned fire not extinguished within 30 minutes of its discovery
- An unplanned ignition or explosion of a blasting agent or an explosive
- An unplanned roof fall in active work areas where roof bolts are in use, or a roof fall that impairs ventilation or impedes passage
- Coal or rock outbursts that cause withdrawal of miners or that disrupt mining activity for more than one hour
- An unstable condition at an impoundment, refuse pile, or culm bank requiring emergency action
- Damage to hoisting equipment in a shaft or slope that endangers an individual or interferes with use of equipment for more than 30 minutes

> **NOTE:** More information about MSHA regulations and procedures is available online at www.msha.gov.

4.4.2 ANSI and ASME Requirements

The American National Standards Institute (ANSI) and the American Society of Mechanical Engineers (ASME) are both private, non-profit organizations that promote uniformity in voluntary standards for engineering.

ANSI accredits standards that are developed by various standards organizations, government agencies, consumer groups, companies, etc. ANSI accreditation signifies that the procedures used by standards-developing organizations meet the Institute's requirements. Using accredited standards helps to ensure that the characteristics and performance of products are consistent, that people use the same definitions and terms, and that products are tested in the same way. ANSI also accredits organizations that perform product or personnel certification in accord with requirements defined in international standards. Additionally, ANSI designates specific standards as American National Standards, or ANS. There are approximately 9,500 American National Standards that carry the ANSI designation.

ASME focuses on developing codes and standards for mechanical devices. Its primary concern is delivering solutions to the practical challenges that engineering professionals encounter on the job. Committees of subject matter experts use an open, consensus-based process to develop ASME standards. ASME has produced approximately 600 codes and standards covering many technical areas, such as cranes and rigging; boilers and pressure vessels; elevators and escalators; piping and pipelines; flanges, fittings, and gaskets; and power plant systems and components. ASME standards are used in more than 100 countries. ASME also has a large technical publishing operation, hosts numerous technical conferences and professional development courses every year, and sponsors many outreach and educational programs that promote engineering and allied sciences.

Government agencies cite many ASME standards as tools to meet their regulatory objectives. ASME standards are voluntary, however, unless the standards have been incorporated into a legally binding business contract or incorporated into regulations that are enforced by an authority that has jurisdiction, such as a federal, state, or local government agency.

> **NOTE**: For more information about ANSI and ASME, visit the websites for these organizations at **www.ansi.org** and **www.asme.org**, respectively.

4.4.3 NFPA Requirements

The National Fire Protection Association (NFPA) is a non-profit US trade association that creates and maintains consensus-based standards and codes for preventing death, injury, and property losses due to fire, electrical, and related hazards. More than 300 ANSI-accredited NFPA codes and standards are available for adoption by local governments and are used throughout the United States as well as in many other countries.

Some of the most widely used NFPA codes include the following:

- *NFPA 1 – Fire Code* (requirements for fire safety and property protection in new and existing buildings)
- *NFPA 70 – National Electrical Code®*
- *NFPA 70B – Recommended Practice for Electrical Equipment Maintenance*
- *NFPA 70E – Standard for Electrical Safety in the Workplace®*
- *NFPA 72 – National Fire Alarm and Signaling Code®*
- *NFPA 101 – Life Safety Code®* (requirements to protect occupants of new and existing buildings from fire, smoke, and toxic fumes)
- *NFPA 704 – Standard System for the Identification of the Hazards of Materials for Emergency Response* (the four-color diamond symbol for identifying the risks posed by hazardous materials)
- *NFPA 921 – Guide for Fire and Explosion Investigations*
- *NFPA 1670 – Standard on Operations and Training for Technical Search and Rescue Incidents*
- *NFPA 1901 – Standard for Automotive Fire Apparatus*

> **NOTE**: All NFPA codes and standards can be viewed (although not printed) free of charge at **www.nfpa.org**

4.4.4 DHS and Coast Guard Requirements

The US Department of Homeland Security (DHS) is a cabinet department of the federal government. In contrast to the Department of Defense, which is charged with military actions abroad, the DHS operates in the civilian sector to protect the United States within, at, and outside its borders. DHS's mission is to prepare for, prevent, and respond to domestic emergencies, particularly terrorism.

Each component agency within DHS has a Designated Safety and Health Official (DSHO) who provides operational program management and oversight for safety and health programs, and develops policy, instructions, standards, requirements and metrics related to safety and health programs within the component.

The US Coast Guard (USCG) operates as a subcomponent of the DHS and is responsible for law enforcement, maritime security, national defense, maritime mobility, and the protection of natural resources. USCG health and safety requirements comply with or exceed applicable OSHA and other federal agency regulations, requirements, and standards. 29 *CFR* Part 1926, *Safety and Health Regulations for Construction* applies to all USCG construction projects and sites.

USCG health and safety policies and requirements are covered in the *USCG Safety and Environmental Health Manual (USCG Commandant Instruction Manual 5100.47A)*. The manual covers a full range of standard health and safety topics (personal protective equipment, hazard communication, confined space entry, etc.) and outlines the notification, analysis, reporting, and recordkeeping requirements for a mishap analysis program.

> **NOTE**
> In USCG terminology, a mishap is essentially the same as an incident. Specifically, the USCG defines a mishap as any unplanned, unexpected, or undesirable event that causes injury, occupational illness, death, material loss, or damage. As explained in the *USCG Safety and Environmental Health Manual*, some mishaps are reportable; others are non-reportable. For more information, visit the Coast Guard website at **www.uscg.mil**.

4.4.5 USACE and NAVFAC Requirements

The US Army Corps of Engineers (USACE, sometimes shortened to CoE) is a federal agency within the Department of Defense. It is one of the world's largest public engineering, design, and construction management agencies. By providing security planning, force protection, research and development, disaster preparedness, and quick response to emergencies and disasters, the USACE supports both the US Department of Homeland Security and the Federal Emergency Management Agency (FEMA).

Safety and health requirements that apply to contractors working on USACE projects are published in the *Engineering Manual (EM) 385-1-1*. *EM 385-1-1* has been adopted by the US Army, Navy, and Air Force, as well as many independent contractors. Generally, EM regulations are in line with OSHA standards, although some EM regulations may be slightly more stringent than what is required by OSHA.

> **NOTE**
> A searchable, electronic version of the EM is available online through the USACE website at **www.usace.army.mil**.

In most cases, before beginning work on a USACE project, contractors must submit an Accident Prevention Plan (APP) that defines how they are going to manage their safety program under the contract. Other required elements within the APP, such as Activity Hazard Analyses (AHAs), are developed and added to the APP during the performance of the contract. AHAs are based on the specific phases of work and the hazards involved and are developed by referencing the additional sections of the EM as needed. *EM 385-1-1* requires contractors to employ at least one competent person at each job site to manage, implement, and enforce the elements of their safety plans.

The Naval Facilities Engineering Command (NAVFAC) is the US Navy's engineering command. It manages the planning, design, and construction of shore facilities around the world for the US Navy and the Marine Corps. Safety and health requirements that apply to contractors working on NAVFAC projects are essentially the same as those for contractors working on USACE projects and are published in *EM 385-1-1*, as described above. Both USACE and NAVFAC require contractors to plan out each phase of a job before actually performing the work.

> **NOTE**
> An APP outline is provided in *Appendix A* of *EM 385-1-1*. Additional electronic tools, forms, templates, and checklists to assist contractors with implementing their health and safety programs are available through an HQ Safety Website link that can be found at the USACE website at **www.usace.army.mil**.

Additional Resources

Basic Safety Administration: A Handbook for the New Safety Specialist, 2003. Fred Fanning, CSP. 2003. Des Plaines, IL: The American Society of Safety Engineers (ASSE).

Construction Project Safety, John Schaufelberger and Ken-Yu Lin. 2014. Hoboken, NJ: John Wiley & Sons.

Construction Safety Planning, David V. MacCollum, P.E., CSP. 1995. Hoboken, NJ: John Wiley & Sons.

The complete *OSHA Safety and Health Regulations for Construction* are located at **www.osha.gov**.

Information about other safety program requirements can be found at the following websites:

American National Standards Institute (ANSI), **www.ansi.org**

American Society of Mechanical Engineers (ASME), **www.asme.org**

Department of Transportation (DOT), **www.dot.gov**

Department of Homeland Security (DHS), **www.dhs.gov**

Environmental Protection Agency (EPA), **www.epa.gov**

Federal Motor Carrier Safety Administration (FMCSA) commercial driver's license (CDL) requirements, **www.fmcsa.dot.gov**

Mine Safety and Health Administration (MSHA), **www.msha.gov**

National Fire Protection Association (NFPA), **www.nfpa.org**

National Institute for Occupational Safety and Health (NIOSH), **www.cdc.gov**

National Occupational Research Agenda (NORA), **www.cdc.gov**

US Coast Guard (USCG), **www.uscg.mil** (includes link to *USCG Safety and Environmental Health Manual [USCG Commandant Instruction Manual 5100.47A]*).

US Army Corps of Engineers (USACE), **www.usace.army.mil** (includes link to *USACE Engineering Manual [EM] 385-1-1*).

US Naval Facilities Engineering Command (NAVFAC), **www.navfac.navy.mil**

The following Code of Federal Regulations references address safety requirements that may apply to construction work:

29 *CFR*, Part 1926 (*Safety and Health Regulations for Construction*).

30 *CFR*, Parts 48, 70, 71, 74, 75, 77, and 90 (*Mine Act standards*).

4.0.0 Section Review

1. The agency that authorizes the enforcement of workplace health and safety standards is _____.
   a. NORA
   b. NIOSH
   c. OSHA
   d. ANSI

2. The EPA law that imposes liability on owners/operators of a facility from which there is a release of hazardous substances into the environment is _____.
   a. CERCLA
   b. CMVSA
   c. CITES
   d. MSHA

3. A valid reason for disqualification from the Commercial Driver's License (CDL) program is _____.
   a. having a serious traffic violation
   b. violating an out-of-service order within a three-year period
   c. using a CMV to transport hazardous materials
   d. leaving the scene of an incident

4. Any construction or maintenance done at a mine or a facility that handles mine products such as portland cement, coal, gravel, and limestone falls under _____.
   a. ASME regulations
   b. MSHA regulations
   c. NIOSH regulations
   d. NAVFAC regulations

Summary

The role of safety technician on a construction site is an important one. The safety technician is responsible for conducting safety inspections; creating job safety plans; organizing and conducting safety training; and investigating and maintaining records of incidents, incidents, injuries, and illnesses. They are responsible for knowing how to prevent incidents and must understand the effects of incident costs on a company. Safety technicians must make sure that workers carefully follow all the regulatory requirements that apply to the given work site.

Review Questions

1. In some cases, a safety technician is responsible for serving as a resource to site management on _____.
   a. emerging market trends
   b. intermodal transportation
   c. operational cost-benefit analyses
   d. environmental regulations

2. The safety term that is defined as a measure of both the probability and the consequences of all hazards of an activity or condition is _____.
   a. harm
   b. incident
   c. safety
   d. risk

3. Incidents that result from the uncontrolled release of energy are classified as _____.
   a. Level I
   b. Level II
   c. Level III
   d. Level IV

4. Incident costs are often classified by the term _____.
   a. net impact
   b. cumulative sum
   c. total incurred
   d. experience modifier

5. An additional safety program element that may be required in order to comply with specific international, federal, state, and local regulations is _____.
   a. management audits
   b. bloodborne pathogens
   c. media relations
   d. cost mitigation

6. As part of their drug and alcohol abuse programs, some larger companies sponsor _____.
   a. smoking cessation clinics
   b. prescription drug disposal services
   c. employee assistance programs (EAPs)
   d. drug/alcohol rehabilitation centers

7. One of the recordkeeping tasks that safety technicians may be required to perform is _____.
   a. reporting workplace illnesses and injuries to OSHA as needed
   b. documenting the company's average EMR per quarter
   c. coordinating OSHA reports with local site safety and health records
   d. using BLS formulas to calculate injury and illness statistics

8. One of the best practices that the Construction Industry Institute (CII) Committee has identified as having the greatest positive impact on safety performance is _____.
   a. management audits and observation
   b. worker involvement and participation
   c. lessons-learned questionnaires
   d. sub-contractor retention bonuses

9. The agency that is authorized to test and certify personal protective equipment is _____.
   a. OSHA
   b. DHS
   c. NIOSH
   d. EPA

10. The program that was developed by NIOSH to facilitate its mission of research collaboration is _____.
    a. NORA
    b. EAP
    c. CDL
    d. SNAP

11. The EPA law that regulates the generation, treatment, storage, and disposal of hazardous waste is _____.
    a. TSCA
    b. SDWA
    c. CAA
    d. RCRA

12. The environmental rules and regulations covering underground petroleum storage tanks fall under _____.
    a. CWA and regional control
    b. RCRA and state control
    c. TSCA and manufacturer control
    d. the Oil Pollution Act of 1990

13. The Commercial Motor Vehicle Safety Act of 1986 established the _____.
    a. minimum national standards that states must meet when licensing CMV drivers
    b. requirement for CMV drivers to obtain a separate federal license
    c. FHWA's testing and licensing standards
    d. requirement for an interstate FHWA/CDL database

14. An employer who knowingly uses a driver to operate a CMV without a valid CDL is subject to a penalty of up to _____.
    a. $2,500
    b. $3,500
    c. $5,000
    d. $10,000

15. The *National Electrical Code*® is a standard that is created and maintained by _____.
    a. ASME
    b. ANSI
    c. NFPA
    d. NTSB

Trade Terms Introduced in This Module

Consensus standards: Standards developed through the cooperation of all parties who have an interest in the use of the standard (e.g., the National Electrical Code®). Consensus standards rely on the expertise of manufacturers, inspectors, craft professionals, maintenance personnel, and safety professionals.

Construction User's Round Table (CURT): CURT describes itself as an autonomous organization that provides a forum for the exchange of information, views, practices and policies of various owners at the national level. Similar groups, called Local User Councils, function at the local level and seek to address problems of cost, quality, safety and overall cost effectiveness in their respective areas.

Culm: Coal refuse.

Direct labor costs: Costs that can be directly related to an incident such as medical costs, workers' compensation insurance, benefits, and liability and property damage insurance payments.

Experience modification rate (EMR): A numeric factor used in determining workers' compensation costs. It rises for contractors with poor incident experience and falls for those with good incident experience.

Gross income: Income before deductions (i.e., taxes).

Gross vehicle weight rating (GVWR): The maximum allowed gross vehicle weight for a vehicle.

Gross unloaded vehicle weight (GUVW): The weight of a liquid cargo trailer without liquids; also known as dry weight.

Job safety analysis (JSA): A careful study of a job or task to find all of the associated hazards and identify methods of safeguarding workers against each hazard.

Net income: Income after deductions (i.e., taxes and expenses).

Profit: Net income over a given period of time.

Release of energy: Events or conditions that release energy from systems, machines, or pieces of equipment.

Surface mines: Mines in which the mining operation is done near the surface of the ground.

Unwarrantable failures: An unwarrantable failure violation (second within 90 days) refers to a situation in which the mine operator knew or should have known that a violation existed and yet failed to take corrective action.

Additional Resources

This module presents thorough resources for task training. The following resource material is suggested for further study.

Basic Safety Administration: A Handbook for the New Safety Specialist, Fred Fanning, CSP. 2003. Des Plaines, IL: The American Society of Safety Engineers (ASSE).
Construction Project Safety, John Schaufelberger and Ken-Yu Lin. 2014. Hoboken, NJ: John Wiley & Sons.
Construction Safety Planning, David V. MacCollum, P.E., CSP. 1995. Hoboken, NJ: John Wiley & Sons.
29 *CFR*, Part 1926 (*Safety and Health Regulations for Construction*).
30 *CFR*, Parts 48, 70, 71, 74, 75, 77, and 90 (*Mine Act standards*).
Extensive safety information is available on the following websites:
American National Standards Institute (ANSI), **www.ansi.org**
American Society of Mechanical Engineers (ASME), **www.asme.org**
The American Society of Safety Engineers (ASSE), **www.asse.org**
Centers for Disease Control and Prevention (CDC), **www.cdc.gov**
Department of Homeland Security (DHS), **www.dhs.gov**
Department of Transportation (DOT), **www.dot.gov**
Environmental Protection Agency (EPA), **www.epa.gov**
Federal Motor Carrier Safety Administration (FMCSA), **www.fmcsa.dot.gov**
Mine Safety and Health Administration (MSHA), **www.msha.gov**
National Fire Protection Association (NFPA), **www.nfpa.org**
US Naval Facilities Engineering Command (NAVFAC), **www.navfac.mil**
OHSAS 18001, Occupational Health and Safety Management, **www.nsai.ie**
OSHA Safety and Health Regulations for Construction, **www.osha.gov**
US Army Corps of Engineers (USACE), **www.usace.mil**
US Coast Guard (USCG), **www.uscg.mil**

Figure Credits

LPR Construction, Module opener
U.S. Department of Labor, Figure 3

Section Review Answer Key

Answer	Section Reference	Objective
Section One		
1. c	1.1.0	1a
2. b	1.2.0	1b
Section Two		
1. c	2.1.0	2a
2. d	2.2.0	2b
Section Three		
1. b	3.1.1	3a
2. c	3.2.0	3b
Section Four		
1. c	4.1.0	4a
2. a	4.2.0	4b
3. d	4.3.3	4c
4. b	4.4.1	4d

NCCER CURRICULA — USER UPDATE

NCCER makes every effort to keep its textbooks up-to-date and free of technical errors. We appreciate your help in this process. If you find an error, a typographical mistake, or an inaccuracy in NCCER's curricula, please fill out this form (or a photocopy), or complete the online form at **www.nccer.org/olf**. Be sure to include the exact module ID number, page number, a detailed description, and your recommended correction. Your input will be brought to the attention of the Authoring Team. Thank you for your assistance.

Instructors – If you have an idea for improving this textbook, or have found that additional materials were necessary to teach this module effectively, please let us know so that we may present your suggestions to the Authoring Team.

NCCER Product Development and Revision
13614 Progress Blvd., Alachua, FL 32615

Email: curriculum@nccer.org
Online: www.nccer.org/olf

❏ Trainee Guide   ❏ Lesson Plans   ❏ Exam   ❏ PowerPoints   Other _____

Craft / Level: _____   Copyright Date: _____

Module ID Number / Title: _____

Section Number(s): _____

Description: _____

_____

_____

_____

Recommended Correction: _____

_____

_____

_____

Your Name: _____

Address: _____

Email: _____   Phone: _____

Positive Safety Communication

OVERVIEW

This module explains how to support positive safety communication on the job site, including communication techniques, motivation, and effective responses to behavioral issues.

Module 75205

Trainees with successful module completions may be eligible for credentialing through NCCER's National Registry. To learn more, go to **www.nccer.org** or contact us at 1.888.622.3720. Our website has information on the latest product releases and training, as well as online versions of our *Cornerstone* magazine and Pearson's product catalog.

Your feedback is welcome. You may email your comments to **curriculum@nccer.org**, send general comments and inquiries to **info@nccer.org**, or fill in the User Update form at the back of this module.

This information is general in nature and intended for training purposes only. Actual performance of activities described in this manual requires compliance with all applicable operating, service, maintenance, and safety procedures under the direction of qualified personnel. References in this manual to patented or proprietary devices do not constitute a recommendation of their use.

Copyright © 2018 by NCCER, Alachua, FL 32615, and published by Pearson, New York, NY 10013. All rights reserved. Printed in the United States of America. This publication is protected by Copyright, and permission should be obtained from NCCER prior to any prohibited reproduction, storage in a retrieval system, or transmission in any form or by any means, electronic, mechanical, photocopying, recording, or likewise. To obtain permission(s) to use material from this work, please submit a written request to NCCER Product Development, 13614 Progress Blvd., Alachua, FL 32615.

75205 V2

From *Safety Technology,* Trainee Guide. NCCER.
Copyright © 2018 by NCCER. Published by Pearson. All rights reserved.

75205
POSITIVE SAFETY COMMUNICATION

Objectives

When you have completed this module, you will be able to do the following:

1. Identify and describe the three key elements for supporting positive safety communication on a job site.
   a. Identify the three types of communication and describe how to use them to communicate effectively with all employees on a job site.
   b. Describe the ARCS model of motivation.
   c. Describe how to respond to behavioral problems.

Performance Task

Under the supervision of your instructor, you should be able to do the following:

1. Communicate safety policies and procedures to all employees on a job site.

Trade Terms

ARCS Model of Motivation
Behavioral-based safety (BBS)
Communication
Diversity
Ethnic groups
Feedback
Jargon
Message
Non-verbal communication

Paraphrase
People-based safety
Receiver
Sender
Verbal communication
Visual communication
Written communication

Industry Recognized Credentials

If you are training through an NCCER-accredited sponsor, you may be eligible for credentials from NCCER's Registry. The ID number for this module is 75205. Note that this module may have been used in other NCCER curricula and may apply to other level completions. Contact NCCER's Registry at 888.622.3720 or go to www.nccer.org for more information.

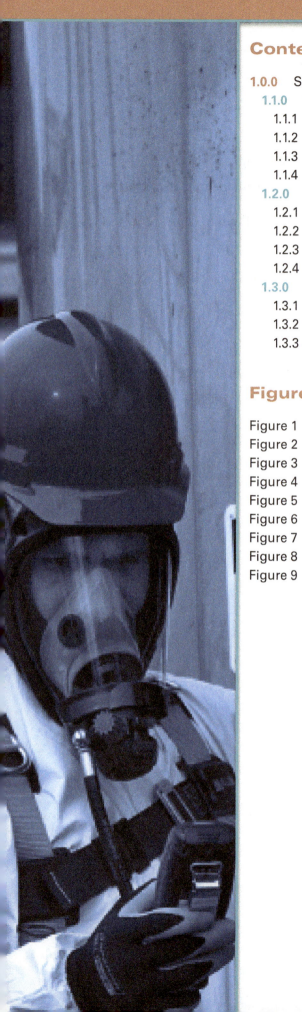

Contents

1.0.0 Supporting Positive Safety Communication 1
    1.1.0 Communication 1
        1.1.1 Verbal Communication 2
        1.1.2 Non-Verbal Communication 3
        1.1.3 Written or Visual Communication 3
        1.1.4 Communication Issues 5
    1.2.0 Motivation 6
        1.2.1 Gaining Attention 7
        1.2.2 Building Relevance 8
        1.2.3 Instilling Confidence 8
        1.2.4 Fostering Satisfaction 8
    1.3.0 Behavior Modification 9
        1.3.1 Poor Attitude Towards the Workplace 12
        1.3.2 Inability to Work with Others 12
        1.3.3 Absenteeism and Turnover 13

Figures

Figure 1   Communication process. 2
Figure 2   Example of written and visual communication. 4
Figure 3   Tailor your message. 6
Figure 4   The ARCS Model of Motivation 7
Figure 5   Safety plaque. 8
Figure 6   Behavior modification matrix 9
Figure 7   Sample safety behavior checklist. 10
Figure 8   Overview report sample. 10
Figure 9   Leadership style and group phase matrix 11

Section One

1.0.0 Supporting Positive Safety Communication

Objective

Identify and describe the three key elements for supporting positive safety communication on a job site.

a. Identify the three types of communication and describe how to use them to communicate effectively with all employees on a job site.
b. Describe the ARCS model of motivation.
c. Describe how to respond to behavioral problems.

Performance Task

1. Communicate safety policies and procedures to all employees on a job site.

Trade Terms

ARCS Model of Motivation: A strategy for motivating people developed by Dr. John Keller. The strategy states that to motivate individuals you must gain their attention, make the issue relevant to them, help them to feel confident that they can be successful, and provide them with a sense of satisfaction once they have achieved their goal.

Behavioral-based safety (BBS): A proactive method of safety management based on psychology. It requires systematic workplace observation, analysis of unsafe behaviors, and resolution of problems, coupled with training and incentives for behavior modification.

Communication: A process by which information is exchanged between individuals through a common system of symbols, signs, or behavior.

Diversity: Differences between individuals, particularly with regard to race, religion, ethnicity, and gender.

Ethnic groups: Large groups of people classed according to common racial, national, tribal, religious, linguistic, or cultural origin or background.

Feedback: The communication that occurs after a message has been sent and received. This communication from the receiver enables the sender to determine whether or not the message was received accurately.

Jargon: Technical terminology known only by people who work directly with the technology being discussed.

Message: The information that the sender is attempting to communicate to the receiver.

Non-verbal communication: Communication achieved through non-spoken means, such as body language, facial expressions, hand gestures, and eye contact.

Paraphrase: A restatement of a text, passage, conversation, or work process that is explained without changing its meaning.

People-based safety: A safety management system centered on empowering the employee to make decisions both for their own safety and the safety of those around them, including giving all workers the authority to stop work in the event of unsafe conditions or behaviors.

Receiver: The person to whom the sender is communicating a message.

Sender: The person who creates the message to be communicated.

Verbal communication: Transfer of information through spoken word. This process involves a sender, receiver, message, and feedback.

Visual communication: Communication through visual aids such as signs, postings, and hand signals.

Written communication: Transfer of information through the written word.

Workers are more likely to work safely if they are motivated, recognized for their good behavior and accomplishments and, when necessary, disciplined fairly, consistently, and properly. Safety technicians must communicate with workers to ensure that they know what is expected of them.

The safety technician has a leadership role on a job site. This means that workers look to them to provide information and guidance about how to get the job done safely. As such, safety technicians need to have the respect of the workers. When an individual is respected and approachable, discipline problems are reduced. This makes it easier to concentrate on keeping the work site and workers safe.

1.1.0 Communication

Effective communication with people at all levels on a job site is an important part of a safety technician's job. In order to be successful, you must develop an understanding of human behavior and have communication skills that enable you to understand and influence others. This includes knowing your audience and adapting the way you dress, speak, etc. accordingly, so that your message will be conveyed most effectively.

There are many definitions for communication. One definition is that communication is the act of accurately and effectively conveying or transmitting facts, feelings, and opinions to another person. Another is that communication is the method of exchanging information and ideas.

Just as there are many definitions for communication, it also comes in more than one form. The different types of communication include the following:

- Verbal
- Non-verbal
- Written

A typical person spends about 80 percent of his or her day communicating through writing, speaking, listening, or using non-verbal communication, such as body language. Of that time, studies suggest that approximately 20 percent of communication is written, and 80 percent involves speaking or listening.

1.1.1 Verbal Communication

Verbal communication refers to the spoken words exchanged between two or more people. It can be done face-to-face or by other means, such as the telephone or two-way radios.

Figure 1 shows the relationship of the five primary components of verbal communication. These components are identified as follows:

- A sender that transfers information
- A message that contains the information to be communicated
- A receiver to accept the message
- Feedback that indicates whether or not the receiver received the desired information
- Distractors of various types that can interfere with or prevent reception or comprehension

The sender is the person who creates the message to be communicated. In verbal communication, the sender actually says the message aloud to the person or persons for whom it is intended. The sender must be sure to speak in a clear and concise manner that can be easily understood by others. This is not an easy task, but with practice anyone can become an effective sender.

Some basic guidelines for becoming an effective sender include the following:

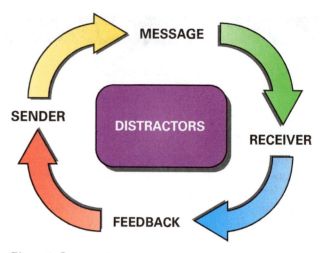

Figure 1 Communication process.

- Avoid talking with anything in your mouth, such as food or gum.
- Find an appropriate rate of speech. Don't talk too quickly or too slowly. This is important, because in some situations people tend to focus on your rate of speech instead of on what you are saying.
- Be aware of your tone. Remember, it's not so much what you say, but how you say it.
- Enunciate to prevent misunderstandings; many letters, such as T, D, B, and P sound similar.
- Don't talk in a monotone. Put some enthusiasm and feeling into your voice.

The message is what the sender is attempting to communicate to the receiver. A message can be a set of directions, an opinion, or a feeling. Whatever its function, a message is an idea or fact that the sender wants the receiver to know.

Before speaking, the sender should determine what it is he or she wants to communicate. The sender should then organize what to say, ensuring that the message is logical and complete. Taking the time to clarify your thoughts prevents rambling, not getting the message across effectively, or confusing the receiver. It also allows you to get to the point quickly.

In delivering the message, the sender should consider the audience. It is important not to talk down to people. Remember that everyone, whether in a senior or junior position, deserves respect and courtesy. Therefore, the sender should use words and phrases that the audience can understand. Try to avoid technical language or slang, when possible. Use short sentences; this gives the audience time to understand and digest one point or fact at a time.

The receiver is the person to whom the message is communicated. For the communication process to be successful, it is important that the receiver understands the message as the sender intended. Therefore, the receiver must listen to what is being said. The first step to becoming a good listener involves realizing that there are many barriers, or distractors, that can get in the way of your listening, particularly on a busy construction job site.

The following are barriers to effective listening:

- Noise, visitors, telephone, or other distractions
- Preoccupation, being under pressure, or daydreaming
- Reacting emotionally to what is being communicated
- Thinking about how to respond instead of listening
- Giving an answer before the message is complete
- Personal biases
- Finishing the sender's sentences

The following tips will help you to overcome these barriers:

- Take steps to minimize or remove distractions; learn to tune out your surroundings.
- Listen for key points.
- Take notes.
- Try not to take things personally.
- Allow yourself time to process your thoughts before responding.
- Let the sender communicate the message without interruption.
- Be aware of your personal biases and try to stay open-minded.

As a receiver, there are many ways to show that you are actively listening to what is being said. This can be accomplished by maintaining eye contact, nodding your head, and taking notes. It may also be accomplished through feedback.

Feedback refers to the communication that occurs after the message has been sent and received. It involves the receiver responding to the message. Feedback is an important part of the communication process because it allows the receiver to communicate how he or she interpreted the message. It also lets the sender know whether the message was understood as intended. In other words, feedback is a way to make sure that the receiver and the sender understand each other.

The receiver can use the opportunity of providing feedback to paraphrase back what was heard. When paraphrasing, it is best to use your own words. This shows the sender that you interpreted the message correctly and could explain it to others if necessary.

In addition, feedback gives the receiver the opportunity to clarify the meaning of the message and request additional information if needed. This is generally done by asking questions.

1.1.2 Non-Verbal Communication

Non-verbal communication refers to things that you can actually see when communicating with others. Examples include facial expressions, body movements, hand gestures, and eye contact. Non-verbal communication can provide an external indication of an individual's inner emotions. It occurs simultaneously with verbal communication, and the sender of the non-verbal communication is often not even aware of it.

Because it can be physically observed, non-verbal communication is just as important as the words used to convey the message. This is true because people are influenced more by non-verbal signals than by spoken words. It is important to be conscious of the non-verbal cues that you send. You don't want the receiver to interpret your message incorrectly based on your posture or an expression on your face. When communicating, try not to carry over issues and incidents from earlier in the day that could affect the current communication exchange.

> **NOTE**
> Standing with your arms crossed over your chest or with your hands on your hips is considered an aggressive or threatening stance. Consider standing with your arms at your sides when speaking to another individual. This helps to ensure that the receiver does not feel threatened or intimidated.

1.1.3 Written or Visual Communication

Much of our communication is written or visual. Written communication or visual communication refers to communication that is documented on paper or transmitted electronically via the computer using words or images. Written or visual communication can also include signs, hand signals, and postings (*Figure 2*).

Many messages on a job have to be communicated in text form. Examples include weekly reports, requests for changes, purchase orders, and correspondence on specific subjects. These items are written because they must be recorded for contractual and historical purposes. In addition, some communication on the job has to be visual because some items that are difficult to explain verbally or by the written word can best be explained through diagrams or graphics. Examples include the plans or drawings used on a job.

6 Easy Steps That Could Save Your Life

How To Don A Harness

1
Hold harness by back D-ring. Shake harness to allow all straps to fall in place.

2
If chest, leg and/or waist straps are buckled, release straps and unbuckle at this time.

3
Slip straps over shoulders so **D-ring is located in middle of back between shoulder blades.**

4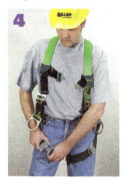
Pull leg strap between legs and connect to opposite end. Repeat with second leg strap. If belted harness, connect waist strap after leg straps.

5
Connect chest strap and position in midchest area. Tighten to keep shoulder straps taut.

6
Snug Fit

After all straps have been buckled, **tighten all buckles so that harness fits snug but allows full range of movement.** Pass excess strap through loop keepers.

Figure 2 Example of written and visual communication.

It is important to understand your audience before writing or creating a visual message. This helps to ensure that the receiver is able to read the message and understand the content. If the receiver is unable to do this, the communication process will be unsuccessful. You need to consider the actual meaning of words or diagrams and how others might interpret them. It's also important for your handwriting to be legible so the message can be understood.

Some basic guidelines for writing include the following:

- Avoid emotion-packed words or phrases.
- Be positive whenever possible.
- Avoid using technical language or jargon.
- Stick to the facts.
- Provide an adequate level of detail.
- Present the information in a logical manner.
- Avoid making judgments, unless asked to do so.
- Check the message for spelling and grammatical errors.
- Make sure the document is legible.
- Be prepared to provide a verbal or visual explanation, if needed.

Some basic guidelines for creating effective visuals include the following:

- Provide an adequate level of detail.
- Make sure that the diagram is large enough to be seen.
- Avoid creating complex visuals.
- Present the information in a logical order.
- Be prepared to provide a written or verbal explanation of the visual, if needed.

1.1.4 Communication Issues

Each person communicates a little differently because we are all unique individuals. The diversity of the workforce can make communication challenging. Keep in mind that your audience consists of individuals from different ethnic groups and cultural backgrounds, with varying levels of education and economic status. You must understand your audience to determine how to effectively communicate with each individual.

The key to effective communication is to acknowledge that people are different. It is important to be able to adjust your communication

Overcoming Communication Barriers in a Diverse Workforce

Today's typical workforce is a diverse mix of people of different generations, genders, races, nationalities, and cultural and ethnic origins. It's easy for communication to get garbled, and for misunderstandings to occur. Overcoming the communication barriers that are likely to exist in a diverse workforce is essentially a matter of applying the standard basic communication techniques for ensuring clear communication and putting particular effort into knowing your audience. Miscommunication often occurs when you mistakenly assume that others think and act as you do. As a result, you can believe you're communicating clearly when you really aren't. A few tips for overcoming this problem include the following:

- Avoid jargon and slang.
- Keep it simple. Use short sentences, clear, simple words, and active voice.
- Use active listening techniques, such as paraphrasing, summarizing, and seeking feedback.
- Be aware of your personal biases and avoid stereotyping, making generalizations, and jumping to conclusions.
- When possible, use the communication media that your audience prefers—phone, texting, email, face-to-face meetings, written reports, visual demonstrations, etc.
- When communicating with people for whom English is not their first language, use a combination of types of communication to help get your message across. For instance, in addition to speaking, use written handouts, visual aids, and demonstrations.
- Be aware that body language, hand gestures, and other non-verbal and visual forms of communication can be misinterpreted due to cultural differences. For instance, in the United States it's considered important to make eye contact with someone who is speaking with you, so the speaker won't get the impression that you're distracted or uninterested. In many Asian countries, however, eye contact can be considered a sign of disrespect or a challenge to authority. Similarly, in the United States and most of Europe, the thumbs-up gesture means approval, but in Islamic countries this gesture is considered rude and offensive.
- To promote better understanding and communication among members of a diverse workforce, encourage workers to teach each other about their differing perspectives, languages, and social conventions.

style to meet the needs of those on the receiving end of your message. This involves relaying your message in the simplest way possible and avoiding the use of words that some may find confusing. Be aware of how you use technical language, slang, jargon, and words that have multiple meanings. Present the information in a clear, concise manner. Avoid rambling, and always speak clearly, using good grammar.

You may have to communicate your message in multiple ways or adjust your level of detail or terminology to ensure that everyone understands your meaning as intended. For instance, you may have to draw a map for a visual person who cannot comprehend directions in a verbal or written form, or you may have to overcome language barriers on the job site by using graphics or visual aids to relay your message.

Figure 3 shows how to tailor your message to your audience.

Workplace stress can also affect communication. Stress has been associated with loss of appetite, ulcers, mental disorders, migraines, difficulty in sleeping, emotional instability, disruption of social and family life, and the increased use of cigarettes, alcohol, and drugs. Stress also affects workers' attitudes and behavior. Some frequently reported consequences of stress are difficulties in communicating, maintaining pleasant relations with co-workers, and judging the seriousness of a potential emergency.

The following suggestions may help to relieve worker stress:

- Educate employees about job stress.
- Address work-related stresses, such as unreasonable work load, lack of readily available resources, or inadequate and unsafe equipment.
- Have regular safety meetings and discussions.
- Establish stress management programs.
- Provide flexibility and innovation by supervisors to create alternative job arrangements.
- Provide an organized and efficient work environment.

Safety technicians should make sure workers are taking positive steps to relieve stress. Safety is likely to suffer if stress is not addressed; quality and productivity can also be affected.

1.2.0 Motivation

The ability to motivate others is a key leadership skill that effective safety technicians must possess. Motivating means getting people to act or perform at a high level consistently. Because safe practices often add steps to a task, time to a job, or paperwork to a process, employees must be motivated to perform their job duties safely. Sometimes the decision to work safely is a choice between doing a task the quick-and-easy way or doing the same task the correct and safe way. If properly instructed and motivated, an employee will choose the correct and safe way every time.

Different people learn in different ways. Be sure to communicate so you can be understood.

Figure 3 Tailor your message.

Motivational techniques are often considered to be limited to praising an employee, giving bonuses and pay raises, scolding someone, or firing an employee. While these may all be effective pieces of a motivational strategy, there is much more to consider. According to Dr. John Keller of Florida State University, there are four major parts to a successful motivational strategy: gaining attention, building relevance, instilling confidence, and fostering satisfaction. *Figure 4* illustrates why Dr. Keller's model is referred to as the ARCS Model of Motivation.

While it is unlikely that you will have the authority to promote someone, give bonuses or raises, or fire someone, you can still be an effective and persuasive motivator by using the ARCS Model.

1.2.1 Gaining Attention

For communication to be effective, the receiver must be open and receptive to the message. To motivate employees, the safety technician must first gain their attention. This can be done in a number of ways, such as having them talk about hazardous situations or near misses they have experienced; citing accurate and remarkable statistics; and showing or describing shocking (but not excessively graphic) images.

When working to get employees' attention, it is useful to have them share actual work-related experiences involving hazardous situations. These accounts are usually of great interest and, because they come from co-workers, are credible. These experiences can be brought up in a conversation or safety meeting by asking for stories about past mishaps relevant to the topic of discussion. For example, if the issue of wearing a hard hat needs to be addressed, ask if anyone has seen an incident where a hard hat saved a life or prevented a serious injury. It is likely that if there are some experienced workers in the group, someone will have a story to tell. When asking employees to share their stories, caution them to not use the names of the companies involved. Also, have ready an example or two of your own in case no one volunteers.

Numbers often help to get someone's attention. You could say, for example, that according to the Bureau of Labor Statistics, construction work is one of the most dangerous occupations, and that every day in America three people die on a construction site. That should grab the attention of any construction worker, who will likely then be eager to listen to the information coming next.

You might want to ask a challenging question. In a discussion about hearing protection, for example, begin with a question like, "What is your hearing worth to you? Can you quantify it or is it priceless? Imagine never hearing the voices of your family again." Then, reinforce the hazard by presenting some statistics. For example, "According to NIOSH, noise-induced hearing loss is the most common work-related injury." You may also wish to tailor your questions to the audience. For example, you might ask older workers, "Would you allow a child to play here?"

Showing photographs of a safety incident can be a very effective way to get someone's attention if it is done using good judgment and care. Pictures that show the nonhuman impact of an incident are often enough for the employees to personalize the image so that you gain their attention. Images of tipped vehicles may be a good way to initialize a conversation about speed limits. Images of a burned-out building would be an effective way to begin a conversation on fire prevention.

You can also use your experience in the safety field to describe incidents that you have witnessed. For example, you might describe how removing the guards on a saw caused a worker to lose a finger.

You can also illustrate how incidents occur by using models, props, or videos. Many equipment manufacturers supply safety videos with animations of specific hazards. You can also use props or models to show how events may turn dangerous. For example, you might have someone sit in a wheeled office chair, and then toss them a rubber ball. The shift in weight and momentum caused by catching the ball causes the chair to move, which can emphasize the importance of wheel chocking.

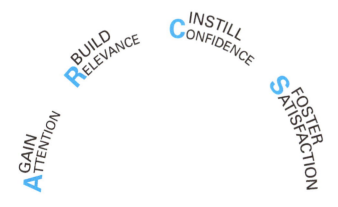

Figure 4 The ARCS Model of Motivation.

1.2.2 Building Relevance

Stories, statistics, images, and illustrations used to gain a person's attention will not have a long-term impact on their own. To effectively motivate people, you must show them how the information is relevant to them personally. You can do this by finding out what is important to the person(s) you're talking to and then help them make the link between what you have to say and what is already important to them.

For example, to begin a safety meeting on hearing protection you might ask a question like, "What is one of your favorite sounds?" This question will almost certainly bring many personal responses. To make the subject of hearing protection relevant, follow these responses with a statement like, "Wearing proper hearing protection could be the difference between losing your hearing and being able to enjoy those sounds for the rest of your life." By making the topic of hearing protection relevant and important on a personal level, it becomes much more likely that they will take the time to obtain and use appropriate hearing protection on the job.

> **NOTE:** If you don't have experience with the task to be performed, find a subject matter expert to help you understand the challenges.

1.2.3 Instilling Confidence

Giving employees the sense that they can be successful in preventing incidents, injuries, and fatalities is a very important key to motivation. If a person expects to fail, he or she will likely not put much effort into a given job. The opposite is true of workers who are confident that they can achieve their goals. They will put in the effort needed to get the job done. Confidence comes from knowing what is expected and having the tools and skills to do it.

> **NOTE:** Praising in public boosts confidence. Disciplining, however, should be done in private as soon as possible after the fact.

Recommendations for building confidence include the following:

- Communicate clear and realistic goals
- Provide employees with the proper tools and equipment
- Provide adequate training
- Reinforce safe behavior

1.2.4 Fostering Satisfaction

For someone to continue to act or perform at a high level, the individual must get some satisfaction from his or her actions. For many employees, there is an internal sense of satisfaction that comes from performing their job well and staying safe on the job. Likewise, fellow employees can be highly effective at motivating their co-workers. However, safety technicians must find ways to create additional reinforcing opportunities for employees by developing recognition programs. Some suggestions for employee recognition programs include the following:

- Many organizations keep track of and set goals for days in operation without a lost-time injury. These provide the whole organization with a goal to attain or a record to break. Once that record is broken, the organization can provide further satisfaction by hosting an event or a party to celebrate.
- To provide satisfaction for individual employees, many organizations provide awards for safe practices. These awards may include plaques or trophies. *Figure 5* shows a typical employee award plaque.
- Monetary awards can also be offered; these might include bonuses, gift certificates, or pay raises.
- Long-term and consistently high performance is sometimes recognized with job opportunities or promotions.

Figure 5 Safety plaque.

It's important to note that these types of incentive programs can turn out to be counterproductive. They can inadvertently discourage reporting and drive incidents underground. When this happens, the problems that caused the incident cannot be resolved. Make sure that there are positive incentives for reporting. This will encourage workers to report safety incidents and help management make the needed repairs or improvements.

1.3.0 Behavior Modification

In a fast-paced, deadline-oriented industry such as construction, there will always be behavioral problems to resolve and decisions to be made. The safety technician has a role in assisting the frontline supervisor in this regard. Behavior modification—influencing worker behavior to create a safe environment—is a large part of everyday work for a safety technician. Achieving this, however, requires good communication skills, a sound understanding of human psychology, and awareness of the fact that managing safety can be a personal issue for people.

Behavioral-based safety (BBS) is a method of applying psychology to workplace safety. It's based on a matrix of options that can be used to modify the behavior of an individual. As shown in *Figure 6*, the options include positive and negative punishment and positive and negative reinforcement. Simply put, the goal of either form of reinforcement is to increase a desired behavior, whereas either form of punishment is used to decrease an undesired behavior.

Positive punishment adds something unpleasant in order to decrease an undesired behavior. For instance, being publicly scolded by the instructor when your cell phone rings during a training session decreases the likelihood of your failing to turn off your phone at the next training session.

Negative punishment is a kind of penalty; it takes away or reduces something an individual enjoys or wants. For example, in football an offending team has to give up yardage.

Negative reinforcement, by contrast, makes a desired behavior more likely by taking away or reducing something that an individual does not enjoy; in other words, it's a relief from some form of unpleasantness. Having to fasten the seat belt to silence an annoying noise when your car's ignition key is turned on is an example of negative reinforcement intended to get you to put the seat belt on.

Positive reinforcement, on the other hand, accelerates or increases a behavior by adding something that an individual desires. Adding praise, for example, is a simple way to increase the practice of employees towards keeping a job site free of debris.

Research has shown that positive reinforcement is the most powerful of the four options for behavior modification. Adding a positive to increase a response not only works better, it also focuses on the positive aspects of the situation. Punishment is often counterproductive because it can engender a negative response such as anger or resentment on the part of the individual who is being punished.

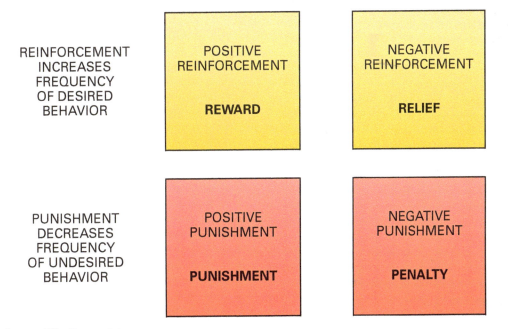

Figure 6 Behavior modification matrix.

> **NOTE:** Keep in mind that managing safety is a personal issue. To be most effective, any positive reinforcement must be closely tailored to the specific individual.

BBS programs require the involvement of everyone in a company. Every level of workforce and management plays a part. The majority of incidents occur at the workforce level, the people who are most likely to be hurt. As a result, they have the strongest interest in being involved in the elimination of unsafe behavior.

BBS is not implemented as a top-down strategy. Instead, both workforce and managers are seen as team members. Employee feedback is welcomed and utilized, not ignored. All workers are involved in the process.

Management must also be committed to the program. BBS requires that observers watch workers to improve their safety performance. Observers must be trained and given sufficient time and materials to complete their tasks. Feedback and positive reinforcement must be provided. BBS is an ongoing program; it cannot work with a half-hearted commitment.

In most industries, there are a few unsafe behaviors that result in many incidents. These behaviors are observable. Studying safety history, such as incidence reports, can help to identify these behaviors. Determining the causes of unsafe behavior requires more analysis. For example, unsafe behaviors can be compiled into a checklist (*Figure 7*) which should be constantly revised. Any worker might enter new, updated unsafe behaviors. Conversely, old checklist items can be removed as behavioral hazards are eliminated.

Note that the Comments column in the sample checklist includes a note stating that names are not to be used. This is because BBS works when behaviors are observed, not individuals. Punishment is not part of BBS because, as previously mentioned, punishment tends to have a negative effect. Instead of encouraging safe behavior, it discourages honest observation. It also builds resentment. Workforce support disappears, and the process becomes meaningless.

The observer's task is to monitor a worker's safety behavior. This must be done on a regular basis. Frequent observation provides more accurate data and it also affects behavior. Workers are more likely to engage in safe behavior when they

SAFETY PRACTICE	Y	N	COMMENTS (NO NAMES!)
EYES ON PATH OR WORK			
PROPER BODY MECHANICS			
CLEAR LINE OF SIGHT			
PINCH POINTS			
WORK PACE			
GETTING HELP			

Figure 7 Sample safety behavior checklist.

OVERVIEW REPORT				
Dates From: 01 Jan 00 To: 01 Apr 00	Total People Observed 575		Total Observations 575	
	SAFES	CONCERNS	% SAFES	SAMPLE
BODY / WORK POSITION				
EYES ON PATH OR WORK	471	28	94%	495
PROPER BODY MECHANICS WHEN LIFTING REACH	203	35	85%	238
CLEAR LINE OF SIGHT	244	32	88%	275
PINCH POINTS	124	14	90%	137
WORK PACE	406	7	98%	413
GETTING HELP	106	5	95%	111
Category Totals:	1554	121	92%	
TOOLS AND EQUIPMENT				
SELECTION AND USE OF EQUIPMENT	120	3	98%	123
FORKLIFT OPERATION	63	16	80%	76

Figure 8 Overview report sample.

are aware of being observed. This creates safer habits even when no observation is taking place.

Decisions can be based on behavioral observation. This requires turning checklist results into measurable qualities. The percentage of observed behaviors safely performed is a common possibility (*Figure 8*). Analyzing trends leads to safety process improvement. The reasons behind unsafe practices often become evident through analysis.

Improvement begins to occur after BBS has been put into action. There's an introductory period of about four weeks as people adjust to their new roles and observers are trained. Unsafe behaviors are identified, checklists are developed, and baseline behavior is established.

After the initial period, goals are set. These are not executive-level goals, but realistic targets set by the workforce. Observation continues, and goals are re-examined regularly. Improvement becomes a continuous process.

Everyone needs to know what's happening in order for BBS to work. Observers should talk to workers on the spot about their behavior. Larger-scale findings can be displayed on posters or discussed in regular meetings. Once at-risk behaviors are identified, request input from employees as to possible causes and solutions. When recommendations are implemented, make follow-up observations to see if there has been a positive change in the percentage of safe observations. As one behavior is corrected, the team can focus on others. This makes BBS a continuous process.

BBS is not a magic bullet. It is a process that must be managed and adjusted on a regular basis. Likewise, you can better influence the behavior of employees if you understand the dynamics of group behavior. The behavior of a group varies depending on which of the three primary phases of group progression exists. The three primary stages, or phases, of group progression are as follows:

- *Strangers* – When a group first meets, members begin to form impressions of where they will fit in. Everyone is interested in making a good impression to some degree and is generally behaving at his or her best. However, some individuals may fear rejection and try to go unnoticed if possible. Group members may be reluctant to speak up and become involved; they tend to be careful about what is said and concerned about how it will be received. There is relatively little trust among the members of the group, although their normal social skills may prevent them from revealing this in any way.

- *Coming together* – In this phase, individuals begin to feel like they are part of the group. Patterns of influence begin to develop, and cliques may form. This is also the time members learn to work together toward common goals. A shared vision develops within the group as a whole. Members are becoming familiar with the strengths and weaknesses of individuals, and informal group leaders are more easily identified as they begin to emerge. More constructive communication begins to take place between the company-designated leader (*e.g.*, supervisor, manager, safety technician) and the group, as well as within the group itself.

- *Cohesiveness* – At this stage, trust has now developed within the group and the individuals feel a common bond. The group now seems more like a team and a single entity instead of a group of strangers. Members of a group in this phase usually possess a greater sense of security and will participate more readily. However, members may still be reluctant to disagree with other members of the group. Informal leaders have emerged and are often sharing leadership within the group, although this may be subtle.

Just as the group goes through phases, so safety technicians should transition through various leadership styles. *Figure 9* graphically illustrates how leadership styles relate to the phases of a group.

During the Stranger phase, you generally need to use a commanding style. Both you and the group members are getting to know one another, and group members have no sense of what to

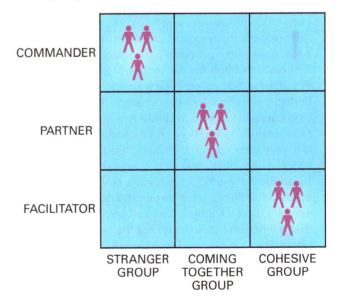

Figure 9 Leadership style and group phase matrix.

do other than follow directions. As a result, you must provide clear direction for the group and concentrate on bringing the group together on a single path. Quickly providing guidance and revealing a common direction to the group eliminates some unknowns and reduces anxiety for individual members.

As the group comes together, it becomes easier to involve shy and reserved individuals, and for you to take on a more supportive Partnering role. Once the individuals form a cohesive group, the Facilitating style of leadership becomes more appropriate.

One of the keys to success in matching leadership style with group phases is to understand that groups often change even after cohesiveness. It is rare that you can transition to a facilitator's role and remain that way until a project's completion. Disruptions of one type or another will occur, and it is up to you to quickly change styles and get the group back on track when it strays off course.

It is likely that a safety technician will encounter both simple and complex behavioral problems and be forced to make decisions about how to respond. A few of the most common employee problems include the following:

- Poor attitude towards the workplace
- Inability to work with others
- Absenteeism and turnover

Human Performance and Practicing Perfection®

On many job sites, the approach to safety is reactive rather than proactive. In other words, it focuses on the response to an incident, such as rescue after a fall, rather than preventing the incident in the first place through the use of proper training and PPE. A proactive safety culture known as Practicing Perfection® is an alternative strategy that seeks to encourage workers to continuously strive for improvement rather than blindly performing tasks that may be unsafe or inefficient. It is based on the concept that most cases of human error can actually be attributed to process, program, or organizational errors. In addition, it focuses on the desire of workers to excel at their jobs and acknowledges that the workers themselves are the best source of information on process improvement.

Source: Practicing Perfection Institute, www.ppiweb.com.

Similar to behavior-based safety, people-based safety also focuses on assessing behavior and providing positive motivation for change. People-based safety is a safety management system centered on empowering the employee to make decisions for both their own safety and the safety of those around them, including giving all workers the authority to stop work in the event of unsafe conditions or behaviors.

1.3.1 Poor Attitude Towards the Workplace

It is not unusual to come across employees who have a poor attitude towards the workplace. Employees with a poor attitude may create safety hazards in two ways. First, they may unintentionally create unsafe working conditions through sloppy or shoddy work, unprofessional behavior, and carelessness. Second, they may intentionally create unsafe working conditions through tool and equipment sabotage, material damage, and fighting.

A poor attitude on the job may be caused by bad relationships with fellow employees, personal problems, negative perceptions of supervision, or dislike for the job in general. The first step in changing an employee's poor attitude is to determine the cause of the problem.

The best way to determine the cause of a poor attitude is to talk with that employee one-on-one, listen to what he or she has to say, and ask questions to uncover information. Once you've had this conversation and assessed the facts, you can determine how to correct the situation and turn the negative attitude into a positive one.

If you discover that a problem stems from factors in the workplace or the surrounding environment, there are options available to remedy the situation. First, the worker can be moved from the situation to a more acceptable work environment. Or, the part of the work environment found to be causing the poor attitude can be changed. Finally, steps can be taken to change the employee's attitude so that the work environment is no longer a negative factor.

1.3.2 Inability to Work with Others

There will be situations in which an employee has a difficult time working with others. This could be the result of personality differences, an inability to communicate, or some other cause. Whatever the reason, the issues must be addressed, and the crew must be able to work together as a team. Teamwork is one of the best ways to prevent incidents and to create a safe working environment.

The best way to determine the reason that individuals don't get along or don't work well together is to talk to the parties involved. Speak with the employees and any other individuals on site to determine the source of the conflict.

Once the reason for the conflict is found, it is easier to determine how to respond. There may be a way to resolve the problem and get the workers to communicate and work as a team again. On the other hand, there may be nothing that can be done that will lead to a harmonious situation. Since this can lead to unsafe working conditions, the safety technician should report the situation to the employee's supervisor or Human Resources Department. In this case, the employee would need to be transferred to another crew or terminated in order to maintain safe working conditions.

1.3.3 Absenteeism and Turnover

Absenteeism and turnover are big problems on construction jobs. It can cause workers to feel pressured to fill in and do jobs for which they are not trained or certified. Absenteeism and turnover may also force an understaffed crew to take on tasks that are too large. Two workers, for example, may find themselves working on a task that requires a crew of three to be done safely. To maintain a safe work environment, therefore, it is important to take steps to reduce absenteeism and turnover.

Absenteeism has many causes, some of which are inevitable. People get sick, and they have to take time off for family emergencies, or to attend family events such as funerals. However, some causes of absenteeism can be prevented.

The most effective way to control absenteeism is to make the company's policy clear to all employees. Companies that do this find that chronic absenteeism becomes less of a problem. New employees should have the policy explained to them, including the number of absences allowed and the reasons for which sick or personal days can be taken. All workers should know how to inform their supervisors when they need to miss work and understand the consequences of exceeding the number of sick or personal days allowed. The safety technician needs to reinforce this policy and emphasize the effect that absenteeism has on the safety of co-workers.

Once the policy on absenteeism is explained to employees, it must be implemented consistently and fairly. If the policy is administered equally, employees will likely follow it. However, if the policy is not administered equally and some employees are given exceptions, it will not be effective. Consequently, the rate of absenteeism will increase.

Despite having a policy on absenteeism, there will always be employees who are chronically late or often miss work. A safety technician who notices an employee abusing the absenteeism policy and causing unsafe working conditions should discuss the situation directly with the employee's supervisor. Confirm that the employee understands the company's policy and is willing to agree to comply with it. If the employee's behavior continues, disciplinary action may be in order.

Turnover refers to the loss of an employee who is fired or leaves the company to work elsewhere. According to the Construction Industry Institute (CII) unwanted employee turnover costs companies over $140 billion annually. This figure reflects the costs for recruiting, training, and administrative costs that are associated with taking on new hires. To calculate the actual cost of losing an employee, consider the following factors:

- The cost of advertising, interviewing, screening, and hiring a new employee.
- The training costs and management time associated with getting a new employee oriented and up to speed. Some sources estimate that in the course of 2 to 3 years, a business likely invests 10 to 20 percent of an employee's salary or more in training.
- The cost of lost productivity. In some cases, it could take a new employee up to 2 years to reach the productivity of an existing employee. Also, in response to high turnover, other employees may tend to disengage and become less productive.
- The cost of errors that new, less-experienced employees often make.

The Society for Human Resource Management (SHRM) estimates that it costs $3,500 to replace one $8.00-per-hour employee when all costs such as recruiting, interviewing, hiring, training, and reduced productivity are considered. SHRM's estimate was the lowest of 17 nationally respected companies who calculate this cost. Other sources estimate that it costs 30 to 50 percent of the annual salary of entry-level employees, 150 percent of mid-level employees, and up to 400 percent for specialized, high-level employees.

Obviously, reducing employee turnover benefits a company's bottom line. One source estimates that a 10-percent reduction in turnover is worth more money than a 10-percent increase in productivity or a 10-percent increase in sales. However, like absenteeism, there are some causes of turnover that cannot be prevented. For instance, it is unlikely that an employee who finds a job elsewhere that pays twice as much as what he or she is earning will stay at the current job.

On the other hand, some employee turnover situations can be prevented. This can be done by ensuring safe working conditions for the crew, treating workers fairly and consistently, and helping to promote good working conditions. Safety technicians need to be alert for any signs of discontent, especially those due to unsafe, or perceived unsafe, working conditions. They must then work with the employees and management to provide resolutions for these issues and minimize turnover. The key to doing so is communication.

Some of the major causes of turnover include the following:

- Non-competitive wages and benefits may lead workers to leave one company or industry for another that pays higher wages and/or offers better benefits.
- A lack of job security may cause workers to leave to find more permanent employment.
- Unsafe project sites will cause workers to leave for safer projects.
- Workers will leave to find a less stressful working environment if they perceive unfair and/or inconsistent treatment.
- Workers will move on if they feel that the working conditions are inadequate.

Essentially, the same actions prescribed to address absenteeism are effective for reducing turnover. Studies have shown that maintaining harmonious relationships on the job site goes a long way toward reducing both turnover and absenteeism. Achieving this goal, however, requires effective leadership on the part of the supervisor and all other management-level employees.

Case History

Making Safety a Core Value

Making safety a core value, not just a priority, is the foundation for creating a positive safety culture. Case in point: Alcoa Inc.

When Paul O'Neill (who would later become Treasury Secretary under President George W. Bush) took over as CEO in 1987, he stated unequivocally that his core value was a zero-injury workplace. That meant he needed to change the company's safety culture.

Because of O'Neill, the automatic routine at Alcoa became that, for any injury, the unit president had to report it to the CEO directly within 24 hours and present a plan for ensuring that this type of injury would never occur again. Those who embraced the system were promoted. Floor employees became supervisors; supervisors became directors; directors became vice presidents—if they committed to zero injuries and embraced the routine of making sure that whenever an incident resulting in injury did occur, all employees learned everything possible from the incident to prevent injuries in the future.

What happened next was astonishing. Not only did Alcoa's safety program change from reactive to proactive, but its entire culture shifted. The keystone safety habits that Paul O'Neill instituted built new corporate habits that streamlined the company's manufacturing process and increased profits (and employee salaries). By the time O'Neill left Alcoa 13 years later, the company's annual net income was five times higher than when he started.

The Bottom Line: Developing positive safety communication starts with buy-in from a company's CEO and top leadership. In the end, the transformation benefits rank-and-file employees as well as the company's profit margin.

Source: "Creating a World-Class Safety Culture, Part 2; Cultural Transformation: Establishing Safety as Everyone's Core Value." Chris Wilson, CAE, Director of Safety; Associated Builders and Contractors, Inc.

Additional Resources

The American Society of Safety Engineers (ASSE), **www.asse.org**

Occupational Safety and Health Organization, **www.osha.gov**

The Psychology of Safety Handbook, E. Scott Geller, Ph.D. 2001. Boca Raton, FL: CRC Press/Lewis Publishers.

The Participation Factor—How to Increase Involvement in Occupational Safety, E. Scott Geller, Ph.D. 2002. Des Plaines, IL: The American Society of Safety Engineer

1.0.0 Section Review

1. The three types of communication are _____.
   a. written, verbal, and digital
   b. verbal, physical, and electronic
   c. verbal, non-verbal, and written/visual
   d. visual/non-verbal, verbal, and virtual

2. The four major parts of a successful motivational strategy are gaining attention, building relevance, instilling confidence, and _____.
   a. establishing hierarchies
   b. fostering satisfaction
   c. incentivizing performance
   d. eliminating distractions

3. The best way to determine the cause of an employee's poor attitude is to _____.
   a. publicly reprimand the employee
   b. move the employee to another job site
   c. send the employee to a job counselor
   d. talk with the employee one-on-one

SUMMARY

Communicating effectively is the foundation for developing positive safety communication among all employees on a job site. The three types of communication are verbal, non-verbal, and written/visual. Safety technicians must understand how to utilize all three forms of communication in order to effectively interact with others.

Communication with site personnel is essential. If you are able to communicate with them, you will be able to provide the recognition, motivation, and feedback they need to keep the workplace safe. You will also be able to motivate workers and modify employee behavior when needed.

Establishing open communication is key. If you are able to speak honestly and respectfully to all workers at all times, even short conversations about safety will be meaningful and easy to understand.

Review Questions

1. Feedback is the component of communication that occurs _____.
    a. when the sender creates the message
    b. before the message reaches the receiver
    c. whenever distractors are present
    d. after the message has been sent and received

2. One form of non-verbal communication is _____.
    a. email
    b. facial expressions
    c. warning labels
    d. piping diagrams

3. What type of communication is documented on paper or transmitted electronically via a computer using words or images?
    a. Verbal
    b. Non-verbal
    c. Written
    d. Formal

4. When seeking to motivate employees, asking them a challenging question is a good way for a safety technician to _____.
    a. gain attention
    b. build relevance
    c. instill confidence
    d. foster satisfaction

5. Which component of the ARCS Model of Motivation is used to show employees how the information that is being conveyed is important to them personally?
    a. Attention
    b. Relevance
    c. Confidence
    d. Satisfaction

6. One way a safety technician can provide further satisfaction for employees to stay safe on the job is by _____.
    a. penalizing individuals for breaking safety policies and procedures
    b. developing recognition programs based on observations
    c. conducting unannounced on-site safety audits
    d. performing periodic task hazard analyses

7. A common tool for determining the causes of unsafe behavior is a(n) _____.
    a. job hazard analysis
    b. experience modification ratio
    c. safety behavior checklist
    d. job safety analysis

8. To determine the cause of an employee's poor attitude towards the workplace, talk with the employee one-on-one, listen to what the employee has to say, and _____.
    a. ask questions to uncover information
    b. interview the employee's co-workers
    c. refer the employee to the Human Resources Department
    d. record interview results on a behavior modification matrix

9. If there is nothing that can be done to resolve the problem when an employee is unable to work with the other members of a crew, the safety technician should _____.
    a. immediately terminate the employee
    b. assign a crew member to mentor the employee
    c. put the employee on a two-week probation
    d. report the situation to the employee's supervisor

10. For a company's policy on absenteeism to be effective it must be _____.
    a. applied flexibly
    b. certified by NIOSH
    c. implemented consistently and fairly
    d. revised annually

Trade Terms Introduced This Module

ARCS Model of Motivation: A strategy for motivating people developed by Dr. John Keller. The strategy states that to motivate individuals you must gain their attention, make the issue relevant to them, help them to feel confident that they can be successful, and provide them with a sense of satisfaction once they have achieved their goal.

Behavioral-based safety (BBS): A proactive method of safety management based on psychology. It requires systematic workplace observation, analysis of unsafe behaviors, and resolution of problems, coupled with training and incentives for behavior modification.

Communication: A process by which information is exchanged between individuals through a common system of symbols, signs, or behavior.

Diversity: Differences between individuals, particularly with regard to race, religion, ethnicity, and gender.

Ethnic groups: Large groups of people classed according to common racial, national, tribal, religious, linguistic, or cultural origin or background.

Feedback: The communication that occurs after a message has been sent and received. This communication from the receiver enables the sender to determine whether or not the message was received accurately.

Jargon: Technical terminology known only by people who work directly with the technology being discussed.

Message: The information that the sender is attempting to communicate to the receiver.

Non-verbal communication: Communication achieved through non-spoken means, such as body language, facial expressions, hand gestures, and eye contact.

Paraphrase: A restatement of a text, passage, conversation, or work process that is explained without changing its meaning.

People-based safety: A safety management system centered on empowering the employee to make decisions for both their own safety and the safety of those around them, including giving all workers the authority to stop work in the event of unsafe conditions or behaviors.

Receiver: The person to whom the sender is communicating a message.

Sender: The person who creates the message to be communicated.

Verbal communication: Transfer of information through spoken word. This process involves a sender, receiver, message, and feedback.

Visual communication: Communication through visual aids such as signs, postings, and hand signals.

Written communication: Transfer of information through the written word.

Additional Resources

This module presents thorough resources for task training. The following resource material is suggested for further study.

The American Society of Safety Engineers (ASSE), **www.asse.org**

Occupational Safety and Health Organization, **www.osha.gov**

The Psychology of Safety Handbook, E. Scott Geller, Ph.D. 2001. Boca Raton, FL: CRC Press/Lewis Publishers.

The Participation Factor—How to Increase Involvement in Occupational Safety, E. Scott Geller, Ph.D. 2002. Des Plaines, IL: The American Society of Safety Engineers (ASSE).

Figure Credits

Honeywell Safety Products, Module opener

Fall protection materials provided courtesy of Miller Fall Protection, Franklin, PA, Figure 2

DIYawards, Figure 5

NuDatum Software, Figure 8

Section Review Answer Key

Answer Section One	Section Reference	Objective
1. c	1.1.0	1a
2. b	1.2.0	1b
3. d	1.3.1	1c

NCCER CURRICULA — USER UPDATE

NCCER makes every effort to keep its textbooks up-to-date and free of technical errors. We appreciate your help in this process. If you find an error, a typographical mistake, or an inaccuracy in NCCER's curricula, please fill out this form (or a photocopy), or complete the online form at **www.nccer.org/olf**. Be sure to include the exact module ID number, page number, a detailed description, and your recommended correction. Your input will be brought to the attention of the Authoring Team. Thank you for your assistance.

Instructors – If you have an idea for improving this textbook, or have found that additional materials were necessary to teach this module effectively, please let us know so that we may present your suggestions to the Authoring Team.

NCCER Product Development and Revision
13614 Progress Blvd., Alachua, FL 32615

Email: curriculum@nccer.org
Online: www.nccer.org/olf

❏ Trainee Guide ❏ Lesson Plans ❏ Exam ❏ PowerPoints Other _____

Craft / Level: _____ Copyright Date: _____

Module ID Number / Title: _____

Section Number(s): _____

Description: _____

_____

_____

_____

Recommended Correction: _____

_____

_____

_____

Your Name: _____

Address: _____

Email: _____ Phone: _____

Hazard Recognition, Environmental Awareness, and Occupational Health

Overview

This module covers job site hazards and unsafe conditions. It explains how to evaluate risks and identify appropriate methods of hazard control. It also discusses environmental concerns that affect the construction industry. An overview of regulations for hazardous materials is included, and the elements of a medical surveillance program are described.

Module 75219

Trainees with successful module completions may be eligible for credentialing through NCCER's Registry. To learn more, go to **www.nccer.org** or contact us at 1.888.622.3720. Our website has information on the latest product releases and training, as well as online versions of our *Cornerstone* magazine and Pearson's product catalog.

Your feedback is welcome. You may email your comments to **curriculum@nccer.org**, send general comments and inquiries to **info@nccer.org**, or fill in the User Update form at the back of this module.

This information is general in nature and intended for training purposes only. Actual performance of activities described in this manual requires compliance with all applicable operating, service, maintenance, and safety procedures under the direction of qualified personnel. References in this manual to patented or proprietary devices do not constitute a recommendation of their use.

Copyright © 2018 by NCCER, Alachua, FL 32615, and published by Pearson, New York, NY 10013. All rights reserved. Printed in the United States of America. This publication is protected by Copyright, and permission should be obtained from NCCER prior to any prohibited reproduction, storage in a retrieval system, or transmission in any form or by any means, electronic, mechanical, photocopying, recording, or likewise. To obtain permission(s) to use material from this work, please submit a written request to NCCER Product Development, 13614 Progress Blvd., Alachua, FL 32615.

75219 V2

From *Safety Technology,* Trainee Guide. NCCER.
Copyright © 2018 by NCCER. Published by Pearson. All rights reserved.

75219
HAZARD RECOGNITION, ENVIRONMETAL AWARENESS, AND OCCUPATIONAL HEALTH

Objectives

When you have completed this module, you will be able to do the following:

1. Identify hazards and unsafe conditions on a job site.
    a. Identify incident types and energy sources.
    b. Describe effective hazard recognition techniques.
2. Explain how to evaluate and control hazards.
    a. Assess risks using a written formula.
    b. Identify the root cause.
    c. Select and use appropriate hazard control methods.
3. Identify environmental concerns on a construction site.
    a. Identify chemical-specific safety programs.
    b. Explain how to comply with environmental laws.
    c. List environmental hazards and their control methods.

Performance Task

Under the supervision of your instructor, you should be able to do the following:

1. Prioritize risks using a written formula.

Trade Terms

Acceptable level of risk
Ambient noise levels
Asbestos
Asbestos containing material (ACM)
Asbestosis
Audible
Baseline
Bioaccumulate
Consequences
Cross-training
Dielectrics
Flaw
Generators
Hazardous waste
Hazardous waste manifest
Light ballasts
Mesothelioma
Polychlorinated biphenyls (PCBs)
Potentially responsible party (PRP)
Probability
Reportable quantity (RQ)
Secondary containment
Silica
Silicosis
Storm water run-off
Superfund
Swales
Target organs
Uncontrolled release of energy
Wetlands

Industry Recognized Credentials

If you are training through an NCCER-accredited sponsor, you may be eligible for credentials from NCCER's Registry. The ID number for this module is 75219. Note that this module may have been used in other NCCER curricula and may apply to other level completions. Contact NCCER's Registry at 888.622.3720 or go to www.nccer.org for more information.

Contents

- 1.0.0 Job Site Hazards and Unsafe Conditions 1
  - 1.1.0 Incident Types and Energy Sources 1
  - 1.2.0 Hazard Recognition Techniques 1
    - 1.2.1 Job Safety Analysis (JSA) 1
    - 1.2.2 Pre-Task Planning 7
    - 1.2.3 Safety Inspections 8
    - 1.2.4 Pre-Job Planning Checklist 8
    - 1.2.5 Job Observations 8
- 2.0.0 Evaluating and Controlling Hazards 12
  - 2.1.0 Risk Assessment 12
    - 2.1.1 Risk Assessment Formula 12
    - 2.1.2 Risk Assessment Matrix 13
  - 2.2.0 Root Cause Identification 14
  - 2.3.0 Hazard Control Methods 15
    - 2.3.1 Engineering/Design 16
    - 2.3.2 Administrative Controls 17
    - 2.3.3 Personal Protective Equipment 17
    - 2.3.4 Safety Devices 17
    - 2.3.5 Job Site Safety Interventions 20
- 3.0.0 Environmental Concerns 23
  - 3.1.0 Chemical-Specific Safety Programs 24
    - 3.1.1 Asbestos 24
    - 3.1.2 Lead 25
    - 3.1.3 PCBs 28
    - 3.1.4 Silica 30
  - 3.2.0 Environmental Laws 30
    - 3.2.1 Clean Water Act 31
    - 3.2.2 Federal Insecticide, Fungicide, and Rodenticide Act 32
    - 3.2.3 Comprehensive Environmental Response, Compensation, and Liability Act 32
    - 3.2.4 Land Disposal Restrictions Program 32
    - 3.2.5 Endangered Species Act 33
    - 3.2.6 National Historic Preservation Act 33
  - 3.3.0 Other Environment-Related Concerns 34
    - 3.3.1 Ambient Noise 34
    - 3.3.2 Weather 34
    - 3.3.3 Mold 35
    - 3.3.4 Electromagnetic Radiation 37
    - 3.3.5 Nanotechnology 37
    - 3.3.6 Chemical Management 39
    - 3.3.7 Hazardous Waste Management 39
    - 3.3.8 Medical Surveillance Programs 42
    - 3.3.9 Training Requirements 46

Figures and Tables

Figure 1   Job safety analysis (JSA) form 2
Figure 2   Alternate JSA form 3
Figure 3   Identifying the sequence of tasks 4
Figure 4   Identifying the hazards for each task 5
Figure 5   Identifying the preventive measures for each hazard 6
Figure 6   What's wrong with this picture? 7
Figure 7   Technology for electronic JSA forms 8
Figure 8   Pre-task planning checklist 9
Figure 9   Safety inspection checklist for scaffolding 10
Figure 10 Calculating risk 13
Figure 11 Risk assessment matrix 13
Figure 12 Using a risk assessment matrix 14
Figure 13 Root causes of incidents 15
Figure 14 Root causes flow chart 16
Figure 15 Hierarchy of controls 17
Figure 16 Personal fall arrest system 19
Figure 17 Chrysotile asbestos 24
Figure 18 Typical ACM found in buildings 26
Figure 19 Typical control zone for an asbestos removal project 27
Figure 20 Posting an asbestos warning sign 27
Figure 21 Lead paint abatement project 28
Figure 22 An approved dust mask 28
Figure 23 Transformer containing PCB oils 29
Figure 24 Microscopic view of crystalline silica particles on a filter 31
Figure 25 Worker using a powered air purifying respirator (PAPR) to
           limit silica dust exposure 31
Figure 26 Tools for control of storm water run-off on a construction site ... 32
Figure 27 Hurricane cleanup and recovery 35
Figure 28 Mold 35
Figure 29 Personal protective equipment for mold cleanup 36
Figure 30 Chemical storage area and secondary containment 40
Figure 31 Hazardous waste containers, properly and improperly labeled 41
Figure 32 NFPA label 41
Figure 33 EPA and DOT labels 42
Figure 34 Uniform hazardous waste manifest 43

Table 1   TSCA Disposal Requirements for Fluorescent Light Ballasts 30
Table 2   Sound Levels of Various Construction Activities 34
Table 3   Recommended Medical Program 44
Table 4   Target Organs and Chemicals 45

SECTION ONE

1.0.0 JOB SITE HAZARDS AND UNSAFE CONDITIONS

Objective

Identify hazards and unsafe conditions on a job site.
  a. Identify incident types and energy sources.
  b. Describe effective hazard recognition techniques.

Trade Terms

Uncontrolled release of energy: Energy that is released as result of an energy source that is uncontrolled. Energy sources can include tools, equipment, machinery, temperature, pressure, gravity, or radiation.

Safety technicians play an important role in identifying and controlling hazards on the job site. They help determine unsafe working conditions, inform workers of identified hazards, and provide safe work alternatives. Safety technicians should advise workers to alert their supervisor about hazardous conditions they see and correct the hazard themselves if they are able to do so. This puts part of the responsibility of hazard recognition and control on each worker as an active member of the safety team.

There are a number of ways to recognize existing and potential hazards on a work site. Some techniques are more complicated than others. An effective technique begins with this basic question: "What could go wrong with this situation or operation?". No matter what technique you use, accurate answers to this question will save lives and protect equipment from damage.

1.1.0 Incident Types and Energy Sources

Incident types and energy sources are considered potential hazard indicators. The best way to determine if a situation or equipment is potentially hazardous is to ask the following questions:

- How can this situation or equipment cause harm?
- What types of energy sources are present that could cause an incident?
- What is the magnitude of the energy?
- What could go wrong that would release the energy?
- How can the energy be eliminated or controlled?
- Will I be exposed to any hazardous materials?

Before you can fully answer these questions, you need to be familiar with the different types of incidents that can happen and the energy sources behind them. Some of the different types of incidents that can cause injuries include the following:

- Falls on the same elevations or falls from elevations
- Being caught in, on, or between equipment
- Coming in contact with acid, electricity, heat, cold, radiation, pressurized liquid, gas, or toxic substances
- Being struck by falling objects
- Being cut by tools or equipment
- Being exposed to high noise levels
- Repetitive motion or excessive vibration

When equipment is the cause of an incident, it is usually because there was an *uncontrolled release of energy*. Types of energy sources that are typically released include the following:

- Mechanical
- Pneumatic
- Hydraulic
- Electrical
- Chemical
- Thermal
- Radioactive
- Gravitational
- Stored energy

1.2.0 Hazard Recognition Techniques

Several techniques have been developed that can help to identify and correct hazards. Some of the classic techniques of hazard recognition are listed as follows:

- Job safety analysis
- Pre-task planning
- Safety inspections
- Pre-job planning checklists
- Recognizing incident types and energy sources
- Job observations

1.2.1 Job Safety Analysis (JSA)

A job safety analysis (JSA), also known as job hazard analysis (JHA), is one approach to hazard recognition. In a JSA, the task is broken down into its individual parts or steps, and then each step is analyzed for its potential hazards. Once a hazard is identified, certain actions or procedures are recommended that will correct that hazard.

Consider the example of a task for which a pump motor must be installed in a tight space. A JSA determines that using a come-along would be safer than having a worker do this manually. This reduces the potential for the worker's hand to be crushed during installation, and the worker is protected from injury. *Figure 1* shows one example of a JSA form.

Figure 2 shows a portion of another JSA form that uses three columns: Sequence of Tasks, Potential Hazards, and Preventive Measures (sometimes referred to as *mitigation*). The column headings mirror the three basic steps in the job safety analysis process.

For example, to conduct a JSA for changing the tire on a vehicle, begin by identifying the individual tasks, in order from first to last, that must be performed to do the job (*Figure 3*).

Next, identify the potential hazards for each task (*Figure 4*).

To complete the analysis, identify the most effective preventive measures for each of the potential hazards (*Figure 5*).

It is important to write legibly when filling out a JSA form; otherwise people won't be able to read it (see *Figure 6*). Increasingly, workers with smartphones or tablets use applications and online forms to complete JSA forms electronically (*Figure 7*).

NOTE

Refer to *Appendix A* for examples of some of the different styles of JSA forms that are available.

JOB SAFETY ANALYSIS FORM

Page __ of __

Job Title:
Job Location:
PPE:
Tools, Materials & Equipment:

Date of Analysis:
Conducted By:
Staffing:
Duration:

Step	Hazard	Quality Concern	New Procedure or Protection

Figure 1 Job safety analysis (JSA) form.

JOB SAFETY ANALYSIS WORKSHEET

Job:

Analyzed by: Date:
Reviewed by: Date:
Approved by: Date:

Sequence of Tasks	Potential Hazards (Energy type and contacts)	Preventive Measures (Barriers)

Figure 2 Alternate JSA form.

JOB SAFETY ANALYSIS WORKSHEET

Job: *Changing tire on a vehicle*

Analyzed by: *Jack Workman* Date: *June 29, 2015*
Reviewed by: Date:
Approved by: Date:

Sequence of Tasks	Potential Hazards (Energy type and contacts)	Preventive Measures (Barriers)
1. Park vehicle.		
2. Get spare tire and tool kit.		
3. Pry off hub cap.		
4. Loosen lug bolts (nuts).		
5. And so on...		

Figure 3 Identifying the sequence of tasks.

JOB SAFETY ANALYSIS WORKSHEET

Job: Changing tire on a vehicle

Analyzed by: Jack Workman Date: June 29, 2015
Reviewed by: Date:
Approved by: Date:

Sequence of Tasks	Potential Hazards (Energy type and contacts)	Preventive Measures (Barriers)
1. Park vehicle.	a) Can be hit by passing traffic. b) Can be hit by vehicle on uneven, soft ground. c) Vehicle may roll on driver.	
2. Get spare tire and tool kit.	a) Lifting spare tire may cause strain.	
3. Pry off hub cap.	a) Hub cap may pop off and hurt the driver.	
4. Loosen lug bolts (nuts).	a) Lug wrench may slip and hurt the driver.	
5. And so on...	a) ...	

Figure 4 Identifying the hazards for each task.

JOB SAFETY ANALYSIS WORKSHEET

Job: Changing tire on a vehicle

Analyzed by: Jack Workman Date: June 29, 2015
Reviewed by: Jill Proctor Date: June 30, 2015
Approved by: Pat Masterson Date: July 2, 2015

Sequence of Tasks	Potential Hazards (Energy type and contacts)	Preventive Measures (Barriers)
1. Park vehicle.	a) Can be hit by passing traffic. b) Can be hit by vehicle on uneven, soft ground. c) Vehicle may roll on driver.	a) Drive to area well clear of traffic. b) Choose a firm, level area. c) Apply the parking break, leave transmission in gear or in PARK, place blocks in front and back of the wheel diagonally opposite to the flat.
2. Get spare tire and tool kit.	a) Lifting spare tire may cause strain.	a) Turn spare into upright position in the wheel well. Using your legs and standing as close as possible, lift spare out of truck and roll to flat tire.
3. Pry off hub cap.	a) Hub cap may pop off and hurt the driver.	a) Pry off hub cap using steady pressure.
4. Loosen lug bolts (nuts).	a) Lug wrench may slip and hurt the driver.	a) Use proper lug wrench; apply steady pressure slowly.
5. And so on...	a) ...	a) ...

Figure 5 Identifying the preventive measures for each hazard.

Job Location: Metal Shop	Analyst: A. Wheeler	Date: 10/10/15

Task Description: Worker reaches into metal box to the right of the machine, grasps a 15-pound casting, and carries it to the grinding wheel. Worker grinds 40 to 50 castings per hour.

Hazard Description: Reaching, twisting, and lifting 15-pound castings from the floor could result in a muscle strain to the lower back.

Hazard Controls:

1. Move castings from the ground to the work zone to minimize lifting. Ideally, place them at waist height or an adjustable platform or pallet.
2. Train workers not to twist while lifting, and reconfigure work stations to minimize twisting during lifts.

Figure 6 What's wrong with this picture?

JSAs can also be used as pre-planning tools. This helps to ensure that safety is planned into the job. When JSAs are used for pre-planning, they contain the following information:

- Tools, materials, and equipment needs
- Staffing or manpower requirements
- Duration of the job
- Quality concerns

1.2.2 Pre-Task Planning

Another approach to hazard recognition is pre-task planning, also called task hazard analysis (THA). Pre-task planning is similar to performing a JSA in that both require workers to identify potential hazards and needed safeguards associated with a job they are about to do. The difference is the form that is used to report the hazard.

During pre-task planning, a pre-printed, fill-in-the-blank checklist, like the one shown in *Figure 8*, is often used to document any hazard found during analysis. The conclusions found during pre-task planning should be discussed with the crew by the first-line supervisor or team leader before work begins. Some companies require workers to sign the completed pre-task planning forms or checklists before they start work. This is so they can document that workers have been told of the potential hazards and safety procedures.

> **NOTE**
> Job safety analyses and pre-task planning are explained in more detail in NCCER Module 75220, *Job Safety Analysis and Pre-Task Planning*.

1.2.3 Safety Inspections

Safety inspections should be performed on a regular basis. Depending on the size of the company and company policies, they may be done on a daily, weekly, monthly, semi-annual, or annual basis.

> **NOTE:** For more information about safety inspections, refer to NCCER Module 75221, *Safety Data Tracking and Trending*.

During a safety inspection, tools, equipment, and the work area are checked for hazards. It is important to plan in advance what will be inspected. Preparing for safety inspections helps to ensure that the inspection will be thorough and successfully identify hazards. The following are questions to consider when preparing for a safety inspection:

- Have there been any reported incidents or near miss incidents?
- Who is responsible for inspecting each area?
- What tools, materials, or equipment need regular inspection?
- What did the last inspection reveal?
- Were the control actions adequate and completed?
- How frequently has the work area been inspected?
- Who will correct any noted deficiencies? How? When?

Problem areas and potential hazards can be dealt with faster and more effectively when inspections are approached in this way. Safety inspections checklists (*Figure 9*) are a good way to make sure safety inspections are done properly and that nothing is missed.

1.2.4 Pre-Job Planning Checklist

Before each job starts, an assessment of the hazards, risks, and associated safety needs for that job should be completed. The following safety pre-planning questions should be considered during the assessment process:

- Are there any special laws or regulations that will affect job site operations?
- Are there any hazardous processes, materials, or equipment that are going to be on site at the same time?
- What personal protective equipment or safety gear is needed for the job?
- Have workers been trained and authorized on any special tools and/or equipment involved in the job?
- Is there equipment maintenance software for tracking workers with certain certifications?
- Are all of the necessary work permits in place?

Knowing the answers to these questions and addressing any potential problems before a job starts will help to prevent incidents and injuries.

1.2.5 Job Observations

Approximately 85 percent of all incidents occur because of the at-risk behavior of workers. Some of these behaviors include wearing personal protective equipment improperly, engaging in horseplay, listening to headphones while working, and using cell phones on the job in non-emergency situations. To address these issues, there should be a program or procedure in place to make sure workers are doing their jobs safely. These are typically called job observation programs. With a job observation program, poor performance and at-risk behavior can be identified by actually watching people as they work.

When observing workers on a job, consider the following questions to make sure everyone is doing their job safely:

- Is the worker following the right procedure for the job?
- Is the worker using the right tools and equipment properly?
- Is the worker using personal protective equipment properly?

Figure 7 Technology for electronic JSA forms.

PRE-TASK PLANNING CHECKLIST				
Name:	Location:			
Signature:	Date:			
		Yes	No	N/A
1. Have underground utilities been located prior to excavation?				
2. In areas where there are known or suspected unexploded ordnance, has the area been cleared by qualified explosive ordnance disposal (EOD) personnel?				
3. Are excavations, the adjacent areas, and protective systems inspected and documented daily?				
4. Are excavations over 5 feet in depth adequately protected by shoring, trench box, or sloping?				
5. When excavations are undercut, is the overhanging material safely supported?				
6. Have methods been taken to control the accumulation of water in excavations?				
7. Are employees protected from falling materials (loose rock or soil)?				
8. Are substantial stop logs or barricades installed where vehicles or equipment are used or allowed adjacent to an excavation?				
9. Have steps been taken to prevent the public, workers, or equipment from falling into excavations?				
10. Are all wells, calyx holes, pits, shafts, etc. barricaded or covered?				
11. Are walkways provided where employees or equipment are required or permitted to cross over excavations?				
12. Where employees are required to enter excavations is access/egress provided every 25 feet laterally?				
13. For excavations less than 20 feet, is the maximum slope 1-1/2 horizontal to 1 vertical?				
14. Are support systems drawn from manufacturer's tabulated data in accordance with all manufacturer's specifications?				
15. Are copies of the tabulated data maintained at the job site?				
16. Are members of support systems securely connected together?				
17. Are shields installed in a manner to restrict lateral or other hazardous movement?				
Comments:				

Figure 8 Pre-task planning checklist.

SAFETY INSPECTION CHECKLIST FOR SCAFFOLDING

1. Are scaffolds and scaffold components inspected before each work shift by a competent person?
   _____ YES   _____ NO

2. Have employees who erect, disassemble, move, operate, repair, maintain, or inspect the scaffold been trained by a competent person to recognize the hazards associated with this type of scaffold and with the performance of their duties related to this scaffold?
   _____ YES   _____ NO

3. Have employees who use the scaffold been trained by a qualified person to recognize the hazards associated with this scaffold and know the performance of their duties related to it?
   _____ YES   _____ NO

4. Is the maximum load capacity of this scaffold known and has it been communicated to all employees?
   _____ YES   _____ NO

5. Is the load on the scaffold (including point loading) within the maximum load capacity of this particular scaffold?
   _____ YES   _____ NO

6. Is the scaffold plumb, square, and level?
   _____ YES   _____ NO

7. Is the scaffold on base plates and are mudsills level, sound, and rigid?
   _____ YES   _____ NO

8. Is there safe access to all scaffold platforms?
   _____ YES   _____ NO

9. Are all working platforms fully planked?
   _____ YES   _____ NO

10. Do planks extend at least 6" and no more than 12" over the supports?
    _____ YES   _____ NO

11. Are the planks in good condition and free of visible defects?
    _____ YES   _____ NO

12. Does the scaffold have all required guardrails and toe-boards?
    _____ YES   _____ NO

13. Are 4:1 (height to width) scaffolds secure to a building or structure as required?
    _____ YES   _____ NO

Figure 9 Safety inspection checklist for scaffolding.

- Is the worker over-reaching or lifting improperly?
- Are others at risk because of this operation?
- Is the area clear of trash and debris?

If you find that workers are in imminent danger, immediately ask them to stop, express concern for their safety, let them know specifically what they are doing wrong, and ask them what steps need to be taken to do their jobs safely. Make an effort to find out why they are doing what they are doing. Try to determine if it is a training issue or a motivational issue. Be sure to discuss your findings with the workers' supervisor(s).

Additional Resources

American Society of Safety Engineers, www.asse.org

Occupational Safety and Health Association, www.osha.gov

Basic Safety Administration: A Handbook for the New Safety Specialist, Fred Fanning, CSP. 2003. Des Plaines, IL: The American Society of Safety Engineers (ASSE).

Construction Project Safety, John Schaufelberger and Ken-Yu Lin. 2014. Hoboken, NJ: John Wiley & Sons.

Construction Safety Planning, David V. MacCollum, P.E., CSP. 1995. Hoboken, NJ: John Wiley & Sons.

1.0.0 Section Review

1. When equipment is the cause of an incident, it is usually because there was an _____.
    a. absence of secondary containment
    b. energy abatement error
    c. inadequate control zone
    d. uncontrolled release of energy

2. A good way to make sure safety inspections are done properly and that nothing is missed is to use a _____.
    a. job safety analysis
    b. safety inspection checklist
    c. risk assessment matrix
    d. pre-task planning checklist

Section Two

2.0.0 Evaluating and Controlling Hazards

Objective

Explain how to evaluate and control hazards.
a. Assess risks using a written formula.
b. Identify the root cause.
c. Select and use appropriate hazard control methods.

Performance Task

Prioritize risks using a written formula.

Trade Terms

Acceptable level of risk: The level of risk that is reasonable when working in hazardous conditions.
Ambient noise levels: Background noise that is related to the jobs done on a work site.
Audible: When a noise or sound is heard or capable of being heard.
Consequences: Something that happens as a result of a set of conditions or actions.
Cross-training: Training workers to do multiple jobs.
Flaw: A part of the design of equipment, parts, or a process that creates a hazard or operational or maintenance difficulties.
Probability: The chance that something will happen.

Risk is a measure of both the probability and consequences of an event. A safe operation is one in which there is an acceptable level of risk. For example, climbing a ladder has risk that is considered to be acceptable if the proper ladder is being used on the job, if it is erected correctly, and if it is in good condition. If any one of these conditions were less than satisfactory, climbing the ladder would not have an acceptable level of risk.

Once hazards have been identified, they must be evaluated. Typically, this is done by determining an acceptable level of risk based on the classification and prioritization of the hazard. These tasks require a combination of good judgment and knowledge about the work process, as well as your company's STOP WORK policy.

Basically, if you are witnessing a hazard, you should stop the task immediately. However, check with management to verify your level of authority. If you see every hazard as life-threatening, or if you shut down a job for a very insignificant hazard, you can cost the company a great deal of money and lose your credibility. On the other hand, if you are able to evaluate the risk and put the identified hazards in proper perspective, you will likely prevent the loss of lives as well as money.

2.1.0 Risk Assessment

An important part of evaluating a hazard is categorizing and prioritizing it. This is helpful because there may be limited resources to handle each and every safety problem. Workers and work sites are safer when the most hazardous conditions are fixed first. That's why it is important to learn how to classify and prioritize hazards. There are two tools you can use to help accomplish this: a risk assessment formula and a risk assessment matrix.

2.1.1 Risk Assessment Formula

Determining the seriousness of a hazard depends on the following three categories:

- *Probability* – The chance that a given event will occur.
- *Consequences* – The results of an action, condition, or event.
- *Exposure* – The amount of time and the degree to which someone or something is exposed to an unsafe condition, material, or environment.

The following formula can be used to help prioritize risks:

Probability + Consequences + Exposure = Risk

The formula works by assigning a numerical value to each hazard category using the following steps:

Step 1 Assign a value of 1 to 4 from the following probability categories:
1. Unlikely to happen
2. Possibly will happen in time
3. Probably will happen in time
4. Likely to happen very soon

> **NOTE:** This step (Step 1) helps to determine the likelihood of an undesired event occurring.

Step 2 Assign a value of 1 to 4 from this list of consequences:
1. Negligible – An injury is not likely.
2. Marginal – Minor illness, injury, or property damage are likely.
3. Critical – Severe illness, injury, or property damage are likely.
4. Catastrophic – Death or permanent disability are likely.

Step 3 Assign a value of 1 to 4 based on the exposure to the hazard:
1. There is little exposure to the hazardous condition or task. Controls are in place to limit exposures and are deemed effective.
2. Exposure is small. A limited number of workers have been exposed to the hazard. Some controls are in place to limit or control exposure.
3. Exposure is still significant. At least four workers have been exposed to the hazard. There are no controls in place to limit or control exposures.
4. Total number of workers exposed to the task or hazardous condition in a day or shift is high, the frequency of exposure or duration of over-exposure to contaminants is high. Or, those exposed, though fewer in number, are not within your control (such as members of the general public, or the employees of another contractor).

Once you have assigned a numerical value for each category, add the numbers together to get a total. Then, compare the total to a specific range of numbers designed to prioritize hazards. The range can vary from 3 to 12. A high number in the range means there is a greater risk of illness, injury, or death. A low number means there is a lower risk. When the risk is high, the hazard should be placed at the top of the priority list and fixed immediately. *Figure 10* shows an example of how risk is calculated for the hazard of tripping over debris in the work area.

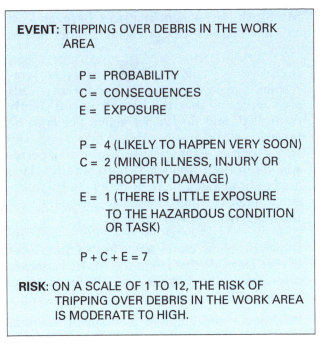

Figure 10 Calculating risk.

ity can be used, but all of these charts are defined by two dimensions that create a grid. The simple example in *Figure 11* is typical. Here, the matrix is formed by plotting frequency (or probability) on the horizontal axis and severity on the vertical axis.

> **NOTE:** These risk values should be considered as guidelines or a management tool. They are not intended to be used as a definitive measurement system.

2.1.2 Risk Assessment Matrix

A risk assessment matrix is another tool for facilitating decision making and prioritizing hazard control. Matrices of varying degrees of complex-

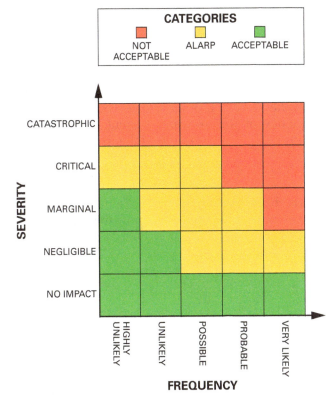

Figure 11 Risk assessment matrix.

Most risk assessment matrices have at least three color-coded areas, or categories, defined as follows:

- *Acceptable* – Typically shown in green; corresponds to low frequency, low severity. This area indicates that the hazard is sufficiently controlled, and no further action needs to be taken.
- *Not acceptable* – Typically shown in red; corresponds to high frequency, high severity. This area indicates that more or better control measures must be implemented to lower the risk that the hazard poses.
- *As low as reasonably possible (ALARP)* – Typically shown in yellow; corresponds to a medium level of risk. Hazards that fall in the ALARP area are not catastrophic, but they usually need to be monitored to keep them at an acceptable level.

Whereas a written risk assessment formula gives a quantitative value, a risk assessment matrix yields qualitative information. In theory, it can be used as a quantitative tool, but in practice it is not. Because a risk matrix uses mostly qualitative descriptions along its axes, it is difficult to assign any real numbers to the matrix and thus use it as a tool for performing calculations. A risk assessment matrix can only give a qualitative score that indicates in which category an event falls (acceptable, unacceptable, or ALARP).

A risk assessment matrix is not intended to be used as a standalone tool for decision making. It is best suited for ranking or prioritizing hazard control needs. For instance, in addition to using the risk assessment formula to calculate the risk for tripping over debris in the work area, you could also apply a risk assessment matrix. With the matrix approach, you would judge the frequency of this unwanted event as "Possible" to "Probable" on the horizontal axis, and then go up to the "Marginal" row on the severity scale. This would place the event in the ALARP zone, or category (*Figure 12*). This helps to confirm that although you need to monitor the hazard to keep the risk level acceptable, you probably don't need to rank it as a high priority hazard that requires immediate correction.

2.2.0 Root Cause Identification

Once a hazard has been identified, the next step is to determine the causes so that corrective actions can be taken. In most cases, there are two types of fixes: the quick fix and the permanent fix. The quick fix, often referred to as an Immediate Temporary Control (ITC), includes marking the hazard,

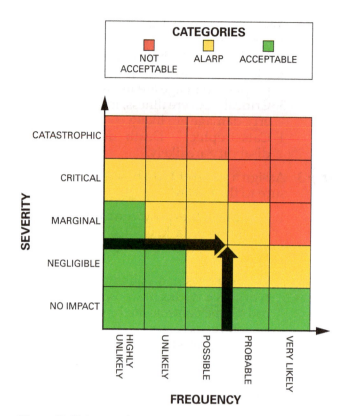

Figure 12 Using a risk assessment matrix.

or repairing or replacing a damaged piece of equipment, such as a broken ladder. The permanent or long-range fix addresses the root cause(s) of a hazard. Root causes are management system failures that failed to detect, correct, or anticipate the unsafe act or condition (*Figure 13*). Properly addressing root causes brings about permanent change and reduces the likelihood of similar incidents occurring.

Hazard analysis flow charts (*Figure 14*) can be helpful in identifying root causes and the associated root actions. The flow chart suggests steps that should be taken to fix this and similar problems. You will notice questions such as "Was the condition discovered during the last inspection? Should it have been?" and "Did anyone report the hazard? Determine why the hazard was not reported." These questions represent the recommended corrective actions or system fixes for the problem. They are designed to make you think of ways to improve hazard control. See *Appendix B* for more examples of hazard analysis flow charts.

Once the recommendations have been made and approved, deadlines and responsibilities for completing the correction must be assigned and communicated to both management and employees. Recommended actions should be tracked until they have been completed. Make sure there is a way to document the corrective actions that have been made and confirm that they are effective. This will be helpful during the recordkeeping

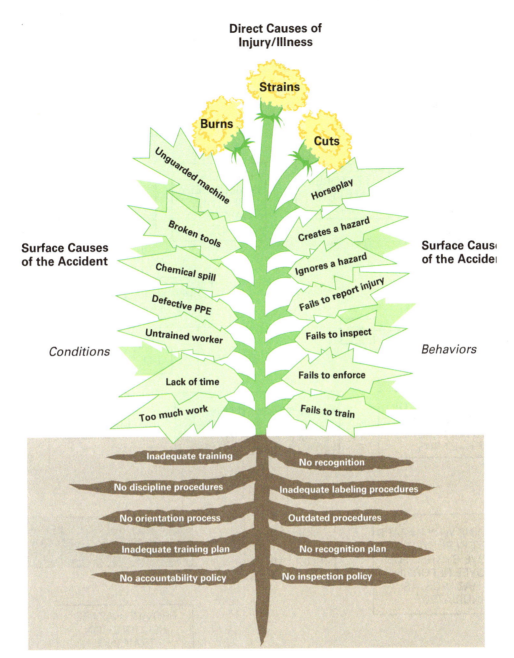

Figure 13 Root causes of incidents.

process. Temporary fixes must also have a deadline by which they must be completed; otherwise, quick fixes may become permanent.

> **NOTE**
> For more information about recordkeeping, refer to NCCER Module 75226, *OSHA Inspections and Recordkeeping*.

2.3.0 Hazard Control Methods

In addition to determining the root causes of hazards and the appropriate corrective action, it is important to understand how to control potential hazards. There are various hazard control methods and different ways of grouping the methods to form a hierarchy of controls (*Figure 15*). In any case, the basic point of a hierarchy of controls is to establish that the higher the severity of the hazard/risk, the more that should be done to eliminate the hazard. Following the hierarchy leads to the implementation of inherently safer systems and substantially reduces the risk of illness or injury.

For the purposes of this module, the methods of controlling hazards are placed into three major categories:

1. Engineering/design
2. Administrative controls
3. Personal protective equipment

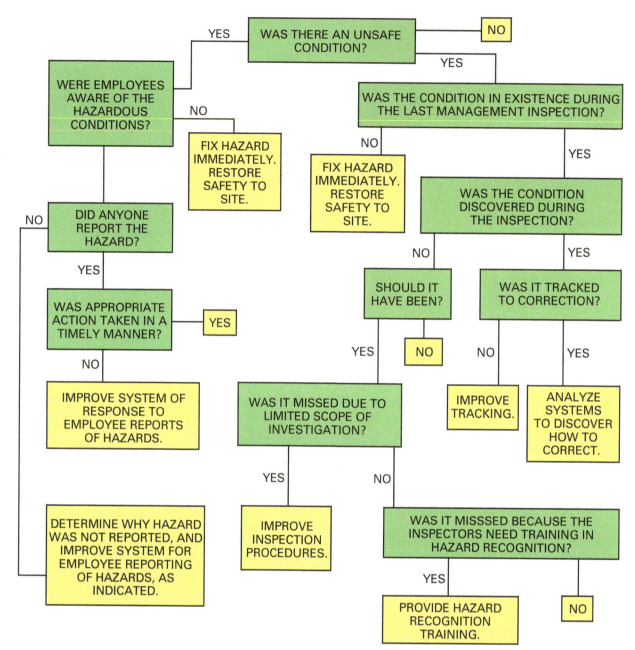

Figure 14 Root causes flow chart.

2.3.1 Engineering/Design

Sometimes equipment is hazardous because of a flaw in the design. NIOSH leads a national initiative called Prevention through Design (PtD). The goal of the initiative is to prevent or reduce occupational injuries, illnesses, and fatalities by eliminating the hazard at the source. For example, many power tools are very loud, creating a hearing hazard that must be managed through the use of hearing protection. The NIOSH Buy Quiet initiative encourages the use of power tools that are designed to operate at a much lower noise level.

Engineering/design can also be used for hazard control by adding features that make it safer to perform a job. This approach controls potential hazards by using engineering or substitution to either remove the hazard from the work process, use a less hazardous or nonhazardous substance or process, or isolate the hazard from the worker. Examples include choosing a nontoxic coating over a volatile coating or building a structure on the ground and then lifting it into place rather than constructing it at a fall-protection height.

Engineering/design controls are generally the most effective method of hazard control because they eliminate a hazard before it comes into contact with a worker. Unlike administrative controls and PPE, engineering/design controls don't require significant effort by the affected employee, and they work even when no one is watching.

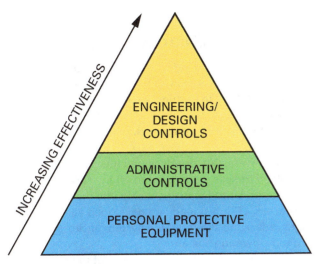

Figure 15 Hierarchy of controls.

Hazards can be controlled by reducing the potential for them to occur. Elements of a hazard potential reducing program usually have engineering/design considerations for the following:

- *Distance* – For example, reducing the chances of a massive fire by relocating refueling areas 50' (15 m) away from flammable materials.
- *Quantity* – For example, replacing one large storage tank with several smaller cylinders. This will reduce the size of a leak in the event of container failure because the amount of hazardous materials has been reduced.

By incorporating these elements into a hazard control program, you are taking the steps necessary to protect workers and equipment.

2.3.2 Administrative Controls

Administrative controls use practices and policies to limit an employee's exposure to a hazard. Safety policies and procedures, operating procedures, and maintenance procedures are examples of administrative controls. They are based on procedures that have been documented and formalized by management. Lockout/tagout, confined-space entry procedures, and work permits are also examples of administrative controls.

Likewise, worker rotation is actually an administrative control. Worker rotation simply involves limiting worker exposure in certain situations. Workers should be rotated if their work involves hot or cold work environments, constant or repetitive motion, or if they are bored. This can be accomplished by cross-training workers for different jobs on the site, adjusting the work schedule, or providing frequent breaks. It's important to rotate workers to ensure that everyone is alert and paying attention to safety.

Limiting the number of personnel exposed to a hazard is also an administrative control. An example of this method is training a small group of five individuals to perform the required hazardous work instead of training every worker on the site. This reduces the number of persons exposed to the hazard.

Warning devices are another form of administrative control. These devices can be audible (horns, bells, and whistles) or visual (signs). They are used to alert workers about hazardous conditions. The effectiveness of audible warning devices is sometimes limited, however, because they are not always distinguishable from other sounds on a work site. In some cases, workers hear alarms or signals so often that they unconsciously tune them out. In order for these warning devices to be effective, they must be distinctive and audible over ambient noise levels. Detailed messages on signs must be simple and easy to understand. Symbols, instead of words, may be used to help workers with limited reading skills or language barriers. Warning devices are more effective as a hazard control method when they are clear and easy to understand. Make sure everyone on site knows what the warning devices look like and/or how they sound.

Administrative controls require training and enforcement to be effective. When administrative controls are specified, they should be audited to verify their effectiveness.

> **NOTE**
> For more information on audits, refer to NCCER Module 75221, *Safety Data Tracking and Trending*.

2.3.3 Personal Protective Equipment

Personal protective equipment is special equipment designed to prevent workers from coming into contact with a hazardous substance through inhalation, ingestion, or absorption. PPE is the most basic method of hazard control. While it protects individuals, it does not necessarily create a safe work environment. Make personal protective equipment an important part of your safety program but understand that it must be used in combination with other hazard controls to be effective. Emphasize to all workers that whenever they use personal protective equipment they need to be aware of and respect its limitations.

2.3.4 Safety Devices

Safety devices can be classified as passive or active. Passive safety devices are essentially engineering/design controls. That is, you don't have

Construction of Marlins Park Baseball Stadium

One of the most notable features of the Marlins Park baseball stadium in Miami, Florida, is its three-section moveable roof. LPR Construction, based in Loveland, Colorado, provided steel erection and construction engineering services to install the roof structures—a $10.4 million project. LPR, in partnership with a structural steel fabricator and a house-moving company, was also responsible for the structural design, fabrication, and installation of a temporary shoring system that supported the trusses for the roof panels during the construction process.

The trusses were built in pieces on skewed ramps atop the shoring towers. Custom-designed heavy-duty rollers were installed between the ramps and the trusses. Then, the assembled trusses were lowered off the shores by rolling the roof panels down the supporting ramps. The concept is similar to launching a ship into the water, except the descent is slow and controlled throughout the entire process. LPR designed and developed all the specifications for the "launching" process, including high-strength rods, specialized jacking system anchorage connections, customized rollers, and a custom hydraulic synchronous lowering system.

Largely due to the innovative construction techniques that were used as well as LPR's commitment to safety, no injuries occurred during the construction process, and the project finished about a month ahead of schedule. Furthermore, on the basis of this project, LPR won the Steel Erectors Association of America Project of the Year award for 2012.

The Bottom Line: With the proper engineering/design controls and a positive safety culture, zero injuries can be achieved even on highly complex, uniquely challenging construction projects.

any control over when the safety device is used. An example of a passive safety device is an air bag in your car. Common types of passive safety devices that are used to control hazards on construction sites include the following:

- *Barricades* – Barricades provide a physical barrier that prevents workers from entering an unsafe area.
- *Safety nets* – Safety nets installed below a high-level work area (typically 6 feet or more) reduce the distance a worker can fall.

Active safety devices, much like personal protective equipment, require action and compliance on the part of workers. An example of an active safety device is the seat belt on a fork truck. The operator must decide to use the belt. When active safety devices are used, the company must develop and enforce policies and procedures mandating their use.

Types of active safety devices commonly used to control hazards on construction job sites include the following:

- *Power transmission guards* – Power transmission guards cover chains, belts, sprockets, or pulleys.

- *Personal fall arrest systems* – A personal fall arrest system (*Figure 16*) consists of an anchorage, body harness, and connectors worn by a worker in a high-level work area (typically 6 feet or more). If the worker falls from the working area, the fall arrest system safely stops (arrests) the fall.
- *Point of operation guards* – Point of operation guards protect workers from the cutting action or movement of a tool. The bottom blade guard on a radial saw is an example of a point of operation guard.
- *Interlocks on safety gates* – Interlocks stop equipment when the safety gate is opened.
- *Hand switches* – Switches are located outside the danger zone and must be touched and/or held down for the machine to operate or cycle. A common example of a hand switch is the safety bar on a power lawn mower.

Exposure Limits

To ensure the health and safety of all workers on a job site, exposure limits must be tracked. The permissible exposure limit (PEL) is the limit set by OSHA as the maximum concentration of a substance that a worker can be exposed to in an 8-hour work shift. A PEL is usually given as a time-weighted average (TWA), although some are short-term exposure limits (STEL) or ceiling limits.

Workers must be trained on the hazards, safeguards, and limitations of the guards or safety devices that are used in their work. This will ensure the effectiveness of safety devices as a hazard control method.

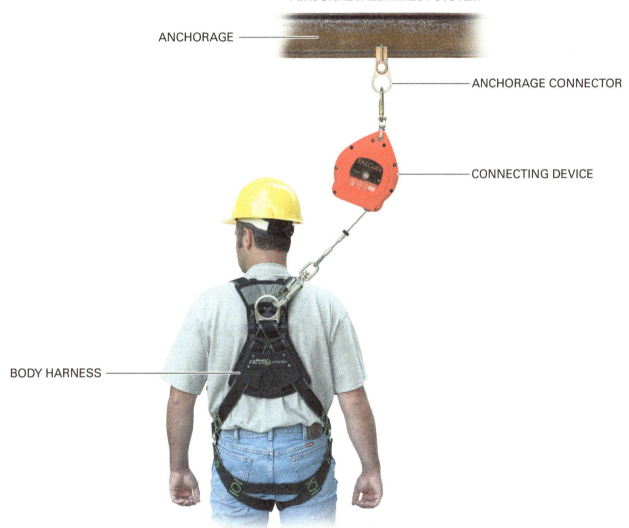

Figure 16 Personal fall arrest system.

> **WARNING!** Make it clear to workers that safety devices should never be destroyed, removed, or modified, no matter how inconvenient the devices might seem.

2.3.5 Job Site Safety Interventions

At-risk behaviors and situations can lead to injuries and incidents if they are not corrected. Current best practices for job site safety interventions in the construction industry are based on behavior modification and employee participation rather than top-down, authoritarian discipline. Basically, behavioral-based safety (BBS) programs make use of two key principles:

- At-risk behaviors are the actions of people, and these actions can be observed.
- At-risk behaviors that are observed can be measured in a meaningful way on a daily basis.

The American Contractors Insurance Group (ACIG) has identified the following ten best practices for improving safety performance:

- Executive-level support
- Pre-task planning for every task
- Management visibility
- Supervision involvement and accountability
- Root cause analysis of incidents
- Measurement and frequent review of key indicators
- Active Risk Management Committee
- Pre-project planning
- Subcontractor safety management
- Employee engagement/involvement

The following recommendations correlate with the ten best practices:

- Improve safety leadership among front-line supervisors, foremen, and other leaders at job sites.
- Align and integrate safety as a core value, working it into the design and planning phases of construction projects and breaking down traditional barriers.
- Optimize management commitment.
- Empower and involve workers.
- Ensure accountability for safety among everyone involved in a construction project (owners, management, safety personnel, supervisors, and workers).
- Improve communication, including reviewing safety policies and procedures to eliminate mixed messages and encouraging supervisors to initiate discussions with their employees.
- Provide supervisors and managers with safety, communication, and leadership skills training to improve the work site safety climate.
- Promote owner/client involvement and encourage owners to adopt a range of best practices.

Safety technicians are likely to encounter both simple and complex behavioral problems and need to make decisions about how to respond. Any safety intervention is more likely to be effective when it is tailored to a specific event. Consider a situation in which a worker is not wearing safety glasses. Simply reminding him or her about the hazards of not wearing eye protection can be sufficient for this relatively minor at-risk behavior. However, a worker who fails to install a barrier to protect personnel below an elevated work area puts more people at risk; this warrants a stronger intervention. Finally, a worker using gasoline as a diluting agent for a crude oil spill presents a seriously unsafe act that could have catastrophic consequences. It is an immediate danger and must be corrected at once.

Reasons for Unsafe Behavior – The first step in any job site safety intervention is to determine the real reason behind the person's at-risk behavior. The best way to do this is usually by talking with the worker directly, listening to what they have to say, and asking questions to uncover information. Having assessed the facts, you can determine how to correct the situation.

People often behave unsafely simply because they have never been hurt while doing their job in an unsafe way. Over time, being free from injury in spite of consistently performing unsafe actions actually reinforces the at-risk behaviors.

Workers are also less likely to follow certain safety policies and procedures if ignoring them is consistently rewarded by an immediate timesaving that achieves extra production. At-risk behavior is sometimes further reinforced by first-line managers turning a blind eye, or even actively encouraging employees to take unsafe shortcuts for the sake of production.

Changing Workers' Behavior – Positive reinforcement is the most powerful form of behavior modification. Praise and social approval are two effective forms of positive reinforcement. Positive, specific praise of safe actions coupled with constructive feedback go a long way in eliminating at-risk behaviors.

Research has shown there is a strong link from behavior change to attitude change. If people consciously change their behaviors, they also tend to re-adjust their associated attitudes and belief systems to fit the new behaviors. Furthermore, if a work crew adopts the norm that thinking

Safety Stand-Downs

A safety stand-down is an event in which normal work is paused and the entire job site focuses on a particular safety issue. Stand-downs can be tailored to specific industries and job sites. These events may be very short—a toolbox talk or a safety huddle about one specific hazard control—or they may be more extensive events that include a range of training activities and information on a variety of hazard controls.

In many cases, employers use stand-downs as interventions to call attention to particular hazards that are present on their individual job sites. The OSHA-hosted National Falls Campaign, by contrast, has adopted the use of the stand-down to draw nationwide attention to fall hazards in the construction industry and reinforce the importance of fall prevention. Contractors, workers, and safety professionals from across the country have participated in these ongoing events. Workers receive training on fall protection, including how to properly inspect and use personal protective equipment. Many companies use these national stand-downs as an opportunity to replace old or worn harnesses and lanyards.

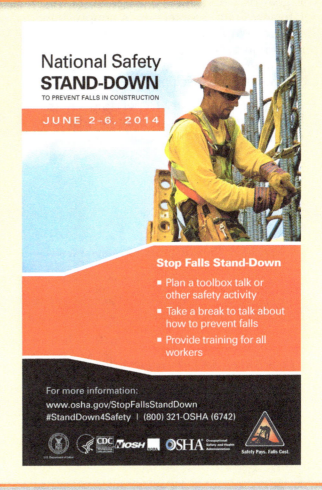

and behaving safely is best for all concerned, the group as a whole will openly disapprove of individuals who deviate from the norm and behave unsafely. In effect, peer pressure from their co-workers can then influence individuals to change their at-risk behavior.

Repeated Unsafe Actions – You may find that certain minor but nonetheless unsafe acts seem particularly difficult to eliminate. Many of these unsafe acts may result from worker carelessness or poor safety attitudes, but most are triggered by deeply ingrained at-risk behavior. Behavioral-based safety addresses these problems by focusing on particular sets of at-risk behavior and making workers more aware of the potential that the behavior has for causing harm. In turn, this serves to motivate workers to control not only their own safety behavior but that of their co-workers as well. Again, teamwork is one of the best ways to prevent incidents and create a safe working environment.

Punishing workers for everyday infractions of safety policies (*e.g.,* not wearing a hard hat) is very difficult to do consistently and does nothing to address the underlying problems (*e.g.,* the hard hat is uncomfortable or gets in the way of doing the job). On the other hand, punishment is a valid option for workers who repeatedly put other people at risk. The bottom line is this: reserve punishment for workers who consistently and deliberately commit unsafe acts. Likewise, use punishment only after you have done everything else in your power to create the safest working environment and provide the most comfortable protective equipment.

When disciplinary action is needed, employers must have a program in place for this. The program must be clear, administered fairly and promptly, and enforced consistently. Many employers use a program of progressive discipline that is completely separate from the behavior-based program they use to educate their employees about unsafe acts. In a progressive discipline program, discipline becomes increasingly harsh with each offense, ultimately leading to termination. Some unsafe acts—such as not using fall arrest protection or not following proper lockout/tagout procedures—may be grounds for immediate discipline or termination, depending on the severity of the hazard(s) associated with the act.

The goal of a disciplinary program, however, is not to terminate workers; it is simply intended to provide employers with an alternative for preventing injuries on the job site.

The following are some important guidelines for administering discipline:

- Discipline in private. Depending on the circumstances, however, it may be advisable for you to have a witness present.

- Clearly explain to the employee the reason(s) for the disciplinary action. Let the person know what will happen if he or she commits subsequent unsafe acts.

- Record each instance whenever you take a disciplinary action. This can include adding a note to an individual's personnel file after issuing a verbal warning.

Additional Resources

American Society of Safety Engineers, www.asse.org

Occupational Safety and Health Association, www.osha.gov

Basic Safety Administration: A Handbook for the New Safety Specialist, Fred Fanning, CSP. 2003. Des Plaines, IL: The American Society of Safety Engineers (ASSE).

Construction Project Safety, John Schaufelberger and Ken-Yu Lin. 2014. Hoboken, NJ: John Wiley & Sons.

Construction Safety Planning, David V. MacCollum, P.E., CSP. 1995. Hoboken, NJ: John Wiley & Sons.

The Psychology of Safety Handbook, E. Scott Geller, Ph.D. 2001. Boca Raton, FL: CRC/Lewis Publishers.

Root Cause Analysis Handbook: A Guide to Effective Incident Investigation, ABSG Consulting. 2005. Brookfield, CT: Rothstein Associates Inc.

2.0.0 Section Review

1. The formula used to prioritize risks assigns numerical values to probability, consequences, and _____.
    a. exposure
    b. training
    c. causes
    d. experience

2. The type of fix that addresses the root cause(s) of a hazard is a(n) _____.
    a. immediate temporary control
    b. long-range or permanent fix
    c. applied contingent remedy
    d. direct and provisional solution

3. Which of the following is a passive safety device?
    a. Personal fall arrest system
    b. Power transmission guard
    c. Barricade
    d. Interlock on a safety gate

Section Three

3.0.0 Environmental Concerns

Objective

Identify environmental concerns on a construction site.
a. Identify chemical-specific safety programs.
b. Explain how to comply with environmental laws.
c. List environmental hazards and their control methods.

Trade Terms

Asbestos: A natural mineral that forms long crystal fibers, used in the past as a fire retardant. It is a known carcinogen.

Asbestos containing materials (ACM): An object that is comprised of asbestos and other compounds.

Asbestosis: Scarring of the lung tissue caused by inhaled asbestos fibers that lodge in the lung's air sacs. This terminal condition is caused by asbestos exposure.

Baseline: In medical surveillance programs, this refers to the initial health status of the person. Subsequent medical reports are compared to the baseline.

Bioaccumulate: The natural process by which chemicals become concentrated in higher levels of the food chain as larger animals consume many smaller contaminated animals or plants.

Dielectrics: Nonconductors of electricity, especially substances with electrical conductivity of less than one-millionth of a siemens.

Generators: Firms that create hazardous waste.

Hazardous waste: A discarded material that has dangerous properties; it may be ignitable, corrosive, toxic, and/or reactive.

Hazardous waste manifest: A manifest is similar to a bill of lading. It is a shipping document that must be used for shipping waste that is considered hazardous by DOT and EPA standards.

Light ballasts: The parts of fluorescent lights that contain electric capacitors, which may contain PCBs.

Mesothelioma: An aggressive and terminal form of cancer that typically develops in the outer lining of the lungs. It is almost exclusively caused by exposure to asbestos.

Polychlorinated biphenyls (PCBs): A group of man-made chemicals, which were widely used as dielectric fluids or additives.

Potentially responsible party (PRP): An individual or firm who may be liable for paying the costs of a Superfund cleanup; a defendant in a Superfund lawsuit.

Reportable quantity (RQ): The amount of a chemical that, when spilled, must be reported to the National Response Center.

Secondary containment: A barrier that collects chemical overflow or spills from their original containers.

Silica: A common mineral often referred to as quartz. It is present in soil, sand, granite, and other types of rocks.

Silicosis: A form of lung disease caused by inhaling crystalline silica dust.

Storm water run-off: Rain that is not absorbed by the soil. Uncontrolled rainwater flows over land and picks up dirt and other contaminants and carries them to the nearest water body.

Superfund: The common name for the Comprehensive Environmental Response, Compensation, and Liability Act.

Swales: Shallow trough-like depressions that carry water mainly during rainstorms or snow melts.

Target organ: A specific organ in the human body most affected by a particular chemical.

Wetlands: Lowland areas, such as marshes or swamps, which are saturated with moisture, especially when regarded as the natural habitat of wildlife.

Since the first environmental laws were passed in 1970, many programs have been created to protect the natural world in which humans and other animals live. Communities and governments are paying more attention to pollution in the air, soil, and water. Safety technicians must be aware of how these programs can affect their job site.

Some of the major environmental issues that affect construction trades include the following:

- Asbestos abatement
- Lead abatement
- Disposal of polychlorinated biphenyls (PCBs)
- Chemical spills
- Water pollution
- Hazardous waste disposal

Improper handling of these materials can result in legal and financial consequences. Soil and water pollution from chemical spills can incur fines, legal actions, and project delays. Waste containing asbestos, PCBs, or other chemicals must be kept separate from any other debris. Segregating these hazardous materials prevents them from contaminating non-hazardous materials or combining with other materials to produce hazardous chemical reactions. Containers of hazardous waste must be properly marked and taken to an approved facility for disposal.

Many general construction firms prefer to hand off these matters to subcontractors who take care of all the details. For example, if an underground chemical storage tank is found while excavating, the contractor may hire an environmental firm to remove it, test soils, and file the proper reports with the government.

Construction company estimators and safety personnel must be familiar with environmental issues. Pre-bid planning can identify possible pitfalls and help to avoid costly mistakes. If your company handles these materials internally, personnel will need additional training and a medical surveillance program. Certification may also be required.

3.1.0 Chemical-Specific Safety Programs

Some environmental programs were created to address just one chemical, the most common being asbestos, lead, PCBs, and silica. Asbestos and PCBs are considered so harmful that they have been banned for most uses. Any current use of them is very restricted. Unfortunately, these compounds were once widely used in construction and may be encountered during the renovation or demolition of older buildings.

3.1.1 Asbestos

Asbestos removal is a very common environmental issue in construction. For many years, asbestos was used in several types of building materials and insulation. *Appendix C* lists products that may contain asbestos. Workers can create a health hazard if they break any asbestos containing material (ACM) during demolition or renovation.

> **WARNING!** Asbestos containing materials discovered during work must be immediately reported to a supervisor, safety technician, or other qualified individual. Craftworkers must not handle any materials they suspect may contain asbestos without proper training and equipment.

Two federal laws regulate asbestos. The general program is the Asbestos Hazard Emergency Response Act (AHERA). Because asbestos exposure is more dangerous for children than adults, the Asbestos School Hazard Abatement Act (ASHAA) law was added to address asbestos removal in school buildings.

Asbestos is a type of mineral. Asbestos crystals form long, thin fibers. It is a good insulator and has been used in building materials as a fire retardant for many years. There are six types of asbestos:

- Chrysotile (white asbestos)
- Amosite (brown asbestos)
- Crocidolite (blue asbestos)
- Tremolite
- Actinolite
- Anthophyllite

Chrysotile (*Figure 17*) makes up 90 to 95 percent of all asbestos found in buildings in the United States. The second most common type is Amosite. Crocidolite is only found in very high-temperature applications. The other types are rare.

Asbestos is hazardous when asbestos dust is inhaled into the lungs. Asbestos that can crumble is known as friable asbestos. When asbestos material is broken, dust is created that contains tiny asbestos fibers. The long, thin fibers are easily airborne. If workers breathe this dust without respiratory protection, the fibers get into the lungs and can cause damage.

The primary diseases caused by asbestos exposure are asbestosis and cancer. Asbestosis is a scarring of the lung tissue that develops slowly over time. The effects may not be seen for 15 to 30 years, at which point it is usually fatal.

Figure 17 Chrysotile asbestos.

Asbestos exposure can also cause lung cancer and mesothelioma, an aggressive type of terminal cancer that develops from asbestos fibers lodged in the outer lining of the lungs. Both lung cancer and mesothelioma have latency periods of over 20 years, and smoking increases the risk. Tar from cigarettes in a smoker's lungs sticks to the asbestos fibers, making it difficult for the body to eliminate the fibers by natural processes.

The known health effects are based on very large exposures to asbestos. In the 1940s, insulators would spray the interiors of ships and tanks with asbestos. The air would be thick with asbestos dust and the workers did not have any respiratory protection. They inhaled a lot of asbestos fibers into their lungs and many later died from asbestosis.

There is not much information on the effects of low-level exposure to asbestos. It is a known carcinogen, and even minimal exposure to a carcinogen can trigger cancer many years later. Because of this and the many asbestosis deaths that have been well-documented, asbestos handling and removal is now tightly regulated.

Asbestos or ACM (*Figure 18*) can be found in many buildings built before 1980. The US government banned production of most asbestos products in the 1970s, but installation of building materials containing asbestos continued into the early 1980s. Asbestos can still be found in these older buildings. Materials in these buildings that may contain asbestos include the following:

- HVAC duct insulation
- Boiler and pipe insulation
- Acoustical ceiling tiles
- Asphalt or vinyl floor tiles
- Exterior siding

New products made with asbestos must be clearly labeled. Building owners must identify existing asbestos and ACM through an asbestos survey and give this information to employees and contractors working in the building. Contractors must determine if there is asbestos or ACM as part of the pre-bid job analysis. If the building owner has not done an asbestos survey, a walk-through of the area should be performed by someone trained in asbestos identification.

An asbestos removal project must follow strict rules for work-zone safety. The OSHA construction standard for asbestos is 29 *CFR* 1926.1101. It covers all phases of asbestos handling and removal including demolition, removal, alteration, repair, maintenance, installation, cleanup, transportation, disposal, and storage. The OSHA standard includes the following requirements:

- A control zone must be created as a barrier to contain dust (*Figure 19*).
- Only trained workers may enter the control zone.
- The air in the control zone must be tested.
- Workers in the control zone must wear respiratory protection.
- A decontamination zone must be used when leaving the control zone.
- Decontamination procedures must be used.
- Access to the area must be limited.
- Warning signs must be posted (*Figure 20*).

Strict safety procedures must be followed when removing asbestos or ACM. All asbestos waste must be bagged in heavy plastic and tagged with specific hazard labels. The waste must be transported and disposed of by a certified firm. Because of the complex procedures involved, many general construction companies subcontract with a certified asbestos abatement firm to remove asbestos.

3.1.2 Lead

Lead is a common metal that is often used for pipes. Years ago, lead was also used as an additive to paints, gasoline, and other chemicals. Lead-based paints were banned in 1978 and it has also been phased out of gasoline. Lead pipes are no longer used for drinking water.

Lead is a toxic metal that can cause serious health problems, especially in children. The OSHA Permissible Exposure Limit (PEL) for lead is no greater than 50 micrograms of lead per cubic meter of air (50 $\mu g/m^3$) averaged over an 8-hour period. The PEL is reduced for shifts longer than 8 hours by the equation PEL = 400/hours worked.

Exposure to lead in young children can damage the brain and kidneys, impair hearing, and cause learning and behavioral problems. It can also cause less severe problems including vomiting, headaches, and appetite loss. Adults are also affected by lead exposure. It can lead to kidney damage, muscle and joint pain, high blood pressure, nerve disorders, and mood changes.

People are exposed to lead in different ways. Very small children have been exposed to lead by eating paint chips that contain lead. Small lead dust particles can be inhaled or swallowed. Lead can seep into water from lead pipes, lead solder on copper pipes, and brass faucets. When lead is swallowed it is carried in the bloodstream, causing damage to the brain, kidneys, and other organs.

(A) WORKER REMOVING ASBESTOS SIDING

(B) ASBESTOS DRAIN PIPES

(C) ASBESTOS INSULATION IN CEILING

(D) ASBESTOS GUTTER

(E) ASBESTOS ROOFING

(F) ASBESTOS TILE

Figure 18 Typical ACM found in buildings.

Figure 19 Typical control zone for an asbestos removal project.

The Federal Residential Lead-Based Paint Reduction Act was enacted to inform homebuyers of possible lead hazards. Sellers and landlords are required to give a lead paint disclosure form to the buyer or renter. In some states, banks will not approve loans for rental units unless lead paints are removed or covered (*Figure 21*). Due to the risk associated with lead, some people want lead pipes and paint removed from their homes.

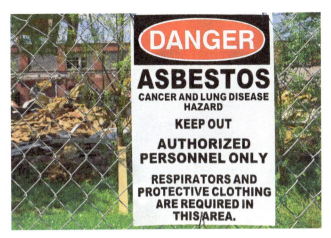

Figure 20 Posting an asbestos warning sign.

Workers may encounter lead-based paint during demolition or renovation of buildings built before 1978. Because dust created from sanding lead-based paints is hazardous, an approved dust mask (*Figure 22*) must be worn to protect the lungs. Any wastes contaminated with lead may be considered hazardous waste, including wood or metal painted with a lead-based paint. Lead piping and solder is usually not considered a hazardous waste. Any waste that may contain lead must be tested. The waste will be regulated as a hazardous waste if it fails a specific test. The proper handling and disposal of hazardous waste is detailed later in this module.

> **NOTE**
>
> Fines for improper disposal of hazardous waste start at $25,000. Your company could become a party to a lawsuit if you dispose of hazardous waste improperly.

Case History

Asbestos Exposure

A contractor spent approximately 15 years cutting and installing asbestos tiles and sheetrock while working on the original World Trade Center from the late 1960s to the early 1980s. Other workers further contaminated the site with asbestos fireproofing, tape, and pipe coverings. When the contractor was diagnosed with mesothelioma nearly 30 years later, he filed and won a lawsuit against the site manager.

The site manager appealed, insisting that he should not be held responsible because he did not pick out the asbestos products that caused the contractor's illness. In 2012, the Supreme Court of New York County upheld the ruling in favor of the contractor. The court maintained that the site manager failed to adequately protect his workers.

The Bottom Line: Employers are responsible for providing a safe and healthful workplace for their employees.
Source: *Asbestos.com., LLC/The Mesothelioma Center*

Figure 21 Lead paint abatement project.

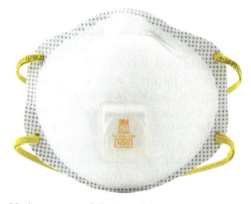

Figure 22 An approved dust mask.

3.1.3 PCBs

Polychlorinated biphenyls (PCBs) are a group of chemicals manufactured from the 1930s to the late 1970s. PCBs are also known by various trade names. More than 1.5 billion pounds of PCBs were manufactured in North America.

PCBs were used in many industrial materials, including sealing and caulking compounds, inks, pesticides, and paint additives. Because of their heat resistance and electrical characteristics, PCBs were used as an additive in hydraulic fluids. They were also used to make coolants, lubricants, and dielectrics for certain kinds of electrical equipment, including transformers and capacitors. They are now banned for most uses, but more than a million tons remain in use in electrical transformers and light ballasts.

While PCBs are not as toxic as many other chemicals, they are persistent. This means that they do not break down or degrade easily. Due to their widespread use and persistence, PCBs can be found in almost every living organism on earth, including humans. PCBs were banned because they do not degrade easily and would eventually reach toxic levels in humans.

In the 1970s, soil and water samples were tested for PCBs. They were found in many areas, including several rivers and in game fish. PCBs are easily transported through groundwater, ocean currents, or by evaporation. If swallowed, PCBs are stored in the body's fat tissue. PCBs bioaccumulate in the food chain. This means that when one animal eats another, it also consumes the PCBs in the body of that animal. As the larger animals eat many smaller animals, the PCBs become concentrated. Because of this, many areas have warnings against eating too many fish from PCB-contaminated waters.

PCBs may cause damage to the immune, nervous, and reproductive systems. Workplace exposure to high levels of PCBs over a long period of time may increase a worker's chance of getting cancer, especially cancer of the liver and kidneys. Some of the short-term health effects are known because workers have been exposed to high levels of PCBs due to incidents on the job site. A concentrated exposure may cause the following conditions:

- A severe form of acne called chloracne
- Swelling of the upper eyelids
- Numbness in the arms and/or legs
- Discoloring of the nails and skin
- Muscle spasms
- Chronic bronchitis
- Problems with the nervous system

In the past, the use of PCBs was very widespread. When it was banned, it was impossible to remove all PCBs at once. They were allowed to remain in some existing electrical equipment. Construction workers may encounter PCBs when decommissioning large electrical equipment or during demolition or renovation of buildings built before 1978.

The US Environmental Protection Agency (EPA) under the Toxic Substances Control Act (TSCA) regulates the handling, transportation, storage, and disposal of highly concentrated PCBs. Many restrictions apply. The waste can only be accepted at approved facilities. Some states regulate smaller concentrations of PCBs, such as those found in light ballasts. It is important to determine if your project has PCBs and how your state regulates them.

PCBs were used in electrical transformers as a fire retardant from 1929 through 1977. The majority of these PCB transformers were installed in apartments, residential and commercial buildings, industrial facilities, campuses, and shopping centers built before 1978. *Figure 23* shows a transformer that contains PCB oils.

Case History

Lead Exposure

In 1991, NIOSH investigators studied lead exposure in twelve workers who were doing abrasive blasting and repainting of a bridge in Kentucky. Blasters wore continuous flow abrasive blasting respirators. Other workers used half-mask, air-purifying respirators with high-efficiency particulate air (HEPA) filters. However, the site had no complete respiratory protection program in place that met OSHA requirements (29 *CFR* 1910.134) and NIOSH recommendations (*NIOSH 1987a; NIOSH 1987b*). There were no coveralls or clean change-rooms available at the site, and no running water.

In the first week of work, the workers' blood lead levels (BLLs) ranged from 5 to 48 micrograms per deciliter (5 to 48 µg/dl). Their BLLs after 1 month of exposure ranged from 9 to 61 µg/dl. Two workers had BLLs exceeding 50 µg/dl. The airborne concentration of inorganic lead ranged from 5 to 29,400 µg/m^3. The BLL measurements clearly showed lead poisoning among the workers, even though they were wearing respirators.

The Bottom Line: To minimize workers' risk of adverse health effects from lead exposure, lead concentrations must be kept as low as possible. All available hazard controls must be used, including engineering controls, safe work practices, and appropriate respiratory protection.

Source: CDC/NIOSH.

Figure 23 Transformer containing PCB oils.

A transformer must have a nameplate attached to one side with the trade name of the dielectric fluid, the approximate weight in pounds, and the amount of fluid (usually in gallons). Since PCBs were marketed under different trade names, the nameplate on a PCB transformer may not carry the specific term "PCB". Trade names for PCBs could include the following:

- Abestol, Aroclor, Askarel, Chlophen
- Chlorextol, DK, EEC-18, Fenclor
- Inerteen, Kennechlor, No-Flamol, Phenoclor
- Pyralene, Pyranol, Saf-T-Kuhl, Solvol
- Non-Flammable Liquid

If the nameplate says PCB or any of the names on this list, the transformer contains PCBs. If the nameplate does not carry any of these labels, or if the label is missing or illegible, the utility company may be able to tell you if the transformer contains PCBs. Otherwise, the only way to be certain is to have the electrical fluid tested by qualified personnel.

Fluorescent light ballasts installed prior to 1978 may also contain PCBs. A small capacitor inside the ballast probably contains PCBs. If it does not contain PCBs, it must be labeled "No PCBs." It may be marked with the percentage of PCBs in the fluid.

If the label is marked >50 ppm (more than 50 parts per million), or if the ballast is damaged or leaking, then it is regulated under TSCA. If it is not leaking or is <50 ppm (less than 50 parts per million) it is not regulated under TSCA. *Table 1* shows the federal disposal requirements. State disposal regulations may be more restrictive.

Any unmarked ballasts must be treated as if they contained PCBs. Testing each ballast is neither practical nor recommended. Testing for PCBs requires destroying the ballast and removing the fluid, creating waste that is even more dangerous. The cost of testing usually outweighs the cost of proper disposal. Improper disposal can lead to very large fines and lawsuits. Federal laws may not regulate disposal of light ballasts, but many state laws do. It is important to understand your local laws and dispose of PCBs properly. Information on federal PCB regulations is available online at **www.epa.gov**. Contact your state Department of Environmental Conservation for information on how PCBs are regulated in your state.

3.1.4 Silica

Silica, often referred to as quartz, is a common mineral. It is found in many materials that are present on construction sites, including sand, concrete, masonry, rock, granite, and landscaping materials. Cutting, grinding, drilling, or otherwise disturbing these materials creates dust that contains crystalline silica particles (*Figure 24*) that are too small to see with the naked eye. Inhaling this silica dust puts workers at risk for developing lung cancer, chronic obstructive pulmonary disease (COPD), and a form of lung disease called silicosis. Silica exposure has also been linked to other forms of cancer as well as kidney disease.

It takes only a very small amount of airborne silica dust to create a health hazard. When it was adopted in 1971, the OSHA PEL for silica for the construction industry was about 250 µg/m^3. In 2017, OSHA enacted a new rule changing the PEL to 50 µg/m^3. This new PEL standard for construction includes the following provisions for employers:

- Measure the amount of silica that workers are exposed to if it may be at or above an action level of 25 µg/m^3, averaged over an 8-hour day.
- Protect workers from respirable crystalline silica exposures above the PEL of 50 µg/m^3, averaged over an 8-hour day.
- Limit workers' access to areas where they could be exposed above the PEL.
- Use dust controls, such as local exhaust ventilation, to capture the dust at its source, and the use of water sprays to suppress the dust where it is produced to protect workers from silica exposure above the PEL.
- Provide respirators to workers when dust controls cannot limit exposures to the PEL (*Figure 25*).
- Offer medical exams, including chest X-rays and lung function tests, every 3 years for workers who are exposed above the PEL for 30 or more days per year.
- Train workers on work operations that create silica exposure and ways to limit exposure.
- Keep records of workers' silica exposure and medical exams.

3.2.0 Environmental Laws

In addition to laws that deal with specific chemicals, there are many general environmental laws to control pollution. These include the Clean Water Act (CWA); the Federal Insecticide, Fungicide, and Rodenticide Act (FIFRA); and the Comprehensive Environmental Response, Compensation, and Liability Act (CERCLA).

Table 1 TSCA Disposal Requirements for Fluorescent Light Ballasts

PCB Capacitor	PCB Fluids	Labeling, Transporting, and Manifesting for Disposal	Disposal Options
No PCB label	Does not contain PCB	Not regulated under TSCA	Not regulated under TSCA. State laws may apply.
Intact and non-leaking	<50 ppm	No labeling or manifesting required.	Municipal solid waste under TSCA. State laws may apply.
Intact and non-leaking	>50 ppm	Is a PCB bulk product waste. No labeling is required. Manifesting may be required.	TCSA incinerator, TCSA/RCRA incinerator, secure landfill or alternate destruction method, risk-based approval.
Leaking	Either <50 ppm or >50 ppm	Is a PCB bulk product waste. No labeling is required. Manifesting may be required.	TCSA incinerator, TCSA/RCRA incinerator, secure landfill or alternate destruction method, risk-based approval.

Figure 24 Microscopic view of crystalline silica particles on a filter.

3.2.1 Clean Water Act

The Clean Water Act (CWA) is a federal law that applies to jurisdictional bodies of water and protects all surface waters. The program covers lakes, rivers, estuaries, oceans, and wetlands. The five major sections of the Act are the following:

- A permit program
- National effluent standards
- Water quality standards
- Oil and hazardous substances
- Storm water discharges

The CWA sets certain water quality standards and prohibits pollution. Any discharge into a body of water must be permitted. These permits are known as NPDES permits, for the National Pollutant Discharge Elimination System.

The first two sections of the CWA apply to companies that have a fixed-pipe discharge into a body of water. A fixed-pipe discharge is known as a point source and must have a permit. The CWA also protects wetlands. Excavation or fill in wetlands or swamp areas is controlled by the CWA. Permits must be obtained for these activities.

The CWA also covers storm water run-off. Storm water that carries any pollutants from a construction site to a body of water requires a permit. This includes suspended particles, even clean soil, that is mixed in the storm water. There are national and state programs for storm water management. Construction firms must control storm water run-off in accordance with local laws.

Figure 25 Worker using a powered air purifying respirator (PAPR) to limit silica dust exposure.

Construction sites that disturb more than one acre of land must obtain a NPDES permit. The permit requires that the site prepare a storm water pollution prevention plan. The plan must include best management practices, including erosion and sediment control and site management or housekeeping practices. The first part requires that storm water, which picks up dirt, be prevented from reaching storm drains or water bodies. The second part requires the prevention of possible contaminants, like oil and chemicals, from getting in the dirt. It is important to be aware of the local storm water rules.

Erosion and sediment controls include the following (*Figure 26*):

- Silt fences and earthen berms
- Swales and collection basins
- Storm drain protection
- Stabilized site entrances

Housekeeping and site management concerns may include the following:

- Truck washout areas
- Waste collection areas
- Temporary fueling areas
- On-site storage of hazardous substances

3.2.2 Federal Insecticide, Fungicide, and Rodenticide Act

The Federal Insecticide, Fungicide, and Rodenticide Act (FIFRA) covers the use of pesticides in the United States. The EPA is responsible for controlling the use of pesticides and for ensuring that they are used properly and do not pose a threat to humans or wildlife.

All pesticides must be registered with the EPA by the manufacturer. The EPA can limit the use of the pesticide in order to protect health and the environment. Pesticides must be marked and used only as approved. The person applying pesticides must be licensed. Make sure that only licensed professionals use pesticides on your site.

3.2.3 Comprehensive Environmental Response, Compensation, and Liability Act

There are hundreds of areas across the United States where chemicals have been dumped, been spilled, or have leaked out of storage. The Comprehensive Environmental Response, Compensation, and Liability Act (CERCLA), commonly known as the Superfund, was created in 1980 to clean up these areas.

The government can force private parties to clean up these contaminated areas. In many cases, the original property owner no longer exists or cannot afford the cleanup. They can find other firms that may have added to the pollution. If the government decides to pursue this action, the accused firm is known as the potentially responsible party (PRP). More than one firm can be responsible. The key is that any company, no matter how much or how little they are at fault, can be held responsible for the entire cost of the cleanup. In most cases, the EPA and the PRPs litigate the cleanup and costs.

The high cost of Superfund actions is a big incentive to make sure that your company does not become a PRP due to improper disposal of hazardous materials or waste. It is important to make sure that your waste is not hazardous, or that any hazardous waste is disposed of properly.

In addition to Superfund actions, CERCLA requires that all spills must be reported. If a company spills or intentionally pours out a hazardous substance into the water or soil, they must file a report with the National Response Center. The minimum amount of a chemical that must be spilled before a report is required is called a reportable quantity (RQ). What constitutes a reportable quantity varies depending on the chemical. For instance, as listed in 40 *CFR* Ch. 1, Table 302.4—*List of Hazardous Substances and Reportable Quantities*, the final reportable quantity for ammonia is 100 pounds (45.4 kilograms), whereas the final reportable quantity for chlorine is 10 pounds (4.54 kilograms).

The specific procedures for spill response also depend on the chemical that has been spilled and how large the spill is. The best practice is to establish a program to report any quantity of spillage.

3.2.4 Land Disposal Restrictions Program

In 1984, Congress created EPA's Land Disposal Restrictions (LDR) program (40 *CFR* 268). The LDR program ensures that toxic constituents present in hazardous waste are properly treated before the waste is land disposed. That is, hazardous wastes must meet mandatory technology-based

(A) SILT FENCE

BMP SUPPLIES, INC. WWW.BMPSUPPLIES.COM

(B) STORM DRAIN PROTECTION

Figure 26 Tools for control of storm water run-off on a construction site.

treatment standards before they can be placed in a landfill. These standards help minimize short- and long-term threats to human health and the environment.

The following basic prohibitions provide the framework for LDR regulations:

- *Disposal prohibition* – Requires that waste-specific treatment standards must be met before a waste can be land disposed.
- *Dilution prohibition* – Ensures that wastes are properly treated and not simply diluted to mask the concentration of hazardous constituents.
- *Storage prohibition* – Prevents the indefinite storage of hazardous wastes instead of treating the waste promptly.

Under 40 *CFR* 268.45, alternative treatment standards are provided for certain types of hazardous debris that are commonly generated in the construction industry—specifically, debris that is often generated when a building or structure undergoes demolition or renovation. Material generated by this type of work may be contaminated with or contain a hazardous waste. For example, scrap piping or tanks may have held hazardous waste and thus be classified as hazardous debris. Construction and demolition (C&D) landfills—which are generally a more affordable disposal solution than sanitary or commercial landfills—may be used for land disposal of this type of debris.

> **CAUTION**
> Many states have significantly reduced or eliminated permitted status for many C&D landfills, chiefly because of cross-contamination of C&D debris by hazardous substances. Check your state and local regulations.

3.2.5 Endangered Species Act

The Endangered Species Act (ESA) was signed into law in 1973. The US Fish and Wildlife Service (USFWS) and the National Marine Fisheries Service (NMFS), along with various state agencies, oversee compliance with this Act. The ESA requires that federally listed species and habitat not be adversely affected during any activity that has federal involvement or is subject to federal oversight—for example, projects that require a NPDES storm water permit for construction. There are approximately 1,930 total plant and animal species listed under the ESA. A related law, the Migratory Bird Treaty Act, protects migratory birds, their nests, and their eggs. Under this Act, construction in nesting areas can constitute a taking of the nest because the habitat has been destroyed.

If a company's activities could affect ESA-protected species or habitats, it may be required to develop mitigation strategies to minimize the impacts. Specific locations have seasonal moratoriums or restrictions to protect wildlife. For example, the Florida Marine Protection Act, along with the ESA, prohibits people from touching or disturbing sea turtles, their hatchlings, or their nests during the nesting season.

Prior to construction, consult the local USFWS office, the National Marine Fisheries Service, and your local conservation agency to determine whether your project could harm endangered or threatened species—and if so, what to do about it. Absent any federal involvement or oversight, private landowners must still ensure that proposed development activities won't result in a take of any listed species, and they may need to develop a habitat conservation plan.

> **NOTE**
> EPA delegates the authority to implement certain regulatory programs to some states. A state may have more stringent environmental requirements than the federal requirements. Always check with your state and local agencies before starting a construction project.

3.2.6 National Historic Preservation Act

Section 106 of the National Historic Preservation Act (NHPA) requires federal agencies to oversee the protection of historic properties. Many states have similar requirements. Contact your local historic preservation office to determine whether your construction activities could have an impact on any historic properties.

> **Did You Know?**
> **Brownfield Program**
> Brownfields are abandoned, idled, or underused industrial and commercial facilities where expansion or redevelopment is complicated by real or perceived environmental contamination. EPA's Brownfield Program provides funding for the assessment, cleanup, and redevelopment of brownfield sites, and leverages public and private investments to help in these efforts.

3.3.0 Other Environment-Related Concerns

Ambient conditions in the environment at the job site need to be evaluated and addressed in order to control hazards. Some of these types of concerns are associated with the following:

- Ambient noise levels
- Extreme or severe weather
- Mold
- Electromagnetic radiation
- Nanotechnology

3.3.1 Ambient Noise

Work performed in the vicinity of large equipment is often subject to excessive noise. If employees are exposed to noise exceeding an 8-hour, time-weighted average sound level of 90 dBA (decibels on the A-weighted scale), feasible administrative or engineering controls must be utilized. In addition, whenever employee noise exposures equal or exceed an 8-hour, time-weighted average sound level of 85 dBA, employers must administer an annual, effective hearing conservation program as described in 29 *CFR* Part 1910.95. NIOSH recommends that the amount of exposure time be cut in half for every 3 dBA above 85 dBA.

The approximate sound levels of various construction activities are shown in *Table 2*. Double protection (both ear plugs and ear muffs) may be required in high-noise areas.

Hearing protection sold in the United States must be labeled by the manufacturer with a noise reduction rating (NRR) per 40 *CFR* Part 211,*Product Noise Labeling*. A single number rating (SNR) is a system determined according to *International Standard ISO 4869*. The tests are carried out by commercial laboratories that are independent of the manufacturers. Like NRRs, SNRs are also expressed in decibels and are used to compare the protection provided by various types of hearing protection.

The NIOSH Buy Quiet initiative seeks to have companies purchase or rent quieter machinery and tools to reduce worker noise exposure. It provides information on equipment noise levels so companies can buy quieter products that make the workplace safer. In addition, by helping to create a demand for quieter products, the initiative motivates manufacturers to design quieter equipment.

You may want to consider implementing a Buy Quiet program at your company. Recommended components of a model program include the following:

- An inventory of your company's existing machinery and equipment with respective noise levels
- A Buy Quiet company policy or procedure
- Educational materials and promotional tools explaining the importance and benefits of Buying Quiet
- Cost-benefit analysis of Buying Quiet

3.3.2 Weather

Extreme weather can turn a job site into a hazardous environment for workers. Safety technicians must anticipate the impact that weather may have on work activities and have a safety plan in place for protecting personnel.

Heat stress is a major hazard that can be even worse for workers wearing protective clothing. The same protective materials that shield the body from chemical exposure also limit the dissipation of body heat and moisture. Personal protective clothing can therefore create a hazardous condition. Depending on the ambient conditions and the work being performed, heat stress can occur very rapidly—within as little as 15 minutes. It can pose as great a danger to worker health as chemical exposure.

Table 2 Sound Levels of Various Construction Activities

Sound Level	Typical Noises and Hazards
0 dB	Lowest audible sound
60 dB	Normal conversation
65 dB	Pickup truck
85 dB	Pneumatic tools, forklift, backhoe, bulldozer, grader: Unprotected exposure of over 8 hours can cause hearing damage.
95 dB	Masonry saw, compressor, truck traffic: Unprotected exposure of over 4 hours can cause hearing damage.
100 dB	Grinders: Unprotected exposure of over 60 minutes can cause hearing damage.
110 dB	Pile drivers: Unprotected exposure of over 15 minutes can cause hearing damage.
140 dB	Jackhammer: Threshold of pain; the danger is immediate.

In its early stages, heat stress can cause rashes, cramps, discomfort and drowsiness, resulting in impaired functional ability that threatens the safety of both the individual and co-workers. Continued heat stress can lead to heat stroke and death. Ways to protect against this hazard include avoiding overprotection; careful training and frequent monitoring of personnel who wear protective clothing; judicious scheduling of work and rest periods; and frequent replacement of water (proper hydration).

At low temperatures and when the wind-chill factor is low, injury from frostbite and hypothermia and impaired ability to work are dangers. To guard against these hazards, make sure that workers wear appropriate clothing; have warm shelter readily available; carefully schedule work and rest periods; and monitor workers' physical conditions. Proper hydration is also important at lower temperatures.

Hurricanes and major storms often result in widespread flooding and damage to property and infrastructure. Cleanup and recovery activities (*Figure 27*) expose people to hazards such as contaminated flood waters and downed power lines which can cause serious injuries or death. Employers are required to protect workers from the anticipated hazards associated with the response and recovery operations that workers are likely to conduct.

Assess the potential for hazardous conditions and/or exposures before allowing workers to perform any cleanup and recovery activities. Based on the initial assessment of hazards, provide workers with the appropriate personal protective equipment, training, and information to safely perform the work. The OSHA fact sheet, "Keeping Workers Safe during Hurricane Cleanup and Recovery" outlines hurricane-related hazards in detail along with the proper precautions to take for each hazard. More information from OSHA about hurricane preparedness and response is available at **www.osha.gov**.

Lightning hazards are another concern on job sites. The use of equipment such as cranes is prohibited whenever there is any indication of a possible lightning strike.

3.3.3 Mold

Molds are the most common forms of fungi found on earth. They can grow on almost any material as long as moisture and oxygen are available. Most molds reproduce by forming microscopic spores that resist drying and can be released into the air.

Airborne mold spores are found indoors as well as outdoors. In fact, mold is present in some form and quantity in all buildings. When the spores land on a suitable moist surface, they begin to grow and release chemicals that digest and eventually destroy the surface and its underlying materials. Molds can also cause adverse health effects in humans.

Molds can be identified by the following characteristics:

- *Sight* – Molds usually appear as distinctly colored wooly mats (*Figure 28*).
- *Smell* – Molds often produce a foul odor, such as a musty, earthy smell.

Figure 27 Hurricane cleanup and recovery.

Figure 28 Mold.

Construction workers are most likely to encounter mold during demolition and renovation projects, but it may be found on new construction projects as well. Workers may also be exposed to mold when performing hurricane cleanup and recovery activities.

In the past few decades, the use of building components such as paper-faced gypsum wallboard and sheathing, paper-faced batt insulation, oriented strand board sheathing (OSB), and cellulose-based fireproofing have become popular. These materials are vulnerable to water damage and are therefore capable of sustaining mold growth.

Health Effects of Mold Exposure – Inhaling certain species of mold spores in large enough concentrations can cause mild to severe health problems in sensitive individuals. The most common health effects are allergic reactions. Typical physical symptoms may include eye, nose, and throat irritation; respiratory complaints; skin irritation; nausea; and dizziness and fatigue. Currently, there are no federal regulations, guidelines, or industry standards establishing acceptable levels of airborne mold.

Mold Prevention – The key to mold prevention is moisture control. When cleaning up spills or leaks, use wet-dry vacuums and fans to dry the area as quickly as possible. Discard water-damaged materials and items that are visibly contaminated with mold. Remove and discard all porous materials—such as carpeting, wallpaper, drywall, paper, wood, etc.—that have been wet for more than 48 hours. Porous materials cannot be cleaned and may remain a source of mold growth.

The only adequate defense against mold infestation in new construction is systematic prevention. This requires using a durable, quality building design and employing construction methods that protect all vulnerable components until the building envelope is complete.

Mold Cleanup and Remediation – Smooth, nonporous surfaces such as metal or glass can usually be decontaminated with appropriate disinfection methods. Molds cannot generally be removed from porous surfaces and cannot be removed from surfaces such as paper gypsum board facers, which actually serve as culturing media (food sources) for mold organisms. Once contaminated by mold, porous or culturing components generally must be removed and replaced. This involves demolition and removal of the affected areas of a structure.

Much like asbestos abatement, the process of exposing and removing mold-contaminated materials disturbs the mold, which can release harmful airborne spores into surrounding areas. The resulting contamination of adjacent surfaces creates additional mold exposure hazards for workers. You may want to consult with an industrial hygienist who is competent in the field of mold abatement to help determine how to address possible mold hazards at your worksite.

Some general guidelines for mold cleanup include the following:

- Ventilate the work area well.
- Place mold-damaged materials in a plastic bag and then discard them.
- Clean mold off hard surfaces and other nonporous materials with detergent and water; allow the surfaces to dry completely, and then disinfect them with a household bleach solution.

> **WARNING!** Never mix bleach with other cleaning products that contain ammonia. This can produce highly toxic chlorine gas.

OSHA recommends using the following personal protective equipment during mold cleanup (*Figure 29*):

- N-95 respirator
- Long gloves (ordinary household rubber gloves when cleaning with water, bleach, and a mild detergent; chemical-resistant gloves if using a disinfectant, biocide, or strong cleaning solution)
- Protective goggles without ventilation holes

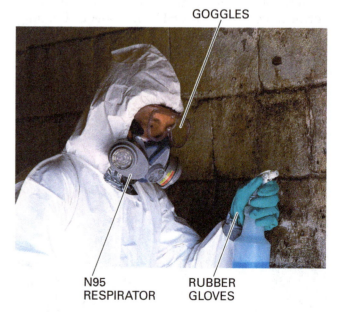

Figure 29 Personal protective equipment for mold cleanup.

> **NOTE:** Additional information concerning mold exposure and cleanup, including specific remediation tactics based on area size, can be obtained from the OSHA website at www.OSHA.gov.

3.3.4 Electromagnetic Radiation

Almost any electronic device that produces or uses radio energy can potentially be a source of electromagnetic interference (EMI), usually in the form of radio frequency (RF) emissions—*i.e.*, radio frequency interference (RFI). A cellphone tower has multiple antennas mounted on it, each of which emits radio signals in a circular pattern around the structure. For this reason, the presence of a cellphone tower near the job site may need to be factored in to your assessment of the hazards that could affect a construction project. Radiated electrical noise from the antennas mounted on the cell tower can interfere with the operation of two-way radios.

Radiated noise is the most common cause of mobile radio interference, and it can cause serious problems with crane operations on construction sites. Crane operations generally depend on a combination of hand signals and two-way radio communication among the crane operator and a signal person and a spotter on the ground. During crane operations, the ability to transmit signals between the operator and the ground personnel must be maintained. If that ability is interrupted, the crane operator must stop operations until signal transmission is reestablished and a proper signal is given. Electromagnetic radiation from the cellphone antennas mounted on a tower could create electronic interference with the two-way radios and disrupt crane operations. The crane communication system must be tested on site before beginning operations to make sure the signal transmission is effective, clear, and reliable.

Whether electromagnetic radiation adversely affects human health is more controversial, and few safety and health organizations have adopted exposure limits. The role that electromagnetic radiation plays in causing cancer is not yet defined. NIOSH has found only weak evidence that RF radiation alone can produce cancer. On the other hand, the chief organs formerly thought to be at risk from exposure to electromagnetic radiation—the skin, eyes, and ears—may not be the only ones affected. In short, there is no consensus yet on the potential hazard of exposure to electromagnetic radiation. Most experts agree that more research on human exposure to this type of radiation is needed.

What is known is that RF emissions from antennas used for wireless transmissions such as cellular and personal communication systems result in exposure levels on the ground that are typically thousands of times less than the safety limits that the Federal Communications Commission (FCC) has adopted. On this basis, there is no reason to believe that cellphone towers could pose a potential health hazard to people nearby.

3.3.5 Nanotechnology

Nanotechnology deals with matter at extremely small dimensions, between approximately 1 and 100 nanometers (nm). A nanometer is one billionth of a meter. Engineered nanomaterials are assembled from nanoscale structures such as carbon nanotubes and filaments or from nanoparticles of materials such as titanium dioxide or cadmium selenide. Nano-products currently used in the construction industry fall into four categories:

1. Cement-bound construction materials (mortar, cement, roofing tiles)
2. Noise reduction and thermal insulation or temperature regulation
3. Surface coatings to improve the functionality of various materials (paints, varnishes)
4. Fire protection

Some nanomaterials may act as chemical catalysts and produce unanticipated reactions, creating a risk of explosions and fires. Products containing these nanomaterials should be handled as possible sources of ignition.

The respiratory tract is the main route of entry for nanoparticles. Inhaled nanoparticles may enter the bloodstream and travel to internal organs. Metal nanoparticles can penetrate flexed, damaged, or diseased skin. The health hazard potential of an exposure depends on the particular nanomaterial involved and the person's exposure level.

Based on current knowledge, if a nanomaterial is permanently bound in a matrix, such as in concrete or in an insulation material, there is a very low or non-existent probability of people being exposed to the nanomaterial or product as long as the material or product is not undergoing destructive treatments such as sanding, milling, or drilling. Further studies under realistic conditions are required to determine with more certainty the exact environmental and health threats of construction products with nanomaterials.

Construction workers are most likely to be exposed to nanomaterials in the following ways:

- When applying a ready-to-use product, such as paint, or a product that is mixed into another material on the job site, such as a nanomaterial additive for concrete
- When performing drilling, sanding, milling, or similar work on a nano-product, thereby exposing the workers to nanomaterials through inhalation, skin contact, or ingestion
- When eating or smoking in the work area (a bad work practice) and unintentionally ingesting nanoparticles

To date, there are no OSHA PELs for nanomaterials. Currently OSHA does recommend the following Occupational Exposure Limits (OELs):

- Worker exposure to respirable carbon nanotubes and carbon nanofibers not to exceed 1.0 micrograms per cubic meter (1.0 µ/m^3) for an 8-hour time-weighted average
- Worker exposure to nanoscale particles of titanium dioxide (TiO2) not to exceed 0.3 milligrams per cubic meter (0.3 mg/m^3) for an 8-hour time-weighted average

No exposure limits exist yet for other nanomaterials.

> **NOTE**
> Nanomaterial use may fall under either OSHA General Industry or Construction standards. OSHA's Nanotechnology Safety and Health Topics page on the OSHA website highlights some of the OSHA standards that may apply to situations where workers handle or are exposed to nanomaterials. The *General Duty Clause, Section 5(a)(1) of the Occupational Safety and Health Act*, may also apply in such situations. States with OSHA-approved state plans may have additional standards that apply to nanotechnology.

Minimizing Worker Exposure: Assessment

Begin by assessing worker exposure to nanomaterials to identify the control measures needed and to determine whether the controls already in place are effectively reducing exposure. This step should include the following:

- Identifying and describing processes and job tasks that could expose workers to nanomaterials
- Determining the physical state of the nanomaterials (dust, powder, spray, or droplets)
- Determining the routes of exposure (inhalation, skin contact, ingestion) of particulates, slurries, suspensions, or solutions of nanomaterials
- Identifying the most appropriate sampling method to determine the quantities, airborne concentrations, durations, and frequencies of worker exposures to nanomaterials
- Determining any additional controls that may be needed based on the exposure assessment results and evaluation of the effectiveness of existing controls

> **NOTE**
> Although the paint and varnish industry has compiled special operational guides on the safe use of their nano-products, many current safety data sheets (SDSs) for nano-products do not provide sufficient data for communicating potential hazards. The International Standards Organization (ISO) recommends that employers provide SDSs for any nanomaterial or nanomaterial-containing product regardless of whether the material is classified as hazardous.

Minimizing Worker Exposure: Hierarchy of Controls

Employers should use a range of hazard control measures and best practices to minimize worker exposure to nanomaterials. The hierarchy of controls approach (*Figure 15*), outlined below, is an efficient way of accomplishing this:

- *Engineering controls* – These are generally the most effective controls:
  - For enclosed operations, use ventilated enclosures and high-efficiency particulate air (HEPA) filters.
  - For unenclosed operations, use containment measures such as local exhaust ventilation (*e.g.*, capture hood, enclosing hood) equipped with HEPA filters and designed to capture the containments at the point of generation or release.
  - Use modifications to reduce airborne nanoparticles. For example, keep raw materials wet if possible.
- *Administrative controls* – These establish best practices and procedures that workers should follow:
  - Provide handwashing facilities and information that encourages the use of good hygiene practices, such as not eating or smoking in the work area.
  - Establish procedures to address cleanup of nanomaterial spills and decontamination of surfaces to minimize worker exposure. For example, prohibit dry sweeping or the use of compressed air for cleanup of dusts that contain nanomaterials. Instead, use wet wiping and vacuum cleaners equipped with HEPA filters.

- Provide at least the following training and information to workers who will be exposed to nanomaterials:
  ~ Identification of the nanomaterials used and the processes in which these materials are used
  ~ Results from any exposure assessments that were conducted at the worksite
  ~ Identification of engineering and administrative controls and personal protective equipment to be used to reduce exposure to nanomaterials
  ~ Use and limitations of personal protective equipment
  ~ Emergency measures to take should a nanomaterial spill or release occur
- Review medical surveillance requirements under OSHA standards (*e.g.*, Cadmium, Respiratory Protection) and, if appropriate, make medical screening and surveillance available for workers who are exposed to nanomaterials.

- *Personal protective equipment* – This is the lowest tier on the hierarchy of controls:
  - Provide workers with the necessary personal protective equipment, such as appropriate respirators, gloves, and protective clothing.

> **NOTE**
> To keep up with current developments in the field of nanotechnology, visit the Nanotechnology page on OSHA's website (**www.osha.gov**) or go to **www.nano.gov**.

3.3.6 Chemical Management

Proper chemical handling is the best defense against chemical spills, exposures, and incidents. A good chemical management program includes information on safety, SDSs, personal protective equipment, storage, spills, and disposal. The program should cover all products that contain hazardous chemicals.

Chemicals should be stored safely and in a central location, if possible. This area should be marked with hazard warnings. Some general rules for good chemical management include the following:

- Keep hazards separate. For example, keep flammables and caustics in separate cabinets.
- Make sure all containers are properly labeled and approved for the chemical being stored.
- Keep containers away from weather or heat extremes.
- Protect containers against damage.
- Have spill cleanup materials at the storage area.
- If chemicals must be transferred to another container, it should also be an approved container labeled with all necessary warnings. Storing chemicals in temporary non-approved containers creates a hazard and may lead to improper disposal.

Pollution prevention and waste minimization are elements of a best practices plan for chemical management. One way to reduce possible pollution is to use less-toxic materials. There are many non-toxic solvents available. Good housekeeping will also minimize waste. Separate chemical or hazardous wastes from general construction debris. Remember, even a small amount of a hazardous substance in a container of waste makes the whole container a hazardous waste.

Secondary containment (*Figure 30*) is an additional barrier between the chemical and the soil. This means that if a chemical is spilled, it falls into a basin or container. It is much easier to clean the basin than the soil. Any areas where chemicals are transferred should have secondary containment. This includes temporary fueling areas.

When chemicals are spilled, all cleanup materials may be hazardous waste. This includes rags, absorbents, and broken containers, and it can also include soil onto which the chemical is spilled. Spill prevention is usually cheaper than hazardous waste disposal.

3.3.7 Hazardous Waste Management

In 1976, the Resource Conservation and Recovery Act (RCRA) was created to handle all aspects of hazardous waste management. Hazardous wastes are tracked from the time they are created to the time they are destroyed. RCRA covers hazardous waste generators, transporters, treatment, storage, and disposal facilities.

All firms that create waste must determine if the waste is hazardous. Firms that create hazardous waste are known as generators. Firms that only produce a small amount of hazardous waste are allowed some exemptions from the rules. They are known as conditionally exempt small quantity generators (CESQGs) and small quantity generators (SQGs). CESQGs create less than 220 pounds of hazardous waste per month. SQGs create 200 to 2,000 pounds per month. Large quantity generators (LQGs) are not exempt and produce more than 2,200 pounds per month. How much hazardous waste your firm creates each month determines which generator category governs your firm.

(A) CHEMICAL STORAGE

(B) SECONDARY CONTAINMENT

Figure 30 Chemical storage area and secondary containment.

The US Environmental Protection Agency (EPA) offers a free booklet, "Managing Your Hazardous Wastes: A Guide for Small Businesses". This guide explains the rules and exemptions for all generators and is available on the EPA website or by phone.

The general RCRA requirements for generators include the following:

- Generators must identify hazardous waste.
- All parties must obtain an EPA identification number.
- Certain records must be kept and reported to EPA.
- Personnel must be trained.
- Hazardous waste can be stored for a limited time.
- Hazardous waste containers must be labeled (*Figure 31*).
- Containers used for hazardous waste must meet international specifications.
- Only permitted carriers and disposal facilities may be used.
- All wastes must be shipped using the hazardous waste manifest system.
- Generators must have a waste minimization plan.

Identifying Hazardous Wastes – The EPA regulates hazardous waste disposal. Most states also have programs for handling hazardous wastes which may differ from the federal program explained here. The state rules must be at least as stringent as the federal standards. Some state programs include additional chemicals such as nickel, PCBs, and oil-contaminated water and soil. It is important to be knowledgeable about the wastes that are regulated in your state. If waste is shipped to other states for disposal, you must be familiar with the regulations for those states as well. Contact your state Department of Environmental Conservation for state regulations. Most disposal facilities will assist you in properly identifying the waste and completing the necessary paperwork.

Identification of hazardous waste can be complex. The EPA rules for properly identifying waste are listed at 40 *CFR* 261: *Identification and Listing of Hazardous Waste*. These rules will help you to determine the proper EPA codes to use on the manifest. Generally, a waste is hazardous if it has hazardous properties or it is listed as a hazardous waste. The following characteristics are hazardous properties:

- Flammable
- Corrosive
- Reactive
- Toxic

The National Fire Protection Association (NFPA) label, generally referred to as a fire diamond, provides information about the specific hazards and the severity of the hazards that may occur during emergency response situations (*Figure 32*). The label consists of four colored sections: red for fire hazards, blue for health hazards, yellow for reactivity hazards, and white for special hazards. The numbers zero through four are used to show the severity of each hazard; the higher the number, the more severe the hazard. Information in the white section describes the

(A) PROPER

(B) IMPROPER

Figure 31 Hazardous waste containers, properly and improperly labeled.

Figure 32 NFPA label.

nature of any special hazard—OX indicates the material is an oxidizer, SA indicates the material is a simple asphyxiant (nitrogen, helium, neon, argon, krypton, or xenon), and W indicates the material is reactive with water.

A drum of gasoline and water would be a flammable hazardous waste. Many chemicals are listed wastes, including cadmium and lead. Any container of waste, even if it has a small amount of a listed chemical, would be a hazardous waste. For example, a discarded bucket of paint that contains cadmium or lead is a hazardous waste.

> **NOTE**
> If you put a small amount of a hazardous waste into a container of general waste, the whole container is considered hazardous waste. Always separate hazardous waste for proper disposal.

In addition to EPA identification, you must also classify the waste according to US Department of Transportation (DOT) rules when you ship it. The DOT rules for shipping hazardous materials (49 *CFR* 170) apply to hazardous waste transportation. You must choose the proper DOT shipping name and hazard class. The containers must be labeled with both EPA and DOT labels and markings, as shown in *Figure 33*.

Most disposal companies that accept hazardous waste will assist you in determining the correct EPA codes and DOT shipping name for your waste.

Shipping Hazardous Waste – A **hazardous waste manifest** (*Figure 34*) is a shipping document that must be used to ship all hazardous wastes. The waste is tracked from where it is produced until it reaches the disposal facility. This is known as cradle-to-grave tracking. Each step in the process is documented. Both DOT and EPA require the manifest for shipping hazardous waste.

The responsibilities of safety technicians related to hazardous waste manifests depends on their company's specific policies and procedures. In some instances, they may be required to complete the manifest. In others, they may simply verify and sign off on the completed manifest before hazardous waste is released for transport to a disposal facility. In most cases, safety technicians are responsible for maintaining the documentation and records associated with the use of hazardous waste manifests.

> **NOTE**
> Some states have their own hazardous waste manifest and do not use the federal form. Be sure to find out which manifest should be used.

The Uniform Hazardous Waste Manifest form is divided into three main sections:

- *Generator* – completed by the generator, it lists the type and quantity of each waste
- *Transporter* – signed by the hazardous waste transporter to acknowledge receipt of the materials

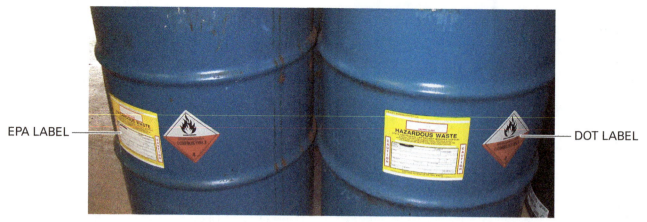

Figure 33 EPA and DOT labels.

- *Facility* – completed by the owner or operator of the disposal facility

The form has multiple copies. Initially, the generator must send a copy to the EPA. Each party that handles the waste signs the manifest and keeps a copy. When the waste reaches the disposal facility, that facility returns a signed copy of the manifest to the generator. This confirms that the waste has been received. The disposal facility also sends a copy to the EPA. The EPA matches the initial copy from the generator with the copy from the disposal facility.

Completing a Hazardous Waste Manifest – The firm that creates the waste must obtain a Generator number from the EPA. This number is limited to a particular site. If the firm creates waste in several places, each site must have its own Generator EPA ID number. Transportation firms that carry hazardous waste must have an EPA ID number. The facility that accepts the waste for disposal must also have an EPA ID number. These are all listed on the top of the manifest.

To complete the section of the form that lists each type of waste, you must use a proper shipping name for each material. All proper shipping names are listed in the DOT Hazardous Materials Table at 49 *CFR* 171.101. You must also list the EPA codes associated with each waste. These are the same EPA codes you found when you initially identified your waste. The RQ and emergency response information must also be listed for each waste.

To complete the generator portion of the form, the manifest must be signed by the generator. This signature is a certification that you have fully and accurately described and packaged the waste. The signature also certifies that the company has a waste minimization plan. This certification is a legal document. You can be held responsible for improper disposal.

> **NOTE**
> Because generators have cradle-to-grave responsibility for the waste they produce, if it is later discovered (*i.e.*, after disposal) that hazardous wastes have been sent to sanitary landfills, and if the source of that waste can be identified, the generator may be held financially responsible for cleanup, treatment, and re-disposal.

3.3.8 Medical Surveillance Programs

Many OSHA and environmental programs such as lead abatement or emergency response actions require medical surveillance. Workers must be tested periodically to make sure that their health has not been affected by exposure to hazardous materials. A medical surveillance program consists of the following elements:

- Surveillance, including pre-employment screening, annual screening, exposure-related testing, and termination examination
- Treatment, both emergency and non-emergency
- Recordkeeping
- Program review

Table 3 is the recommended medical program from NIOSH, OSHA, EPA, and the US Coast Guard for hazardous waste site operations.

Medical Testing – Every employee should have a medical evaluation at critical stages in his or her employment. The first screening takes place during pre-employment. The second is done at annual checkups. Screening is also done if an incident occurs. The final screening is performed when the employee stops working for the company.

The pre-employment screening is a series of medical tests that workers must have before they can start the job. These tests serve two purposes. First, it ensures that the worker is physically qualified to perform the work. This is especially

Please print or type. (Form designed for use on elite (12-pitch) typewriter.) Form Approved. OMB No. 2050-0039

UNIFORM HAZARDOUS WASTE MANIFEST	1. Generator ID Number	2. Page 1 of	3. Emergency Response Phone	4. Manifest Tracking Number

GENERATOR

5. Generator's Name and Mailing Address — Generator's Site Address (if different than mailing address)

Generator's Phone:

6. Transporter 1 Company Name — U.S. EPA ID Number

7. Transporter 2 Company Name — U.S. EPA ID Number

8. Designated Facility Name and Site Address — U.S. EPA ID Number

Facility's Phone:

9a. HM	9b. U.S. DOT Description (including Proper Shipping Name, Hazard Class, ID Number, and Packing Group (if any))	10. Containers No.	10. Containers Type	11. Total Quantity	12. Unit Wt./Vol.	13. Waste Codes			
	1.								
	2.								
	3.								
	4.								

14. Special Handling Instructions and Additional Information

15. **GENERATOR'S/OFFEROR'S CERTIFICATION:** I hereby declare that the contents of this consignment are fully and accurately described above by the proper shipping name, and are classified, packaged, marked and labeled/placarded, and are in all respects in proper condition for transport according to applicable international and national governmental regulations. If export shipment and I am the Primary Exporter, I certify that the contents of this consignment conform to the terms of the attached EPA Acknowledgment of Consent. I certify that the waste minimization statement identified in 40 CFR 262.27(a) (if I am a large quantity generator) or (b) (if I am a small quantity generator) is true.

Generator's/Offeror's Printed/Typed Name — Signature — Month Day Year

INT'L

16. International Shipments ☐ Import to U.S. ☐ Export from U.S. Port of entry/exit:

Transporter signature (for exports only): Date leaving U.S.:

TRANSPORTER

17. Transporter Acknowledgment of Receipt of Materials

Transporter 1 Printed/Typed Name — Signature — Month Day Year

Transporter 2 Printed/Typed Name — Signature — Month Day Year

DESIGNATED FACILITY

18. Discrepancy

18a. Discrepancy Indication Space ☐ Quantity ☐ Type ☐ Residue ☐ Partial Rejection ☐ Full Rejection

Manifest Reference Number:

18b. Alternate Facility (or Generator) — U.S. EPA ID Number

Facility's Phone:

18c. Signature of Alternate Facility (or Generator) — Month Day Year

19. Hazardous Waste Report Management Method Codes (i.e., codes for hazardous waste treatment, disposal, and recycling systems)

1.	2.	3.	4.

20. Designated Facility Owner or Operator: Certification of receipt of hazardous materials covered by the manifest except as noted in Item 18a

Printed/Typed Name — Signature — Month Day Year

EPA Form 8700-22 (Rev. 3-05) Previous editions are obsolete. DESIGNATED FACILITY TO DESTINATION STATE (IF REQUIRED)

VOID

Figure 34 Uniform hazardous waste manifest.

Table 3 Recommended Medical Program

Component	Recommended	Optional
Pre-Employment Screening	• Medical history • Occupational history • Physical examination • Determination of fitness to work wearing protective equipment • Baseline monitoring for specific exposures	• Freezing pre-employment specimen for later testing (limited to specific situations)
Periodic Medical Examinations	• Yearly update and occupational history • Yearly physical examination • Testing based on examination results, exposures, and job class and task	• Yearly testing with routine medical tests
Emergency Treatment	• Provide first aid on site • Develop liaison with local hospital and medical specialties • Arrange for decontamination of victims • Arrange in advance for transport of victims • Transfer medical records; give details of incident and medical history to next care provider	
Non-Emergency Treatment	• Develop mechanism for non-emergency health care	
Recordkeeping and Review	• Maintain and provide access to medical records in accordance with OSHA and state regulations • Report and record occupational injuries and illnesses • Review Site Safety Plan regularly to determine if additional testing is needed • Review program periodically; focus on current site hazards, exposures, and industrial hygiene standards	

Source: NIOSH/OSHA/USCG/EPA

important if the worker will be wearing a respirator. Some firms require extensive testing for work in extreme environments or where medical services are not readily available.

Second, pre-employment screening will document the initial health of the worker. This is known as establishing a baseline. Any changes in the worker's health can be compared to the baseline. One of the most common examples of baseline use is audiometric testing for how exposure to noise may have affected an individual's hearing. This test establishes a baseline to compare with the employee's post-employment hearing level. It is important to have a baseline to monitor an employee's health changes. Without the baseline, it would be difficult to determine whether any health issues arose during the person's employment.

Workers who handle many types of hazardous chemicals must have an annual physical. Chemical exposure will be tested through blood or other samples. Medical tests cannot directly check for a specific chemical exposure; rather, the tests check the status of target organs. Most chemicals affect a particular organ, such as the lungs or liver. *Table 4* lists the corresponding target organs for several chemicals that construction workers may encounter on the job. Annual screening results are compared to the baseline to see if there is any significant change. General trends in a worker's health will also show possible exposure or stress.

Any employee who is injured, becomes ill, or develops signs or symptoms due to possible overexposure to a hazardous substance must receive

Table 4 Target Organs and Chemicals

Chemical	Target Organs
Ammonia	Eyes, skin, respiratory system (lungs, bronchi, trachea, pharynx, and nasal passages)
Asbestos	Respiratory system (lungs, bronchi, trachea, pharynx, and nasal passages), eyes
Asphalt fumes	Eyes, respiratory system (lungs, bronchi, trachea, pharynx, and nasal passages)
Carbon monoxide	Cardiovascular system (heart, arteries, veins, blood, blood vessels), lungs, central nervous system (brain and spinal cord)
Chlorine dioxide	Eyes, respiratory system (lungs, bronchi, trachea, pharynx, and nasal passages)
Chlorodiphenyl (54% chlorine; also called polychlorinated biphenyl [PCB])	Skin, eyes, liver, reproductive system (sex organs)
Chromic acid and chromates	Blood, respiratory system (lungs, bronchi, trachea, pharynx, and nasal passages), liver, kidneys, eyes, skin
Dimethyl methane	Central nervous system (brain and spinal cord)
Hydrogen sulfide	Eyes, respiratory system (lungs, bronchi, trachea, pharynx, and nasal passages), central nervous system (brain and spinal cord)
Lead	Eyes, gastrointestinal tract (esophagus, stomach, small and large intestines), central nervous system (brain and spinal cord), kidneys, blood, gingival tissue (gums, mucosal tissue surrounding the teeth)
Manganese compounds and fumes (as Mn)	Respiratory system (lungs, bronchi, trachea, pharynx, and nasal passages), central nervous system (brain and spinal cord), blood, kidneys
Silica, crystalline	Eyes, respiratory system (lungs, bronchi, trachea, pharynx, and nasal passages)

a medical exam as soon as possible. This includes illness or symptoms due to health hazards from an emergency response or hazardous waste operation. Follow-up treatment and medical evaluations may be required.

All workers should have an exit physical to document the state of their health when they leave the job.

Ergonomics – Some of the most common injuries in the construction industry result from job demands that push the human body beyond its natural limits. On the job, workers often lift, stoop, kneel, twist, grip, stretch, reach overhead, or work in awkward positions. This puts them at risk of developing musculoskeletal disorders such as back problems, tendinitis, sprains, and strains.

Applying ergonomics, the scientific study of people at work, is a way to design the job to fit the worker instead of forcing the worker to fit the job. Ergonomics covers all aspects of a job, from physical stresses on joints and muscles to environmental factors that can affect hearing, vision, and general health and comfort. The overall goal is to reduce stress and eliminate injuries and disorders associated with the overuse of muscles, bad posture, and repeated tasks. This goal can be accomplished by designing tasks, work areas, tools, equipment, and so on to fit the worker's physical capabilities and limitations.

Many ergonomics experts recommend that you develop your own ergonomics programs to analyze risk factors at your work site and find the most effective solutions. Start by asking the following questions:

- What kinds of activities are most likely to cause injuries?
- How can we minimize the risk of workers being injured while performing these activities?

Then, review the cost, quality, and site-specific information you obtain to make sure the solutions you choose will best meet your needs.

Often just a small change in tools, equipment, or materials can make a big difference in preventing injuries. There are many simple and inexpensive ways to make construction tasks easier, more comfortable, and better suited to the needs of the human body. For example, consider the following:

- Use tools with extension handles that let workers stand upright while doing floor level tasks (such as autofeed screw guns or rebar tying tools that accommodate extension handles).
- Use mechanical lifts or person-lifts to make overhead work safer and more comfortable.

An ergonomics program can be a valuable way to reduce injuries, improve worker morale, and lower workers' compensation costs. These programs often increase productivity as well.

Recordkeeping and Program Evaluation – Records must be kept of any employment-related physical exams. OSHA and state regulations require certain restrictions on access to these records. OSHA also requires reporting and recordkeeping for incidents and illnesses. Medical records on any exposed worker must be kept for the duration of employment plus 30 years. This requirement is waived if the employee works for the contractor for less than one year and receives a copy of his or her medical records upon leaving the company.

Every medical surveillance program should be evaluated regularly. It should be updated for any changes in legal requirements and proven to be effective in protecting worker health.

3.3.9 Training Requirements

Many of the environmental laws covered in this module establish training requirements which are listed in the regulations. Some programs have a set number of hours of training; for example, the Hazardous Waste Operations (HAZWOPER) has a 40-hour initial course. Most training programs include annual or periodic refresher courses. It is important to keep accurate records of worker training. OSHA, insurance companies, and other government agents may inspect these training records.

There are many professional safety firms that offer training. These programs are available in a variety of formats, from standard lecture to online courses. Verify that the training meets the specific program requirements; some training may overlap.

There are different training needs depending on job tasks. Most programs have several levels of training. The first level is for employees whose daily tasks are not directly related to any hazard. The second level is for those who may encounter a hazard. The third level is for those who work around the hazard every day. The final level is for those workers who are responsible for safety or emergency response.

Several programs offer certification or licenses. The following list presents some of the training programs that may be required:

- *Asbestos handlers*: 29 *CFR* 1926.1101
- *Confined space entry*: 29 *CFR* 1910.146
- *DOT HM-126*: Hazardous materials handling and transportation, 49 *CFR* 172
- *Hazard communication*: 29 *CFR* 1910.1200
- *HAZWOPER*: Hazardous Waste Operations and Emergency Response, 29 *CFR* 1910.120
  - *HAZWOPER 40-hour course*: Designed for workers who are involved in cleanup operations, emergency response, and the storage, disposal, or treatment of hazardous substances or uncontrolled hazardous waste sites.
  - *HAZWOPER 24-hour course*: Covers broad issues pertaining to hazard recognition at work sites.
  - *HAZWOPER 8-hour course*: An annual online refresher course for those with previous HAZWOPER training.
- *Commercial pesticide applicators*: FIFRA laws managed by state agencies
- *Respiratory protection*: 29 *CFR* 1910.134

Additional Resources

Basic Safety Administration: A Handbook for the New Safety Specialist, Fred Fanning, CSP. 2003. Des Plaines, IL: The American Society of Safety Engineers (ASSE).

Construction Project Safety, John Schaufelberger and Ken-Yu Lin. 2014. Hoboken, NJ: John Wiley & Sons.

Construction Safety Planning, David V. MacCollum, P.E., CSP. 1995. Hoboken, NJ: John Wiley & Sons.

Field Safety, NCCER. Current edition. New York, NY: Pearson.

Hazardous Materials Behavior and Emergency Response Operations, Denis E. Zeimet, Ph.D., CIH and David N. Ballard. 2000. Des Plaines, IL: The American Society of Safety Engineers (ASSE).

Managing Your Hazardous Wastes: A Guide for Small Businesses, US Environmental Protection Agency. 2001. Washington, DC.

NIOSH Construction Sector, www.cdc.gov/niosh

NIOSH Ergonomics for Construction Workers, DHHS (NIOSH) Publication No. 2007-122, August 2007.

NIOSH Pocket Guide to Chemical Hazards, DHHS (NIOSH) Publication No. 2005-149.

Occupational Safety and Health Administration (OSHA), www.osha.gov

Environmental Protection Agency (EPA), www.epa.gov

3.0.0 Section Review

1. An environmental hazard that construction workers may encounter when decommissioning large electrical equipment from the late 1970s is _____.

    a. asbestos
    b. lead
    c. PCBs
    d. fungicides

2. The EPA's Land Disposal Restrictions (LDR) program is covered by _____.

    a. 29 *CFR* 1926
    b. 30 *CFR* 77
    c. 40 *CFR* 268
    d. 49 *CFR* 195.428

3. Medical records for an employee who has had a chemical exposure must be kept for the duration of employment plus _____.

    a. 5 years
    b. 10 years
    c. 20 years
    d. 30 years

SUMMARY

Important aspects of incident prevention include hazard recognition, evaluation, and control, with attention given to the various environmental issues that affect the construction industry. Companies must comply with current environmental and health and safety laws and regulations. Safety technicians need to stay informed of new developments that will affect environmental conditions and the health and safety of the workers at their job sites. When incidents are prevented, the overall health and well-being of every employee on site is improved, and operating costs remain affordable.

Review Questions

1. Incident types and energy sources are considered potential _____.
   a. emerging indicators
   b. intermodal indicators
   c. hazard indicators
   d. regulatory indicators

2. The type of energy involved in an incident in which worker is struck by a falling object is _____.
   a. hydraulic
   b. pneumatic
   c. electrical
   d. gravitational

3. Which of the following is used to check the tools, equipment, and work area at a job site for potential hazards?
   a. A job safety analysis
   b. A task safety analysis
   c. A safety inspection
   d. A risk assessment checklist

4. Watching people as they work with the goal of identifying poor performance and at-risk behavior is the purpose of a _____.
   a. root cause analysis
   b. job observation program
   c. pre-job planning inspection
   d. tailgate talk

5. In the risk assessment formula, a probability value of 2 means the chance that a given event will occur is _____.
   a. possible in time
   b. probable in time
   c. likely very soon
   d. unlikely

6. In the risk assessment formula, a consequences value of 3 means the results of an action, condition, or event is _____.
   a. critical
   b. negligible
   c. catastrophic
   d. marginal

7. The cause of an incident that resulted from management system failures that failed to detect, correct, or anticipate the unsafe act or condition is a(n) _____.
   a. indirect flaw
   b. root cause
   c. near miss
   d. Level I mishap

8. The questions on a hazard analysis flow chart represent recommended corrective actions for _____.
   a. immediate temporary controls
   b. indirect causes
   c. administrative controls
   d. root causes

9. Warning devices are used to _____.
   a. implement operating procedures
   b. provide active or passive hazard control
   c. correct equipment design flaws
   d. alert workers about hazardous conditions

10. Which of the following is an active safety device?
    a. Safety net
    b. Barricade
    c. Hand switch
    d. Control zone

11. Asbestos fibers can cause diseases such as cancer and asbestosis when the fibers contact a person's _____.
    a. skin
    b. eyes
    c. lungs
    d. liver

12. A common environmental hazard in construction is _____.
    a. lead-based paint
    b. rain gardens
    c. nitrogen blankets
    d. swales

13. To comply with RCRA requirements hazardous waste generators must have a(n) _____.
    a. lead abatement program
    b. waste minimization plan
    c. sediment control plan
    d. NPDES permit

14. The DOT rules for shipping hazardous materials and transporting hazardous waste can be found in _____.
    a. 29 *CFR* 1926
    b. 30 *CFR* 77
    c. 40 *CFR* 261
    d. 49 *CFR* 170

15. A medical surveillance program should include _____.
    a. best management practices and site maintenance
    b. annual screening and recordkeeping
    c. respiratory protection and secondary containment
    d. baseline treatment of target organs

Trade Terms Introduced in This Module

Acceptable level of risk: The level of risk that is reasonable when working in hazardous conditions.

Ambient noise levels: Background noise that is related to the jobs done on a work site.

Asbestos: A natural mineral that forms long crystal fibers, used in the past as a fire retardant. It is a known carcinogen.

Asbestos containing material (ACM): An object that is comprised of asbestos and other compounds.

Asbestosis: Scarring of the lung tissue caused by inhaled asbestos fibers that lodge in the lung's air sacs. This terminal condition is caused by asbestos exposure.

Audible: When a noise or sound is heard or capable of being heard.

Baseline: In medical surveillance programs, this refers to the initial health status of the person. Subsequent medical reports are compared to the baseline.

Bioaccumulate: The natural process by which chemicals become concentrated in higher levels of the food chain as larger animals consume many smaller contaminated animals or plants.

Consequences: Something that happens as a result of a set of conditions or actions.

Cross-training: Training workers to do multiple jobs.

Dielectrics: Nonconductors of electricity, especially substances with electrical conductivity of less than one-millionth of a siemens.

Flaw: A part of the design of equipment, parts, or a process that creates a hazard or operational or maintenance difficulties.

Generators: Firms that create hazardous waste.

Hazardous waste: A discarded material that has dangerous properties; it may be ignitable, corrosive, toxic, and/or reactive.

Hazardous waste manifest: A manifest is similar to a bill of lading. It is a shipping document that must be used for shipping waste that is considered hazardous by DOT and EPA standards.

Light ballasts: The parts of fluorescent lights that contain electric capacitors, which may contain PCBs.

Mesothelioma: An aggressive and terminal form of cancer that typically develops in the outer lining of the lungs. It is almost exclusively caused by exposure to asbestos.

Polychlorinated biphenyls (PCBs): A group of man-made chemicals, which were widely used as dielectric fluids or additives.

Potentially responsible party (PRP): An individual or firm who may be liable for paying the costs of a Superfund cleanup; a defendant in a Superfund lawsuit.

Probability: The chance that something will happen.

Reportable quantity (RQ): The amount of a chemical that, when spilled, must be reported to the National Response Center.

Secondary containment: A barrier that collects chemical overflow or spills from their original containers.

Silica: A common mineral often referred to as quartz. It is present in soil, sand, granite, and other types of rocks.

Silicosis: A form of lung disease caused by inhaling crystalline silica dust.

Storm water run-off: Rain that is not absorbed by the soil. Uncontrolled rainwater flows over land and picks up dirt and other contaminants and carries them to the nearest water body.

Superfund: The common name for the Comprehensive Environmental Response, Compensation, and Liability Act.

Swales: Shallow trough-like depressions that carry water mainly during rainstorms or snow melts.

Target organ: A specific organ in the human body most affected by a particular chemical.

Uncontrolled release of energy: Energy that is released as result of an energy source that is uncontrolled. Energy sources can include tools, equipment, machinery, temperature, pressure, gravity, or radiation.

Wetlands: Lowland areas, such as marshes or swamps, which are saturated with moisture, especially when regarded as the natural habitat of wildlife.

Appendix A

Examples of Job Safety Analysis Forms

JOB HAZARD ANALYSIS FORM

Page __ of __

Job Title:
Job Location:
PPE:
Tools, Materials & Equipment:

Date of Analysis:
Conducted By:
Staffing:
Duration:

Step	Hazard	Quality Concern	New Procedure or Protection

© PSA, Inc. / NCCER / JSA Proc./Forms

STEVEN P. PEREIRA, CSP

Job Title:	Job Location:	Analyst:	Date:

Task #:	Task Description:

Hazard Type:	Hazard Description:

Consequence:	Hazard Controls:

Rational or Comment:

U.S. DEPARTMENT OF LABOR

JOB SAFETY ANALYSIS WORKSHEET

Job:

Analyzed by: Date:
Reviewed by: Date:
Approved by: Date:

Sequence of Tasks	Potential Hazards (Energy type and contacts)	Preventive Measures (Barriers)

STEVEN P. PEREIRA, CSP

Appendix B

Hazard Analysis Flow Charts

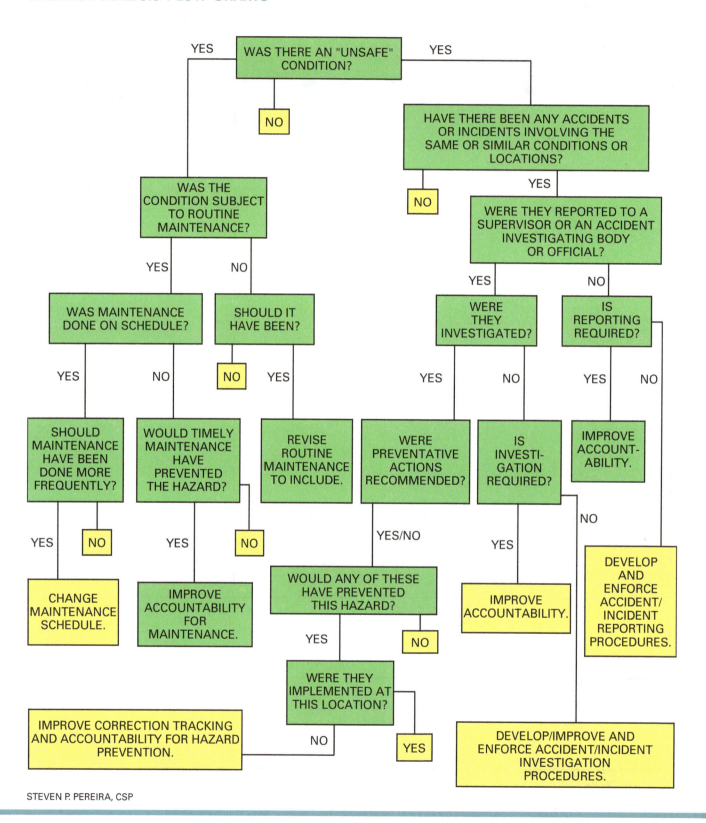

STEVEN P. PEREIRA, CSP

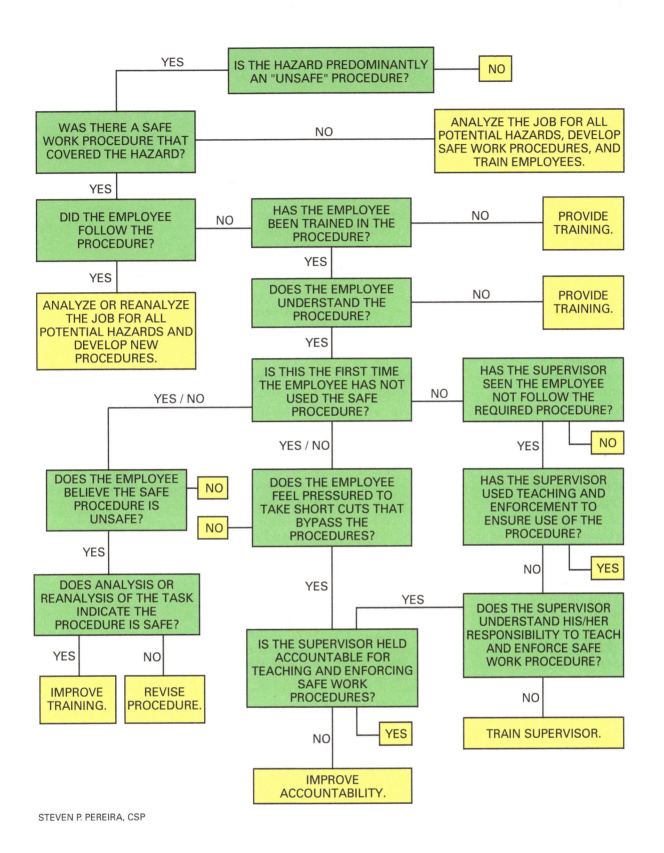

Module 75219 Hazard Recognition, Environmental Awareness, and Occupational Health

Appendix C

ASBESTOS CONTAINING MATERIALS

Products and materials that may contain asbestos include the following:

> **NOTE:** This list does not include every product or material that may contain asbestos. It is intended as a general guide to show which types of materials may contain asbestos.

- Acoustical plaster
- Adhesives
- Asphalt floor tile
- Base flashing
- Blown-in insulation
- Boiler insulation
- Breaching insulation
- Caulking and putties
- Ceiling tiles and lay-in panels
- Cement pipes
- Cement siding
- Cement wallboard
- Chalkboards
- Construction mastics (floor tile, carpet, ceiling tile, etc.)
- Cooling towers
- Decorative plaster
- Ductwork flexible fabric connections
- Electrical cloth
- Electrical panel partitions
- Electrical wiring insulation
- Elevator brake shoes
- Elevator equipment panels
- Fire blankets
- Fire curtains
- Fire doors
- Fireproofing materials
- Flooring backing
- Heating and electrical ducts
- High temperature gaskets
- HVAC duct insulation
- Joint compound
- Laboratory gloves
- Laboratory hoods or table tops
- Pipe insulation (corrugated air-cell, block, etc.)
- Roofing felt
- Roofing shingles
- Spackling compounds
- Spray-applied insulation
- Taping compounds (thermal)
- Textured paints or coatings
- Thermal paper products
- Vinyl floor sheeting
- Vinyl floor tile
- Vinyl wall coverings
- Wallboard

Additional Resources

This module presents thorough resources for task training. The following resource material is suggested for further study.

Basic Safety Administration: A Handbook for the New Safety Specialist, Fred Fanning, CSP. 2003. Des Plaines, IL: The American Society of Safety Engineers (ASSE).
Construction Project Safety, John Schaufelberger and Ken-Yu Lin. 2014. Hoboken, NJ: John Wiley & Sons.
Construction Safety Planning, David V. MacCollum, P.E., CSP. 1995. Hoboken, NJ: John Wiley & Sons.
Field Safety, NCCER. Current edition. New York, NY: Pearson.
Hazardous Materials Behavior and Emergency Response Operations, Denis E. Zeimet, Ph.D., CIH and David N. Ballard. 2000. Des Plaines, IL: The American Society of Safety Engineers (ASSE).
Managing Your Hazardous Wastes: A Guide for Small Businesses, US Environmental Protection Agency. 2001. Washington, DC.
NIOSH Construction Sector, **www.cdc.gov/niosh**
NIOSH Ergonomics for Construction Workers, DHHS (NIOSH) Publication No. 2007-122, August 2007.
NIOSH Pocket Guide to Chemical Hazards, DHHS (NIOSH) Publication No. 2005-149.
The Psychology of Safety Handbook, E. Scott Geller, Ph.D. 2001. Boca Raton, FL: CRC/Lewis Publishers.
Root Cause Analysis Handbook: A Guide to Effective Incident Investigation, ABSG Consulting. 2005. Brookfield, CT: Rothstein Associates Inc.
American Society of Safety Engineers (ASSE), **www.asse.org**
Environmental Protection Agency (EPA), **www.epa.gov**
Occupational Safety and Health Administration (OSHA), **www.osha.gov**

Figure Credits

Miller and Long Co., Inc./elcosh.org, Module opener
Steven P. Pereira, CSP, Figures 1, 14, Appendix A01A, A02A, A02B
GoCanvas.com, Figure 7
U.S. Army Corps of Engineers, Figure 8
U.S. Department of Labor, Figures 9, SA02, Appendix A01B
LPR Construction, SA01
Fall protection materials provided courtesy of Miller Fall Protection, Franklin, PA, Figure 16
© Tony Rich Asbestorama.com, Figure 17
U.S. Air Force photo by Airman 1st Class Ryan Callaghan, Figure 18A
© Toa555/ Dreamstime.com, Figure 18B
NASA, Figure 18C
© Lcswart/ Dreamstime.com, Figure 18D
© Eleni Seitanidou/Dreamstime.com, Figure 18E
© Bernard Maurin/Dreamstime.com, Figure 18F
© Karin Hildebrand Lau/Shutterstock.com, Figure 19
© Bronwyn Photo/Shutterstock.com, Figure 20
© iStockphoto.com/James Mullineaux, Figure 21
Photo courtesy of 3M Personal Safety Division, Figure 22
© Yobidaba/ Dreamstime.com, Figure 23
U.S. Geological Survey, Figure 24
© Glen Jones/Dreamstime.com, Figure 25
Topaz Publications, Inc., Figure 26A
BMP Supplies Inc., www.bmpsupplies.com, Figure 26B
Andre R. Aragon/FEMA, Figure 27
Marilee Caliendo/FEMA, Figure 28
MSA The Safety Company, Figure 29
Eagle Manufacturing Company, Figure 30
University of California, Riverside, Figures 31A, 33
© Brandon Bourdages/Shutterstock.com, Figure 31B
U.S. Environmental Protection Agency, Figure 34

Section Review Answer Key

Answer	Section Reference	Objective
Section One		
1. d	1.1.0	1a
2. b	1.2.3	1b
Section Two		
1. a	2.1.1	2a
2. b	2.2.0	2b
3. c	2.3.4	2c
Section Three		
1. c	3.1.3	3a
2. c	3.2.4	3b
3. d	3.3.8	3c

NCCER CURRICULA — USER UPDATE

NCCER makes every effort to keep its textbooks up-to-date and free of technical errors. We appreciate your help in this process. If you find an error, a typographical mistake, or an inaccuracy in NCCER's curricula, please fill out this form (or a photocopy), or complete the online form at **www.nccer.org/olf**. Be sure to include the exact module ID number, page number, a detailed description, and your recommended correction. Your input will be brought to the attention of the Authoring Team. Thank you for your assistance.

Instructors – If you have an idea for improving this textbook, or have found that additional materials were necessary to teach this module effectively, please let us know so that we may present your suggestions to the Authoring Team.

NCCER Product Development and Revision
13614 Progress Blvd., Alachua, FL 32615

Email: curriculum@nccer.org
Online: www.nccer.org/olf

❏ Trainee Guide ❏ Lesson Plans ❏ Exam ❏ PowerPoints Other _____

Craft / Level: _____ Copyright Date: _____

Module ID Number / Title: _____

Section Number(s): _____

Description: _____

_____

_____

_____

_____

Recommended Correction: _____

_____

_____

_____

_____

Your Name: _____

Address: _____

Email: _____ Phone: _____

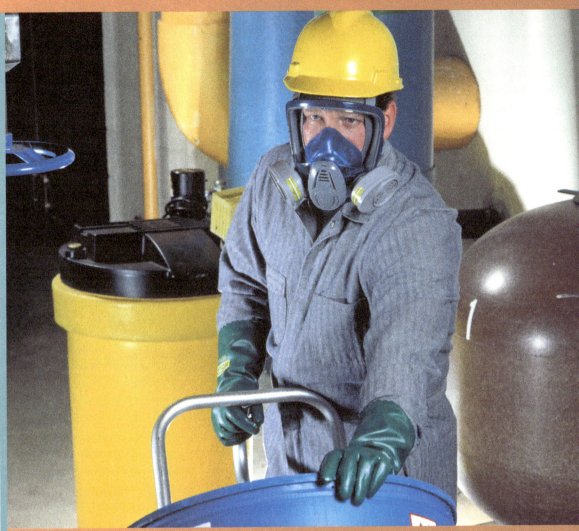

Job Safety Analysis and Pre-Task Planning

Overview

This module provides guidance on safety performance analysis and employee coaching. It also explains how to conduct a job safety analysis and perform pre-task planning.

Module 75220

Trainees with successful module completions may be eligible for credentialing through NCCER's National Registry. To learn more, go to **www.nccer.org** or contact us at 1.888.622.3720. Our website has information on the latest product releases and training, as well as online versions of our *Cornerstone* magazine and Pearson's product catalog.

Your feedback is welcome. You may email your comments to **curriculum@nccer.org**, send general comments and inquiries to **info@nccer.org**, or fill in the User Update form at the back of this module.

This information is general in nature and intended for training purposes only. Actual performance of activities described in this manual requires compliance with all applicable operating, service, maintenance, and safety procedures under the direction of qualified personnel. References in this manual to patented or proprietary devices do not constitute a recommendation of their use.

Copyright © 2018 by NCCER, Alachua, FL 32615, and published by Pearson, New York, NY 10013. **All rights reserved.** Printed in the United States of America. This publication is protected by Copyright, and permission should be obtained from NCCER prior to any prohibited reproduction, storage in a retrieval system, or transmission in any form or by any means, electronic, mechanical, photocopying, recording, or likewise. To obtain permission(s) to use material from this work, please submit a written request to NCCER Product Development, 13614 Progress Blvd., Alachua, FL 32615.

75220 V5

From *Safety Technology,* Trainee Guide. NCCER.
Copyright © 2018 by NCCER. Published by Pearson. All rights reserved.

75220
JOB SAFETY ANALYSIS AND PRE-TASK PLANNING

Objectives

When you have completed this module, you will be able to do the following:

1. Identify the factors involved in analyzing performance.
    a. List the factors that contribute to incidents.
    b. Describe techniques for coaching and counseling workers who have performance problems.
    c. Explain how to use the ABC model to encourage safe behavior.
2. Explain how to analyze the hazards of specific jobs and tasks.
    a. Identify the benefits of a job safety analysis (JSA).
    b. Describe how to conduct a JSA.
    c. Identify the differences between a JSA and pre-task planning.
    d. Describe how to conduct pre-task planning.

Performance Tasks

Under the supervision of your instructor, you should be able to do the following:

1. Conduct a job safety analysis (JSA).
2. Conduct pre-task planning.

Trade Terms

Causal links
Consequences
Ergonomics
Job
Pattern
Reinforcement
Recognition
Structured on-the-job training (SOJT)
Task
Trend

Industry Recognized Credentials

If you are training through an NCCER-accredited sponsor, you may be eligible for credentials from NCCER's Registry. The ID number for this module is 75220. Note that this module may have been used in other NCCER curricula and may apply to other level completions. Contact NCCER's Registry at 888.622.3720 or go to www.nccer.org for more information.

Contents

- 1.0.0 Performance Analysis 1
  - 1.1.0 Factors That Contribute to Incidents 1
    - 1.1.1 The Behavioral Law of Effect 1
    - 1.1.2 Observation and Reinforcement 2
    - 1.1.3 Performance Barriers 2
    - 1.1.4 Impairment Factors 3
    - 1.1.5 Employee Misconduct 4
    - 1.1.6 Chances and Risks 4
    - 1.1.7 Inattention, Distraction, and Fatigue 6
  - 1.2.0 Coaching and Counseling 7
    - 1.2.1 Dealing with At-Risk Behavior 7
    - 1.2.2 Dealing with Safe Behavior 7
  - 1.3.0 The ABC Model 7
    - 1.3.1 Activator 8
    - 1.3.2 Behavior 8
    - 1.3.3 Consequences 8
- 2.0.0 Job and Task Hazard Analysis 10
  - 2.1.0 Reasons for Conducting a Job Safety Analysis 10
    - 2.1.1 Recognize and Reduce the Risk of Potential Hazards 11
    - 2.1.2 Improve Hazard Awareness 11
    - 2.1.3 Develop Standardized Work Practices 11
    - 2.1.4 Facilitate Job Training 11
  - 2.2.0 Conducting a Job Safety Analysis 11
    - 2.2.1 Selecting Jobs or Tasks to be Analyzed 11
    - 2.2.2 Preparing for a Job Safety Analysis 12
    - 2.2.3 Collecting Data 12
    - 2.2.4 Identifying Hazards and Risk Factors 14
    - 2.2.5 Developing Solutions 14
    - 2.2.6 Common Errors When Conducting a JSA 15
  - 2.3.0 Pre-Task Planning 15
  - 2.4.0 Conducting Pre-Task Planning 15

Figures and Tables

- Figure 1 Horseplay is at-risk behavior 2
- Figure 2 Example of an impaired worker 4
- Figure 3 Worker using required PPE, causing fatigue 6
- Figure 4 ABC model 8
- Figure 5 The job/task relationship 10
- Figure 6 Example of a job safety analysis form 13
- Figure 7 Hazard analysis chart 15
- Figure 8 Hierarchy of controls 15
- Figure 9 Sample pre-task planning forms 17

Section One

1.0.0 Performance Analysis

Objective

Identify the factors involved in analyzing performance.
a. List the factors that contribute to incidents.
b. Describe techniques for coaching and counseling workers who have performance problems.
c. Explain how to use the ABC model to encourage safe behavior.

Trade Terms

Consequences: The final outcome of actions and behaviors.

Job: A regular activity performed to achieve some end.

Reinforcement: Supporting or strengthening of a behavior or action.

Recognition: Something that is given or returned verbally, through body language, or in writing for good or bad behavior.

Task: A discrete step or portion of a job.

Safety technicians are responsible for identifying at-risk behavior. They must learn how to assess the behavior of workers by accurately observing them and understanding barriers and impairment issues. They must make sure that workers are aware of the consequences of at-risk behavior and should be prepared to enforce those consequences.

As you develop these skills, you will become increasingly adept at creating a safe and productive work environment, and the value you bring to your company will increase as well.

> **NOTE:** For more information on risk analysis, refer to NCCER Module 75219, *Hazard Recognition, Environmental Awareness, and Occupational Health*.

1.1.0 Factors That Contribute to Incidents

Most incidents are the result of at-risk behavior (*Figure 1*). It is important to understand human behavior and the human factors that lead to incidents. The following concepts need to be understood in order to accurately analyze performance problems:

- The Behavioral Law of Effect
- Observation and reinforcement
- Performance barriers
- Impairment factors
- Unpreventable employee misconduct
- Chances and risks
- Inattention, distraction, and fatigue

1.1.1 The Behavioral Law of Effect

The Behavioral Law of Effect states: "Behavior that brings recognition is repeated, whereas behavior that does not bring recognition or is punished is not repeated." This Law does not specify whether the behavior is good or bad; it simply states that if the behavior is recognized in a positive way, it will be repeated. Behavior that is followed by discomfort or a negative consequence is less likely to be repeated.

This Law also does not indicate the source of the recognition, which could be positive or negative. Consider the example of a crew experiencing a mechanical problem with some heavy equipment. The operator quickly diagnoses the problem and fixes it. A while later, the supervisor tells the operator that he did a great job of keeping the unit on line. The operator feels good that his efforts were recognized and will likely fix the same problem in the same way in the future.

Suppose that the same operator facing the same equipment problem does not follow the correct procedure when making the repair. As a result, the worker injures his hand. Because of the unpleasant consequence, the worker is now less likely to repeat the incorrect behavior.

If every at-risk or hazardous act resulted in an injury, the incident itself would reinforce the need to follow safety procedures. However, since literally thousands of at-risk or hazardous acts

NIOSH FACE Program

The National Institute of Occupational Safety and Health (NIOSH) conducts investigations of fatal occupational injuries through its Fatality Assessment and Control Evaluation (FACE) program. By conducting these investigations, NIOSH can identify factors that contributed to the fatalities and then use this information to develop comprehensive recommendations for preventing similar deaths. Full texts of the fatality investigation reports are available at the NIOSH website, www.cdc.gov/niosh/FACE/.

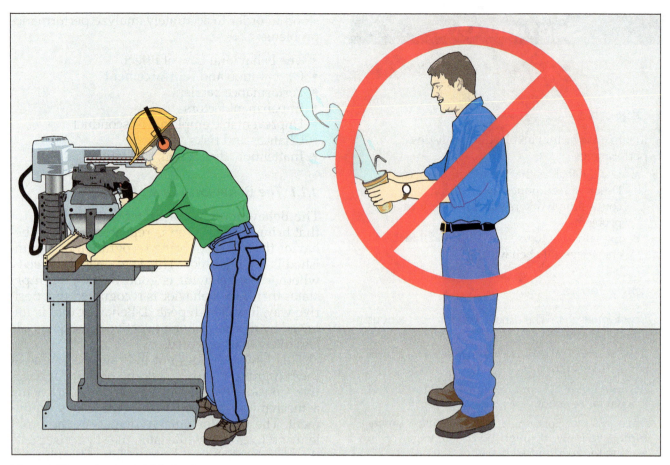

Figure 1 Horseplay is at-risk behavior.

are committed each day with no adverse consequences, at-risk behavior is encouraged. Safety technicians should offer positive recognition whenever they observe a worker doing something right.

1.1.2 Observation and Reinforcement

Observing a worker confirms and reinforces that person's behavior, regardless of whether it's safe or at-risk. This is because the worker being observed assumes that his or her behavior is acceptable if it is done openly without any adverse consequences. Overlooking or ignoring a safety violation essentially gives approval to the violation and implies permission for the condition to exist.

There will be times when you will be asked to overlook at-risk behavior. However, overlooking a few simple safety violations compromises your entire attitude towards safety. Never do this. Always let workers know that you mean what you say, and that if you observe an at-risk act or condition you will ask them to correct it. This will help to reinforce safe behavior.

1.1.3 Performance Barriers

Often, there are obstacles that prevent workers from performing their job safely. These obstacles are called performance barriers. Common performance barriers include the following:

- Physical barriers
- Knowledge barriers
- Execution barriers

Physical Barriers – Physical barriers cause a worker to be physically unable to perform a job safely. Some examples of physical barriers include the following:

- Poor eyesight
- Loss of hearing
- Lack of individual strength or agility
- Illness

A worker with a physical barrier should be assigned a different job, or the specific job task should be restructured, if possible.

Knowledge Barriers – Knowledge barriers present a problem when the worker does not know or remember how, when, or why to do an assigned task. The following factors should be considered when trying to determine if a knowledge barrier exists:

- Education
- Training
- Skill level
- Ability to understand written or oral instructions
- Language barriers

Training and practice are possible solutions for correcting knowledge barriers.

Execution Barriers – An execution barrier exists when a worker is physically and mentally capable of completing a job task but does not do it. This type of behavior may be the result of either a task barrier or an incentive barrier.

Task barriers result when the worker is not provided with a proper work environment, the appropriate number of people for the task, or the right tools or equipment to do the job. A lack of material can also be a barrier. Task barriers are most easily fixed by providing what is needed to get the job done safely.

> **NOTE**
> Failure to use personal protective equipment can be a task barrier. For instance, workers may not be using safety glasses as required because they consider the glasses unattractive or uncomfortable. It is relatively easy to fix the problem by providing safety glasses that are more attractive or comfortable, so workers will want to use them.

Incentive barriers can be created when a worker has been recognized for doing the job incorrectly; has been punished even though the job was done correctly; or has never received feedback for either positive or negative behavior. For example, even though a worker may have used the wrong equipment to complete a task, the supervisor is pleased that the task was completed on time. As a result, it is likely that the worker will continue this at-risk behavior until there is an incident. Positive feedback for negative behavior encourages the behavior.

Positive feedback can be in the form of recognition. Negative feedback can be in the form of disciplinary actions or lost privileges. Incentive barriers can be fixed by noting the performance problem and providing the appropriate positive or negative feedback.

1.1.4 Impairment Factors

Workers can sometimes get confused, distracted, or disoriented and do things they otherwise would not have done (*Figure 2*). This behavior can be the result of fatigue, illness, intense or prolonged stress, or substance abuse. These are considered impairment factors. Impairment factors are different from performance barriers because they affect the worker in such a way that it would be difficult to do any other type of work; in contrast, performance barriers only affect one task.

It can be easy to say that an individual just forgot to follow the correct procedure and leave

---

Case History

Fatal Fall

On April 19, 2012, a laborer fell approximately 13.5 feet (over 4 meters) from a residential roof to a concrete driveway. The laborer was a member of a crew of eight workers for a construction subcontractor. The workers were replacing shingles on a roof that was accessed by a ladder. When the incident occurred, the laborer was on the garage side of the home and out of sight of his co-workers. The co-workers heard the laborer hit the ground, rushed to his aid, and called 911. Emergency Medical Services subsequently pronounced the laborer dead at the scene.

The Bottom Line: Based on its identification of the contributing factors for this incident, NIOSH concluded that this fatality could have been prevented by the following:

- Developing a comprehensive safety program that provides fall prevention training, complies with OSHA standards, and is in a language and at a literacy level that the employees can understand.
- Having a competent person inspect the worksite before the work begins to identify fall hazards, determine the appropriate fall prevention systems for the workers, and ensure that personal fall arrest systems for the workers are assembled, installed, and maintained properly.

Source: NIOSH FACE Report 2012-02

Figure 2 Example of an impaired worker.

it at that. But if the procedure suffered because of prescription drug use or substance abuse, the behavior cannot be ignored or overlooked. Pay attention to the worker's behavior so that you can accurately judge the nature of the impairment and take the appropriate actions to correct the situation.

Impairment factors such as fatigue, stress, illness, and the effect of medications are less clearly identifiable. These factors are sometimes only identified during an incident investigation. If an incident investigation is being done, interview those directly involved to determine if any of these factors may have contributed to the incident. It may also be helpful to explore these issues with witnesses or with people close to those directly involved in the incident. It may even be necessary to contact a doctor to establish and/or confirm the effects of the suspected impairment factor on behavior.

1.1.5 Employee Misconduct

Employee misconduct is behavior that happens when workers clearly disregard policies, procedures, and regulations. Be aware of behavior that is inappropriate and correct it as needed. Before classifying a worker's behavior as employee misconduct, first confirm that other factors did not contribute to the behavior.v

1.1.6 Chances and Risks

Some people are more willing than others to accept a higher level of risk. As an example, picture two aquariums, each holding a rattlesnake. In one of the aquariums is a $100 bill, and the other has $1,000. Some people would take the chance of a snake bite for $1,000; others would do it for $100; and some wouldn't take the chance at all. Why do you think people take these kinds of risks? The answer is that there are consequences that influence both real and perceived risk. People generally move towards consequences that are soon, certain, and positive.

Personal Risk Acceptance – Personal riswk acceptance is the level of risk at which a worker is willing to accept the consequences of his or her actions. This type of risk is categorized by the following two factors of personal acceptance:

- *It won't happen to me.* This is a common belief among workers who take risks. Past experience has led the worker to believe there is a relatively low level of risk. They say to themselves,

"I've done it this way a thousand times and never had an incident." And they are right. They haven't been hurt and therefore will continue to ignore the risk.

- *Real versus perceived risk.* The actual risk associated with a specific hazard or behavior is determined by the likelihood and severity of a possible incident, not a worker's previous experience with the same situation. Very often, it's easy to overlook this and take a risk based on personal experience rather than the actual risk factor. For example, the risk of dying in a car wreck while traveling to and from work is much higher than the risk of dying due to radiation exposure from a nearby nuclear power plant. Yet it is likely that workers wouldn't volunteer to go near the power plant but they would still travel to and from work every day.

Factors That Influence Perceived Risk – Everyone's perception of risk is different. Keep this in mind when you are having performance problems with a worker. This knowledge will be helpful in determining a solution. The following are factors and situations that influence perceived risk:

- *Voluntary vs. mandatory exposure.* Hazards can be either voluntary or mandatory. Voluntary hazards include activities such as motorcycling, snowmobiling, and skiing. These activities are generally perceived as low risk because they are enjoyable. For example, you might consider it low risk to ski down a mountain, but if you were required to work from a small platform raised 25 feet (8 meters) off the ground, you would probably perceive that risk as much greater. It's important to recognize that individuals perceive risks differently. This will influence a worker's willingness to complete a task.
- *Familiar hazards are considered to be lower risk.* The more that is known about a hazard, the more risk one is willing to take. For example, when you first learned to drive, you probably drove slowly and carefully. As you became more familiar with the vehicle and its hazards, you were willing to drive faster and with less caution. The same familiarity occurs with tools and equipment; over-confidence leads to incidents. Therefore, it is important to observe workers who have a lot of experience as much as you would observe those with little experience.
- *Controllable vs. uncontrollable.* If a worker feels that a hazard can be controlled, it is often perceived as less risky. If the hazard is considered uncontrollable, as in the case of severe weather, it is considered more dangerous. Workers can become complacent when it comes to controllable hazards, because as they learn more about how to control a hazard, their respect for it diminishes. For example, when workers first learn to work with sulfuric acid, they are extremely cautious and wear the prescribed personal protective equipment. After working with the chemical for several years without a major incident, they may become complacent. They know that if they get to water quickly, the burn can be minimized significantly. As a result, they often don't treat the acid with the respect it deserves. Monitoring and observation can help to eliminate this problem.
- *Risk compensation.* When people are provided with personal protective equipment or safety devices, they often become careless. For example, a group of workers was asked to drive forklifts through a practice course. Their speed was clocked each time they completed the course. When seat belts were required, the drivers increased their speed considerably. It was concluded that personal protective equipment sometimes gives users a false sense of security. It was also concluded that knowledge of the risks and proper safety training can prevent this from happening.
- *No one is watching.* Some workers only follow safety policies and procedures when they are being watched. They are often willing to take risks only when they know they can get away with it. Make sure workers know the consequences of their actions even when no one is watching.
- *Peer pressure and social conformity.* Peer pressure is an extremely powerful motivator. It can be positive or negative. Very few workers want to be different; most want to be part of a group and will change their behavior to fit into that group. For example, many people in the workplace recognize the need for and the value of hard hats and other forms of personal protective equipment, but often don't use them because their co-workers don't. Make sure everyone is following all safety policies and procedures at all times, and always wear appropriate PPE yourself to set a good example.
- *Acceptable consequences.* If the consequences of an action are soon, certain, and positive, they are likely to be repeated. For example, consider a worker named Jim, who is required to grind a piece of metal using a bench grinder approximately ten times a day. The task of grinding takes only 30 to 45 seconds. When Jim is grinding, he doesn't use the face shield because it is dirty and needs to be readjusted. Jim feels confident that an incident will not happen because

the task takes less than a minute and he is wearing safety glasses. He has done it this way for six years without an incident; as far as he is concerned, the consequences have been acceptable. Jim has a false sense of safety because he has always completed the job without an incident. But this doesn't mean that an incident couldn't happen the next time the grinder is used. One way to stop this behavior is to make workers accountable for their actions. In this case, Jim should be appropriately disciplined and made aware of the consequences of his actions.

1.1.7 Inattention, Distraction, and Fatigue

Loss of focus is a situation in which a worker has difficulty doing a job because his or her mind is wandering. The worker becomes temporarily distracted, fails to pay attention, and may commit a hazardous act. The hazardous act puts the worker at risk and can sometimes cause an incident. This loss of focus may be due to daydreaming, concern about off-the-job problems, or a sudden distraction. If you suspect a worker is distracted or inattentive, talk to him or her. Find out what may be causing the distraction and figure out a solution, such as rotating the worker to a different job task. Aggressive work schedules with overtime, shift work, and extended work shifts all put workers at risk for fatigue. The fatigue that can arise from strenuous work schedules can be compounded by physical and environmental conditions at the job site. It's important to take a pre-emptive approach to address worker fatigue and thereby prevent potential incidents or injuries.

The length of work shifts is a chief risk factor for fatigue. Research indicates that as work shift length increases, the risk of incidents and injuries also rises. The risk further increases for night shifts and with an increase in the length of the work week. Additional factors that increase the risk of fatigue include a disrupted sleep schedule; exposure to environmental, physical, and/or chemical hazards; and the use of certain types of required personal protective equipment (*Figure 3*). The most proactive way to address worker fatigue is to adopt an approach that assesses and controls each identified risk factor in proportion to the hazard it presents at a given work site. This is what a fatigue management plan or program is designed to accomplish.

A fatigue management plan uses three key procedures—assessment of risk factors, selection of controls, and evaluation of control effectiveness—to develop site-specific plans for minimizing worker fatigue. Basically, implementation of

Figure 3 Worker using required PPE, causing fatigue.

the plan involves first assessing the risks associated with the tasks to be performed and the conditions under which the tasks will be carried out; then, selecting controls (policies, procedures, and work practices) to minimize these risks; and finally, following up by regularly performing evaluations to judge how effective the controls are in mitigating worker fatigue.

Applying a hierarchy of controls, select the work practices, policies, and resources needed to address the most significant risk factors. Different routines of work and variations in workload will impact cumulative fatigue over a single shift and throughout a work rotation. In some cases, a single control, such as limiting the length of a work shift, will be feasible. In other cases, you may need to use a combination of controls—such as monitoring for signs and symptoms of fatigue (practices), providing rest breaks and mandatory days off (policies), and adding workers and/or rotating workers through jobs during a work shift (resources)—to offset the physical demands of the task and the operational need for extended shifts.

Working for seven days or more without a day off is both physically and mentally fatiguing and should be avoided when possible. In addition, moving workers between shifts can result in fatigue as their bodies require several days to adjust. Maintaining a consistent schedule with frequent breaks and time off will help to keep workers alert and safer on the job site.

> **Did You Know?**
>
> ## The Downside of Energy Drinks
>
> Caffeine and sugar are the two main ingredients in most energy drinks. The high caffeine content can cause headaches, insomnia, nervousness, rapid heartbeat, and increased blood pressure. Both caffeine and sugar contribute to the energy crash that some people experience after consuming energy drinks. The resulting low blood sugar levels have been shown to correlate with poor decision-making ability, mental fatigue, and impaired anger management. Consuming excessive amounts of sugar also increases the risk for obesity. Mega-doses of B vitamins, particularly B12, are often found in energy drinks as well. Excessive intake of B vitamins may result in liver or nerve damage.
>
> Taurine is another common ingredient in energy drinks. Research from Weill Cornell Medical College suggests that instead of acting as a stimulant, taurine actually has more of a sedative effective on the brain. Scientists suspect that taurine, as well as caffeine and sugar, may play a role in the crash that often follows the initial stimulation that occurs when consuming energy drinks.
>
> In addition, some energy drinks contain alcohol, which can impair workers on the job. Some job sites prohibit the use of energy drinks.

1.2.0 Coaching and Counseling

Safety technicians who understand the causes of performance problems also need to learn how to coach or counsel those who are placing themselves or others in a hazardous or at-risk situation. The skills of coaching and counseling are necessary tools for communicating with workers and correcting at-risk behavior. When workers feel that they can communicate openly, it builds trust and respect between both parties. When you have earned the confidence of an individual, they will feel secure enough to communicate and will be more open to the suggestions, ideas, and actions that will reduce at-risk behavior.

1.2.1 Dealing with At-Risk Behavior

At-risk behavior affects everyone on the job site, not just the worker taking risks. Keep in mind that workers observe and often imitate unsafe behaviors. When you have determined that someone's behavior is too risky, you must do the following:

- Without startling the person, stop what they are doing.
- Take the worker aside and speak to them privately.
- Express concern for his or her safety and well-being.
- Identify the action or behavior that was incorrect or at risk.
- Explain why the action or behavior was incorrect or at risk.
- Try to find out if there are any barriers or impairment factors.
- Ask the worker to explain or demonstrate the proper way to perform the job but be prepared to show or explain how the task should be done.
- Leave on a positive note.

Use the same technique whether you are dealing with co-workers, visitors, or supervisors. It's important to remember that everyone deserves to be treated with dignity and respect. No one wants to be publicly embarrassed. Do your coaching and counseling privately.

1.2.2 Dealing with Safe Behavior

People desire recognition and thrive on compliments. When you see someone performing a job safely, provide positive feedback. Remember, behavior that is recognized is repeated. A good job that is not recognized (or is punished) is not repeated. When you observe a worker performing a job safely, you should do the following:

- Praise publicly.
- Specifically state the correct behavior observed.
- Compliment the worker on a job well done.

1.3.0 The ABC Model

The ABC model is a tool that will help you understand human behavior and thereby encourage safe behavior. Its main objective is to find ways to change behavior through positive recognition.

It also provides insight about why people behave in a certain way and helps determine the types of consequences associated with the behavior.

The ABC model assigns the following terms to A, B, and C:

- *A* – Activator (or antecedent)
- *B* – Behavior
- *C* – Consequences

By understanding the ABC model (*Figure 4*), including the relationship between the activators and behavior, behavior and consequences, and consequence and activator, you can apply the model to change the behavior.

1.3.1 Activator

An activator is any condition that prompts behavior. For example, if the phone rings, you answer it. The ringing phone is considered the activator. Sometimes activators are clear and other times they are not. Some examples of activators that are easy to point out and correct include:

- Incentive programs
- Policies
- Signs

When activators are not clear, it is difficult to find an immediate cause for at-risk behavior. If you find that the activator is not clear and an investigation must be done, use the following questions to determine what prompted the at-risk behavior:

- Is there a policy or procedure that addresses the behavior?
- Is the policy or procedure specific to the behavior?
- Is the policy or procedure known, understood, and accepted as valid?
- Is the behavior activated by practice, habit, and peer influence or job requirements?

Once you answer these questions, you will be able to determine the cause of the behavior and immediately correct the problem. Most workers consider basic safety as common sense rather than a specific procedure that must be followed. If you notice that a safety procedure is not being followed, provide concrete examples of injuries that have resulted from failure to follow the specific policy or procedure.

1.3.2 Behavior

Behavior is the second part of the ABC model and probably the most easily explained. In the context of this model, behavior refers to the specific behavioral response to the activator. For example, a safety policy exists stating that safety glasses must be worn. The policy is the activator and the wearing of safety glasses is the behavior. The behavior is both observable and measurable.

1.3.3 Consequences

Consequences are the results of the behavior, whether positive or negative. Workers must be made aware of both the positive and negative consequences of their behavior. For example, a hard hat that is uncomfortable may not be worn. The immediate consequence of this action is greater comfort. This is perceived as positive by the worker, but it is actually negative because of the injuries the worker can receive. To encourage workers to wear hard hats, some companies color code them based on experience. A green hard hat might indicate an inexperienced worker, and a brown one would be worn by someone with more experience. In this case, using a piece of safety equipment as a badge of honor makes wearing it a positive experience.

In the context of the ABC model, consider the following factors when examining the consequences of at-risk behavior:

- Is the recognition pleasurable or positive?
- Is the recognition certain?
- Is the recognition soon?
- Is the recognition valuable to the employee?

If the answer to any of these questions is yes, then it is likely that the at-risk behavior will continue unless corrected. Training and strong leadership can help workers understand the consequences of their behavior.

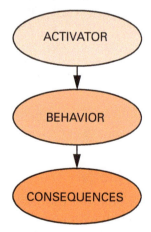

Figure 4 ABC model.

Additional Resources

American Society of Safety Engineers, **www.asse.org**

Occupational Safety and Health Organization, **www.osha.gov**

The Participation Factor—How to Increase Involvement in Occupational Safety, E. Scott Geller, Ph.D. 2002. Des Plaines, IL: The American Society of Safety Engineers (ASSE).

The Psychology of Safety Handbook, E. Scott Geller, Ph.D. 2001. Boca Raton, FL: CRC/Lewis Publishers.

Root Cause Analysis Handbook: A Guide to Effective Incident Investigation, ABSG Consulting. 2005. Brookfield, CT: Rothstein Associates, Inc.

1.0.0 Section Review

1. Lack of individual strength is an example of which type of performance barrier?
    a. Execution
    b. Motivation
    c. Knowledge
    d. Physical

2. Which of the following is an appropriate way to deal with a worker who is exhibiting at-risk behavior?
    a. Use a raised tone of voice to scold the worker publicly.
    b. Gather the entire crew and explain to them why the behavior is risky.
    c. Take the worker aside and discuss the issue privately.
    d. Ignore the immediate behavior but speak to the worker's supervisor later.

3. In the ABC model, the letters A, B, and C stand for _____.
    a. attitude, behavior, and change
    b. activator, behavior, and consequences
    c. antecedent, barrier, and condition
    d. aptitude, barrier, and correction

Section Two

2.0.0 Job and Task Hazard Analysis

Objective

Explain how to analyze the hazards of specific jobs and tasks.

a. Identify the benefits of job safety analysis (JSA).
b. Describe how to conduct a JSA.
c. Identify the differences between a JSA and pre-task planning.
d. Describe how to conduct pre-task planning.

Performance Tasks

1. Conduct a job safety analysis (JSA).
2. Conduct pre-task planning.

Trade Terms

Causal links: A relationship between two events or occurrences in which one is the cause of the other.

Ergonomics: The applied science of equipment design, as for the workplace; intended to maximize productivity by reducing operator fatigue and discomfort.

Pattern: An indication of how predictable the reoccurrence of an event is.

Structured on-the-job training (SOJT): Training that takes place during the performance of the job with the guidance of documentation or a facilitator.

Trend: The tendency to take a particular direction.

Developing methods and controls to prevent incidents from occurring can dramatically increase the safety of a job or task. *Figure 5* shows the relationship between a job and a task. Conducting job safety analyses (JSA) and pre-task planning can help site personnel reduce the risk of incidents. A job safety analysis (*e.g.*, welding a pipe) is a careful study of a job to find all of the associated hazards. Pre-task planning is a detailed study of a task to find all of the associated hazards (*e.g.*, beginning with driving to the job and then breaking the task into its component subtasks, such as turning on the torch).

Both job safety analyses and pre-task planning also identify procedures and practices for reducing the risks associated with the jobs and tasks. Time spent doing JSA and pre-task planning should be considered an investment. Each helps to reduce the number of incidents, improve productivity, and aid in developing job training.

> **NOTE:** OSHA has extensive information on job safety analyses and pre-task planning on their website at **www.osha.gov**.

2.1.0 Reasons for Conducting a Job Safety Analysis

There are several reasons to conduct JSAs. Most of the benefits are directly related to safety. Other benefits are in the areas of productivity and training. JSAs benefit the employees, the company, and the workers. JSAs also help future planning, identifying areas of needed improvement, and development of safety processes.

> **NOTE:** The terms *job safety analysis (JSA)* and *job hazard analysis (JHA)* are often used interchangeably, depending on the company or job site. Each involves a methodical review of the specific steps required to complete a task, along with identifying the anticipated hazards and the safety precautions for avoiding those hazards. In some cases, a JHA will also include information on the probability or severity of the listed hazards. *Appendix A* shows an example of a job hazard analysis (JHA).

JOB: ELECTRICIAN TASK: CHANGE BURNED OUT LIGHT BULB IN A DESK LAMP
STEP 1: LOCATE BURNED OUT BULB.
STEP 2: DETERMINE BULB TYPE.
STEP 3: UNPLUG THE LAMP.
STEP 4: REMOVE BURNED OUT BULB.
STEP 5: INSTALL WORKING BULB.
STEP 6: RESTORE POWER TO THE LAMP.

Figure 5 The job/task relationship.

2.1.1 Recognize and Reduce the Risk of Potential Hazards

The main benefit to conducting a JSA is that it provides information on potential hazards associated with a job or task. It is used to train inexperienced workers and remind experienced workers of safety precautions they might have forgotten. Besides identifying hazards, a JSA will also help determine ways to eliminate or minimize unsafe or unhealthy conditions. Identifying hazards and determining the appropriate steps, equipment, and controls needed to reduce those hazards will minimize incidents and injuries.

2.1.2 Improve Hazard Awareness

Unknown or unforeseen hazards, such as poor housekeeping, may be uncovered through the careful study and analysis involved in a JSA. As JSAs are conducted on individual jobs, you will begin to develop a larger picture of the overall hazard level at the site. This information can be used to prevent incidents or illnesses and to reduce the level of risk throughout.

2.1.3 Develop Standardized Work Practices

JSAs also provide a means for standardizing work practices. When conducting a JSA, a job is studied in great detail based on prior experience. Several workers may be observed while performing a job. They may also be interviewed to determine all of the steps involved in performing the job. This study produces a large volume of information on those practices that work efficiently and safely and those that do not. Documentation of this information can be used to develop standardized work practices.

2.1.4 Facilitate Job Training

Observing and analyzing a job and the tasks that go into performing that job will yield documentation and information that can be useful when developing training for the job. This training could be job-specific safety training, process training, or structured on-the-job training (SOJT). Because the research that goes into developing technical and safety training often involves a similar type of job analysis, the data provided by a JSA may easily be adapted for this purpose.

2.2.0 Conducting a Job Safety Analysis

A JSA should be done before the work is underway. Depending on the job to be analyzed, conducting a JSA can be a rigorous process. Always do as much in advance as possible. A thorough JSA involves five basic steps:

Step 1   Select the job to be analyzed.

Step 2   Prepare for the JSA.

Step 3   Collect data.

Step 4   Identify hazards and risk factors.

Step 5   Develop solutions.

2.2.1 Selecting Jobs or Tasks to be Analyzed

A priority rating system must be used to rank jobs for analysis. It is suggested that jobs be rated based on the following criteria:

- History of incidents or injuries
- Potential for serious injuries
- Newness of the job or the equipment used
- A change in procedure or routine
- High turnover rates

Using these criteria along with your experience and professional judgment will help you to select jobs for analysis that will provide the greatest impact on work site safety.

History of incidents or injuries – A job that is known to frequently result in incidents or injuries should be strongly considered for a JSA. This data can come from previously completed jobs. For example, if workers who receive and open shipments of materials are frequently cutting themselves with box cutters or razors, a JSA may be in order to determine the causes and develop solutions.

Potential for serious injuries – A job that is more likely to result in serious injury or a fatality should be rated a higher priority for a JSA than one in which the injuries are less severe. If, for example, a job may result in an incident involving high-voltage electrocution or a fall from a great height, that job should receive a higher priority for a JSA than a job where the most likely incidents typically result in minor injuries.

Newness of the job or the equipment used – When a new job or process is started, it should be thoroughly analyzed to determine what risk factors it introduces to the work site and the workers. If a new piece of equipment is introduced, all of the jobs for which that equipment will be used should be subjected to a JSA to make sure that the new equipment does not create new or different hazards.

A change in procedure or routine – Changing a procedure or routine may introduce new risks or hazards because workers are not used to these changes. Also, the new procedure or routine may create new hazards that were not present before, or it may aggravate currently existing or likely hazards.

High turnover rates – Jobs with high turnover rates warrant a JSA for two reasons. First of all, the unsafe condition or perceived unsafe condition of the job may be a cause for the high turnover rate. Also, a high flow of new people through a job means that the workers have little experience and are more likely to have an incident or cause one. Clearly documented work processes and safety procedures help to reduce the occurrence of such incidents.

2.2.2 Preparing for a Job Safety Analysis

In order to work most efficiently, some preparation needs to go into the JSA process before data collection begins. First, take a look at the general conditions under which the job is being performed and develop a list of questions to be answered or a checklist of items to evaluate while conducting the JSA. Some standard questions to ask include the following:

- Is there adequate lighting?
- Are there line-of-fire hazards (*e.g.*, the potential for contacting energized equipment)?
- Are there live electrical hazards or the potential for the release of energy (thermal, electrical, pneumatic, etc.)?
- Are there fire hazards?
- Are there chemical hazards?
- Are there fall hazards?
- Are there excavation/engulfment hazards?
- Are there confined space hazards?
- Is there a potential for an environmental hazard or weather change?
- Is there a possibility of workers being struck by/pinched by/or caught in-between moving equipment?
- Are there tools or equipment that need repair?
- What is the availability of personal protective equipment?

It is useful to create a form on which to record your observations before conducting the JSA. Use of a form will simplify data collection, documentation, and recordkeeping. *Figure 6* shows a typical example of a completed job safety analysis form.

2.2.3 Collecting Data

There are several methods of collecting data, including the following:

- Documentation review
- Direct observation of prior work activities that are incorporated into the JSA to meet working conditions
- Group discussion

When conducting a JSA, it is not necessary to use each method of data collection. However, the more techniques used, the better the overall picture.

Documentation Review – The first resource for collecting data is the documentation associated with the job being analyzed. The documentation includes the following:

- Reports on incidents or near misses occurring on the job (an OSHA injury/near miss form is provided in *Appendix B*)
- Job descriptions
- Equipment and tool maintenance logs
- Procedure manuals
- Safety data sheets (SDSs) for materials used on the job
- Previous observations

Study and analysis of this data will provide important background information on how the job should be done and the necessary equipment and safeguards. This research may also reveal a certain trend, causal links, or a common pattern associated with incidents. Reviewing the associated documentation will not typically provide all of the necessary details, nor is it likely to provide solutions. It will, however, make the other aspects of data collection far more efficient. It is a key method of data collection.

Direct Observation – Direct observation is simply going to the job site, watching workers do the job, and asking them about the job and its hazards. Direct observation is the preferred method of data collection because it gives the best perspective on the job, the conditions under which it is performed, the tasks that need to be performed, and the risks associated with the job and its tasks.

JOB SAFETY ANALYSIS	JOB TITLE: <u>Casting Finisher</u> Page 1 of 1	JSA No. 123	DATE: February 15, 2016	NEW ☐ REVISED ☐
	TITLE OF PERSON WHO DOES JOB: Grinder Operator	SUPERVISOR: Rusty Bridger	ANALYSIS PERFORMED BY: JJ Morales	
ORGANIZATION: Central Castings	LOCATION: Metal Shop	DEPARTMENT: Grinding	REVIEWED BY: Chris Tran, EH&S	
SEQUENCE OF BASIC JOB STEPS	POTENTIAL HAZARDS	RECOMMENDED ACTION OR PROCEDURE		
1. Reach into right box and select casting	• Could strike hand on wheel when reaching into box located beneath wheel • Could tear hand on sharp corners of casters	• Relocate box to side of wheel • Wear leather gloves		
2. Grasp casting, lift, and position	• Could strain shoulder/elbow by lifting casting from low box with elbow extended • Casting could slip from hand during positioning and fall on toe	• Place box on a pallet to reduce the lift distance • Wear steel-toed safety shoes/boots		
3. Push casting against wheel and grind burr	• Small wheel guard could allow hand to strike against wheel • Wheel could explode if cracked or installed incorrectly • Wheel friction with caster causes flying sparks/chips • Caster metal and wheel metal generate respirable dust • Loose sleeves could get caught in machinery	• Provide larger wheel guard with tongue guard and work rest • Check rpm rating of wheel • Inspect wheel for cracks • Wear safety goggles • Provide local exhaust system • Wear close-fitting, long-sleeved shirts with buttoned cuffs		
4. Place finished casting into box	• Could strike hand on finished castings due to buildup of completed stock	• Routinely remove excess buildup of completed stock		

Figure 6 Example of a job safety analysis form.

Direct observation can be used to evaluate and fine-tune existing JSAs. The drawbacks to direct observation are that it can be distracting to the workers and may cause them to change the way they work because they know they are being observed. The best way to work around these issues is to keep a low profile as much as possible.

Group Discussion – A group discussion is a structured interview with people involved in the job being analyzed. This may include workers who perform the job, their supervisors, and engineers who design the processes that the workers use. A group discussion may be used in advance of safety meetings to prepare for observation, or it may be used to follow up on an observation. The group discussion should be conducted away from the immediate work site, preferably in a conference room or classroom. Take detailed notes so you can revisit the data at a later time. Structure the questions to be asked in advance. These questions should be generated by the review of documentation or direct observation. If possible, use two people to run the group discussion. One person should ask the questions, encourage discussion, and document the generated feedback on flipcharts. The other person should quietly observe the discussion and take careful notes. This will help to ensure that the meeting runs quickly and smoothly and that none of the important data gets lost or goes uncaptured.

2.2.4 Identifying Hazards and Risk Factors

When identifying hazards and risk factors, carefully consider and analyze each task within the job, each tool or piece of equipment used, and all of the materials used in order to determine what risks they pose to the workers. This is done to determine all of the safety hazards presented by a job and to develop solutions, preventive measures, and procedures to eliminate or reduce those hazards.

Hazard and risk identification should be done in two steps. First, while making the initial observation, take notes about possible risks or hazards. Ask the following questions about each task:

- Is there a danger of striking against, being struck by, or otherwise making harmful contact with an object?
- Can employees be caught in, on, by, or between objects?
- Is there a potential for a slip, trip, or fall? If so, will it be on the same elevation or to a different elevation?
- Can workers strain themselves by pushing, pulling, lifting, bending, or twisting?
- Is the environment hazardous to anyone's safety or health?

These initial observations should be researched further and explored more thoroughly by qualified individuals or in a group discussion. After all of the potential hazards and risks have been identified, each should be thoroughly analyzed and documented. This may be done using a chart similar to the one shown in *Figure 7* or by writing a report covering the job, its tasks, and the associated risks.

A severity and frequency score should be assigned to each identified hazard. The severity score indicates the expected severity of an injury resulting from that hazard. For example, if the hazard is a high-voltage electrical shock that could result in severe burns or a fatality, the severity score would be high. The frequency score indicates how probable it is that the hazard would result in an incident. These scores assist in assigning priorities to the hazards so that the analysis can be used to its best advantage.

> **NOTE**
> More information on risk assessment and the use of a risk matrix is provided in NCCER Module 75219, *Hazard Recognition, Environmental Awareness, and Occupational Health* and Module 75222, *Site-Specific Safety Plans*.

2.2.5 Developing Solutions

The final step in conducting a JSA is to develop practical methods of making the job safer. The goal is to prevent the occurrence of potential incidents and minimize the risk and severity of injury should an incident occur. This is best achieved by applying a hierarchy of controls (*Figure 8*). With this approach, the greater the severity of the hazard/risk, the more effort should be applied to eliminate it. The major methods of controlling hazards generally fall into three major categories:

1. Engineering/design
2. Administrative controls
3. Personal protective equipment

Following the hierarchy leads to the implementation of inherently safer systems and substantially reduces the risk of illness or injury. The principal solutions are described as follows:

- Find a safer way to do the job.
- Change the physical conditions that create the hazard.
- Change work methods or procedures to eliminate hazards.
- Try to reduce the necessity or frequency of doing the job or some of its tasks.

Job:	Hazard	Severity (0 – 3)	Frequency (0 – 3)
TASK 1			
TASK 2			
TASK 3			
TASK 4			
TASK 5			

Figure 7 Hazard analysis chart.

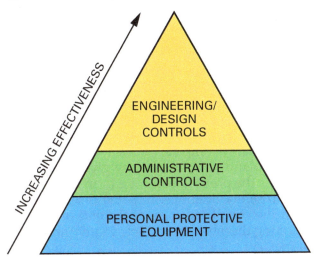

Figure 8 Hierarchy of controls.

The higher the severity and frequency scores of a hazard, the greater degree of attention is required to eliminate the risk. For example, if cleaning out a chemical tank is likely to result in death due to asphyxiation, then every effort should be made to minimize the risks associated with that job. These efforts may include purchasing automated equipment to clean the tank rather than risking the safety of workers, providing better job safety training, or acquiring better personal protective equipment.

Developing solutions should be done with a team that includes safety experts, process engineers, and other experts such as ergonomists (experts in the field of ergonomics), fire experts, or medical doctors. Most changes will also require a commitment on the part of supervisory and management personnel. For this reason, it is a good idea to involve them at an appropriate point in the process.

2.2.6 Common Errors When Conducting a JSA

When conducting JSAs, there are some common errors that should be avoided. Some of the most common errors are the following:

- Making the breakdown so detailed that an unnecessarily large number of tasks are listed
- Making the job so general that all potential hazards are not covered
- Failing to identify the education and experience levels of the target audience
- Not prioritizing JSAs by severity

The best way to avoid these mistakes is to take the time up front to make a clear plan on how the JSA will be conducted, who will be a part of the JSA process, and what the likely outcomes and uses will be for the results of the JSA.

2.3.0 Pre-Task Planning

Pre-task planning is a process for identifying and evaluating potential hazards associated with a given task or work assignment. Once the hazards have been identified, methods of eliminating or controlling the hazards should be developed and incorporated into the job plan. Pre-task planning reduces incidents and near misses through hazard analysis and improves both productivity and communication by clearly identifying work tasks.

2.4.0 Conducting Pre-Task Planning

Pre-task planning is typically completed just prior to starting the task. The evaluation may be performed individually or collectively. The ideal situation is to have the entire work group involved in the process. Some companies mandate that the supervisor coordinate the process, while others encourage individual crew members to lead the process. When employees are working alone, they are encouraged to personally conduct pre-task planning on their own.

The steps in pre-task planning typically include the following:

Step 1 Identify the task to be performed on a pre-task planning form. Most companies provide a pocket-size form for pre-task planning with a list of items to consider. Two sample forms are shown in *Figure 9*. Potential hazards are checked off or noted on the form.

Step 2 Identify the step-by-step procedure required to complete the task.

Step 3 Identify the potential hazards for each step and indicate the level of risk.

Step 4 Identify the action(s) that will be taken to control the safety, efficiency, and operational risks at each step. Actions can include engineering controls, proper training, off-shift/shutdown work, and safety precautions, such as PPE.

Step 5 Identify an emergency action plan.

Step 6 Review the plan and have all workers sign off to indicate that they understand their roles and responsibilities in safely completing the task.

Step 7 Post the plan at the job site and communicate the impacts of the expected work to other crews on the job site.

Some pre-task planning forms incorporate the JSA process, which requires listing each step, the associated hazards, and the appropriate safeguards. Completed pre-task planning forms are typically reviewed by the supervisor at the start of the workday or during the day. The forms are also used to audit safe work practices during the day. Upon completion of the task, completed pre-task planning forms are usually collected and retained for a pre-determined length of time.

Pre-task planning is conducted each time a new task is assigned. It is possible for a craft professional to complete or review two or more pre-task planning forms on any given day. Some firms use the pre-task planning form for pre- and post-job safety briefings. In some cases, workers initial the completed form at the end of the day, indicating the job was completed safely and without incident.

PRE-JOB SAFETY BRIEFING / TASK HAZARD ANALYSIS

Task/work to be performed _____
_____

Tools/equipment/materials involved
_____
_____

Physical Hazards

- Falls on same elevation
- Falls from elevation
- Pinch points
- Rotating/moving equipment
- Electrical hazards
- Hot/cold substances/surfaces
- Strains/sprains/repetitive motion
- Struck by falling/flying objects
- Sharp objects

Hazardous Chemicals/Substances

- Flammable materials
- Reactive materials
- Corrosive chemicals
- Toxic chemicals
- Oxidizers
- Hazardous wastes
- Biohazards
- Radiation hazards
- Other

Energy Sources

- Where is the energy?
- What is the magnitude of the energy?
- What could happen or go wrong to release the energy?
- How can it be eliminated or controlled?
- What am I going to do to avoid contact?

PRE-JOB SAFETY BRIEFING / TASK HAZARD ANALYSIS

Permits Required

- Welding & Burning
- Lockout/Tagout
- Excavation
- Line Entry
- Electrical Hot Work
- Confined Space Entry
- Critical Lift
- Vehicle Entry
- Other

PPE Requirements

- Safety Glasses
- Goggles
- Face Shield
- SCBA
- Respirator
- Gloves
- Chemical Suit
- Head Protection
- Safety Shoes/Boots
- Full Body Harness
- Lanyard
- Lifeline
- Other
- Special Precautions

Employee Signatures

_____
_____
_____
_____

Figure 9 Sample pre-task planning forms.

Additional Resources

American Society of Safety Engineers, www.asse.org

Occupational Safety and Health Organization, www.osha.gov

Job Hazard Analysis: A Guide to Identifying Risks in the Workplace, George Swartz. 2007. Rockville, MD: Government Institutes, Inc.

Root Cause Analysis Handbook: A Guide to Effective Incident Investigation, ABSG Consulting. 2005. Brookfield, CT: Rothstein Associates, Inc.

2.0.0 Section Review

1. Which of the following is a benefit of doing a JSA?

    a. Randomized risk perception
    b. Increased turnover rates
    c. Minimized training requirements
    d. Standardized work practices

2. The two steps involved in identifying hazards and risk factors are _____.

    a. asking the workers what the hazards are and asking their supervisor what the risks are
    b. making notes of likely hazards while observing and then following up with thorough analysis and documentation
    c. preparing your questions in advance and jotting down quick notes to answer the questions
    d. holding a group discussion and reviewing the documentation and related research

3. Once pre-task planning has identified and evaluated potential hazards for a given task or assignment, methods of eliminating or controlling the hazards should be developed and _____.

    a. incorporated into the job plan
    b. reported to the local OSHA office
    c. reviewed by Human Resources
    d. run through a cost-benefit analysis

4. Completed pre-task planning forms are typically reviewed by the supervisor at the start of a workday and are also used during the day to _____.

    a. assess worker attitudes
    b. cross-reference JSA records
    c. audit safe work practices
    d. mandate worker rotations

SUMMARY

Understanding human behavior is a key element to successful risk analysis and assessment. Once you know why workers behave the way they do and the factors that influence their behavior, you will be better equipped to help prevent incidents and injuries.

In this module, you learned the factors involved in human behavior, as well as techniques to improve your observation skills. You learned how to communicate the consequences of human behavior to workers, thereby making everyone on site accountable for safety.

You also learned about the documentation and paperwork, such as job safety analysis forms, that safety technicians make use of when identifying and evaluating the hazards associated with specific jobs and tasks. If performed correctly, JSAs and pre-task planning are important tools not only for increasing productivity and reducing waste, but also for creating and maintaining a safe work environment.

Review Questions

1. Which type of performance barrier causes a worker not to complete a task even though he or she is physically capable of doing so?
   a. Physical
   b. Knowledge
   c. Impairment
   d. Execution

2. An example of an impairment factor is _____.
   a. defective safety equipment
   b. fatigue
   c. lack of training
   d. lack of PPE

3. The belief that "it won't happen to me" is a factor of _____.
   a. the ABC model
   b. unpreventable employee misconduct
   c. personal risk acceptance
   d. the Behavioral Law of Effect

4. When dealing with at-risk behavior by a worker, the safety technician should _____.
   a. publicly discipline the worker to set an example for the rest of the crew
   b. inform the worker's supervisor and direct him or her to punish the worker
   c. attempt to elicit the desired behavior by changing the worker's attitude
   d. try to find out if there are any barriers or impairment factors

5. Safety technicians should compliment workers when they are observed performing jobs safely because _____.
   a. praise is the least expensive incentive
   b. behavior that is recognized is repeated
   c. you can't correct what you can't observe
   d. compliments counteract perceived risks

6. To discourage at-risk behavior the consequences of safe behavior must be _____.
   a. reportable, repeatable, and reinforcing
   b. deferred, flexible, and strict
   c. soon, certain, and positive
   d. objective, reasonable, and adjustable

7. In the ABC model, consequences are the _____.
   a. behavior that causes the condition
   b. specific response to the condition
   c. condition that prompts behavior
   d. positive or negative results of a behavior

8. One reason for conducting a job safety analysis is _____.
   a. to prevent an OSHA inspection
   b. because surprise inspections help keep employees in line
   c. to determine which employees are engaging in at-risk behavior
   d. to help facilitate job training

9. JSAs help reduce risk by _____.
   a. reporting hazardous situations to OSHA
   b. revealing hazardous situations to the media
   c. identifying hazards and determining appropriate steps needed to reduce them
   d. identifying ways to conceal unsafe conditions

10. When selecting jobs for analysis, the highest priority should be given to jobs with the _____.
    a. largest number of affected workers
    b. highest potential for serious injuries
    c. greatest impact on productivity
    d. longest history of performance

11. When preparing to do a JSA, a safety technician should first _____.
    a. look at the general conditions under which the job is performed
    b. ask management what they would like you to find
    c. break the job down into as many tasks as possible
    d. determine which risks you'd like to find

12. Reviewing documentation _____.
    a. is the preferred method of data collection
    b. will typically provide all of the necessary job detail
    c. provides important background information
    d. should include reviewing employee medical records

13. A common error when conducting a JSA is _____.
    a. spending time reviewing documentation
    b. placing too much emphasis on risky jobs
    c. calling in other experts to help develop solutions
    d. breaking the job into too many tasks

14. The purpose of pre-task planning is to _____.
    a. map out the work schedule and finalize work crew shift rotations
    b. observe and document instances of at-risk behavior on a job site
    c. identify and evaluate potential hazards associated with a given task
    d. devise an equitable allocation of resources on a construction project

15. Pre-task planning should be conducted _____.
    a. each time a new task is assigned
    b. soon after starting work on a task
    c. before initiating a job safety analysis
    d. as required by OSHA regulations

Trade Terms Introduced in This Module

Causal links: A relationship between two events or occurrences in which one is the cause of the other.

Consequences: The final outcome of actions and behaviors.

Ergonomics: The applied science of equipment design, as for the workplace; intended to maximize productivity by reducing operator fatigue and discomfort.

Job: A regular activity performed to achieve some end.

Pattern: An indication of how predictable the reoccurrence of an event is.

Reinforcement: Supporting or strengthening of a behavior or action.

Recognition: Something that is given or returned verbally, through body language, or in writing for good or bad behavior.

Structured on-the-job training (SOJT): Training that takes place during the performance of the job with the guidance of documentation or a facilitator.

Task: A discrete step or portion of a job.

Trend: The tendency to take a particular direction.

Appendix A

JOB HAZARD ANALYSIS EXAMPLE

JOB HAZARD ANALYSIS	
20141219.1: 20150218-001	

Task: Replace Light Bulb	**Date:** 02/18/2015
Department: ABC Division	**Analysis Developed By:** John W. Garcia
Location(s): 3333 Wilshire Blvd., Suite 820, Los Angeles, CA 90010	**Analysis Reviewed By:**
Person(s) Performing This Job: Maintenance Worker	**Supervisor:** John Smith
Job Start Date: 02/19/2015	**Duration:** Frequently performed task

Task/Step	Potential Hazards	Recommended Safe Job Procedures
1. Get Ladder and new light bulb	1. Ladders (portable, fixed)	Get proper ladder to accommodate light fixture height
2. Turn off light switch	1. Hand tools	Lock out light switch in room.

Revised: 03/30/2015

Task/Step	Potential Hazards	Recommended Safe Job Procedures
3. Change out light bulb	1. Electrical Wiring 2. Hand tools	Carefully remove light cover and replace light fixture. Make sure to use proper electrical insulated tools.
4. Put ladder back in storage	1. Ladders (portable, fixed)	Carefully stow ladder and tools away

POTENTIAL PHYSICAL HAZARDS OF THIS JOB			
Physical Hazards	Prob.	Sev.	Consequences
Electrical Wiring			Electrocution or shock
Hand tools			Falling (< 6 feet), tripping, or slipping
Ladders (portable, fixed)			Falling (> 6 feet)

HAZARD CONTROL MEASURES USED FOR THIS JOB	
Administrative Controls: Lockout/tagout	**Required Training:** Electrical safety Ladders, stairways, and other working surfaces Tools
Engineering Controls:	**Required PPE:** Personal protective equipment Safety glasses
Required Permit(s):	**Other Information:**

Revised: 03/30/2015

Appendix B

OSHA Report of Injury/Near Miss Form

Employee's Report of Injury Form

Instructions: Employees shall use this form to report <u>all</u> work related injuries, illnesses, or "near miss" events (which could have caused an injury or illness) – *no matter how minor*. This helps us to identify and correct hazards before they cause serious injuries. This form shall be completed by employees as soon as possible and given to a supervisor for further action.

I am reporting a work related: ☐ Injury ☐ Illness ☐ Near miss	
Your Name:	
Job title:	
Supervisor:	
Have you told your supervisor about this injury/near miss? ☐ Yes ☐ No	
Date of injury/near miss:	Time of injury/near miss:
Names of witnesses (if any):	
Where, exactly, did it happen?	
What were you doing at the time?	
Describe step by step what led up to the injury/near miss. (continue on the back if necessary):	
What could have been done to prevent this injury/near miss?	
What parts of your body were injured? If a near miss, how could you have been hurt?	
Did you see a doctor about this injury/illness? ☐ Yes ☐ No	
If yes, whom did you see?	Doctor's phone number:
Date:	Time:
Has this part of your body been injured before? ☐ Yes ☐ No	
If yes, when?	Supervisor:
Your signature:	Date:

Supervisor's Accident Investigation Form

Name of Injured Person _____

Date of Birth _____   Telephone Number _____

Address _____

City _____   State _____   Zip _____

(Circle one)   Male   Female

What part of the body was injured? Describe in detail. _____
_____

What was the nature of the injury? Describe in detail. _____
_____
_____

Describe fully how the accident happened. What was employee doing prior to the event? What equipment, tools were being used? _____
_____
_____

Names of all witnesses:

_____   _____

_____   _____

Date of Event _____   Time of Event _____

Exact location of event: _____

What caused the event? _____
_____
_____

Were safety regulations in place and used? If not, what was wrong? _____
_____

Employee went to doctor/hospital? Doctor's Name _____

Hospital Name _____

Recommended preventive action to take in the future to prevent reoccurrence.
_____
_____
_____

_____   _____
Supervisor Signature   Date

2

Incident Investigation Report

Instructions: Complete this form as soon as possible after an incident that results in serious injury or illness. (Optional: Use to investigate a minor injury or near miss that *could have resulted in a serious injury or illness*.)

This is a report of a:	❑ Death ❑ Lost Time ❑ Dr. Visit Only ❑ First Aid Only ❑ Near Miss
Date of incident:	This report is made by: ❑ Employee ❑ Supervisor ❑ Team ❑ Other_____

Step 1: Injured employee (complete this part for each injured employee)

Name:	Sex: ❑ Male ❑ Female	Age:
Department:	Job title at time of incident:	

Part of body affected: (shade all that apply)	Nature of injury: (most serious one) ❑ Abrasion, scrapes ❑ Amputation ❑ Broken bone ❑ Bruise ❑ Burn (heat) ❑ Burn (chemical) ❑ Concussion (to the head) ❑ Crushing Injury ❑ Cut, laceration, puncture ❑ Hernia ❑ Illness ❑ Sprain, strain ❑ Damage to a body system: ❑ Other _____	This employee works: ❑ Regular full time ❑ Regular part time ❑ Seasonal ❑ Temporary
		Months with this employer
		Months doing this job:

Step 2: Describe the incident

Exact location of the incident:	Exact time:

What part of employee's workday? ❑ Entering or leaving work   ❑ Doing normal work activities
  ❑ During meal period   ❑ During break   ❑ Working overtime   ❑ Other_____

Names of witnesses (if any):

Number of attachments:	Written witness statements:	Photographs:	Maps / drawings:

What personal protective equipment was being used (if any)?

Describe, step-by-step the events that led up to the injury. Include names of any machines, parts, objects, tools, materials, and other important details.

Description continued on attached sheets: ❑

Step 3: Why did the incident happen?

Unsafe workplace conditions: (Check all that apply)
- ❑ Inadequate guard
- ❑ Unguarded hazard
- ❑ Safety device is defective
- ❑ Tool or equipment defective
- ❑ Workstation layout is hazardous
- ❑ Unsafe lighting
- ❑ Unsafe ventilation
- ❑ Lack of needed personal protective equipment
- ❑ Lack of appropriate equipment / tools
- ❑ Unsafe clothing
- ❑ No training or insufficient training
- ❑ Other: _____

Unsafe acts by people: (Check all that apply)
- ❑ Operating without permission
- ❑ Operating at unsafe speed
- ❑ Servicing equipment that has power to it
- ❑ Making a safety device inoperative
- ❑ Using defective equipment
- ❑ Using equipment in an unapproved way
- ❑ Unsafe lifting
- ❑ Taking an unsafe position or posture
- ❑ Distraction, teasing, horseplay
- ❑ Failure to wear personal protective equipment
- ❑ Failure to use the available equipment / tools
- ❑ Other: _____

Why did the unsafe conditions exist?

Why did the unsafe acts occur?

Is there a reward (such as "the job can be done more quickly", or "the product is less likely to be damaged") that may have encouraged the unsafe conditions or acts?   ❑ Yes ❑ No
If yes, describe:

Were the unsafe acts or conditions reported prior to the incident?   ❑ Yes ❑ No

Have there been similar incidents or near misses prior to this one?   ❑ Yes ❑ No

4

Step 4: How can future incidents be prevented?

What changes do you suggest to prevent this incident/near miss from happening again?

☐ Stop this activity ☐ Guard the hazard ☐ Train the employee(s) ☐ Train the supervisor(s)

☐ Redesign task steps ☐ Redesign work station ☐ Write a new policy/rule ☐ Enforce existing policy

☐ Routinely inspect for the hazard ☐ Personal Protective Equipment ☐ Other: _____

What should be (or has been) done to carry out the suggestion(s) checked above?

Description continued on attached sheets: ☐

Step 5: Who completed and reviewed this form? (Please Print)

Written by:

Title:

Department:

Date:

Names of investigation team members:

Reviewed by:

Title:

Date:

Additional Resources

This module presents thorough resources for task training. The following reference material is recommended for further study.

American Society of Safety Engineers, **www.asse.org**

Occupational Safety and Health Organization, **www.osha.gov**

Job Hazard Analysis: A Guide to Identifying Risks in the Workplace, George Swartz. 2007. Rockville, MD: Government Institutes, Inc.

The Participation Factor—How to Increase Involvement in Occupational Safety, E. Scott Geller, Ph.D. 2002. Des Plaines, IL: The American Society of Safety Engineers (ASSE).

The Psychology of Safety Handbook, E. Scott Geller, Ph.D. 2001. Boca Raton, FL: CRC/Lewis Publishers.

Root Cause Analysis Handbook: A Guide to Effective Incident Investigation, ABSG Consulting. 2005. Brookfield, CT: Rothstein Associates, Inc.

Figure Credits

MSA The Safety Company, Module opener, Figure 3

Steven P. Pereira, CSP, Figure 9

ehs International, Inc., www.ehsinc.org, Appendix A

U.S. Department of Labor, Appendix B

Section Review Answer Key

Answer	Section Reference	Objective
Section One		
1. d	1.1.3	1a
2. c	1.2.1	1b
3. b	1.3.0	1c
Section Two		
1. d	2.1.3	2a
2. b	2.2.4	2b
3. a	2.3.0	2c
4. c	2.4.0	2d

NCCER CURRICULA — USER UPDATE

NCCER makes every effort to keep its textbooks up-to-date and free of technical errors. We appreciate your help in this process. If you find an error, a typographical mistake, or an inaccuracy in NCCER's curricula, please fill out this form (or a photocopy), or complete the online form at **www.nccer.org/olf**. Be sure to include the exact module ID number, page number, a detailed description, and your recommended correction. Your input will be brought to the attention of the Authoring Team. Thank you for your assistance.

Instructors – If you have an idea for improving this textbook, or have found that additional materials were necessary to teach this module effectively, please let us know so that we may present your suggestions to the Authoring Team.

NCCER Product Development and Revision
13614 Progress Blvd., Alachua, FL 32615

Email: curriculum@nccer.org
Online: www.nccer.org/olf

❑ Trainee Guide ❑ Lesson Plans ❑ Exam ❑ PowerPoints Other _____

Craft / Level: _____ Copyright Date: _____

Module ID Number / Title: _____

Section Number(s): _____

Description: _____

_____

_____

_____

Recommended Correction: _____

_____

_____

_____

Your Name: _____

Address: _____

_____

Email: _____ Phone: _____

Safety Data Tracking and Trending

Overview

This module discusses the role of the safety technician in conducting safety inspections, audits, and employee safety observations. It presents both traditional and predictive methods of performance measurement and explains how to analyze safety data in order to prevent future incidents.

Module 75221

Trainees with successful module completions may be eligible for credentialing through NCCER's Registry. To learn more, go to **www.nccer.org** or contact us at 1.888.622.3720. Our website has information on the latest product releases and training, as well as online versions of our *Cornerstone* magazine and Pearson's product catalog.

Your feedback is welcome. You may email your comments to **curriculum@nccer.org**, send general comments and inquiries to **info@nccer.org**, or fill in the User Update form at the back of this module.

This information is general in nature and intended for training purposes only. Actual performance of activities described in this manual requires compliance with all applicable operating, service, maintenance, and safety procedures under the direction of qualified personnel. References in this manual to patented or proprietary devices do not constitute a recommendation of their use.

Copyright © 2018 by NCCER, Alachua, FL 32615, and published by Pearson, New York, NY 10013. All rights reserved. Printed in the United States of America. This publication is protected by Copyright, and permission should be obtained from NCCER prior to any prohibited reproduction, storage in a retrieval system, or transmission in any form or by any means, electronic, mechanical, photocopying, recording, or likewise. To obtain permission(s) to use material from this work, please submit a written request to NCCER Product Development, 13614 Progress Blvd., Alachua, FL 32615.

75221 V2

From *Safety Technology*, Trainee Guide. NCCER.
Copyright © 2018 by NCCER. Published by Pearson. All rights reserved.

75221
Safety Data Tracking and Trending

Objectives

When you have completed this module, you will be able to do the following:

1. Explain how to conduct safety inspections, audits, and employee observations.
    a. Describe how to conduct a safety inspection.
    b. Describe how to conduct a safety audit.
    c. Describe how to perform a safety observation.
2. Identify methods of measuring safety performance.
    a. Identify traditional methods (lagging indicators) of measuring safety performance.
    b. Identify predictive methods (leading indicators) of measuring safety performance.
    c. Explain how to analyze safety data in order to prevent future incidents.

Performance Tasks

Under the supervision of your instructor, you should be able to do the following:

1. Conduct two or more of the following:
    a. Safety inspection
    b. Safety audit
    c. Employee safety observation
    d. Near miss report

Trade Terms

Audit
Behavioral-based safety (BBS)
Direct cause
Experience modification rate (EMR)
Hazardous conditions
Incidence rate
Indirect cause
Inspection
Job safety analysis (JSA)

Lagging indicators
Leading indicators
Near miss
Observation
Pre-task plan
Root cause
Safety audit
Safety inspection
Unsafe behavior

Industry Recognized Credentials

If you are training through an NCCER-accredited sponsor, you may be eligible for credentials from NCCER's Registry. The ID number for this module is 75221. Note that this module may have been used in other NCCER curricula and may apply to other level completions. Contact NCCER's Registry at 888.622.3720 or go to www.nccer.org for more information.

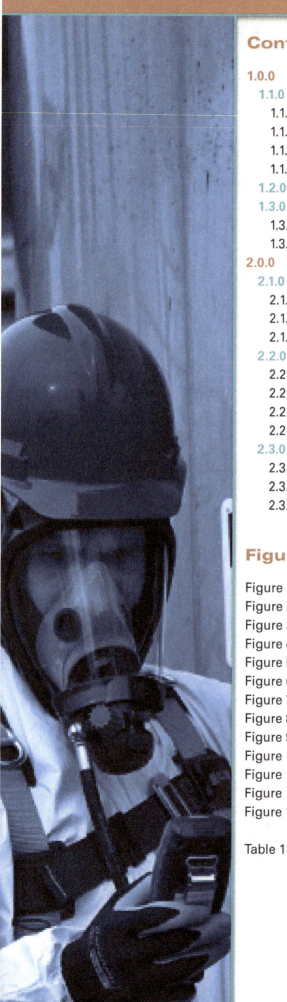

Contents

- **1.0.0** Safety Inspections, Audits, and Employee Observations 1
  - **1.1.0** Safety Inspections 4
    - 1.1.1 Types of Safety Inspections 5
    - 1.1.2 Items Included in a Safety Inspection 5
    - 1.1.3 Safety Inspection Time Frame 6
    - 1.1.4 Safety Inspection Findings 6
  - **1.2.0** Safety Audits 8
  - **1.3.0** Safety Observations 8
    - 1.3.1 Performing a Safety Observation 8
    - 1.3.2 Safety Observation Findings 9
- **2.0.0** Measuring Safety Performance 12
  - **2.1.0** Traditional Methods of Measuring Safety Performance 12
    - 2.1.1 Recordable Incidence Rates 13
    - 2.1.2 Analyzing Injuries 16
    - 2.1.3 Workers' Compensation Experience Modification Rate (EMR) ... 17
  - **2.2.0** Predictive Methods for Measuring Safety Performance 17
    - 2.2.1 Behavior Observations 17
    - 2.2.2 Safety Audits and Inspections 17
    - 2.2.3 Near Miss/Near Hit Reporting 18
    - 2.2.4 Training 21
  - **2.3.0** Analysis of Safety Data 22
    - 2.3.1 Incident Analysis 22
    - 2.3.2 Job Safety Analysis 22
    - 2.3.3 Safety Management 22

Figures and Tables

Figure 1 Construction industry's Fatal Four 1
Figure 2 Causes of deaths in construction for 2013 2
Figure 3 Why worry about safety? 2
Figure 4 Heinrich's triangle 3
Figure 5 Hazard control process 4
Figure 6 Safety audit documentation 9
Figure 7 Employee observation documentation 10
Figure 8 Excerpt from BLS incidence rate statistics 15
Figure 9 Excerpt from injury report by cause 16
Figure 10 Sample audit checklist 18
Figure 11 Root and indirect causes of incidents 21
Figure 12A Sample JSA excerpt (1 of 2) 23
Figure 12B Sample JSA excerpt (2 of 2) 23

Table 1 Sample Lockout/Tagout Procedure Checklist 6

SECTION ONE

1.0.0 SAFETY INSPECTIONS, AUDITS, AND EMPLOYEE OBSERVATIONS

Objective

Explain how to conduct safety inspections, audits, and employee observations.
a. Describe how to conduct a safety inspection.
b. Describe how to conduct a safety audit.
c. Describe how to perform a safety observation.

Performance Task

1. Conduct two or more of the following:
   a. Safety inspection
   b. Safety audit
   c. Employee safety observation
   d. Near miss report

Trade Terms

Audit: To review safety policies and procedures to see if they are adequate and being used.

Hazardous conditions: Circumstances or objects that cause injury or illness. Most hazardous conditions arise as a result of unsafe (at-risk) behaviors.

Inspection: The act of checking an area to identify, report, and correct hazards to workers, materials, and equipment.

Near miss: An unplanned event that did not result in injury, illness, or damage but had the potential to do so. Also called a near hit or a close call.

Observation: Watching workers as they perform a job for the purpose of determining whether they are working safely or committing any unsafe acts.

Safety audit: A predictive method of safety management requiring observation and reporting by an unbiased individual.

Safety inspection: A predictive method of safety management requiring observation and reporting by a workforce supervisor.

Unsafe behavior: Action taken or not taken that increases risk of injury or illness. Also called at-risk behavior.

The construction industry reported 874 work-related deaths in 2014. That's about 20 percent of all workplace deaths in the country for that year. On any given day, two or more construction workers died from on-the-job injuries and 537 construction workers were injured.

The following four conditions, which are commonly referred to as the construction industry's Fatal Four (*Figure 1*), are the cause of the majority of construction fatalities:

- Falls from elevations
- Struck by
- Caught in/between
- Electrical shock

Figure 2 shows a breakdown of the causes of deaths in the construction industry for 2013.

In addition to deaths and immediate injuries to workers, unsafe conditions and practices can cause other potential problems, including the following:

- Damage or destruction of materials and equipment
- Scheduling delays or work shutdowns
- Loss of productivity
- Future health problems for the workers
- Damage to the environment
- Loss of morale

There are other less-obvious areas that are affected by unsafe conditions/practices and incidents and injuries. They are related to the financial well-being of the company and can determine whether or not your company wins contract bids. These financial considerations include the following:

- High cost of workers' compensation insurance
- Occupational Safety and Health Administration (OSHA) fines
- Legal action

Figure 1 Construction industry's Fatal Four.

Module 75221   Safety Data Tracking and Trending   1

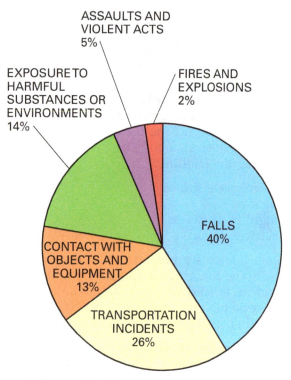

Figure 2 Causes of deaths in construction for 2013.

In the 1930s, H. W. Heinrich introduced the concept of what has come to be known as Heinrich's triangle (*Figure 4*), also called the safety triangle or the safety pyramid. The triangle illustrates Heinrich's theory that for every fatality, there are a great many more no-injury, near miss incidents. By Heinrich's reasoning it follows that reducing the overall frequency of relatively minor workplace injuries will produce an equivalent reduction in the number of severe injuries. This has, in fact, become the underlying principle of modern industrial workplace safety: changing worker behavior is the key to reducing the number and severity of workplace incidents.

Safety is part of everyone's job. Individuals are responsible for not only their own safety but also for the safety of everyone around them. Safety technicians are responsible for coordinating site safety activities. They help to find and correct safety hazards. As a safety technician, sometimes you'll be working under the direction of the safety manager. More often, you'll be working on your own. Either way, you'll often perform the following safety tasks:

- Inspection
- Audit
- Observation

To do these tasks well, you'll need to develop skills in the following areas:

What does all this mean to you and your co-workers? The obvious answer is possible death or physical disability, but injuries and incidents also cost money (*Figure 3*). Your company will need to pass the increased cost on to their customers. That can mean losing bids, which means less work and lower pay for you.

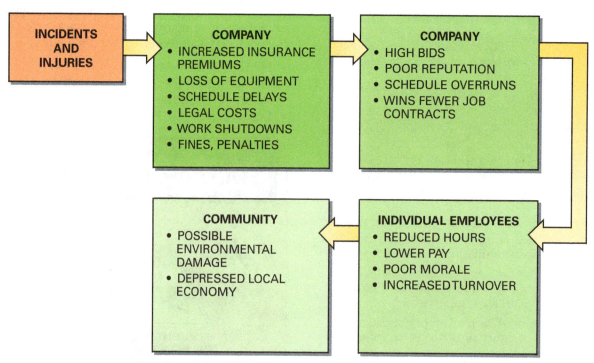

Figure 3 Why worry about safety?

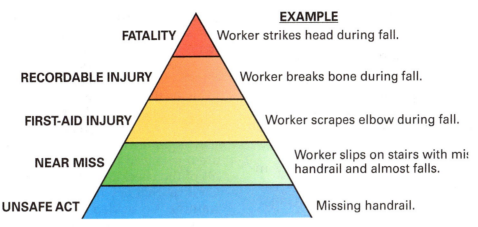

Figure 4 Heinrich's triangle.

- Serve as a resource to site personnel for safety and health matters.
- Review your company's past safety performance, including illness and injury statistics.
- Understand the hazards associated with the type of work that is performed on site.
- Learn your company's policies, procedures, and practices for the work performed.
- Make intelligent decisions to develop solutions to hazards and unsafe practices.
- Develop people skills so that you can talk to personnel on site, including workers, supervisors, and superintendents.

The main reason for conducting any safety inspection, audit, or observation is to identify and correct unsafe conditions and practices before an incident or injury can occur. This requires performing the following steps:

Step 1 Identify the unsafe condition or practice.

Step 2 If the condition or practice is an immediate danger to workers or property, provide an immediate temporary solution.

Step 3 Determine the cause of the unsafe condition or practice.

Did You Know?

The Swiss Cheese Model

Catastrophic incidents are often the result of a chain of failures rather than a single event. The Swiss Cheese Model of accident causation illustrates that most incidents can be prevented if any one of the protections is in place. Each system, such as safety procedures and PPE use, is shown as a slice of Swiss cheese stacked against other systems as a barrier between a worker and an incident. The risk of a hazard becoming reality is mitigated by the various layers stacked behind one another. In theory, holes in one system do not allow a risk to materialize because other defenses exist behind them. These holes vary in size and importance. However, a failure is inevitable when a hole in each slice of cheese momentarily aligns to allow a hazard to pass through all the slices.

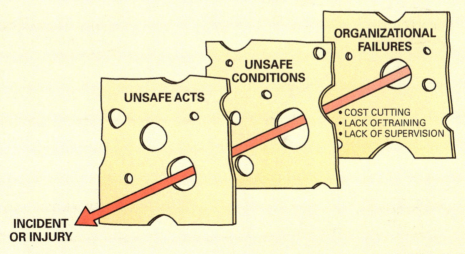

Step 4 Take steps to correct the condition or practice.

Step 5 Follow up to be sure that the corrective action actually fixed the problem.

Step 6 Document the problem, the solution, and how and when you followed up; confirm that the solution really fixed the problem.

Figure 5 shows the hazard control process.

A good safety program, including safety inspections, audits, and observations, will save your company more money than it costs to implement them.

1.1.0 Safety Inspections

Safety inspections involve checking work areas to identify, report, and correct hazards to workers, materials, and equipment. Every day, you and other workers on construction sites perform informal safety inspections. You do so when you carry out basic tasks such as visually examining scaffolds for damage, wear, or missing components; making sure that fire extinguishers are in good working order; or when you check the position and stability of a ladder used to exit a trench on a job site.

Maintaining good housekeeping on the job site is also a form of safety inspection. When you see an unsafe condition, you fix it—for example, by cleaning oil spots off the floor or returning an abandoned tool to storage. This type of informal inspection is continuous and ongoing.

Safety technicians also perform formal safety inspections. These inspections need to be documented. The use of a written checklist is generally recommended so that nothing is forgotten or missed. A written record is also needed to prove that your company has an ongoing safety program. A properly prepared inspection checklist serves a dual purpose: it provides a list of items or areas to be inspected and it also serves as a permanent record of the inspection.

The safety technician needs to periodically evaluate the checklist to ensure that it is still valid and useful. If a checklist is out of date, it will not be a useful tool. Update the checklist whenever new equipment, materials, and work procedures are introduced to your work site. Make sure that all of the checklist components are still relevant. Confirm that any new SDSs are in place and that training is provided on the new equipment.

On the other hand, beware of overreliance on checklists. Keep in mind that the experience of the inspector, the quality of the inspection, and the condition at the time of inspection all impact the inspection process.

> **NOTE**
> Check the OSHA website (www.osha.gov) or your local OSHA office for examples of checklists.

On most sites, you'll need more than one checklist because some areas or equipment will need to be inspected more or less frequently than others. For example, fire extinguishers may need to be inspected monthly, but a ladder or scaffold should be inspected before each use. Keep safety checklists close to where they will be needed. All completed safety checklists must be stored in a safe location, such as in a fireproof safe, and retained according to company policy.

As specified in 29 *CFR* 1926.20(B), frequent regular inspections of job sites, materials, and critical equipment such as scaffolding should be made by a competent person designated by the employer.

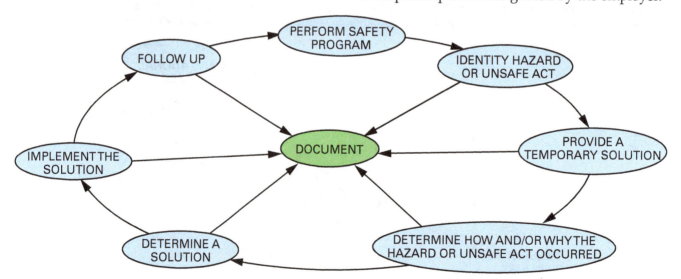

Figure 5 Hazard control process.

1.1.1 Types of Safety Inspections

In addition to continuous inspections, there are three types of scheduled inspections. They are described as follows:

- *Periodic inspections* – These inspections are made weekly, monthly, semi-annually, or at other set intervals. They are for specific items and equipment such as storage areas or heavy machinery.
- *Intermittent inspections* – These inspections are not performed at a set interval. They are sometimes called special inspections. You might need to do a special inspection when you notice that a large number of near misses have been occurring at your work site.
- *General inspections* – These inspections include areas not covered by periodic inspections. They might include walkways, tool rooms, or maintenance shops.

1.1.2 Items Included in a Safety Inspection

The items to be included in a safety inspection vary. It depends on the type of work, the hazards and risks involved, and the number of people at the site. Some of the items that may need to be included are listed here, as follows:

- Environmental factors such as lighting and ventilation
- Hazardous supplies and materials
- Production and related equipment
- Power source equipment
- Electrical equipment
- Hand and power tools
- Personal protective equipment (PPE)
- First aid supplies and equipment
- Fire protection equipment
- Walkways, sidewalks, and roadways
- Elevators, stairways, and personnel lifts
- Working surfaces
- Transportation equipment
- Warning and signaling devices
- Containers
- Storage facilities
- Structural openings such as doorways or windows
- Building and structures
- Grounds and storage areas
- Loading and shipping platforms

Table 1 shows a sample checklist for lockout/tagout procedures. Any "No" answers result in immediate work stoppage. You can find many other examples of checklists on the OSHA website (**www.osha.gov**). Maintain completed checklists according to company policy. Note that electronic checklists and software applications are also available.

Case History

Fatal Fall

A worker was climbing down a 400-foot (122 meter) telecommunications tower. He had not received proper training on the ladder safety device he was using, which consisted of a carabiner, carrier rail, safety sleeve, and body harness. Moreover, the body harness he used was not a component of the manufacturer's ladder safety device. The weight of the worker, his tools, and the equipment was more than the 310-pound (141-kilogram) rating of the body harness. The safety sleeve was connected to the harness at the chest D-ring instead of to the navel D-ring as specified by the manufacturer of the ladder safety device.

When the worker lost his footing, the ladder safety device or system failed to arrest his fall. The safety sleeve did not activate correctly to stop the fall. The chest D-ring ripped out of the body harness, and he plunged 90 feet (27 meters) to his death.

After the incident, the manufacturer issued a safety notice instructing users to tie off to a shock-absorbing lanyard in addition to using the ladder safety device.

The Bottom Line: A competent person should inspect protective equipment each time to make sure all of its components are compatible and in good working order before the equipment is used. Workers must be adequately trained in the proper use of the specific personal protective equipment required for the jobs they'll be performing.

Source: OSHA Fatal Facts, No. 7 – 2014

1.1.3 Safety Inspection Time Frame

The frequency of safety inspections depends on your work site. In some cases, inspections are done on a regularly scheduled basis. Sometimes they may be done randomly and without notice. Other inspections are performed only when needed. For example, cranes and rigging equipment may need to be checked before each use. Most of the time, this will take place during normal work hours.

More detailed inspections and tests are usually needed at monthly and yearly intervals. Safety technicians should constantly evaluate the frequency of their inspections. If your work site starts seeing an increase in minor incidents or near misses, you may need to conduct frequent, unannounced inspections. Ask yourself the following questions to help you decide how often you need to perform inspections:

- What is the potential for injury?
- What is the potential for loss?
- What is the lifespan of the part or equipment?
- What is the safety history of the work site?
- Does OSHA or another agency require inspections?

1.1.4 Safety Inspection Findings

When you find an unsafe condition or work practice that presents an immediate danger to workers or property, have it fixed right away. This may include removing equipment from service or erecting a barrier to prevent access to hazards by unauthorized workers. Any delay increases the risk that someone will get hurt or that equipment will be damaged.

For instance, say you discover that a piece of equipment is leaking oil onto the floor. What are some of the things you could do? Your first actions will probably be immediate and temporary, but sufficient enough to prevent a serious

Table 1 Sample Lockout/Tagout Procedure Checklist

Yes/No/NA	Lockout/Tagout Procedures
	Is all machinery or equipment capable of movement required to be de-energized or disengaged and locked out during cleaning, servicing, adjusting, or setting up operations, whenever required?
	Where the power disconnecting means for equipment does not also disconnect the electrical control circuit: 　Are the appropriate electrical enclosures identified? 　Are means provided to ensure the control circuit can also be disconnected and locked out?
	Are the appropriate electrical enclosures identified?
	Are means provided to ensure the control circuit can also be disconnected and locked out?
	Is the locking out of control circuits in lieu of locking out main power disconnects prohibited?
	Are all equipment control valve handles provided with a means for locking out?
	Does the lockout procedure require that stored energy (mechanical, hydraulic, air, etc.) be released or blocked before equipment is locked out for repairs?
	Are appropriate employees provided with individually keyed personal safety locks?
	Are employees required to keep personal control of their key(s) while they have safety locks in use?
	Is it required that only the employee exposed to the hazard place or remove the safety lock?
	Is it required that employees check the safety of the lockout by attempting a startup after making sure no one is exposed?
	Are employees instructed to always push the control circuit stop button immediately after checking the safety of the lockout?
	Is there a means provided to identify any or all employees who are working on locked-out equipment by their locks or accompanying tags?
	Are a sufficient number of accident preventive signs or tags and safety padlocks provided for any reasonably foreseeable repair emergency?
	When machine operations, configuration, or size requires the operator to leave his or her control station to install tools or perform other operations, and when part of the machine could move if accidentally activated, is such element required to be separately locked or blocked out?
	In the event that equipment or lines cannot be shut down, locked out, and tagged, is a safe job procedure established and rigidly followed?

incident. An appropriate response could include the following actions:

- Assess the area for additional hazards.
- If the equipment is powered on, turn it off.
- Attach a "Danger – Do Not Operate" tag to the equipment.
- Set up safety cones to alert workers that the floor may be slippery.
- Arrange for the floor to be cleaned.

After the immediate hazard has been fixed, follow up with the equipment operator's supervisor to make sure that the equipment is being repaired properly. Also, check to see if this is an isolated problem or a symptom of a bigger issue.

Not all safety issues will be as simple as this one. Some problems can be complicated and will require in-depth analysis to find the cause and determine a solution. When you find an unsafe condition or work practice that is outside of your area of responsibility, notify the responsible person. Often, this is the job superintendent.

Always document your findings, being careful to write clearly. Record the date and time of the inspection. Make a note of any problems that were found and what was done to fix them. Record who was notified, including the date and time when this was done. Once the fix is verified to have been successful, record that the hazard has been taken care of.

After you perform a safety inspection, it's important to inform management, job site supervisors, and co-workers of the results. Let them know your positive findings (*e.g.*, appropriate use of PPE) as well as negative findings (*e.g.*, poor housekeeping). Management will be able to use this information to make informed decisions in the future.

Case History

Fatal Electrocution

A laborer who primarily spoke Spanish was working in a five-acre materials storage yard in a rural area. A three-phase, 7,200-volt overhead power line was located directly above the area where steel augers and steel pipe had been stored. As the laborer was guiding an auger that was being lifted by a truck-mounted crane onto a truck, the truck boom moved, causing the crane boom or load line to contact the power line. The electricity flowed through the laborer's body to ground through his feet. Two workers employed by another subcontractor who were assisting the laborer were also shocked and knocked to the ground by the electric current.

The crane operator was not shocked. He reversed the boom sufficiently to move it away from contact with the lines and got off the crane to evaluate the situation. He then re-entered the crane to lower the load to the ground before returning to the injured men. Finding that the laborer had no apparent pulse and did not appear to be breathing, the crane operator began cardiopulmonary resuscitation (CPR). One of the workers who had been shocked ran to a nearby building and called 911, while the other injured worker waited for Emergency Medical Services (EMS).

The EMS ambulance traveled from a nearby town to the incident site, but had difficulty finding the remote storage yard and took about 20 minutes to arrive. Once EMS personnel were at the incident site, they continued CPR on the laborer, who remained unresponsive. He was transported to a nearby hospital, where he was pronounced dead. The two other injured workers were transported to another hospital in the area where they were examined. One of the workers was released that day, and the other was admitted to the hospital and released two days later.

The Bottom Line: Don't let language barriers prevent comprehensive safety training for all workers. Use a translator when possible or have written manuals translated. Post pictorial representations of key safety elements in hazardous locations. For grass roots (remote) job sites, coordinate with the authority that has jurisdiction for providing emergency medical services to identify and address barriers to the timely response of medical emergencies.

Source: NIOSH In-house FACE Report 2005-01

1.2.0 Safety Audits

A *safety audit* is a review of safety policies and procedures to see if they are accurate and being used. Audits are performed on workers, equipment, and procedures, and can be used to identify positive conditions and behaviors as well as areas for improvement (coaching opportunities).

Safety audits are usually performed on a periodic basis. They can be done more or less frequently depending on who is doing the audit. For example, a site audit performed by the corporate office could be done on a quarterly basis. If the audit is performed by the site safety technician, it could be done on a weekly basis. The size of a project and the company's safety requirements have an impact on how often safety audits are done.

During an audit, you'll review applicable safety policies and procedures as well as safety and health regulations. To determine whether your policies and procedures are accurate and up-to-date, ask the following questions:

- Does this policy or procedure still apply to the work site?
- Is the procedure accurate?
- Do the policies and procedures address the hazards and required safety precautions?
- Is the procedure being followed on the work site?

Safety technicians need to develop a good working relationship with all personnel. Workers are an excellent resource when performing audits. During the audit process, use perception surveys to identify gaps between management and workers. Review the safety policies and procedures that apply to your areas of responsibility. Discuss any ideas for improvement you might have with site personnel and your safety manager.

When your work site starts to use new equipment or materials, don't wait for the next scheduled safety audit. Review the applicable policy and procedure to make sure that it's still applicable.

Be sure to document that you performed a safety audit. Your findings and the control actions taken should also be recorded. Either typed or handwritten documentation is acceptable. *Figure 6* shows an example of a documented safety audit.

1.3.0 Safety Observations

A safety observation is the act of watching a worker or group of workers during the performance of their job. The purpose of the observation is to determine whether workers are performing their job safely or committing any unsafe acts. Always reinforce safe behavior by telling workers what they are doing right.

Your company will provide you and your co-workers with written policies and procedures to ensure a safe working environment. Safety observations performed by safety technicians not only ensure that the policies and procedures are carried out, it also ensures that the policies and procedures themselves are safe and practical. Stay up-to-date; make sure you know your company's safety policies and procedures as well as the applicable OSHA regulations.

1.3.1 Performing a Safety Observation

To perform a safety observation, a safety technician must do the following:

- Observe the worker in the performance of his or her job.
- Validate training and job instructions.
- Note safe and at-risk behavior.

Stay alert and continuously monitor the worker's activities for unsafe practices. Also, monitor the job site for unsafe acts and conditions, and be ready to mitigate any hazards to protect the workers on shift.

Unsafe Acts – You must be able to recognize when a worker's behavior is unsafe. The following is a list of the most common unsafe acts found on a job site:

- Lack of proper and appropriate training
- Failing to use appropriate personal protective equipment
- Failing to communicate potentially hazardous conditions or unsafe behavior to co-workers
- Failing to follow instructions or procedures
- Using incorrect or defective tools or equipment
- Lifting improperly
- Taking an improper working position
- Making safety devices inoperable
- Operating equipment at improper speeds
- Operating equipment without authority or proper training
- Servicing, repairing, or adjusting equipment while it is in motion or energized
- Loading or placing equipment or supplies improperly or dangerously
- Using equipment improperly
- Working under a suspended load or in an obviously dangerous area
- Working while impaired by alcohol or drugs (either legal or illegal)
- Engaging in horseplay

SAFETY AUDIT REPORT LOG

April 4, 2015 –
Performed safety audit on using the paint booth safety policy and procedure. Found that the recommended respirator filter does not filter particle size for new metal primer paint. Recommended to Ralph Brown, Safety Manger, that we change filters. He agreed. Ordered new filters.

H. White, Site Safety Tech

April 5, 2015 –
Received filters. Inked change into current procedure and sent it to the office for word processing. Notified all workers at site to use the new filters. Hung sign on paint booth supply cabinet as a reminder.

H. White, Site Safety Tech

Figure 6 Safety audit documentation.

> **NOTE:** The terms *unsafe behavior* and *at-risk behavior* are sometimes used interchangeably.

Unsafe Conditions – Unsafe conditions are physical conditions that are different from acceptable, normal, or correct conditions. The following is a list of the most common unsafe conditions:

- Poor housekeeping
- Poor illumination
- Poor ventilation
- Congested workplaces
- Defective tools, equipment, or supplies
- Extreme temperatures
- Excessive noise/pressure
- Fire and explosive hazards
- Hazardous atmospheric conditions
  - Gases
  - Dusts/fibers
  - Fumes
  - Vapors
- Inadequate supports or guards
- Inadequate warning systems
- Weather hazards

1.3.2 Safety Observation Findings

When you see a worker performing an unsafe act or practice that presents the immediate risk of an incident or injury, stop the work immediately. This action is expected as part of a STOP WORK program. Then, talk to the employee about the unsafe practice, keeping the following guidelines in mind:

- Stay calm. Try not to startle the worker.
- Express your concern for his or her well-being and safety.
- State the nature of the unsafe act or practice that you observed; be specific.
- Inform the worker of the appropriate safety policy, practice, or procedure.
- Emphasize the importance of following the safety policy or procedure.
- Remember to exercise the coaching aspect of the interaction. Encourage the worker to think critically about safe work practices by asking questions such as, "What should you do?"
- Enlist the aid of on-site experts, such as journey-level workers, to help coach on policies and procedures. Have them demonstrate the correct procedure if necessary.
- Let the worker practice the procedure with your guidance, or under the supervision of a journey-level worker if needed. Make sure that he or she has the knowledge required to do the job properly. If necessary, schedule retraining before the worker is allowed to return to the job.

> **NOTE:** Training a worker to do something right the first time is vital; it is very difficult to relearn a task once it has been performed the wrong way.

- Most safety technicians have a tablet or smartphone on the job site. Use it to share videos and other information with the worker when teachable moments occur.
- Always close an observation on a positive note.

> **NOTE:** Recurring or serious problems should be documented and handled according to your company's disciplinary policy.

Don't blame workers for unsafe acts when other factors are involved. Help them become more aware of their surroundings. If they can recognize unsafe conditions on their own, it will make your job easier. Safety-conscious workers recognize hazards and take steps to mitigate or eliminate them. This could involve stopping to talk to a co-worker about their at-risk or hazardous behavior. Point out that this must be done in a non-threatening way to be effective.

As the safety technician, in addition to telling workers when they do something wrong, it's important to let them know when they're doing a good job. You don't need to pat them on the back all the time, but positive feedback is needed. Remember that behavior rewarded is behavior repeated. Behavior that is punished or ignored is not repeated.

Be sure to document the results of your safety observations (*Figure 7*). You may need these records to identify workers with poor safety habits.

> **NOTE**
> Consider OSHA requirements as the minimum safety standards that must be met. State, local, client, and/or facility requirements may be more stringent. Always follow the most stringent requirements.

EMPLOYEE OBSERVATION REPORT LOG

April 3, 2015 -
During a routine employee observation, I noticed that John Smith didn't have eye protection on while using an angle grinder. All other guards were in place. I stopped Smith and explained the importance of eye protection while using the grinder. Smith said he understood and would wear eye protection from now on.

H. White, Site Safety Tech

Figure 7 Employee observation documentation.

OSHA Inspections

Safety technicians may be required to escort OSHA officials on an inspection tour of the work site. Be aware that your company does not need to grant OSHA officials access to your work site except as noted here. By law, OSHA can be required to obtain search warrants to conduct inspections. However, few employers feel the need to exercise this right. To ensure compliance with OSHA standards, inspections are made where and when considered advisable by the agency. Inspections are made without prior notice to the company being inspected. It's important to know your company's policy and adhere to it when dealing with OSHA or any other state, city, local, or federal agency.

- Some exceptions that allow OSHA inspectors immediate access to a work site are as follows:
- A fatality or a catastrophic incident that resulted in the hospitalization of one or more people, amputation, or the loss of an eye
- Verbal or written complaints received by OSHA concerning an immediate hazard
- Routine inspections of sites that are very hazardous
- Routine inspections of other industries
- Follow-up inspections to see if corrections for hazards found during an initial inspection have been put into place
- Referrals from local law enforcement agencies or other federal agencies
- If a safety inspector unexpectedly appears at your work site, remember the following guidelines:
- Be polite. Remember, inspectors are just doing their job.
- If your company's policy is to have you or a supervisor escort the inspector on the tour, or if you are unsure of your company's policy about outside inspections, call your safety manager or construction superintendent.
- Don't leave the inspector standing in the cold, rain, or hot sun while you are contacting your supervisor.
- Show the inspector where to get water. Offer coffee if it's available at your site.
- For more information about OSHA inspections, refer to NCCER Module 75226, **OSHA Inspections and Recordkeeping**.

Other Federal Agencies

OSHA is not the only federal agency affecting worker safety and industry standards. Other administrations involved in safety efforts include the Mine Safety and Health Administration (MSHA), the Environmental Protection Agency (EPA), and the Consumer Product Safety Commission (CPSC). Whether or not you need to comply with the policies and procedures of these agencies depends on the location of your company, the type of work, and the specific hazards at your job site.

Additional Resources

American Society of Safety Engineers, www.asse.org

Occupational Safety and Health Organization, www.osha.gov

1.0.0 Section Review

1. The items that should be included in a safety inspection depend on the type of work, the hazards and risks involved, and the _____.
   a. shift rotation at the time of inspection
   b. skill levels of the work crews
   c. number of people at the site
   d. physical limitations of the inspector

2. During a safety audit, the safety technician _____.
   a. reviews safety policies and procedures for accuracy
   b. checks the work area for hazards and unsafe conditions
   c. observes whether a worker performs unsafe acts
   d. escorts OSHA inspectors around the work site

3. Performing an employee safety observation consists of observing a worker as he or she performs a job, validating training and job instructions, making note of both safe and at-risk behavior, and providing _____.
   a. a monetary award for safe behavior
   b. immediate feedback to the worker
   c. a summary report to OSHA
   d. motivation for attitudinal change

Section Two

2.0.0 Measuring Safety Performance

Objective

Identify methods of measuring safety performance.

a. Identify traditional methods (lagging indicators) of measuring safety performance.
b. Identify predictive (leading indicators) of measuring safety performance.
c. Explain how to analyze safety data in order to prevent future incidents.

Trade Terms

Behavioral-based safety (BBS): A proactive method of safety management based on psychology. It requires systematic workplace observation and analysis of unsafe behaviors, resolution of problems, and is coupled with training and incentives for behavior modification.

Direct cause: The immediate cause of an injury or illness, not accounting for any underlying unsafe behaviors or conditions.

Experience modification rate (EMR): A lagging indicator of illness and injury rates based on insurance claims and predicted claims over a three-year period. It is most often applied by the insurance industry.

Incidence rate: A lagging indicator of illness and injury rates based on a Bureau of Labor Statistics (BLS) formula. It is measured in annual incidents per 100 workers.

Indirect cause: The underlying cause of an injury or illness. Categories include hazardous conditions and unsafe (at-risk) behaviors.

Job safety analysis (JSA): A method for studying a job to identify hazards and potential incidents associated with each step and developing solutions that will eliminate, minimize, and prevent hazards and incidents. Also called a job hazard analysis (JHA).

Lagging indicators: A measure of performance based only on historical reporting.

Leading indicators: A measure of performance based on predictive observations and analysis.

Pre-task plan: A process for identifying and evaluating potential hazards associated with a given task or work assignment.

Root cause: The deepest level, system-related cause of an injury or illness. Requires in-depth analysis to discover. Also referred to as a basic cause.

Safety technicians use various methods of performance measurement for the tracking and trending of safety- and health-related issues. They often use lagging indicators or leading indicators. These performance measurements provide information that will help to determine effective ways to prevent incidents.

2.1.0 Traditional Methods of Measuring Safety Performance

The effectiveness of a safety program is difficult to measure directly because when a safety program is effective there aren't any incidents. However, there are indicators that are used to show performance trends. Traditional safety indicators include the following:

- The incidence rate, including the Total Recordable Incidence Rate (TRIR), Lost Time Rate (LTR), and Days Away, Restricted, or Transferred (DART)
- Insurance company loss runs and loss ratios
- The workers' compensation experience modification rate (EMR)
- Workers' compensation costs in dollars and cents per man-hour

---

Did You Know?

Deaths and Injuries in the Workplace

In 2014:

- There were 4,679 workplace fatalities due to unintentional injuries.
- There were 3.3 deaths per 100,000 workers.
- On the job, nearly 3 million American workers suffered injuries.
- Transportation incidents accounted for 1,891 of the 4,679 workplace fatalities.
- The construction industry accounted for 874 deaths.

Source: Bureau of Labor Statistics

These methods are all known as lagging indicators. The data is analyzed after an incident occurs. These methods provide a good way to study past performance. Much can be gained from studying past outcomes, but lagging indicators are not always reliable or definitive. For instance, based on the size of the company, lagging indicators are not always a true representation. Likewise, incidence rates show trends but are not always a true indicator of future safety performance. Decisions based solely on past performance may lag well behind current and future problems. To best determine performance trends, lagging indicators should be used along with leading indicators.

2.1.1 Recordable Incidence Rates

OSHA requires the collection and recording of certain data. An organization's injury and illness record can be evaluated using this data. Types of recordable incidence rates include the following:

- Total injury and illness cases
- Lost workday injury and illness cases
- Lost workday injury-only cases
- Lost workday illness-only cases
- Injury and illness cases without lost workdays
- Injury-only cases without lost workdays
- Illness-only cases without lost workdays

These rates can show safety trends within a single organization. You can compare them to standards in your industry or across different industries to highlight problem areas. You can also use this data to prevent work-related injuries and illnesses.

OSHA offers a free computer program that calculates incidence rates. The Safety Pays program can be downloaded from their website at

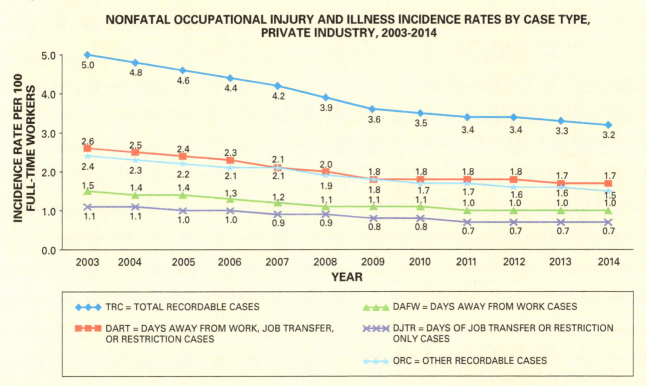

How the Bureau of Labor Statistics (BLS) Tracks Incidence Rates

The BLS produces data that track the number of occupational injuries, illnesses, and fatalities. The Census of Fatal Occupational Injuries (CFOI) publishes a complete count of all deaths that result from injuries at work in all sectors of the economy. The Survey of Occupational Injuries and Illnesses (SOII) uses employer-reported data to estimate counts and incidence rates of nonfatal injuries and illnesses for the private, state government, and local government sectors. The SOII includes summary-level industry data for OSHA-recordable injury and illness cases and also includes data on the detailed case circumstances and worker characteristics for cases involving days away from work.

www.osha.gov. The program calculates incidence rates and other statistical data. This information can be used as one aspect of analyzing your company's safety performance.

Calculating incidence rates – Incidence rates are computed using a Bureau of Labor Statistics (BLS) method. The BLS uses a basis of 200,000 hours worked. This is based on 100 employees working fifty 40-hour weeks per year. This standardizes the rate for all industries, regardless of size, so they can be compared directly. Incidence is considered in terms of a rate per 100 workers.

The calculation is performed in the following manner:

Step 1 Count the total number of OSHA recordable injuries and illnesses for one year. This information should be available from OSHA Form 300A, lines G, H, I, and J.

Step 2 Calculate the number of hours actually worked in that year. Use payroll records. Do not include any paid time off.

Step 3 Incidence rate = (number of injuries and illnesses × 200,000) ÷ (annual hours worked)

The same formula is used to calculate particular incidence rates. For example, you might need to know the incidence rate of lost workday injury-only cases. In that case, you would count only those specific cases in Step 1.

Evaluating incidence rate statistics – Incidence rate data is compared to the previous year's rates. To some personnel, this provides a quick measure of the success of a company's health and safety program. If the incidence rate shows a decline, then health and safety practices can be considered successful. If there is no change or an increase in rates, then health and safety practices need to be revisited.

In spite of the fact that OSHA incidence rates are one of the most widely used measures of safety performance, most safety professionals agree that they are actually poor indicators. Safety technicians

---

Case History

Struck-by-Object Fatality

A primarily Spanish-speaking carpenter's helper was working with an English-speaking crane operator to disassemble a lattice boom on a truck-mounted crane. The carpenter's helper had received no specific training, either in English or Spanish, about safety hazards associated with disassembling booms and had never helped to disassemble a crane boom before. The crane operator had limited experience with the disassembly and assembly of booms and had never taken the lead in disassembling a boom. In the urgency to get the task done, he forgot to make sure that the boom was supported.

The crane operator easily knocked out the first lower pin on the boom, but the second lower pin would not budge. The carpenter's helper indicated with hand motions that they could use the first pin that had been removed as a punch to drive out the second bottom pin. So, while the carpenter's helper held the first pin against the second bottom pin, the crane operator repeatedly pounded on the first pin with a sledgehammer. The crane operator motioned to the carpenter's helper to move so that the boom would not fall on him, but instead of moving to the side of the boom he went under the lower boom section. After a few blows with the sledgehammer, the second bottom pin came out. When this happened, the boom hinged on the top two pins and the lower boom section fell on the carpenter's helper and pinned him to the ground.

Workers at the incident site called 911 and used a nearby hydraulic crane to lift the boom off the carpenter's helper. The two pins that had been removed were then driven back into place to secure the boom sections. EMS personnel arrived within about 5 minutes of being called. They transported the carpenter's helper to the hospital, where he was pronounced dead.

The Bottom Line: This incident underscores the importance of providing comprehensive safety training in language(s) and literacy level(s) that workers can understand, as well as the importance of ensuring that all crew members have a basic understanding of the correct procedures involved in hazardous operations. If the victim had fully appreciated the potential for uncontrolled movement, he might have positioned himself to the side of the boom rather than underneath it, which would have reduced the risk of being struck by the falling boom sections.

Source: NIOSH In-house FACE Report 2006-01

must operate in an ethical manner, never misrepresenting or overstating the accuracy or validity of any safety performance measurements.

The BLS reports data according to the North American Industry Classification System (NAICS). The NAICS codes for construction are 236, 237, and 238. BLS provides rates for a wide variety of workplace environments (*Figure 8*). Statistics are broken down by type of industry and number of employees.

BLS data also shows how rates are distributed through an industry. The BLS table includes data titled First Quartile, Median, and Third Quartile. This indicates what proportion of comparable firms have incidence rates lower than yours.

INCIDENCE RATES OF TOTAL RECORDABLE CASES OF NONFATAL OCCUPATIONAL INJURIES AND ILLNESSES, BY QUARTILE DISTRIBUTION AND EMPLOYMENT SIZE, 2014

Industry, NAICS code, and establishment employment size	Average incidence rates for all establishments: (mean)	One-quarter of the establishments had a rate lower than or equal to: (1st quartile)	One-half of the establishments had a rate lower than or equal to: (median)	Three-fourths of the establishments had a rate lower than or equal to: (3rd quartile)
Construction				
Total all sizes	3.6	(4)	(4)	(4)
1 – 10	2.9	(4)	(4)	(4)
11 – 49	4.2	(4)	(4)	6.5
50 – 249	3.8	.9	2.8	5.7
250 – 500	2.5	.8	1.7	3.5
1,000+	1.4	.2	.5	2.0
Construction of buildings (NAICS 236000)				
Total all sizes	3.3	(4)	(4)	(4)
1 – 10	3.3	(4)	(4)	(4)
11 – 49	4.2	(4)	(4)	6.5
50 – 249	2.9	(4)	1.7	4.6
250 – 500	2.1	.5	1.4	3.0
1,000+	1.0	.2	.6	2.0
Residential building construction (NAICS 236100)				
Total all sizes	4.1	(4)	(4)	(4)
11 – 49	4.8	(4)	(4)	7.1
50 – 249	3.9	.9	3.2	4.6
250 – 999	5.9	1.7	3.2	14.6
Nonresidential building construction (NAICS 236200)				
Total all sizes	2.7	(4)	(4)	(4)
1 – 10	2.0	(4)	(4)	(4)
11 – 49	3.7	(4)	(4)	6.5
50 – 249	2.6	(4)	1.5	4.1
250 – 500	1.5	.4	1.2	2.1
1,000+	1.0	.2	.6	2.0

Figure 8 Excerpt from BLS incidence rate statistics.

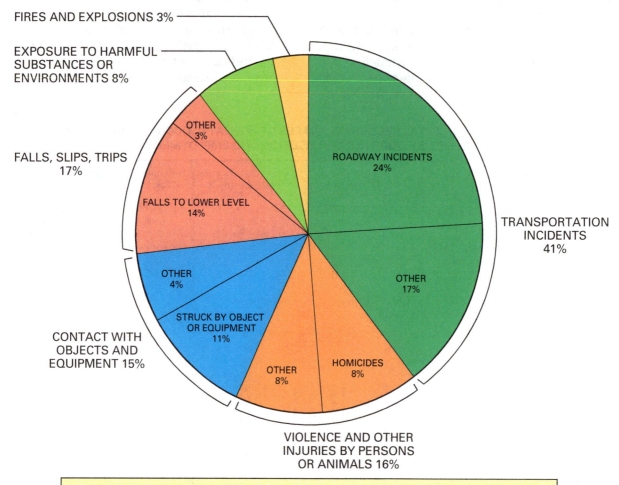

Figure 9 Excerpt from injury report by cause.

2.1.2 Analyzing Injuries

Incidence statistics help to determine historical and cross-industry trends. More specific information is also often available. Reports dealing with the nature of injuries may prove useful. Often these reports deal with the type of injury or part of the body injured. These forms may be available through BLS or state safety boards. You can create them for your firm using a chart or graph, such as the one shown in *Figure 9*.

These reports provide indicators of the types of safety measures needed. As you can see in the chart, slips, trips, and falls are common injury types. Therefore, the safety program should target slip, trip, and fall hazards. These hazards could be reduced by using different footwear or employing better housekeeping practices. Analyzing the nature of an injury often leads to identification of types of personal protective equipment that can help to prevent similar injuries.

The sample chart was compiled with data from BLS analysis of OSHA incidence records—fatality, severity, and recordability. Similar tables are also available with other breakdowns. For example, injuries may be examined as a function of a particular craft or of worker age. These tables can span several industries.

A wide variety of statistics must be included in internal safety histories; incidence reports are not sufficient. Each incident reported requires capturing as much data as possible, including the worker's age, type and nature of injury, work accomplished, and environmental conditions. Careful study of past indicators can provide valuable assistance in preventing future mishaps.

2.1.3 Workers' Compensation Experience Modification Rate (EMR)

A firm's safety record can be analyzed using the EMR. This is an insurance industry indicator used to calculate the premiums that firms pay for workers' compensation insurance. The National Council on Compensation Insurance (NCCI) typically provides the EMR.

The EMR compares actual losses in insurance claims against expected losses. A three-year period, not including the current year, is used for the calculations. Injuries that lead to insurance claims increase the EMR. A clean record over a period of time will decrease it.

When safety is compared across similar industries, the average EMR is always 1.0. An EMR higher than 1.0 means a firm has higher than average losses in claims. A better than average safety record results in an EMR of less than 1.0. Clients use EMR as a pre-qualifier for bid awards.

The EMR is used to calculate workers' compensation premiums. The industry standard rate is multiplied by the firm's EMR. An EMR of 0.7 means a 30 percent discount on the premium. At 1.25, a firm is paying a 25 percent surcharge.

In addition to insurance premiums, the EMR is used as a general safety indicator over a period of years. A decreasing EMR points to a successful safety program. If the EMR rises, improving safety must become a priority.

EMR provides a measure of safety program success across industries. It is an industry average that shows how you compare to similar firms. Many contracts set a threshold for the EMR level that the contractor must meet in order to bid.

The use of EMR as an indicator of industry safety has become common, but it is not without problems. The EMR uses data from past years, so it is immediately dated. It does not take fraudulent claims into account and it does not consider incidents that could not have been prevented. Some companies attempt to keep the EMR low by paying for some smaller medical claims themselves. If the insurance company does not pay a claim, the EMR is not affected.

The EMR is simply one indicator to check. It is not the only or primary measure of safety performance.

2.2.0 Predictive Methods for Measuring Safety Performance

Modern methods for measuring safety take a predictive approach. Some analysis may be based on past performance, but the data is used differently. Past observations may well lead to future changes. Leading indicators are processes designed to identify, prevent, and control incidents and injuries before they occur. Examples of leading indicators include the following:

- Behavior observations
- Audits
- Near miss/near hit reporting
- Training

Leading indicators look at potential causes of safety mishaps as well as prevention tools. Hazardous conditions identify those things or objects that lead to injury. Unsafe behaviors, also called at-risk behaviors, reveal those actions taken or not taken that lead to injury. Reducing hazardous conditions and unsafe actions can prevent injury and illness.

2.2.1 Behavior Observations

Behavioral-based safety (BBS) is a method of applying psychology to workplace safety. Unsafe behavior causes at least 80 percent of all workplace incidents. Some estimates go as high as 96%. This indicates that an improvement in safety performance can be achieved by focusing on unsafe behavior.

BBS is best suited for actions directly controlled by an individual. It includes actions such as ducking under a ladder or engaging in horseplay. It also includes failure to act, such as not properly stowing equipment or not wearing a hard hat.

2.2.2 Safety Audits and Inspections

Safety issues, especially recurring problems, can be revealed through safety audits and inspections. Frequent in-house inspections, coupled with independent audits, provide an excellent means of detecting hazardous trends.

Safety audits – A safety audit is a systematic examination of workplace behavior and safety practices (*Figure 10*). The auditor checks that policies and procedures have been implemented and are adequate. Auditors should be independent, either a separate agency within the company or an outside firm. Government agencies or insurance carriers can also order independent audits.

An audit compares current practices with other standards as well as with past practices within the same company using historical audit data. The audit also compares the company to other companies. BLS data can identify standard risks associated with a task across a range of industries.

An audit should include every component of the safety process. This can be done collectively or one component at a time. At a minimum,

ELEVATED WORK AUDIT			UNGUARDED 8' OR ABOVE			
PLANT LOCATION _____						
AUDITOR _____ DATE __/__/__ SHIFT 1 2 3						
OBSERVATION	1	2	3	4	5	
* EMPLOYEE USING FALL PROTECTION SYSTEM						
FULL BODY HARNESS USED						
LIFELINE/LANYARD NOT TOO LONG						
LIFELINE/LANYARD HOOKED NEAR EMPLOYEE						
LIFELINE/LANYARD PROPERLY TIED OFF						
LIFELINE/LANYARD SECURED TO ANCHOR POINT						
ANCHOR POINT WILL SUPPORT WEIGHT						
ANCHOR POINT DOES NOT HAVE SHARP EDGES						
RETRACTABLE DEVICE WITHIN CERTIFICATION DATE						
WAS PPE USED TO MOVE TO ELEVATED WORK?						
MARK: YES = Y NO = N NOT APPLY = N/A (*) CIRCLE SIGNIFICANT VIOLATIONS = 0	COMMENTS:					
CONFORMANCE = OBSERVATIONS (−) VIOLATIONS (×) 100 / OBSERVATIONS						

Figure 10 Sample audit checklist.

management, training, and operating procedures should be covered. The short-term goal of an audit is to reveal strengths and weaknesses of the whole program. The long-term goal is to reduce losses through incident prevention activities.

An audit needs to consider historical safety data first, and it needs to verify that problems uncovered by previous audits and inspections have been resolved. Once the audit begins, don't allow the checklist to run the audit. Well-trained auditors keep their eyes open for much more than what is written on the clipboard.

A successful audit examines unsafe behavior and hazardous conditions as well as policies and procedures. It's not enough to only note physical hazards. Unsafe acts in the workplace must also be considered. An audit must involve more than a cursory examination. Direct questioning of workers is frequently the best tool for gathering data.

The safety technician is an auditor. First-line supervisors should implement safety policies and procedures in their areas and the safety technician audits the implementation and the results. For example, the safety technician may audit compliance with lockout/tagout, confined-space entry, elevated work, or fall protection procedures.

Safety inspections – Work supervisors normally carry out safety inspections. A safety inspection tends to follow a set checklist format. The checklist may include OSHA violations and will also likely include unsafe conditions and physical hazards.

The inspection process works hand-in-hand with less-frequent audits. An audit may uncover an item requiring improvement, and then inspections can be used to monitor progress toward that improvement.

First-line supervisors are key players in the inspection process. They are familiar with the physical work area and the immediate dangers to workers. An auditor may work closely with the workforce for one part of an audit. The supervisor is there on a daily basis and is best prepared to recognize old and new hazards.

Middle and upper management also need to play a role. Inspections only work when everyone is behind them. Line managers must work to implement changes suggested through the inspection process. Executive-level decisions need to take safety audit and inspection findings into account.

2.2.3 Near Miss/Near Hit Reporting

A near miss—also called a near hit or a close call—is an unplanned event that did not result in injury, illness, or damage but had the potential to do so. Only a fortunate break in the chain of events prevented any injury, fatality, or damage.

Near miss incidents often precede loss-producing events but may be overlooked because they don't result in an injury, illness, or damage. History has shown, however, that loss-producing incidents, both serious and catastrophic, were preceded by warnings or near miss incidents. Recognizing and reporting near miss incidents can significantly improve worker safety and enhance an organization's safety culture.

Near miss incidents can result from many circumstances, such as site conditions, worker behavior, machinery failure, etc. Whatever caused the near miss, a root cause analysis must be conducted to identify the defect in the system that resulted in the error and the factors that may either make the result worse or better. If you capture near misses and act on their causes, you can reduce or eliminate the occurrence of incidents. In fact, near miss reporting has been shown to increase employee relationships and encourage teamwork towards creating a safer work environment. It also demonstrates management's commitment to safety, and it captures data for statistical analysis, correlation studies, and trend detection. In short, near miss reporting is a useful leading indicator.

Many lost-time injuries and fatalities likely had previous unreported near miss incidents relating to the process. Such injuries and illnesses can be prevented and lives saved by implementing a near miss reporting program. Near miss reporting and investigation identify and control safety or health hazards before they cause a more serious incident. The goal of a near miss reporting program is not to lay blame but to find out what happened, determine the cause(s), and put barriers in place to prevent reoccurrence.

Ideally, a near miss reporting system includes both mandatory reporting for incidents with high loss potential and voluntary, non-punitive reporting for minor and near miss incidents. Near miss reports should describe what witnesses observed throughout the event, plus the factors that prevented loss from occurring. This information can then be used to identify and eradicate the root cause of the near-miss incident.

The process is actually simple as well as very effective. A near miss management program consists of seven essential steps:

Step 1 Identification of a near miss

Step 2 Disclosure of the observed near miss

Step 3 Distribution of communication about the near miss

Step 4 Root cause analysis of the near miss

Step 5 Solution identification:
- Find out what happened and determine the causes of the incident
- Rethink the safety hazard
- Introduce ways to prevent reoccurrence
- Establish training needs

Step 6 Dissemination of investigation conclusions

Step 7 Resolution of the problem

Identification – The purpose of this step is to uncover hazards that were either missed earlier or that require new controls (engineering/design controls; administrative controls, including policies and procedures; or personal protective equipment). Conduct an investigation as soon as possible following the near miss to gather all the necessary facts, determine the true cause, and develop recommendations to prevent a reoccurrence.

Disclosure – Employees should not be punished for disclosing a near miss; rather they should receive positive reinforcement for reporting the incident. Make it clear to employees at every level of the organization that safety is a crucial issue and explain how they can actively and effectively participate in the workplace safety program. Consistently communicate safety performance expectations. Spell out the goals and objectives in terms of reducing the cost and frequency of incidents and injuries. Emphasize that in your company's safety culture, reporting near misses is both important and necessary.

Distribution – Establish a process for transferring the information from the individual making the report to those who will make decisions about the preventive actions that need to be taken. This will enable senior management to make immediate decisions on corrective actions. Clearly define what the expected results are from a particular recommendation and set priorities for each action. Identify which actions should be completed before operations resume.

Use two levels of communication for incident investigations. First, issue an official, limited distribution incident investigation report of the most important information that was discovered. Second, issue a widely-distributed flyer (hard copy and/or electronic) to communicate contributing factors of the near miss and review the most important lessons learned from the incident.

Create an action plan to make sure that recommendations for corrective or preventive actions receive prompt attention. Follow up to be sure that the recommendations were implemented and are effective.

Root cause analysis – When analyzing a near miss, you must assess the various causes that contributed to its occurrence. You must also determine corrective actions or solutions to rectify the root cause so that reoccurrence is less likely.

Depending on the potential severity or complexity of a near miss, cause determination can occur informally between the reporting individual and his or her direct supervisor, or it may require formation of an investigative team for a thorough analysis with resulting recommendations.

The initial analysis of a safety incident will show the *direct cause* of the injury. This is the actual action that caused the injury. Behind every direct cause is an *indirect cause*. Indirect causes include hazardous conditions and unsafe behaviors. Steps can usually be taken immediately to prevent both types of indirect causes. There is often a deeper cause behind an indirect cause. A root cause is an inadequacy in the safety system or process. Identifying a root cause is more difficult but removing one can improve a safety record faster than removing many indirect causes.

Consider the following example. A worker unintentionally splashes caustic soda on his face. The direct cause of the injury is the chemical's reaction with his skin. The indirect cause may be a hazardous condition, like an improperly closed container. It could also be behavioral. Perhaps the worker did not practice safe handling techniques. The root cause might then be a company-wide lack of training on chemical handling.

The relationship between the various causes of an incident can be seen in a weed analogy (*Figure 11*). The direct causes are at the flowering top, impossible not to notice. Indirect causes form the stem and leaves, quickly observable to the trained eye. Root causes run deeper and require digging to uncover and correct them. Correcting an indirect cause without touching a root cause is like mowing a dandelion. The untouched roots will simply spring up again with a new problem.

Root causes always precede indirect causes. A poorly designed safety system leads to unsafe conditions and acts. Some unsafe conditions and acts include obvious safety-related factors such as training programs, hazard identification, and accountability. They also include procedures, budgets, and processes that are not typically associated with safety. You need to examine all the possibilities to find root causes.

Root causes intertwine throughout an organization. Eliminating a root cause that affects one hazard may reduce other hazards in unexpected ways. The flip side is that there may be more than one root cause tied to an indirect cause. Several roots may have to be eliminated before the unsafe condition is reduced.

All levels of the workforce must participate in root cause analysis. Craft workers are often the best source for information. Interviews may reveal, for example, that unsafe practices are followed because of a misperception about company policy.

Because root causes can be policy issues, upper management ultimately must be responsible for implementing change. Feedback and continuous analysis must be conducted. A change in procedure to get rid of the root cause for one hazard may well create another. Every policy change must be examined in practice.

Solution – Identify corrective action(s) that are certain to be effective in preventing a reoccurrence of the problem. Solutions should be within your control and meet your goals and objectives without introducing new unforeseen problems. Implement and observe the recommended root cause correction(s) to verify that they are effective.

Dissemination – Provide near miss investigation conclusions to all employees with related job functions to identify the fundamental reasons why the near miss incident occurred and the associated root cause(s). This makes all employees aware of the issues relating to the near miss and helps to find opportunities for eliminating potential risks in the future.

Resolution – Once a near miss incident or hazard is identified, track that the correction needed to resolve the problem is made and report the correction. Resolving near misses is not only important to ensure potential incidents don't occur, it is also essential to the success of your overall near miss program. If employees don't think near misses are acted upon, they will stop reporting them. If near misses continue to be reported, they do not require continuous rewards; a simple "good catch" provides ample reinforcement.

It is also important to audit the performance of a near miss program. This provides a view of what is working well and what needs to be improved upon. All levels of employees should be included in the information/data collection and analysis for the audit. This is the only way to find out if management expectations match the employee expectations, which is critical for an effective near miss program.

Near miss management system – Use of a management system can make it easier to implement and maintain a near miss reporting program. The following are three key components of such a system:

- Management process including a near miss management team to design the system, provide training, and monitor system performance; and a near miss management oversight team to provide high-level guidance and monitor near miss management team performance
- Tools, such as an electronic tracking system to manage the system, including tracking the progression of the seven steps in the process
- Employee training for two groups of employees: those who will manage the reporting system and those who will be using the system to report near misses

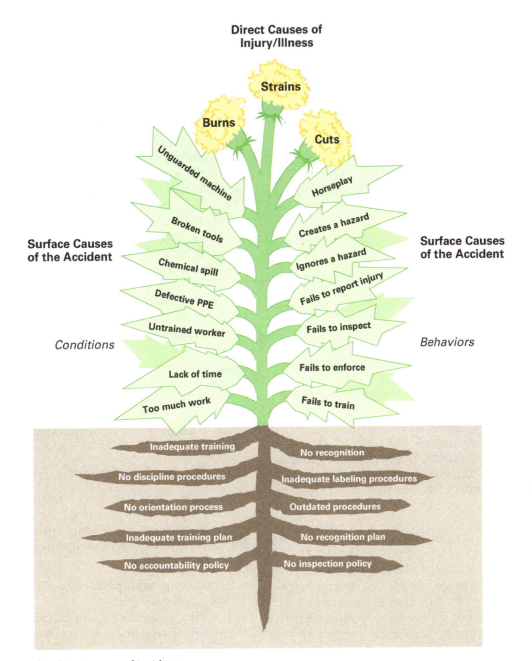

Figure 11 Root and indirect causes of incidents.

2.2.4 Training

Training workers to behave safely and use equipment correctly can be considered a good leading indicator because it correlates to the safety performance at a job site. Basically, the greater the number of workers that are effectively trained, the safer the job site will be.

Training as a leading indicator, however, is more than simply signing employees up for a class. First, set a goal for the training, such as training everyone on the company's lockout/tagout policies and procedures. Second, track the training to be sure it is being conducted and everyone is attending. Third, follow up to verify that the training is effective. If you observe

---

Think About It

OSHA Fatality Report

Two remodeling construction employees were building a wall. One of the workers was killed when he was struck by a nail fired from a powder-actuated tool. The tool operator, while attempting to anchor plywood to a 2 × 4 stud, fired the tool. The nail penetrated the stud and the plywood partition before striking the victim.

What was the direct cause of this fatal incident?
What are the potential indirect causes?
What are the potential root causes?

workers continuing to commit unsafe acts after they have received training, investigate further to determine the root cause of the performance problem.

2.3.0 Analysis of Safety Data

There are many ways of gathering and examining safety data. Both historical (lagging) and predictive (leading) indicators should be used. Data analysis must be done before making procedural changes. Changes are proven effective when injury and illness rates drop.

2.3.1 Incident Analysis

Incident analysis is a reactive examination of the factors that led to an injury or illness. An incident can be caused by a variety of factors, requiring separate but interconnected study. Injury, event, and system analysis are all necessary.

Injury analysis focuses on the injury itself. The initial concentration is on how the injury was sustained (the direct causes of the injury). Root cause analysis may be used to assess underlying causes for the injury.

Event analysis examines the indirect, or surface, causes of an incident. Hazardous conditions and unsafe behaviors are uncovered at this stage. The conditions and behaviors may be corrected based on this analysis, but it may also point to a deeper problem.

This is where system analysis takes over. It uses the facts and observations gathered from injury and event analyses. Basic, or root, causes are uncovered. Indirect causes are traced back to problems in the system. This is the most challenging stage of analysis, because it takes a critical look at company policies and requires thorough examination of all the details.

An incident generally occurs as a result of many factors coming together. Studies have shown that there may be, on average, anywhere from 10 to 27 separate contributing factors. Any analysis of workplace injuries must acknowledge this level of complexity. Only by understanding the relationships between contributing factors can analysis contribute to better safety practices.

2.3.2 Job Safety Analysis

A job safety analysis (JSA), or job hazard analysis (JHA), is a predictive method for studying a job before it begins. Specifically, hazards and potential incidents are identified prior to breaking ground. A completed JSA describes a number of ways to avoid incidents, as shown in the excerpt of a completed JSA in *Figure 12A* and *Figure 12B*.

> **NOTE:** A sample JSA form for insulation/scaffolding is provided in the *Appendix* of this module.

Completing a JSA involves first gathering data about the job. Historical data is analyzed to determine the risks associated with the type of work. All of the key people involved in the job need to be identified as they will each have a say in the JSA development.

Hazards are identified based on everyone's feedback, followed by safe practices in response to those hazards. Clear communication is essential here. Do not say, "Use hand tool," when "Use pliers," is more accurate.

The JSA should be open to continuous alteration and improvement. If, for example, the environment changes, then the agreed-upon content may also have to change.

> **NOTE:** Job safety analysis is covered in more detail in NCCER Module 75220, *Job Safety Analysis and Pre-Task Planning*.

2.3.3 Safety Management

Predictive safety measures require that action be taken before incidents occur. All prescribed methods require continuous support. A successful safety management program will include the following activities:

- Scheduling and performing inspections, audits, and observations
- Tracking inspections, audits, and observations (ensuring that they are performed as scheduled)
- Tracking the status of open action items resulting from incident investigations and safety inspections
- Tracking and trending near miss and first aid reports
- Tracking the promptness of incident reporting and investigations
- Verifying that every required JSA/JHA, pre-task plan, or pre-job safety checklist is being used as intended
- Reporting on the completeness and accuracy of the incident investigation reports and JSAs.
- Making full use of employee perception surveys

A perception survey is a means of measuring leading indicators of safety performance by seeking feedback about a company's safety program

Figure 12A Sample JSA excerpt (1 of 2).

Job Steps in Sequence	Hazards Identified	Hazard Control
1. Setting up scaffold	Struck by falling scaffolding Improper set-up	Have a competent person supervise the erection, alteration, moving, or dismantling of the scaffold. Before erecting the scaffold, analyze work area to identify existing hazards (overhead lines, elevated temperatures, structural stability, soil condition, etc.) that could affect personnel as they erect, alter, move, dismantle, or work on the scaffold. Use at least two people to install the scaffold. Install the scaffold according to manufacturer's requirements and in conformance with 29 CFR 1926, subpart L.
2. Climbing up scaffold	Struck by falling scaffolding Fall injury to self Fall injury to others	Have a competent person who is qualified to recognize the hazards associated with the type of scaffold being used. Train each person who will perform work while on the scaffold. Inspect scaffold on a regular basis. Only use the scaffold for its intended purpose. Provide a safe means of access to the scaffold. Properly plank/deck the working surfaces of the scaffold. Have personnel use required fall protection if they will be working above 6 feet (2 meters). Require workers to maintain three points of contact while climbing up the scaffold. Do not allow persons who are not required for the job to stand under or around the scaffold.
3. Working while on a scaffold	Fall injury to self Fall injury to others	Have a competent person inspect the scaffold befor each work shift, whenever the scaffold is to be used, and before work resumes after the scaffold has been altered or moved. Document the inspection on a Scaffold Inspection Tag and attach the tag to the scaffold. Have personnel use required fall protection if they will be working above 6 feet (2 meters). Remind workers to look before they walk while working on the scaffold. Require workers to maintain three points of contact while climbing up the scaffold. Do not allow persons who are not required for the job to stand under or around the scaffold.

Figure 12B Sample JSA excerpt (2 of 2).

Case History

Caught-In Object Fatality

A laborer at a residential construction site was cleaning a portable mortar mixer at the end of his shift to prepare the mixer for the following day. He began with the drum turned upside down and used a garden hose to spray the paddles and inside of the mixer, allowing the residue to drain out. The laborer then rotated the drum upright and continued spraying water into it. During this procedure the engine of the mixer was running, the mixing blades were rotating, the guard lifter was disconnected, and the drum guard was fully open.

A painter working nearby heard the laborer calling for help and saw that the laborer's arm was caught in the mixer and his body was being pulled into the rotating paddles. The painter tried to turn off the mixer but could not disengage the gears. A bricklayer who was working in another area responded to the painter's calls for help and disengaged the mixing paddles by moving the clutch lever upward and then turning the engine off. A neighbor called 911, and EMS personnel responded within minutes. Rescue workers dismantled the drive mechanism to reverse the mixing paddles and extricated the laborer, who was pronounced dead at the scene.

The Bottom Line: Adequate training to identify unsafe conditions and adequate oversight to avoid unsafe behaviors might have saved this worker's life. Never remove safety guards.

Source: NIOSH In-house FACE Report 2003-13

from employees at all levels within the organization. The questions on the survey ask respondents to evaluate key elements such as top management commitment; employee involvement; and worker safety policies and procedures, including safety and health training, hazard prevention and control, and site conditions. The results of the survey provide a snapshot of the current safety culture through the perceptions of both senior and mid-level management as well as craft workers from the field.

Perception surveys should be conducted annually, and individual responses should be kept confidential. Through analysis of the survey responses, you can quickly see any gaps between the perceptions of field employees versus management-level employees. The gaps identify safety issues that need to be addressed. A common perception gap that can show up in a survey is differing expectations of training. For example, craft workers in the field may think that they are not getting adequate or relevant training, while management believes the company provides plenty of training opportunities.

Work on reducing the biggest gaps first. As you do so, compare the leading indicator measurements from the survey to lagging indicators. By combining data from perception surveys with lagging indicators, you can identify trends. To maintain a positive safety culture, this data-driven process must be supported with management involvement in safety observations and decisions.

Using Lagging Indicators to Identify Trends

Lagging indicators, which depict past safety performance, have long been used to measure safety performance in the construction industry. In fact, safety regulatory agencies and insurance companies require companies to report lagging indicator data. This data includes metrics such as the OSHA recordable injury rate (RIR); the days away, restricted work, or transfer (DART) injury rate; and the experience modification rate (EMR) on workers' compensation. Because lagging indicators such as incidence rates are calculated according to standard formulae, the indicators can easily be used to make comparisons between companies and industries.

Likewise, because lagging indicators occur regularly over the life of construction projects, these indicators can be used to detect changes in performance over time—in other words, trends. For example, analysis of BLS statistics for workplace falls to a lower level from 2004-2013 shows a decreasing trend (improvement) during that time period for the United States construction industry as well as for all industries in the United States.

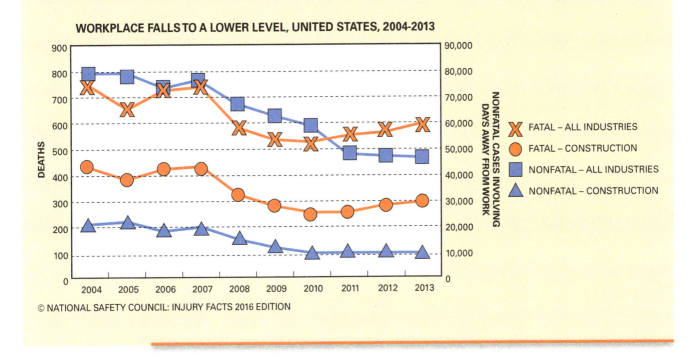

Additional Resources

American Society of Safety Engineers, www.asse.org

Occupational Safety and Health Organization, www.osha.gov

Encyclopedia of Occupational Health and Safety, International Labor Office. 1998. Albany, NY: Boyd Printing.

Fundamentals of Industrial Hygiene, Edited by Barbara A. Plog, MPH, CIH, CSP, and Patricia J. Quinlan, MPH, CIH. 2002. Itasca, NY: National Safety Council.

Root Cause Analysis Handbook: A Guide to Effective Incident Investigation, ABSG Consulting. 2005. Brookfield, CT: Rothstein Associates Inc.

2.0.0 Section Review

1. The BLS method uses an incidence rate based on _____.
   a. one worker
   b. 10 workers
   c. 100 workers
   d. 1,000 workers

2. Which of the following is an indirect cause of an incident?
   a. An inadequate training plan
   b. An unguarded machine
   c. Being cut by a jagged edge
   d. Falling through an opening

3. Incident analysis of the factors that led to an injury or illness requires separate but interconnected analyses of the _____.
   a. person, plan, and performance
   b. consequence, attitude, and resources
   c. motivation, activation, and resolution
   d. injury, event, and system

SUMMARY

As a safety technician, your primary goal is to help find and correct safety hazards before an incident or injury can occur. This involves performing safety inspections, audits, and employee observations and properly documenting them. You will need to keep current on the regulatory agencies that govern the work you do, especially OSHA regulations. You'll also use performance measurements, including both historical (lagging indicator) and predictive (leading indicator) data, for the tracking and trending of safety- and health-related issues.

All of these tasks are designed to decrease the frequency and severity of incidents. Careful analysis of all available data, coupled with continuous dynamic support for safety measures, helps to reduce workplace injury statistics. A good safety record at your job site promotes the well-being of every worker and contributes to the financial health of your company as well.

Review Questions

1. Safety inspections _____.
   a. validate the documentation of hazard mitigation
   b. identify, report, and correct hazards to workers, materials, and equipment
   c. verify the accuracy of safety policies and procedures
   d. are the responsibility of the safety manager

2. Safety technicians need to communicate the results of a safety inspection to job site supervisors, workers, and _____.
   a. site management
   b. the local OSHA office
   c. the company's legal team
   d. Human Resources

3. Safety audits are usually conducted _____.
   a. bi-weekly
   b. monthly
   c. annually
   d. periodically

4. The main reason for performing safety observations is to _____.
   a. determine whether workers are working safely or committing any unsafe acts
   b. discipline employees who violate safety practices
   c. have observations on record for OSHA requirements
   d. make employees feel accountable and empowered

5. Which of the following is an unsafe act that safety technicians commonly find during employee safety observations?
   a. Hazardous atmospheric conditions
   b. Congested workplace
   c. Improper lifting
   d. Excessive noise

6. When talking with an employee who has been observed performing an unsafe act, the safety technician should _____.
   a. criticize the worker for endangering others
   b. specifically state the nature of the unsafe act
   c. ask questions to put the worker off guard
   d. end with a cautionary warning

7. Insurance premiums are traditionally calculated using _____.
   a. the experience modification rate
   b. interviews of front-line workers
   c. observations and inspections
   d. on-site general safety audits

8. Leading indicators attempt to identify, prevent, and control incidents _____.
   a. through training
   b. by modeling behaviors
   c. before they occur
   d. that are unique to the construction industry

9. The fields on a JSA include the organization name, identified hazards, and _____.
   a. injury report
   b. other recordable cases
   c. NAICS code
   d. sequence of job steps

10. A safety management program typically includes tracking _____.
    a. inspections, audits, and observations
    b. BBS interventions per employee
    c. the status of project completion
    d. non-job related medical claims

Trade Terms Introduced in This Module

Audit: To review safety policies and procedures to see if they are adequate and being used.

Behavioral-based safety (BBS): A proactive method of safety management based on psychology. It requires systematic workplace observation and analysis of unsafe behaviors, resolution of problems, and is coupled with training and incentives for behavior modification.

Direct cause: The immediate cause of an injury or illness, not accounting for any underlying unsafe behaviors or conditions.

Experience modification rate (EMR): A lagging indicator of illness and injury rates based on insurance claims and predicted claims over a three-year period. It is most often applied by the insurance industry.

Hazardous conditions: Circumstances or objects that cause injury or illness. Most hazardous conditions arise as a result of unsafe (at-risk) behaviors.

Incidence rate: A lagging indicator of illness and injury rates based on a Bureau of Labor Statistics (BLS) formula. It is measured in annual incidents per 100 workers.

Indirect cause: The underlying cause of an injury or illness. Categories include hazardous conditions and unsafe (at-risk) behaviors.

Inspection: The act of checking an area to identify, report, and correct hazards to workers, materials, and equipment.

Job safety analysis (JSA): A method for studying a job to identify hazards and potential incidents associated with each step and developing solutions that will eliminate, minimize, and prevent hazards and incidents. Also called a job hazard analysis (JHA).

Lagging indicator: A measure of performance based only on historical reporting.

Leading indicator: A measure of performance based on predictive observations and analysis.

Near miss: An unplanned event that did not result in injury, illness, or damage but had the potential to do so. Also called a near hit or a close call.

Observation: Watching a worker during the performance of his or her job for the purpose of determining whether the worker is working safely or committing an unsafe act.

Pre-task plan: A process for identifying and evaluating potential hazards associated with a given task or work assignment.

Root cause: The deepest level, system-related cause of an injury or illness. Requires in-depth analysis to discover. Also referred to as a basic cause.

Safety audit: A predictive method of safety management requiring observation and reporting by an unbiased individual.

Safety inspection: A predictive method of safety management requiring observation and reporting by a workforce supervisor.

Unsafe behavior: Action taken or not taken that increases risk of injury or illness. Also called at-risk behavior.

Appendix

INSULATION/SCAFFOLD JOB SAFETY ANALYSIS FORM

STARCON — a CIANBRO company
The Proactive Approach to Reducing Risk

JSA Grading — Score Using (2, 1, or 0) Maximum = 10   Total Score: _____
- Were Specific Hazards Identified Correctly?
- Were Detailed Job Steps Listed Correctly?
- Were Specific Hazards Mitigated Properly and How?
- Is JSA Filled Out Completely and Signed?
- Can TM's Tell Me The Location of the Eye Wash, Safety Shower and Evacutaion Area?

JSA Audited By _____
Date: _____
Time: _____
JSA Notes: _____

Insulation / Scaffold Job Safety Analysis (JSA)

Date: _____   Time: _____   T/M Name: _____   Permit #: _____
Supervisor: _____   Plant: _____
Plant Rep: _____   Unit Name: _____
Emergency Phone #: _____   Location: _____
Job Description: _____

STOP WORK OBLIGATION

THE BIG 5:
- RESPONSIBILITY
- ACCOUNTABILITY
- INTERVENTION
- VALUE
- PRIDE

IF YOU SEE IT YOU OWN IT

Emergency Information

1. What Direction is the Wind Coming From? _____
2. Reviewed All Emergency Alarms / Phone Numbers with TM's: Yes ___ No ___
3. Escape Routes Reviewed Yes ___ No ___   List Escape Routes Below
   Primary: _____
   Secondary: _____
4. Location of Eye Wash Stations / Safety Showers: _____
5. My Evacuation Assembly Point Is: _____
6. Total Number of Team Members on Job: _____
   Supervisors Approval: _____
   Supervisors Review Date: _____ Time: _____

Team Member Signature and TM ID | Initial Changes

1. _____ ID# _____
2. _____ ID# _____
3. _____ ID# _____
4. _____ ID# _____
5. _____ ID# _____
6. _____ ID# _____
7. _____ ID# _____
8. _____ ID# _____
9. _____ ID# _____
10. _____ ID# _____
11. _____ ID# _____
12. _____ ID# _____

Job Scope Change / Modifications Discussion

Team Member: _____
Changes Made: _____

Supervisors Approval of Changes: _____
Date: _____ Time: _____
New JSA Required: Yes ___ No ___

PAUSE / WHAT IS DIFFERENT?

1. Scope — Yes ___ No ___
2. Unexpected Events — Yes ___ No ___
3. Local Activities — Yes ___ No ___
4. People — Yes ___ No ___
5. Pace — Yes ___ No ___
6. Fatigue — Yes ___ No ___
7. Area Conditions — Yes ___ No ___
8. Other — Yes ___ No ___

HAZARD MANAGEMENT

- **IDENTIFY**
- **ASSESS**
- **CONTROL**
- **RECOVER**

Job Steps in Sequence | Hazards Identified | Hazard Control

Job Steps in Sequence	Hazards Identified	Hazard Control
1.		
2.		
3.		
4.		
5.		
6.		
7.		
8.		
9.		
10.		
11.		
12.		
13.		
14.		
15.		

Hazards Identified in Work Area	Hazard Control Methods	PPE Equipment Utilized
☐ Thermal Burns (Tracing/Hot Piping/Steam)	☐ 360 Degree Awareness	☐ Safety Glasses
☐ Asbestos	☐ Safety Nets Installed	☐ Goggles / Spoggles
☐ Overhead Power Lines	☐ Utilized Salt / De-Ice terials	☐ Side Shields (Prescription Glasses)
☐ Congested Area / Other Workers	☐ Abated / Encapsulated	☐ Safety Harness
☐ Insufficient Lighting	☐ Proper Distance Identified	☐ Double Retractable Device
☐ Heat Stress / Over Exertion	☐ Spotter Assigned	☐ Cheater Strap - Beam Strap
☐ Slips, Trips, Falls	☐ Twist and Release Method Used	☐ Radio (Battery Charged / Working Properly)
☐ Pinch Points	☐ Switch Covers Installed	☐ Hearing Protection
☐ Line of Fire	☐ Proper PPE Utilized for the Job	☐ Double Hearing Protection
☐ Snow, Ice, Rain, High Wind	☐ Lock Out / Tag Out in Place, Box#_____	☐ Leather Gloves
☐ Rotating Equipment	☐ Proper Barricades in Place / Signs Posted	☐ Cut Resistance Gloves (Kevlar)
☐ Rail Road Crossing / Tracks	☐ Continuous Monitor Utilized	☐ Nitrile Gloves (Alky Use)
☐ Leading Edge	☐ Proper Communication in Place	☐ Kevlar Sleeves (Tracing)
☐ Cuts, Punctures	☐ Proper Lifting Ergonomics Utilized	☐ H2S Monitor ☐ Tested
☐ Loud Noises	☐ Clean Work Area (House Keeping)	☐ Hard Hat
☐ Pump Switches / Electrical Equipment	☐ Tools Inspected and Safe for Use	☐ Steel Toe Boots
☐ Radiation Hazards	☐ Safety Equipment Inspected Prior to Use	☐ Rubber Boots
☐ Particles In Eyes	☐ Proper Equipment Grounding	☐ Face Shield (Z87 +)
☐ Dropping Materials	☐ Scaffold Inspected & Updated, Harness Inspected	☐ Fresh Air Equipment
☐ Chemical Exposure	☐ Respirator with Proper Cartridge	☐ (Vest, Horn, 5 Min Pack, 30 Min Pack, etc.)
☐ Sprains / Strains	☐ Proper donning / doffing of PPE	☐ Lock Out / Tag Out Locks in Place
☐ Elevated Work	☐ Other_____	☐ FR Clothing
☐ Sharp Objects	☐ Other_____	☐ Body Protection Other: (Bunker, Tyvek, Rain Suit)
☐ Other:_____	☐ Other_____	☐

Operations / Projects / Maintenance

1. Equipment Preparation and Transfer Complete   Yes ____ No ____ N/A ____
2. Job Scope Reviewed / Understood   Yes ____ No ____
3. Proper Safety Equip. On Job Site   Yes ____ No ____
4. Proper Permits Utilized: Yes ____ No ____
   Cold Work _____ Confined Space _____ Alky _____ Hot Work _____
5. Special Permit _____   Yes ____ No ____
6. Joint Job Site Visit Conducted (JJSV)   Yes ____ No ____
7. Confined Space, Bottle Watch, Hole Watch, Fire Watch Attendant Assigned
   Yes ____ No ____ N/A ____
8. Proper Communications Set in Place   Yes ____ No ____
9. All Valves / Switches / Tubing / Blinds: Closed, De-Energized, and Verified:
   Yes ____ No ____ N/A ____
10. Atmospheric Testing Completed   Yes ____ No ____ N/A ____
11. Confined Space Procedure / Rescue Plan Reviewed With Certified Safety
    Attendant   Yes ____ No ____ N/A ____
12. Proper Tools and Equipment for Job Inspected   Yes ____ No ____
13. MSDS Reviewed and Understood   Yes ____ No ____ N/A ____
14. Material Transport Complete   Yes ____ No ____ N/A ____
15. Other: _____   Yes ____ No ____ N/A ____

System Breach Documentation / Variance

1. Physically Verified That No Hydrocarbon / Hazardous Material is Trapped
   Yes ____ No ____ N/A ____
2. Equipment in Place to Collect and Disperse any Spills
   Yes ____ No ____ N/A ____
3. Is There Hot Work Within a 50 foot Radius Horizontally or Vertically
   Yes ____ No ____ N/A ____
4. Safe Provisions for Egress From Work Site   Yes ____ No ____
5. Emergency Response Plan in Place Required   Yes ____ No ____
6. Safety Watch in Place   Yes ____ No ____ N/A ____
7. All Appropriate PPE in Place for Protection in Case of Leakage
   Yes ____ No ____ N/A ____
8. Process and Mechanical Witnessed the Sniff Testing
   Yes ____ No ____ N/A ____
9. System Breach Checklist Completed   Yes ____ No ____ N/A ____
10. Field Verification Completed   Yes ____ No ____ N/A ____
11. All Signatures Acquired   Yes ____ No ____ N/A ____
12. System Breach Variance Required   Yes ____ No ____ N/A ____
13. Variance Documentation Complete and Approval Acquired
    Yes ____ No ____ N/A ____

Note: If Any Answer Above is a "No" Approval to Proceed Must be Obtained by Company and Site Leadership

Job Task Completion Review

1. Work Site Cleaned / Barricades Removed   Yes ____ No ____
2. Red Tags / LOTO Removed and Signed Off   Yes ____ No ____
3. Permits Pulled and Closed Out   Yes ____ No ____
4. Unit Sign-In / Sign-Out Logs Verified   Yes ____ No ____
5. Post JJSV Required / Completed   Yes ____ No ____
6. Open Manways / Holes Covered   Yes ____ No ____
7. Any Safety Involvement / Concerns   Yes ____ No ____

Concerns: _____

VISION for Safety at STARCON

Contractors and Customers / Owners by ensuring that our enabling and sustaining systems are in place and that our climate / culture supports our vision for safety as a core value. We will see all TM's Sub-Contractors, and Customer / Owners exhibiting positive behaviors that are reinforced and demonstrated through our commitment to Leading With Safety by:

- Focusing Resources Where the Higher Risk Occur
- Working with TM's Sub-Contractors, and Customer / Owners to Implement Lasting Improvements
- Identifying Hazards and Mitigating the Risk
- Intervening **(IF YOU SEE IT YOU OWN IT - TAKE ACTION)**
- Verifying Safety is Implemented Throughout All Work Phases
- Consistently Demonstrating and Supporting Safety Leadership at All Levels
   (Credibility, Action, Resolve, and Engagement)
- Continuously Investing in the Development of Safety Leaders

No One Gets Hurt!

SAFETY AND QUALITY ARE OUR CORE VALUES

Additional Resources

This module presents thorough resources for task training. The following resource material is suggested for further study.

American Society of Safety Engineers, **www.asse.org**

Occupational Safety and Health Organization, **www.osha.gov**

Encyclopedia of Occupational Health and Safety, International Labor Office. 1998. Albany, NY: Boyd Printing.

Fundamentals of Industrial Hygiene, Edited by Barbara A. Plog, MPH, CIH, CSP, and Patricia J. Quinlan, MPH, CIH. 2002. Itasca, IL: National Safety Council.

Root Cause Analysis Handbook: A Guide to Effective Incident Investigation, ABSG Consulting. 2005. Brookfield, CT: Rothstein Associates Inc.

Figure Credits

J. Vinton Schafer & Sons, Inc. and CCBC Catonsville/elcosh.org, Module opener

U.S. Department of Labor, Figure 1, SA02, SA03, SA05, SA06

Bureau of Labor Statistics, SA04, Figures 8, 9

Brett Richardson, Figure 12, Appendix

National Safety Council, SA07

Section Review Answer Key

Answer	Section Reference	Objective
Section One		
1. c	1.1.2	1a
2. a	1.2.0	1b
3. b	1.3.1	1c
Section Two		
1. c	2.1.1	2a
2. b	2.2.3	2b
3. d	2.3.1	2c

NCCER CURRICULA — USER UPDATE

NCCER makes every effort to keep its textbooks up-to-date and free of technical errors. We appreciate your help in this process. If you find an error, a typographical mistake, or an inaccuracy in NCCER's curricula, please fill out this form (or a photocopy), or complete the online form at **www.nccer.org/olf**. Be sure to include the exact module ID number, page number, a detailed description, and your recommended correction. Your input will be brought to the attention of the Authoring Team. Thank you for your assistance.

Instructors – If you have an idea for improving this textbook, or have found that additional materials were necessary to teach this module effectively, please let us know so that we may present your suggestions to the Authoring Team.

NCCER Product Development and Revision
13614 Progress Blvd., Alachua, FL 32615

Email: curriculum@nccer.org
Online: www.nccer.org/olf

❏ Trainee Guide ❏ Lesson Plans ❏ Exam ❏ PowerPoints Other _____

Craft / Level: _____ Copyright Date: _____

Module ID Number / Title: _____

Section Number(s): _____

Description: _____

_____

_____

_____

Recommended Correction: _____

_____

_____

_____

Your Name: _____

Address: _____

_____

Email: _____ Phone: _____

Site-Specific Safety Plans

Overview

This module discusses the use of pre-bid checklists to identify hazards and develop a site-specific safety plan. It explains the importance of an emergency action plan and covers the basic components of a safety program. Several regulatory requirements relevant to the construction industry are identified and described.

Module 75222

Trainees with successful module completions may be eligible for credentialing through NCCER's Registry. To learn more, go to **www.nccer.org** or contact us at 1.888.622.3720. Our website has information on the latest product releases and training, as well as online versions of our *Cornerstone* magazine and Pearson's product catalog.

Your feedback is welcome. You may email your comments to **curriculum@nccer.org**, send general comments and inquiries to **info@nccer.org**, or fill in the User Update form at the back of this module.

This information is general in nature and intended for training purposes only. Actual performance of activities described in this manual requires compliance with all applicable operating, service, maintenance, and safety procedures under the direction of qualified personnel. References in this manual to patented or proprietary devices do not constitute a recommendation of their use.

Copyright © 2018 by NCCER, Alachua, FL 32615, and published by Pearson, New York, NY 10013. All rights reserved. Printed in the United States of America. This publication is protected by Copyright, and permission should be obtained from NCCER prior to any prohibited reproduction, storage in a retrieval system, or transmission in any form or by any means, electronic, mechanical, photocopying, recording, or likewise. To obtain permission(s) to use material from this work, please submit a written request to NCCER Product Development, 13614 Progress Blvd., Alachua, FL 32615.

75222 V2

From *Safety Technology,* Trainee Guide. NCCER.
Copyright © 2018 by NCCER. Published by Pearson. All rights reserved.

75222
SITE-SPECIFIC SAFETY PLANS

Objectives

When you have completed this module, you will be able to do the following:

1. Explain how to use existing pre-bid planning checklists to identify specific job site hazards and requirements.
    a. Describe how to use a risk assessment matrix to determine the risk of a given situation.
    b. Describe how to modify an existing site safety plan to meet specific job conditions.
    c. List the requirements of a site-specific safety program.
2. Explain how to develop an emergency action plan.
    a. Identify the elements of an emergency action plan.
    b. Describe how to plan for specific types of emergencies.
    c. Describe how to communicate with traditional and social media during an emergency.
3. Identify the basic components of a safety program.
    a. Identify essential safety program policies and procedures.
    b. Identify effective practices for providing safety orientation and training.
4. Identify regulatory requirements.
    a. Identify Occupational Safety and Health Administration (OSHA) requirements.
    b. Identify Environmental Protection Agency (EPA) requirements.
    c. Identify Department of Transportation (DOT) requirements.
    d. Identify other regulatory requirements.

Performance Tasks

Under the supervision of your instructor, you should be able to do the following:

1. Use a risk assessment matrix to determine the risk of a given situation.
2. Modify an existing site safety plan to meet specific job conditions.

Trade Terms

Contractor pre-qualification
Defibrillator
Immediately dangerous to life and health (IDLH)
Intrinsically safe
Pre-bid checklist
Probability of occurrence
Process Safety Management (PSM)
Risk assessment
Universal precautions

Industry Recognized Credentials

If you are training through an NCCER-accredited sponsor, you may be eligible for credentials from NCCER's Registry. The ID number for this module is 75222. Note that this module may have been used in other NCCER curricula and may apply to other level completions. Contact NCCER's Registry at 888.622.3720 or go to www.nccer.org for more information.

Contents

- 1.0.0 Pre-Bid Safety Planning 1
  - 1.1.0 Hazard Identification 1
    - 1.1.1 Risk Assessment 2
    - 1.1.2 Hazard Control 2
  - 1.2.0 Preparing a Site Safety Plan 2
  - 1.3.0 Site Safety Plan Requirements 3
- 2.0.0 Emergency Action Plans 4
  - 2.1.0 Elements of an Emergency Action Plan 5
    - 2.1.1 Chain of Command 5
    - 2.1.2 Communications 6
    - 2.1.3 Accounting for Personnel 6
    - 2.1.4 Emergency Response Teams 6
    - 2.1.5 Training 8
    - 2.1.6 Personal Protection 10
    - 2.1.7 Medical Assistance 15
    - 2.1.8 Security 15
  - 2.2.0 Pre-Planning for Specific Types of Emergencies 15
    - 2.2.1 Trapped Workers 15
    - 2.2.2 Severe Weather 16
    - 2.2.3 External Threats 16
    - 2.2.4 Fire Emergencies 16
  - 2.3.0 Dealing with the Media 18
- 3.0.0 The Site Safety Plan 20
  - 3.1.0 Site Location and Layout 20
    - 3.1.1 Traffic Patterns 20
    - 3.1.2 Adjacent Hazards 21
    - 3.1.3 Scope of Work 21
    - 3.1.4 Personnel Health and Safety 22
    - 3.1.5 Safety Coordination 22
  - 3.2.0 Administration 23
    - 3.2.1 Training 23
- 4.0.0 Regulatory Requirements 25
  - 4.1.0 OSHA Requirements 25
  - 4.2.0 EPA Requirements 26
  - 4.3.0 DOT Requirements 28
  - 4.4.0 Other Regulatory Requirements 28
    - 4.4.1 MSHA Requirements 28
    - 4.4.2 ANSI and ASME Requirements 30
    - 4.4.3 NFPA Requirements 30
    - 4.4.4 DHS and Coast Guard Requirements 31
    - 4.4.5 USACE and NAVFAC Requirements 31
    - 4.4.6 Training Requirements 32

Figures and Tables

Figure 1 Risk assessment matrix. 2
Figure 2 Emergency communications equipment. 6
Figure 3 Types of fire extinguishers. 7
Figure 4 Firefighting risk assessment. 9
Figure 5 Personal protective equipment. 11
Figure 6 Four types of respirators. 12
Figure 7 Procedures to follow before using a respirator. 14
Figure 8 Standpipe with Siamese fire department connections. 17
Figure 9 Excavation site traffic. 20
Figure 10 Standard underground markers. 21
Figure 11 A plant covered under OSHA PSM rules. 25
Figure 12 EPA and DOT labels. 28
Figure 13 Hazardous waste manifest. 29
Figure 14 NFPA 704 diamond. 31

Table 1   Evacuate or Fight? 8
Table 2   Particulate Respirators 13

Section One

1.0.0 Pre-Bid Safety Planning

Objective

Explain how to use existing pre-bid planning checklists to identify specific job site hazards and requirements.

a. Describe how to use a risk assessment matrix to determine the risk of a given situation.
b. Describe how to modify an existing site safety plan to meet specific job conditions.
c. List the requirements of a site-specific safety program.

Performance Tasks

1. Use a risk assessment matrix to determine the risk of a given situation.
2. Modify an existing site safety plan to meet specific job conditions.

Trade Terms

Contractor pre-qualification: A process of screening contractors to allow them to bid on jobs for a specific company.

Pre-bid checklist: A list of questions or issues that aids contractors in gathering the information needed to prepare a bid package.

Probability of occurrence: The likelihood that a specific event will occur, usually expressed as the ratio of the number of actual occurrences to the number of possible occurrences.

Risk assessment: The process of qualifying or ranking hazards given their probability and consequences.

Planning safety into a project is just as important as planning for materials and equipment. Pre-bid safety planning can increase productivity; for example, making plans to install stairs earlier in the schedule so that they are available when needed. Failure to plan for safety may result in costly work slowdowns or stoppage due to serious incidents, injuries, or citations from regulatory agencies.

A site analysis performed by a safety technician can potentially be used in the bidding process. The analysis must be completed before a bid is submitted, so all expenses and other provisions are included. Most contractor pre-qualification programs include safety requirements. An example of a contractor pre-qualification form is provided in *Appendix A*.

In the process of pre-bid safety planning, it is essential to include every important aspect of site operations. The use of a pre-bid checklist will help to ensure that nothing is missed. Several groups have developed pre-bid checklists to aid their membership. You may need to modify a standard checklist to fit your operation. Note that specific checklists may be needed for different phases of a project. For example, you may use one checklist for initial site construction and another for finish work. An example of a pre-job planning safety checklist is shown in *Appendix B*.

Pre-bid safety planning should be done for every job. Planning should take into consideration the work being performed along with the necessary safety precautions. Develop incident prevention measures for each phase and component of the work. The plan should also include recommended hazard control measures. Because some phases of the work may be subcontracted, the plan should include provisions for the safety of subcontractors as well.

Planning should identify all potential areas for loss, including the following:

- People (employees, contractors, and the general public)
- Buildings or structures
- Equipment
- The environment

Pre-job safety planning should follow this general format:

1. Identify the hazards
2. Evaluate the consequences
3. Rank the hazards by risk
4. Control or minimize the hazards

1.1.0 Hazard Identification

In order to plan safety into a job, certain types of information must be obtained and analyzed. The information that your plan requires can be extensive. Some of the information that needs to be included in the plan will include the following:

- Site location and layout
- Scope of work
- Client and contract requirements
- Regulatory permits and requirements

1.1.1 Risk Assessment

Once the hazards are identified, assess the associated risks. The two aspects of risk assessment are consequences and probability. You need to determine what would happen in case of an incident. The consequences can be categorized as follows:

- I – Catastrophic: death or loss of facility
- II – Critical: severe illness, injury, or property damage
- III – Minor illness, injury, or property damage
- IV – Low hazard but a violation of standards

Next, determine the potential for the event to happen. This is known as probability of occurrence. These can be ranked as follows:

- A – Likely to happen now or very soon
- B – Probably will happen in time
- C – Possibly will happen in time
- D – Unlikely to happen

These two factors can be used to determine the overall risk using a matrix similar to the one shown in *Figure 1*.

Poor housekeeping is an example of a lesser hazard that would rate as a moderate risk. A fuel tank explosion could have catastrophic results, but it is unlikely to happen. Therefore, this would also rate as a moderate risk. The consequences and probability of different incidents must be determined according to the specific situation.

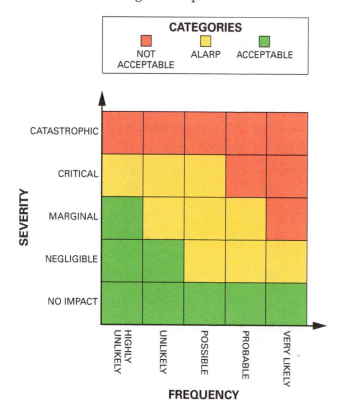

Figure 1 Risk assessment matrix.

> **NOTE:** For more information on risk assessment, refer to NCCER Module 75219, *Hazard Recognition, Environmental Awareness, and Occupational Health* and Module 75220, *Job Safety Analysis and Pre-Task Planning*.

1.1.2 Hazard Control

After identifying the hazards, the safety technician must establish proper safeguards to minimize them. The hazards should be addressed in the order of priority determined by the risk assessment. Any catastrophic hazard needs to be eliminated entirely. Major hazards considered critical or serious must be minimized. Moderate, minor, and negligible hazards must also be addressed in order to reduce the possibility of an incident.

These hazards can be minimized or eliminated in several ways. In order of preference, hazard controls include the following:

- Engineering controls or substitution to eliminate or reduce the hazard (*e.g.*, guarding or isolating the hazard)
- Administrative controls (training, procedures, signs, alarms, instruction, or personnel rotation)
- Use of personal protective equipment

Hazard controls should be balanced with the risk assessment. The best solution will eliminate the hazard, which is the goal for catastrophic or critical risks. The personal protective equipment choice of hazard control merely protects workers from the hazard. It should only be used with minor or negligible risks.

1.2.0 Preparing a Site Safety Plan

A site safety plan is sometimes referred to as an environmental safety and health (ES&H) or safety, health, and environment (SH&E) plan. The site safety plan should be prepared using the information gathered in the pre-bid checklist or other tools. Safety/scheduling software is also available for use on smartphones, tablets, and personal computers (PCs). These applications (apps) often include safety checklists for each bid item. The company's existing safety program or Safety and Loss Prevention Manual can be modified to address specific safety issues on your site.

There are many professional safety firms that offer assistance for preparing a site safety plan. State and federal OSHA offices offer online guidance. The California DOSH offers a model program for Incident and Illness Prevention;

these guidelines can be downloaded from **www.dir.ca.gov/dosh/puborder.asp**. Always be familiar with the current state and federal laws that apply to your project.

Another way to prepare a site-specific plan is to list site-specific information and use the existing procedures indicated in the work plan. Copies of relevant standard operating procedures must be appended. Typical sections generally include the following:

- Site information (including site security and layout)
- Scope of work
- Key contacts
- Existing on-site and adjacent hazards
- Standard procedures used (HazCom, fall protection, welding, etc.)
- Additional procedures
- Emergency response procedures and muster points (meeting areas)
- Training verification, such as confined space, Hazardous Waste Operations and Emergency Response (HAZWOPER), etc.

A sample outline for a generic site safety plan is provided in *Appendix C*.

1.3.0 Site Safety Plan Requirements

A site safety plan will vary in size and complexity depending on the scope and location of the project. At a minimum, the plan must include information on safety responsibilities and emergency procedures. It should have provisions for hazard communication, incident prevention, inspections, grounded electrical systems, recordkeeping, personal protective equipment, elevated work, and housekeeping. Operation-specific safety procedures (for example, hoisting and rigging, or demolition) can be appended to a site safety plan.

The site safety plan must be coordinated with all groups on a project (the client and all subcontracting companies, including all workers). The client or general contractor will also have a safety plan. Roles, responsibilities, and procedures must be clear before the job starts. Some federal and state programs have strict policies for safety coordination. In addition, the site plan needs to be coordinated with any local, state, and/or federal emergency response services that you may rely on.

In order for any plan to be effective, there must be adequate oversight and review. The site plan needs to include inspections, auditing, and reporting. The plan must be reviewed as the project progresses to make sure that the safety goals are met.

Additional Resources

American Society of Safety Engineers, **www.asse.org**

California DOSH Incident and Illness Prevention Program, **www.dir.ca.gov/dosh/puborder.asp**

Occupational Safety and Health Organization, **www.osha.gov**

1.0.0 Section Review

1. Which of the following is a method of minimizing hazards?
    a. Engineering controls
    b. Pre-bid checklists
    c. Chain of command
    d. Negative reinforcement

2. One way to prepare a site safety plan is to use the assistance of a(n) _____.
    a. experienced subcontractor
    b. local emergency response team
    c. professional safety firm
    d. ASME-affiliated engineer

3. Operation-specific safety procedures that are appended to a site safety plan could include _____.
    a. medical monitoring programs
    b. demolition procedures
    c. pre-bid checklists
    d. communication methods

Section Two

2.0.0 Emergency Action Plans

Objective

Explain how to develop an emergency action plan.

a. Identify the elements of an emergency action plan.
b. Describe how to plan for specific types of emergencies.
c. Describe how to communicate with traditional and social media during an emergency.

Trade Terms

Defibrillator: An electronic device that administers an electric shock of preset voltage to the heart through the chest wall in an attempt to restore the normal rhythm of the heart during ventricular fibrillation.

Immediately dangerous to life or health (IDLH): A situation that poses a threat of exposure to airborne contaminants when that exposure is likely to cause death or immediate or delayed permanent adverse health effects or prevent escape from such an environment.

Intrinsically safe: An electric tool or device that is UL-rated to be explosion-proof under normal use. For example, a fuel pump must be intrinsically safe to prevent explosions sparked by static electricity.

Universal precautions: A set of precautions designed to prevent transmission of human immunodeficiency virus (HIV), hepatitis B virus (HBV), and other blood-borne pathogens when providing first aid or health care. Under universal precautions, blood and certain body fluids of all patients are considered potentially infectious for HIV, HBV, and other blood-borne pathogens.

Any construction site or industrial site can experience an emergency. The types of catastrophic emergencies that commonly occur include the following:

- Personal injuries
- Release of toxic gases
- Chemical spills
- Fire
- Explosion
- Structural collapse
- Trench cave-in
- Confined-space incidents
- Incidents at extreme heights
- Natural disasters (earthquakes, hurricanes, tornadoes, tsunamis, etc.)
- Weather emergencies
- External threats/terrorism

Having an emergency action plan can reduce the response time to job site emergencies. In many cases, advance planning can minimize or reduce the severity of the incident. For example, when the names, addresses, and phone numbers of the nearest medical, fire, police, and emergency response agencies are posted conspicuously at each job site phone and provided to employees, it will be easier to get help when you need it.

As part of the emergency action plan, everyone should be familiar with the site emergency reporting and response procedures. Prompt access to safety data sheets (SDSs) is necessary to ensure that proper first aid and follow-up medical care can be given. Emergency action plans should require that each job site train one or more persons in basic first aid, cardiopulmonary resuscitation (CPR), and the use of an automated external defibrillator (AED). It is always best to train as many people as possible. Lives are saved when all of these elements are incorporated into emergency action plans.

The effectiveness of response during emergencies depends on the amount of planning and training that have been done. Management must show support for safety programs and understand the importance of emergency planning. If management is not committed to protecting employees and minimizing property loss, it will be difficult to promote a safe workplace. It is therefore management's responsibility to work with personnel in the field to develop and implement an emergency action plan that can be adapted to meet job site conditions. The input and support of all employees must be obtained to ensure an effective program. Emergency action plans should be developed for each job and be comprehensive enough to effectively address any type of emergency. When required by a particular OSHA standard, the emergency action plan must be in writing.

2.1.0 Elements of an Emergency Action Plan

As stated in 29 *CFR* 1926.35(b), emergency action plans must have, at a minimum, the following elements:

- Emergency escape procedures and emergency escape route assignments
- Procedures to be followed by employees who remain to perform (or shut down) critical operations before they evacuate
- Procedures to account for all employees after emergency evacuation has been completed
- Rescue and medical duties for those employees who are to perform them
- The preferred means for reporting fires and other emergencies
- Names or regular job titles of persons or departments to be contacted for further information or explanation of duties under the plan

The emergency action plan should address all potential emergencies that can be expected in the workplace. A pre-job hazard assessment is necessary to determine what types of emergencies could reasonably occur. This should be done during the pre-job planning phase. For work done at existing facilities, the contractor and the host employer should coordinate their emergency reporting and response procedures before the job begins.

For information on the safe use of chemicals, contact the manufacturer or supplier to obtain SDSs. These forms describe the hazards that a chemical may present; list precautions to take when handling, storing, or using the substance; and outline emergency and first aid procedures.

The emergency action plan should list in detail the procedures for those employees who must remain behind to care for essential operations or until their evacuation becomes absolutely necessary. These employees can include crane and mobile equipment operators, riggers, and other personnel who may be needed during the emergency.

For emergency evacuation, the plan should include floor plans or workplace maps that clearly show the emergency escape routes and safe or refuge areas. All employees must be told what actions they are to take in case of emergency situations that may occur in the workplace.

The emergency action plan is a living document that must be monitored during all phases of construction to account for changes on the site. The job site emergency action plan should be reviewed with employees initially when the plan is developed, whenever the employees' responsibilities under the plan change, and whenever the plan is altered.

In addition to the elements previously discussed, effective emergency action plans must provide detailed information about the following:

- Chain of command
- Communications
- Accounting for personnel
- Emergency response teams
- Training procedures
- Personal protection
- Medical assistance
- Security
- Recordkeeping

2.1.1 Chain of Command

A chain of command must be established so that employees know who has the authority to make decisions. At most existing industrial facilities, an emergency response team coordinator is selected to coordinate the work of the emergency response team. That person is first in the chain of command. In larger organizations, there may be a plant coordinator in charge of plant-wide operations, public relations, and ensuring that outside aid is called in. In order to ensure that these important functions are taken care of, additional backup must be arranged so that trained personnel are always available in case of an emergency. The duties of the emergency response team coordinator include the following:

- Assessing the situation and determining whether an emergency exists that requires activating the emergency procedures
- Directing all efforts in the area, including evacuating personnel and minimizing property loss
- Ensuring that outside emergency services such as medical aid and local fire departments are called in when necessary and are aware of specific hazards that may require specialized training or equipment
- Directing the shutdown of plant operations when necessary

The contractor must implement a chain of command as part of the emergency action plan. This chain of command must be flexible and work in conjunction with local law enforcement and emergency response personnel.

On multi-employer work sites, the construction manager (CM) or the general contractor (GC) controls the site but must communicate the plan to all workers on the site, assigning control and responsibility as appropriate.

2.1.2 Communications

During a major emergency involving a fire or explosion, it may be necessary to evacuate all portions of the site, including offices. Normal services such as electricity, water, telephones, and internet may be unavailable. Under these conditions, it may be necessary to designate an alternate area to which employees can report or that can act as a focal point for incoming and outgoing calls. Because time is an essential element for adequate response, the person designated as being in charge should make this the alternate headquarters so that he or she can be easily reached.

Emergency communications equipment (*Figure 2*) such as portable radio units (walkie-talkies), cell phones, or public-address systems should be on hand for notifying employees of the emergency and for contacting local authorities, such as law enforcement officials, the fire department, ambulance services, and emergency response contractors. Some workplaces require intrinsically safe (explosion-proof) communication devices.

A method of communication is needed to alert employees to an evacuation or to take other action as required in the plan. Alarms should be audible or visible by everyone on site and should have an auxiliary power supply in the event electricity is affected. The alarm should be distinctive and recognizable as a signal to evacuate the work area or perform actions designated under the emergency action plan. Emergency contact information for each employee should be assembled and stored in two places—one on-site and one off-site.

The employer should explain to each employee the means for reporting emergencies. Emergency phone numbers should be posted on or near telephones, on employees' notice boards, or in other obvious locations. The warning plan should be in writing and management must make sure that each employee knows what it means and what action is to be taken.

Keep an updated written list of key personnel, listed in order of priority. Site personnel should be provided with the phone numbers of key personnel in the main office, including daytime and after-hours numbers.

2.1.3 Accounting for Personnel

In an emergency, site management will need to know when all personnel have been accounted for. This can be difficult on a construction project. Someone on site should be appointed to account for personnel and to inform police or emergency response personnel of those persons believed missing. Electronic timekeeping systems can be

Figure 2 Emergency communications equipment.

useful in tracking employees. Morning safety meeting rosters can also be used.

2.1.4 Emergency Response Teams

Emergency response teams are the first line of defense in emergencies. Before assigning personnel to these teams, the contractor must ensure that they are physically capable of performing the duties that may be assigned to them. Depending on the size of the facility, there may be one or several teams trained in the following areas:

- Use of various types of fire extinguishers (*Figure 3*)
- First aid, including CPR and AED
- Confined-space rescues
- Evacuation procedures
- Chemical spill control procedures
- Use of self-contained breathing apparatus (SCBA)
- Emergency rescue procedures for trenches and elevated locations

> **WARNING!** Only individuals with the proper training should attempt rescue operations. Anyone who has not been trained to do so should not attempt a rescue. Make sure emergency response teams are prepared for the expected hazards.

The type and extent of the emergency will depend on the site conditions. The response will vary according to the type of emergency, the material handled, the number of employees, and the availability of outside resources. Emergency response teams should be trained in the types of

EXTINGUISHER TYPE	TYPE OF FIRE
	**Ordinary Combustibles** Fires in paper, cloth, wood, rubber, and many plastics require a water-type extinguisher labeled A.
 CO₂ OR  Dry Chemical	**Flammable Liquids** Fires in oils, gasoline, some paints, lacquers, grease, solvents, and other flammable liquids require an extinguisher labeled B. **Electrical Equipment** Fires in liquids, fuse boxes, energized electrical equipment, computers, and other electrical sources require an extinguisher labeled C.
	**Ordinary Combustibles, Flammable Liquids, or Electrical Equipment** Multi-purpose dry chemical extinguishers are suitable for use on Class A, B, and C fires.
	**Metals** Combustible metals such as magnesium and sodium require special extinguishers labeled D.

Figure 3 Types of fire extinguishers.

possible emergencies and the emergency actions to be performed. They should be informed about special hazards, such as the storage and use of flammable materials, toxic chemicals, radioactive sources, and water-reactive substances, to which they may be exposed during fire and other emergencies. Make sure that SDSs are readily available for all of the chemical hazards on the site.

> **NOTE**
> On a typical construction site, contractors may not have organized emergency response teams. Instead, they may rely on off-site emergency response personnel. An exception would be on some large projects where contractors may be asked to serve on emergency response teams.

Three basic options for fighting small fires are outlined in 29 *CFR* 1910.157, as summarized in *Table 1*. Option 1 of totally evacuating the work site rather than fighting the fire will most effectively minimize the potential for fire-related injuries to employees. This option also makes it unnecessary to train employees to use portable fire extinguishers. However, any extinguishers on site must be properly inspected and maintained. Other factors such as the availability of a public fire department or the possibility of blocked exit routes from the job site must also be considered when selecting the best option for your workplace.

Regardless of the chosen option for fighting a small fire, emergency response teams must be trained to determine when it is best to not to intervene if a fire breaks out. For example, team members must know how to recognize if a fire is too large for them to handle or whether emergency rescue procedures should be performed (*Figure 4*). If there is any possibility that the team members could receive injuries, they should wait for professional firefighters or emergency response groups. Always err on the side of caution and order a localized or site-wide evacuation if the conditions warrant it. All fires should be reported to the site supervisor, regardless of severity.

2.1.5 Training

Training is essential for a successful emergency action plan. Before implementing the plan, a sufficient number of persons must be trained to assist in the safe and orderly evacuation of employees. Training for each type of disaster response is necessary so that employees know what actions are required in each case. There have been many emergencies in which would-be rescuers died when they failed to follow orders to evacuate.

Table 1 Evacuate or Fight?

	Option 1	Option 2	Option 3
Evacuation	Total evacuation of employees from workplace immediately when alarm sounds.	Partial evacuation of employees	Some employees may be evacuated, as set out in the Emergency-Action and Fire Prevention Plan.
Employees who are authorized to use portable fire extinguishers to fight fires	Not required	Designated employees only; all others must evacuate immediately when alarm sounds	Designated employee
Employee training in accordance with Emergency-Action and Fire Prevention Plan	Required	Required	Required
Care of fire extinguishers in workplace	Any portable fire extinguishers in the workplace must be regularly inspected, tested, and maintained.	Meet all general fire extinguisher requirements. Regularly inspect, test, and maintain all portable fire extinguishers in the workplace.	Meet all general fire extinguisher requirements. Regularly inspect, test, and maintain all portable fire extinguishers in the workplace.
Additional employee firefighting training requirements	Not required	Designated employees must be trained annually on the use of portable fire extinguishers.	Designated employees must be trained annually on the use of portable fire extinguishers.

Alarm Sounds

A wide range of sounds can be used as audible alarms—continuous tones, tones that sweep from low to high or vice versa, whooping tones, warbles, chimes, voice commands, etc. The sounds can be produced by various devices such as bells, horns, sirens, and loudspeaker systems. Alarm sounds are user-selected; that is, there's no one universal set of alarm sounds. The same sound could mean different things from one job site to another. This is why it is critical that everyone on a given job site be trained to identify and correctly respond to the specific alarm sounds used on that site.

One alarm sound that is virtually universal for all construction sites is the sound produced by a backup beeper. 29 *CFR* 1926.601(b)(4) specifies that vehicles with obstructed rear views must have either a reverse signal alarm audible above the surrounding noise level, or a human spotter. Backup beepers usually produce a single-tone backup alarm with a typical volume of 97–112 decibels (dB) at the source.

The *National Fire Alarm and Signaling Code* (*NFPA 72®*), only permits bells, horns, and sirens in existing fire alarm systems. New systems must use temporal (time-coded) signals or voices. Temporal signals are accomplished by interrupting a steady sound in an On/Off pattern for specific intervals in a repeating cycle.

Likewise, *American National Standard Institute (ANSI) Standard S.341, Audible Emergency Evacuation Signal*, recommends a standard temporal signal pattern for notifying personnel of the need to evacuate a site. The signal consists of three successive On phases, lasting 0.5 second each, separated by 0.5 second of Off time. At the completion of the third On phase, there must be 1.5 seconds of Off time before the full cycle is repeated. The total cycle lasts 4.0 seconds. This signal is reserved, however, for total evacuation situations. A different signal pattern should be designated for emergencies that require relocating to a safe area or sheltering in place.

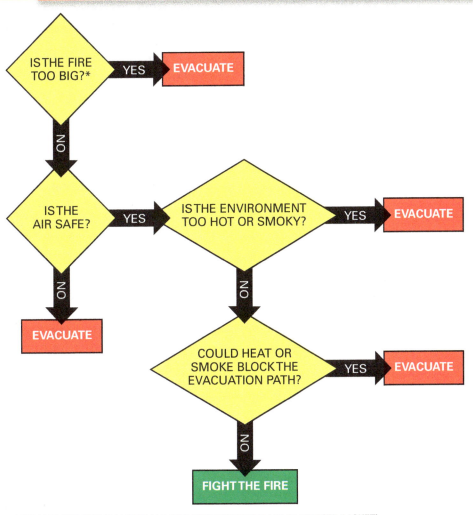

*SPREADING TO OTHER MATERIALS AND NOT CONTAINED (AS IN A WASTE BASKET)

Figure 4 Firefighting risk assessment.

In addition to the specialized training for emergency response team members, all employees should be trained in the following:

- Evacuation plans
- Alarm systems
- Reporting procedures for personnel
- Shutdown procedures
- Types of potential emergencies

These training programs should be provided as follows:

- Initially when the plan is developed
- For all new employees
- When new equipment, materials, or processes are introduced
- When procedures have been updated or revised
- When exercises show that employee performance must be improved
- At least annually

The emergency control procedures should be written in concise terms and made available to all personnel. A drill should be held at random intervals at least annually, and an evaluation of performance made immediately by management and all employees. When possible, drills should include groups supplying outside services, such as fire and police departments. On jobs with multiple contractors, the emergency plans should be overseen by the general contractor or construction manager and should be coordinated with the other contractors and employees on site. The emergency plan should be reviewed at least annually and updated to maintain adequate personnel response and program efficiency.

> **NOTE**
> On small or short-term jobs, drills may not be practical. However, it is a good idea to have local law enforcement and fire department personnel visit your site early in the project for pre-planning purposes. General evacuation procedures should be outlined during site safety orientation.

2.1.6 Personal Protection

Effective personal protection is required for any person who may be exposed to potentially hazardous substances. In emergency situations, employees may be exposed to a wide variety of hazards, including the following:

- Chemical splashes or contact with toxic materials
- Falling objects and flying particles
- Falls from elevations
- Unknown atmospheres that may contain toxic gases, vapors or mists, or inadequate oxygen to sustain life
- Fires and electrical hazards

It is extremely important that employees be adequately protected in these situations. Some of the safety equipment that may be used includes the following (*Figure 5*):

- Safety glasses, goggles, or face shields for eye protection
- Hard hats and safety shoes for head and foot protection
- Respiratory protection for breathing protection
- Chemical-resistant suits, gloves, hoods, and boots for body protection from chemicals
- Body protection for abnormal environmental conditions, such as extreme temperatures
- Personal fall-arrest systems

The equipment selected must meet or exceed the standards of the appropriate governing body. Governing bodies include the American National Standards Institute (ANSI) standards cited by OSHA, Mine Safety and Health Administration (MSHA), and the National Institute for Occupational Safety and Health (NIOSH). The choice of proper equipment is not a simple matter; health and safety professionals should be consulted before making any purchases. Manufacturers and distributors of health and safety products may be able to answer questions if they are given enough information about the potential hazards involved.

Respirators – A consultation with a qualified safety professional/industrial hygienist may be needed in order to select adequate respiratory protection for the concentration level and type of contaminant (based on air monitoring tests and historical data, if necessary). The equipment manufacturer can also supply information on equipment selection and fit tests. Respiratory protection is necessary for toxic atmospheres of dusts, mists, gases, or vapors, and for oxygen-deficient atmospheres. There are four basic categories of respirators (*Figure 6*):

- Air-purifying devices such as filters, gas masks, and chemical cartridges remove contaminants from the air but cannot be used in oxygen-deficient atmospheres.
- Air-supplied respirators should not be used in atmospheres that are **immediately dangerous to life or health (IDLH)** unless operated in the positive pressure mode and equipped with an escape bottle.

Figure 5 Personal protective equipment.

- A self-contained breathing apparatus (SCBA) is required for unknown atmospheres, oxygen-deficient atmospheres, or atmospheres immediately dangerous to life or health.
- Escape masks are temporary devices used to make an emergency exit from an unsafe atmosphere.

Respirators must be selected based on the type of expected hazard and the required protection factor. *Table 2* lists types of respirators and their assigned protection factors.

Before assigning or using respiratory equipment, the following conditions must be met:

- A medical evaluation must be made to determine if the employee is physically able to use the respirator.
- Written procedures (*Figure 7*) must be prepared covering the safe use and proper care of the equipment. Employees must be trained in these procedures and in the use and maintenance of the respirators.
- A fit test must be made to ensure a proper match between the facepiece of the respirator and the face of the wearer. This testing must be repeated annually or as needed. Training must provide the employee with an opportunity to handle the respirator, have it fitted properly, test its facepiece-to-face seal, wear it in normal air for a familiarity period, and wear it in a test atmosphere.
- A regular maintenance program must be instituted including cleaning, inspecting, and testing of all respiratory equipment. Respirators used for emergency response must be inspected after each use and at least monthly to assure that they are in satisfactory working condition. A written record of inspection must be maintained.
- Distribution areas for equipment used in emergencies must be readily accessible to employees.

(A) SELF-CONTAINED BREATHING APPARATUS (SCBA)

(B) SUPPLIED AIR MASK

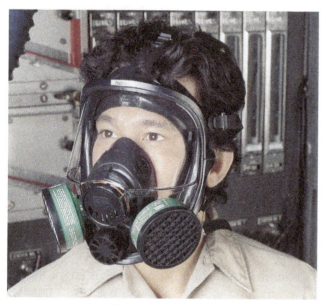
(C) FULL FACEPIECE MASK

(D) HALF MASK

Figure 6 Four types of respirators.

Table 2. Particulate Respirators

Assigned Protection[1] Factor	Type of Respirator
5	Quarter mask respirator
10	Any air-purifying elastomeric half-mask respirator equipped with appropriate type of particulate filter.[2]
	Appropriate filtering facepiece respirator.[2,3]
	Any air-purifying full facepiece respirator equipped with appropriate type of particulate filter.[2]
	Any negative pressure (demand) supplied-air respirator equipped with a half-mask.
25	Any powered air-purifying respirator equipped with a hood or helmet and a high efficiency (HEPA) filter.
	Any continuous flow supplied-air respirator equipped with a hood or helmet.
50	Any air-purifying full facepiece respirator equipped with N-100, R-100, or P-100 filter(s).
	Any powered air-purifying respirator equipped with a tight-fitting facepiece (half or full facepiece) and a high-efficiency filter.
	Any negative pressure (demand) supplied-air respirator equipped with a full facepiece.
	Any continuous flow supplied-air respirator equipped with a tightfitting facepiece (half or full facepiece).
	Any negative pressure (demand) self-contained respirator equipped with a full facepiece.
1,000	Any pressure-demand supplied-air respirator equipped with a half mask.
2,000	Any pressure-demand supplied-air respirator equipped with a full facepiece.
10,000	Any pressure-demand supplied-air respirator equipped with a full facepiece.
	Any pressure-demand supplied-air respirator equipped with a full facepiece in combination with an auxiliary pressure-demand self-contained breathing apparatus.

[1] The protection offered by a given respirator is contingent upon (1) the respirator user adhering to complete program requirements (such as the ones required by 29 *CFR* 1910.134), (2) the use of NIOSH-certified respirators in their approved configuration, and (3) individual fit testing to rule out those respirators that cannot achieve a good fit on individual workers.

[2] Appropriate means that the filter medium will provide protection against the particulate in question.

[3] An assigned protection factor of 10 can only be achieved if the respirator is qualitatively or quantitatively fit tested on individual workers.

BEFORE USING A RESPIRATOR YOU MUST DETERMINE THE FOLLOWING:

1. WHETHER THE INDIVIDUAL HAS BEEN MEDICALLY CLEARED FOR RESPIRATOR USE
2. THE TYPE OF CONTAMINANT(S) FOR WHICH THE RESPIRATOR IS BEING SELECTED
3. THE CONCENTRATION LEVEL OF THAT CONTAMINANT AND THE PROTECTION FACTOR OF THE RESPIRATOR
4. WHETHER THE RESPIRATOR CAN BE PROPERLY FITTED ON THE WEARER'S FACE

YOU MUST READ AND UNDERSTAND ALL RESPIRATOR INSTRUCTIONS, WARNINGS, AND USE LIMITATIONS CONTAINED ON EACH PACKAGE BEFORE USE.

Figure 7 Procedures to follow before using a respirator.

A SCBA offers the best protection to employees involved in controlling emergency situations. It should have a minimum service life rating of 30 minutes. Conditions that require an SCBA include the following:

- Leaking cylinders or containers, smoke from chemical fires, or chemical spills that indicate high potential for exposure to toxic substances
- Atmospheres with unknown contaminants or unknown contaminant concentrations, confined spaces that may contain toxic substances, or oxygen-deficient atmospheres

Confined spaces – Confined spaces include tanks, vaults, pits, sewers, pipelines, vessels, crawl spaces, and similar settings. Entry into confined spaces can expose workers to a variety of hazards, including toxic gases, flammable atmospheres, reduced atmospheric pressures, oxygen deficiency, electrical hazards, and hazards created by mixers and impellers that have not been deactivated and locked out. Personnel should never enter a confined space under normal circumstances unless the atmosphere has been tested for adequate oxygen, combustibility, and toxic substances. Conditions in a confined space must be considered immediately dangerous to life and health unless proven otherwise. A competent employee is required to evaluate each confined space for expected hazards. Continuous monitoring is required; the suspension of a permit may result due to changing entry conditions.

Emergency situations may involve entering confined spaces to rescue employees who are overcome by hazardous atmospheres. If a confined space must be entered in an emergency, these precautions must be followed:

- All lines containing inert, toxic, flammable, or corrosive materials must be disconnected or isolated before entry.
- All impellers, agitators, or other moving equipment inside the vessel or confined space must be locked out.
- Employees must wear appropriate personal protective equipment when entering a confined space.
- Rescue procedures must be specifically designed for each entry. When there is an atmosphere immediately dangerous to life or health, or a situation that has the potential for causing injury or illness to an unprotected worker, a trained confined space attendant must be present. The attendant must maintain communication with all workers within the confined space and be prepared to summon rescue personnel if necessary. The confined space attendant cannot enter the confined space. Instead, this person should assist workers leaving the space, including the use of hoisting equipment to retrieve the entrants from outside the space. This is called an external rescue.
- First responders must give advance notice if they are called out to another emergency and are unavailable.
- All rescue team members must be CPR/first aid certified.

The following resources provide more complete descriptions of procedures to follow while working in confined spaces:

- 29 *CFR* 1926 Subpart AA, *Confined Spaces in Construction*
- *NIOSH, Publication Number 80-106, Criteria for a Recommended Standard: Working in Confined Spaces* (September 1979)
- *NIOSH, Publication Number 87-113, A Guide to Safety in Confined Spaces* (July 1987)

> **NOTE:** The subject of confined spaces is also covered in NCCER Module 75224, *Permits and Policies*.

2.1.7 Medical Assistance

In a major emergency, time is a critical factor if injuries are to be minimized. Most contractors do not have a formal medical program, but they are required to have the following medical and first aid services:

- In the absence of an infirmary, clinic, or hospital in close proximity to the workplace that can be used for the treatment of any injured employee, the employer must ensure that personnel are adequately trained to render first aid.
- Where the eyes or body of any employee may be exposed to harmful corrosive materials, eye washes or suitable equipment for quick drenching or flushing must be provided in the work area for immediate emergency use. Employees must be trained to use the equipment.
- The employer must ensure the ready availability of medical personnel for advice and consultation on matters of employee health. This does not mean that health care must be provided, but rather that if health problems develop in the workplace, medical help will be available to resolve them.

To fulfill these requirements, the following actions should be considered:

- Survey the medical facilities near the job site and make arrangements to handle routine and emergency cases. A written emergency medical procedure must be prepared for handling incidents with a minimum of confusion.
- If the job is located far from medical facilities, at least one (and preferably more than one) employee on each shift must be adequately trained to render first aid. The American Red Cross, some insurance carriers, local safety councils, fire departments, and others may be contacted for this training.
- Inform all personnel about the hazards of blood-borne pathogens and the need to follow universal precautions when exposed to blood or other bodily fluids.
- First aid supplies should be provided for emergency use. This equipment should be ordered through consultation with a physician.
- Emergency phone numbers must be posted in conspicuous places near or on telephones and also provided to all employees. On the emergency phone list, post the physical address for the job site with cross streets and roads and/or GPS coordinates.
- Sufficient ambulance service should be available to handle any emergency. This requires advance contact with ambulance services to ensure they become familiar with your location, access routes, and hospital locations.
- Every job site should have at least one AED and everyone on the site must know where it is located. Safety technicians are responsible for maintaining and regularly inspecting AEDs.

2.1.8 Security

During an emergency, it is often necessary to secure the area in order to prevent unauthorized access and protect vital records and equipment. An off-limits area must be established using barricades and signs. It may be necessary to notify local law enforcement personnel or on-site personnel (owner or host employer) to secure the area and prevent the entry of unauthorized personnel.

Certain records also may need to be protected, such as essential accounting files, legal documents, and lists of employees' relatives to be notified in case of emergency. These records should be stored in duplicate outside the plant/job site and in protected secure locations on site.

2.2.0 Pre-Planning for Specific Types of Emergencies

Pre-planning is an essential part of an emergency action plan. It helps to ensure that all personnel know what to do during specific types of emergencies. Some of the emergencies that can occur on a site include the following:

- Trapped workers
- Severe weather
- External threats
- Fire emergencies

2.2.1 Trapped Workers

Rescuing trapped workers requires specially trained and equipped personnel. Some examples of the types of conditions and work locations in which workers can get trapped include the following:

- Extreme heights (scaffolds, structural steel, upper floors)
- Confined spaces (sewers, manholes, process vessels, tanks, etc.)
- Trench cave-ins
- Equipment in contact with energized power lines
- Structural collapses
- Explosions
- Extreme weather

Did You Know?
AEDs Can Save Lives During Cardiac Emergencies

AEDs improve the odds of survival in the event of an out-of-hospital cardiac arrest. Having an AED nearby allows for administration of time-critical treatment. Providing defibrillation as quickly as possible greatly improves the victim's chance of survival. Having an AED appropriately located in a business or workplace improves the survival rate of people experiencing a cardiac crisis.

Why should employers make AEDs available to employees?

There are 300,000–400,000 deaths per year in the United States from cardiac arrest. Most cardiac arrest deaths occur outside the hospital. Current survival rates are 10.6 percent if not witnessed, but 31.4 percent if witnessed and action is taken immediately. Jobs with shift work, high stress, and exposure to certain chemicals and electrical hazards increase the risks of heart disease and cardiac arrest.

What causes cardiac arrest, and how does an AED improve survival rates?

Abnormal heart rhythms, with ventricular fibrillation (VF) being the most common, cause cardiac arrest. Treatment of VF with immediate electronic defibrillation can increase the number of victims surviving to more than 90 percent. With each minute of delay in defibrillation, 10 percent fewer victims survive.

Is AED equipment expensive?

The average initial cost for an AED currently ranges from $1,000 to $2,000.

Are AEDs difficult to use?

AEDs are easy to use and provide clear audio and visual instructions. In a mock cardiac arrest, untrained sixth-grade children were able to use AEDs without difficulty.

The Bottom Line: AEDs are effective, easy to use, and relatively inexpensive.

Source: Occupational Safety and Health Administration (OSHA) and the American Heart Association

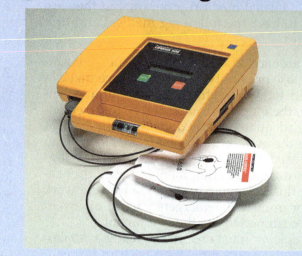

Many contractors erroneously assume that the local fire department can handle all of these emergencies. Many times, they cannot. If your job could conceivably involve any of these types of emergencies, it is critical that plans be worked out in advance. Find out what the local first responders can and cannot do and develop your plans accordingly.

2.2.2 Severe Weather

An emergency action plan should cover all severe weather conditions that are likely to affect your job. These can encompass a wide range of conditions, including extreme heat and cold, thunderstorms, lightning, tornadoes, hurricanes, flooding, mudslides, blizzards, earthquakes, and tsunamis.

If your job is in an area where major thunderstorms or tornadoes often occur, your plan must include a means of notifying all site personnel when there is a storm approaching. You will also need to identify suitable shelter locations. On very remote sites, a weather alert radio may be needed.

2.2.3 External Threats

Don't overlook the potential for bomb threats, civil disturbances, terrorist activity, and protests in the vicinity of your job site. On jobs with labor unrest, bomb threats often occur. Make sure your emergency action plan addresses these issues. Failure to do so can result in disorder on the site and potential injuries or deaths.

2.2.4 Fire Emergencies

Fires are one of the more common types of emergencies on construction projects. Failure to prepare for fires can result in catastrophic consequences. As covered in 29 *CFR* Subpart F, the OSHA construction standards call for a fire protection and prevention plan throughout all phases of construction, repair, alteration, or demolition work. Fire protection and suppression equipment is required. An adequate number of portable fire extinguishers are mandated. Fire extinguishers can be supplemented with fire hoses, but training is necessary if personnel are expected to use the equipment.

Some key OSHA requirements for fire protection at construction sites are identified as follows:

- Access to all available firefighting equipment shall be maintained at all times.
- All firefighting equipment shall be regularly inspected and maintained in operating condition. Defective equipment shall be immediately replaced.
- As warranted by the project, the employer may be required to provide a trained and equipped firefighting organization (Fire Brigade) to ensure adequate protection of life.
- A temporary or permanent water supply of sufficient volume, duration, and pressure required to properly operate the firefighting equipment shall be made available as soon as combustible materials accumulate.
- Where underground water mains are to be provided, they shall be installed, completed, and made available for use as soon as practicable.
- A fire extinguisher rated not less than 2A shall be provided for each 3,000 square feet (279 square meters) of the protected building area, or major fraction thereof. Travel distance from any point of the protected area to the nearest fire extinguisher shall not exceed 100 feet (30 meters). The location of the extinguisher should be clearly marked.
- One or more fire extinguishers rated not less than 2A shall be provided on each floor. In multistory buildings, at least one fire extinguisher shall be located adjacent to stairways.
- A fire extinguisher rated not less than 10B shall be provided within 50 feet (15 meters) of wherever more than five gallons (19 liters) of flammable or combustible liquids or five pounds (2.3 kilograms) of flammable gas are being used on the job site.
- Portable fire extinguishers shall be inspected periodically and maintained in accordance with *Standard for Portable Fire Extinguishers, NFPA No. 10-2013*.
- During demolition involving combustible materials, charged hose lines supplied by hydrants, water tank trucks with pumps, or equivalent shall be made available.
- If the facility being constructed includes the installation of automatic sprinkler protection, the installation shall closely follow the construction and be placed in service as soon as applicable laws permit following completion of each story.
- During demolition or alterations, existing automatic sprinkler installations shall be retained in service as long as reasonable.
- In all structures in which standpipes are required, or where standpipes exist in structures being altered, they shall be brought up as soon as applicable laws permit and shall be maintained as construction progresses in such a manner that they are always ready for fire protection use. The standpipes shall be provided with Siamese fire department connections on the outside of the structure at the street level, which shall be conspicuously marked (*Figure 8*). There shall be at least one standard hose outlet at each floor.
- An alarm system shall be established by the employer whereby employees on the site and the local fire department can be alerted for an emergency.
- The alarm code and reporting instructions shall be conspicuously posted throughout the job site.

Figure 8 Standpipe with Siamese fire department connections.

2.3.0 Dealing with the Media

In the immediate aftermath of an emergency or incident, it is likely that the news media will try to get the story for newspapers, television reports, and online media. They will want the following information:

- What happened?
- Was anyone hurt?
- Why did it happen?
- How will the community be affected?
- How is the construction company going to fix it?

As the safety technician, you may not have the answers to all of these questions immediately. If this is the case, it is acceptable to tell the media that an investigation is currently underway, and the appropriate company representative will report the findings as soon as they are known.

Employees should be restricted from posting information about the situation to social media. Make certain to provide a written policy and have all employees sign it. Only authorized personnel should communicate with the media. Depending on the circumstances, such information could become part of a criminal investigation governed by law enforcement or other authorities.

Additional guidelines for dealing with the media include the following:

- Know your company policy. Only provide information that you are allowed to give. If there is a company spokesperson, politely advise the press to speak to that person.
- Express concern for the safety and well-being of any injured personnel and their families.
- Be prepared if you are to provide information.
- Provide only the facts; do not lie or offer your opinion.
- Say you don't know when you don't know.
- Be polite and helpful.
- Maintain control; don't answer leading questions.
- Be professional; don't make jokes.
- Never make off-the-record statements.
- Never release the names of workers involved.
- Avoid discussion of legal questions.
- Do not accept responsibility or admit liability for the incident.
- Stress your company's commitment to safety.
- Keep reporters out of hazardous areas.
- Keep track of the persons you talked to.
- End the interview when you have covered the subject.

Additional Resources

American Society of Safety Engineers, **www.asse.org**

Occupational Safety and Health Organization, **www.osha.gov**

29 *CFR* 1926 Subpart AA, *Confined Spaces*

29 *CFR* 1926 Subpart F, *Fire Protection and Prevention*

NIOSH, Publication Number 80-106, Criteria for a Recommended Standard: Working in Confined Spaces, September 1979.

NIOSH, Publication Number 87-113, A Guide to Safety in Confined Spaces, July 1987.

2.0.0 Section Review

1. The first line of defense in an emergency is _____.
   a. the contractor
   b. local law enforcement
   c. the emergency response team
   d. site security

2. In areas where major thunderstorms or tornadoes often occur, the emergency action plan should describe how to notify all site personnel and _____.
   a. how to contact first responders
   b. provide guidelines for preventing unauthorized access
   c. state who is in charge of community and media relations
   d. identify suitable shelter locations

3. When dealing with the media after an emergency or incident occurs, _____.
   a. answer all questions immediately
   b. use social media to dispel rumors
   c. do not offer your opinions
   d. never say you don't know

Section Three

3.0.0 The Site Safety Plan

Objective

Identify the basic components of a safety program.

a. Identify essential safety program policies and procedures.
b. Identify effective practices for providing safety orientation and training.

Trade Term

Process Safety Management (PSM): The Process Safety Management of Highly Hazardous Chemicals, 29 CFR 1910.119. An OSHA standard that covers certain chemical plants and refineries.

A site-specific safety plan includes safety procedures to address conditions at a particular job site. It is usually prepared by the safety manager and implemented with assistance from the safety technician. It can be a single document or a collection of documents and references. *ANSI A10-38-200* sets the standard for the elements that a good site safety plan must have.

The plan must be sufficiently detailed to cover the anticipated hazards on the job site. Each site has unique features that create safety challenges. Remember that a job can be done in several phases, each with its own particular safety hazards.

A site-specific safety plan helps to create a safe work environment. It demonstrates a commitment to safety to employees, clients, and government inspectors. Safety briefings are more effective when they address specific conditions at a particular job site. A clear plan will list safety issues that the safety technician can address at the appropriate time.

In some situations, a site safety plan is mandatory. A general contractor may require site-specific safety plans from subcontractors. A site safety plan may also be required when working in plants covered by OSHA's **Process Safety Management (PSM)** rules.

3.1.0 Site Location and Layout

The site safety plan must take into consideration the site location and layout. It is important to know whether the area is rural, residential, commercial, or industrial. Each of these areas present different safety concerns. Pedestrian and vehicular traffic are of greater concern in urban or commercial areas. In rural areas, emergency services may be limited or distant.

Site location also affects site security. You need to know what safeguards are needed to protect the public, the client's employees, and other contractors from hazards on the job site. Likewise, you need to be aware of other contractors or client activities that will affect the safety of your personnel.

Material and tool storage is another feature of site security. Storage areas should be safe as well as secure. Access must be controlled both for safety and to reduce theft. Loading areas are high-hazard areas. You need to consider site access and material storage areas to minimize hazards from traffic and loading operations. Storage of chemicals and flammables needs special consideration; this includes on-site vehicle fueling areas.

3.1.1 Traffic Patterns

In addition to loading areas, the general traffic patterns are an important part of site safety. Consider where employees will park and how vehicles and equipment will move through the site (*Figure 9*). Moving vehicles are a major safety hazard. Traffic

Figure 9 Excavation site traffic.

control is the most important feature of site layout. If possible, equipment and vehicles should move through the site in a one-way traffic pattern to reduce the chance of vehicle collisions.

Site access is also a factor. Job sites in congested city areas present hazards not found on a rural site. Consider how vehicles and equipment will access the site. Public roads or sidewalks may need to be blocked. Permits are usually required. If you are on private property, the client's site may have restrictions which must be incorporated into the plan.

The placement of a field office and employee break areas also affects safety. These areas tend to have a lot of pedestrian traffic. Walking routes may need to be designated in addition to roadways.

3.1.2 Adjacent Hazards

The site location may include nearby hazards that will impact your site safety plan. These hazards can be natural or man-made. Significant slopes, water bodies, and other topological features must be considered. This is especially important during excavation and hazardous-materials handling. A hill leading down to a river can turn a minor chemical spill into a major environmental problem.

Utilities can be a major safety issue. Ensure that the locations of all utilities are marked (*Figure 10*). If possible, the site safety plan should include contact information for all major utilities, including the following:

- Water
- Electricity
- Gas
- Sewer
- Telephone and fiber-optic lines

All activities on the client's site or adjacent areas can affect site safety. Therefore, you need to be aware of area hazards and consider interactions in both directions. How will your actions affect existing operations? How will existing operations affect your work? Construction activities can create noise, air, water, or ground pollution problems. These hazards must be minimized.

You will also need to determine whether the job site is subject to environmental concerns. Are there waterways that must be protected? Will a National Pollutant Discharge Elimination System (NPDES) permit be required? Will your work activities adversely affect any legally protected wildlife species or their habitats?

3.1.3 Scope of Work

The next significant consideration is the scope of work. It's important to consider the safety precautions for different aspects of the project. The work may be done in phases with very different safety considerations for each phase. Safety must be carefully planned for high-hazard work, which can include the following:

- Blasting
- Excavation
- Demolition
- Work at extreme heights
- Work in confined spaces
- Work over water
- Handling asbestos, chemicals, hazardous waste, or other regulated substances

Generally, high-hazard work is a smaller portion of a larger project. The timing of the various phases of the work plan must be included in the site safety plan. These operations may require

Figure 10 Standard underground markers.

Case History

Laborer Killed in Fall through Roof

An employer was demolishing the roof of a warehouse. Work was done at night because the coal tar would release hazardous gases if disturbed during the day. The site had adequate lighting. None of the workers on the job were using fall protection.

After the roofing material was removed, 4' × 8' (1.2 m × 2.4 m) sheets of plywood were exposed. Workers were replacing damaged plywood. The helper's job was to follow the workers, pick up the damaged sheets, and dispose of them in a chute.

One worker removed a sheet of damaged plywood but had run out of nails to attach the new plywood. He walked away to get more nails and left the opening where the damaged plywood had been removed unguarded. The crew was not informed that the opening was temporarily unguarded. The opening was covered by silver-colored insulation inside the roof.

The helper came along, picked up the sheet of damaged plywood, and headed for the chute. He stepped into the opening, ripped through the insulation, and fell approximately 27' (8 m) to the floor below and died.

The Bottom Line: Fall protection was not used because an event like this may have been considered unlikely to happen. The result was catastrophic. Use a risk assessment matrix to prioritize risks and minimize hazards.

---

special tools, equipment, and permits. Training, special precautions, and inspections should be scheduled in parallel with the work plan.

Some special tools or equipment may be needed at various times during the job. Fall-protection equipment, intrinsically safe electrical equipment, and air handling or testing equipment are a few examples. These should be specified in the site safety plan.

3.1.4 Personnel Health and Safety

The next phases of site safety planning combine site as well as task considerations. Safety technicians must identify all the safety equipment needed, similar to a foreman preparing a materials and tools breakout. Include both standard and specialized safety equipment. Standard safety equipment generally includes the following:

- Hard hats
- Safety glasses
- Gloves
- Hearing protection
- Boots

Specialized equipment may include the following items:

- Fire-retardant clothing
- Chemical-resistant gloves or aprons
- Safety harnesses and fall-protection gear
- Debris and personnel nets
- Respiratory equipment

The site safety plan should identify who will supply the personal protective equipment (PPE). Some clients may provide this equipment, but extra supplies may be needed. The site safety plan should specify where specialized PPE, such as hazmat suits, would be required. OSHA now requires employers to provide basic job-specific PPE. Employees may have to purchase other standard PPE, such as boots and fire-retardant clothing. PPE storage locations need to be designated. The plan may include a system to issue gear and restock disposable supplies.

In addition to personal protective equipment, consider basic sanitary needs. If toilet facilities and drinking water are not available on site, they should be included in the site safety plan. Dehydration is a common hazard on job sites. Shower, changing, or decontamination areas may be needed for the duration or only for a specific phase of the project.

3.1.5 Safety Coordination

The site location and scope of work will provide a solid foundation for a site safety plan. In order to be effective, you must coordinate your plan with other parties, including the following:

- The client
- Other contractors
- Emergency services

If the project is on an existing industrial site, the client will have a health and safety program. If several contractors are working on the site or if subcontractors are hired, you must coordinate your plans. Contractors must be aware of the client's site safety plan. The client's employees must be informed of new procedures that will affect their work.

Many of the site safety features must be coordinated with other parties. Coordination is essential for the following areas:

- Site security
- Vehicle and pedestrian traffic
- Communications
- Chemical handing
- Waste disposal
- Emergency response
- Work permit systems
- Lockout/tagout procedures
- Critical lifts
- Site-specific orientation and training

OSHA's policy on multi-employer worksites applies and holds the GC or CM responsible. Your firm can be held responsible if other employees or subcontractors are harmed by your actions. The site is only as safe as the least-safe worker.

Emergency procedures – Coordination is especially important in emergency response. A serious incident or fire can affect the entire site. All safety personnel must work together in an emergency. Procedures will need to be tailored to site conditions.

The site safety plan must be coordinated with other parties. Often the most difficult aspect of an emergency is coordination and communication. All parties must understand each other's programs. You should consider drills for long-term projects. It is also a good idea to invite fire or emergency responders most likely to respond to an emergency on your site to tour the project on a regular basis. This has been shown to dramatically improve response time.

All parties must understand general procedures for the following situations:

- Emergency lighting and loss of power
- Emergency reporting and notifications
- Emergency communication systems
- Security-related issues
- On-site medical facilities
- Off-site medical facilities
- Emergency assembly and evacuation routes
- Hazardous materials emergency response
- Fire procedures

It is imperative that all personnel understand any alarm systems. The site safety plan needs to identify any emergency alarms, buzzers, or bells that will be used.

Emergency contact numbers should be on a posted list. A more extensive list that includes key personnel for all firms can be included as well.

3.2.0 Administration

The site safety plan is of no effect sitting on a desk. It must be put into action through training, notifications, and inspections. The plan must be communicated to the workers in order to be effective.

A safety bulletin board is a common tool that can be used for communicating safety information. The board must include mandatory OSHA posters and notices as well as Department of Labor (DOL) posters. It can also include emergency information and the location of the safety plan and SDSs. Posting new information on a regular basis will increase its effectiveness. Posting notices of high-hazard activities or temporary safety precautions is a good way to encourage workers to check it regularly.

Safety meetings, audits, and inspections should be scheduled consistent with the work plan. The results of audits or inspections can also be posted to encourage good performance.

The site safety plan must also include any mandatory reporting requirements. These include safety incidents, near misses, and reports to regulatory agencies. The client or general contractor may also need to be notified. The contract should be reviewed for specific reporting requirements.

3.2.1 Training

The site safety plan must include a section on training. Training is required at several stages of employment, including initial, periodic, and task-specific training. Records must be kept of all training. Some clients will need copies of these training records.

At a minimum, the plan should include safety orientation for new employees. Current employees who are new to the job site should also receive a basic orientation.

Some clients or projects will require a medical monitoring program. Workers must be tested when they are hired, annually, and when they stop working for the firm. Any work-related medical evaluation should also be included. For example, if the worker received first aid for an injury on the job, there must be a record of this. There are strict requirements for maintaining medical records that vary by state. You need to be aware of your state's requirements.

Additional Resources

American Society of Safety Engineers, **www.asse.org**

U.S. Environmental Protection Agency, **www.epa.org**

Occupational Safety and Health Organization, **www.osha.gov**

3.0.0 Section Review

1. An example of standard safety equipment is _____.
   a. a respirator
   b. chemical-resistant gloves
   c. hearing protection
   d. debris nets

2. A safety bulletin board should include Department of Labor (DOL) Equal Employment Opportunity (EEO) posters as well as _____.
   a. mandatory OSHA posters and notices
   b. safety meeting agendas
   c. pre-bid checklist findings
   d. crew rotation schedules

Section Four

4.0.0 Regulatory Requirements

Objective

Identify regulatory requirements.
a. Identify Occupational Safety and Health Administration (OSHA) requirements.
b. Identify Environmental Protection Agency (EPA) requirements.
c. Identify Department of Transportation (DOT) requirements.
d. Identify other regulatory requirements.

Safety technicians have a responsibility to know which local, state, and federal requirements apply to the work being done on their site. Three major government agencies that provide regulatory requirements applicable to the construction industry are the following:

- Occupational Safety and Health Administration (OSHA)
- Environmental Protection Agency (EPA)
- Department of Transportation (DOT) Commercial Driver's License Program

Depending on the specific project, numerous other requirements may apply as well. Safety technicians must be familiar with all of the requirements and regulations that apply to the work being performed at their job site.

4.1.0 OSHA Requirements

Safety technicians are required to know and follow the regulations established by OSHA. OSHA offers several resources on standards and policies related to the construction industry on their website at **www.osha.gov**.

For additional information about OSHA and the services that this agency provides, contact the OSHA or State Plan Office in your area. The complete *OSHA Safety and Health Regulations for Construction* are located at **www.osha.gov** and covered in 29 *CFR* 1926. Take the time to learn which OSHA regulations apply to the work that is being done on your site. Failure to do so could mean serious injuries, loss of lives, or equipment loss or damage.

The Occupational Safety and Health Act also created the National Institute for Occupational Safety and Health (NIOSH). NIOSH is part of the Centers for Disease Control and Prevention (CDC) within the US Department of Health and Human Services (DHHS).

Unlike OSHA, NIOSH does not issue standards that are enforceable under US law. Rather, NIOSH's mandate is to develop recommendations for health and safety standards, in addition to conducting research and providing information, education, and training to help ensure safe and healthful working conditions. NIOSH may also conduct on-site investigations to determine the toxicity of materials used in workplaces, and to test and certify personal protective equipment and hazard measurement instruments.

Two federal laws regulate asbestos. The general program is the Asbestos Hazard Emergency Response Act (AHERA). Because asbestos exposure is more dangerous for children than adults.n additional law was created to deal with asbestos removal in school buildings, the Asbestos School Hazard Abatement Act (ASHAA). The OSHA Construction Standard for asbestos is 29 *CFR* 1926.1101. It covers all phases of asbestos handling and removal, including demolition, removal, alteration, repair, maintenance, installation, cleanup, transportation, disposal, and storage.

Firms covered under OSHA's Process Safety Management (PSM) regulations also have special requirements for site safety (*Figure 11*). *The Process Safety Management of Highly Hazardous Chemicals* standard can be found at 29 *CFR* 1910.119. Generally, these regulations cover industrial facilities that handle large quantities of chemicals listed as highly hazardous. Chemical processing plants and refineries are two examples of these facilities.\

Construction contractors who work in a PSM facility may require site-specific training. Certain procedures must be followed for contractors who

Figure 11 A plant covered under OSHA PSM rules.

Third-Party Safety Program Evaluators

Maintaining safety, insurance, quality, and regulatory information on contractors and suppliers takes time and money that can strain a company's internal resources. For this reason, some owners use third-party safety program evaluators, such as ISNetworld, PEC Premier, PICs, BROWZ, and I-Square-Foot, to name a few. The third-party evaluators provide services that standardize contractor management across multiple sites and geographic regions by communicating requirements and expectations to contractors and suppliers and exchanging data with other internal systems. Using third-party evaluators generally simplifies the pre-qualification process and may also improve safety standards for the client company.

Third-party evaluators typically collect self-reported information from contractors and suppliers and maintain this information in a centralized database. The evaluators review the information to verify that it meets the requirements of both the client and the relevant regulatory agencies. Then the evaluators are able to connect clients with suitable contractors and suppliers. Third-party evaluators may also provide safety training to contractor personnel.

work in or around hazardous processes. General requirements include the following:

- The host company must pre-screen all contractors.
- Contractors must verify that their employees are trained for the tasks they will perform.
- The contractors must inform the host company of hazards they may bring onto the site, including supplying any SDSs.
- The host employer must inform the contractors of any hazards near where they will be working.
- The host employer must control contractors' access to hazard areas.
- Contractors must be informed of site emergency response procedures.
- If an incident involving highly hazardous chemicals occurs, a contractor representative must participate in the incident investigation.

On a PSM site, the host employer and the contractor must work together to identify training needs. Records must be kept of training and plant safety orientation.

4.2.0 EPA Requirements

It is important to understand how work being done on a construction site affects the environment. This includes knowing the major environmental regulations covering the release, treatment, storage, and disposal of potentially hazardous materials into air, water, and soil. Safety technicians need to have a working knowledge of the following EPA laws:

- *Comprehensive Environmental Response, Compensation, and Liability Act (CERCLA)* – This law, commonly known as the Superfund, imposes liability on owners or operators of a facility from which there is a release of hazardous substances into the environment. In addition to Superfund actions, CERCLA requires that all spills must be reported. The specific procedures for spill response also depend on the chemical that has been spilled and how large the spill is. In any case, the best practice is to establish a program to report any quantity of spillage.
- *Resource Conservation and Recovery Act (RCRA)* – This law regulates the generation, transport, treatment, storage, and disposal of hazardous waste. RCRA requires the owner/operator of a facility to undertake corrective action to clean up a facility used for the treatment, storage, or disposal of hazardous waste. There is also a complex permit program with which the owner/operator must comply. The EPA rules for properly identifying waste are listed at 40 *CFR 261, Identification and Listing of Hazardous Waste.*
- *Underground storage tank regulations (under RCRA and state control)* – This law imposes operating, reporting, financial assurance, and potential cleanup obligations on persons and companies owning and operating underground petroleum storage tanks (USTs). The rules and regulations covering USTs are quite detailed and prescriptive. In addition, many of the states have their own regulations and enforcement policies.
- *Clean Air Act (CAA) of 1990* – This act is a complex, multi-faceted statute that is designed to regulate air emissions from stationary and mobile sources.
- *Clean Water Act (CWA)* – This act prohibits the discharges of pollutants from point sources and storm water into navigable waters of the United States without a permit. The act imposes liability on the person who is responsible for the operation and/or equipment that results in a discharge. Any discharge into a body of water must be permitted. These permits are

known as NPDES permits, for the National Pollutant Discharge Elimination System. The CWA also covers storm water run-off. Storm water that carries any pollutants from a construction site to a body of water requires a permit. There are national and state programs for storm water management. Construction firms must control storm water run-off in accordance with local laws. Construction sites that disturb more than one acre of land must obtain a NPDES permit. The permit requires that the site prepare a storm water pollution prevention plan. It is important to be aware of local storm water rules.

> **NOTE**
> EPA delegates the authority to implement certain regulatory programs to some states. A state may have more stringent environmental requirements than the federal requirements. Always check with your state and local agencies before starting a construction project. Allowable waste quantities and timelines vary from state to state and can result in large fines if not reported accurately.

- *Safe Drinking Water Act (SDWA)* – This act imposes federal drinking water standards on virtually all public water systems. The act requires the establishment of drinking water standards for maximum contaminant levels (MCLs).
- *Toxic Substances Control Act (TSCA)* – This act governs the manufacture and use of chemical products. The importing and exporting of chemicals are regulated under the act. In addition, there are specific regulations regarding the use, management, storage, and disposal of highly concentrated polychlorinated biphenyl (PCB) materials. You can find information about federal PCB regulations online at **www.epa.gov**. Contact your state Department of Environmental Conservation to learn how PCBs are regulated in your state.

> **Did You Know?**
> **Reforming the TSCA**
>
> When the TSCA was passed in 1976, there were about 62,000 toxic substances already in use in the United States. All of these substances were presumed to be safe, and Congress grandfathered them in without testing. Since 1976, roughly another 20,000 chemicals have come on the market, and very few have been tested. At the time of this writing, Congress is debating about reforming the TSCA to strengthen the regulation.

- *Oil Pollution Act of 1990* – This Act imposes strict liability on responsible parties for removal costs and damages resulting from discharges of oil into navigable waters of the United States. An owner/operator of an onshore facility that results in a discharge is considered a responsible party.
- *Federal Insecticide, Fungicide, and Rodenticide Pollution Act (FIFRA)* – This Act covers the use of pesticides in the United States. The EPA is responsible for controlling the use of pesticides and for ensuring that they are used properly and do not pose a threat to humans or wildlife. Make sure that only licensed professionals use pesticides on your site.
- *Land Disposal Restrictions (LDR) Program* (40 *CFR* 268) – The LDR program ensures that toxic constituents present in hazardous waste are properly treated before the waste is land disposed. That is, hazardous wastes must meet mandatory technology-based treatment standards before they can be placed in a landfill. Under 40 *CFR* 268.45, alternative treatment standards are provided for certain types of hazardous debris that are commonly generated in the construction industry—specifically, debris that is often generated when a building or structure undergoes demolition or renovation. Material generated by this type of work may be contaminated with or contain a hazardous waste. For example, scrap piping or tanks may have held hazardous waste and thus be classified as hazardous debris. Construction and demolition (C&D) landfills—which are generally a more affordable disposal solution than sanitary or commercial landfills—may be used for land disposal of this type of debris.

> **CAUTION**
> Many states have significantly reduced or eliminated permitted status for many C&D landfills, chiefly because of cross-contamination of C&D debris by hazardous substances. Check your state and local regulations.

- *Endangered Species Act (ESA)* – The US Fish and Wildlife Service (USFWS) and the National Marine Fisheries Service (NMFS), along with various state agencies, oversee compliance with this Act. The ESA requires that federally listed species and habitat not be adversely affected during any activity that has federal involvement or is subject to federal oversight. A related law, the Migratory Bird Treaty Act, protects migratory birds, their nests, and their eggs. Under this Act, construction in nesting areas can constitute a taking of the nest because the habitat has been destroyed. In addition, nesting predatory birds have special protections. Any nesting predator cannot be disturbed until the nest is abandoned. Prior to construction, consult your local USFWS office, the National Marine Fisheries Service, and your local conservation agency to determine whether your project could harm any endangered or threatened species—and if so, what to do about it. Absent any federal involvement or oversight, private landowners must still ensure that proposed development activities won't result in a take of any listed species, and they may need to develop a habitat conservation plan.

4.3.0 DOT Requirements

Safety technicians must ensure that any worker operating heavy equipment or other commercial vehicle has a current Commercial Driver's License (CDL). They must know the regulatory requirements established by the Department of Transportation's CDL Program and the standards set by the Commercial Motor Vehicle (CMV) Safety Act of 1986.

> **NOTE**
> For more information about CDL requirements, visit the DOT's Federal Motor Carrier Safety Administration (FMCSA) website at www.fmcsa.dot.gov.

DOT rules for shipping hazardous materials (49 *CFR* 170) apply to hazardous waste transportation. You must choose the proper DOT shipping name and hazard class. The containers must be labeled with both EPA and DOT labels and markings (*Figure 12*). Both DOT and EPA require a shipping document for shipping hazardous waste. A hazardous waste manifest is shown in *Figure 13*.

> **NOTE**
> Allowable waste quantities and timelines vary from state to state and can result in large fines if not reported in a timely manner.

4.4.0 Other Regulatory Requirements

Safety technicians need to be familiar with requirements and recommendations from other agencies and organizations that can apply to the construction industry. These include requirements from any of the following:

- US Mine Safety and Health Administration (MSHA)
- American National Standards Institute (ANSI)
- American Society of Mechanical Engineers (ASME)
- National Fire Protection Association (NFPA)
- US Department of Homeland Security (DHS)
- US Coast Guard (USCG)
- US Army Corps of Engineers (USACE)
- US Naval Facilities Engineering Command (NAVFAC)

Make sure you are familiar with all of these agencies and their regulations. It could cost your company a great deal of money in fines and lawsuits if you are not.

4.4.1 MSHA Requirements

The US Mine Safety and Health Act of 1977 established the Mine Safety and Health Administration (MSHA). The mission of MSHA is to ensure safe and healthful working conditions for miners and any other workers doing work at a mine, rock quarry, or mine material-processing facility. It is important for safety technicians to understand MSHA requirements because any construction or maintenance done at a mine or facility that

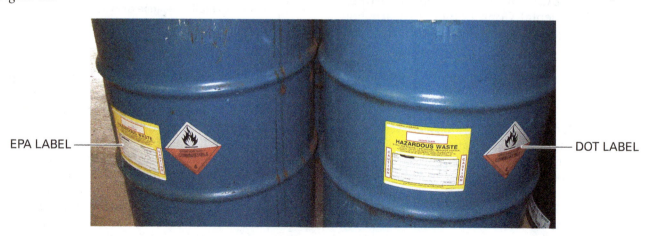

Figure 12 EPA and DOT labels.

Please print or type. (Form designed for use on elite (12-pitch) typewriter.) Form Approved. OMB No. 2050-0039

GENERATOR

UNIFORM HAZARDOUS WASTE MANIFEST	1. Generator ID Number	2. Page 1 of	3. Emergency Response Phone	**4. Manifest Tracking Number**

5. Generator's Name and Mailing Address — Generator's Site Address (if different than mailing address)

Generator's Phone:

6. Transporter 1 Company Name — U.S. EPA ID Number

7. Transporter 2 Company Name — U.S. EPA ID Number

8. Designated Facility Name and Site Address — U.S. EPA ID Number

Facility's Phone:

9a. HM	9b. U.S. DOT Description (including Proper Shipping Name, Hazard Class, ID Number, and Packing Group (if any))	10. Containers No.	10. Containers Type	11. Total Quantity	12. Unit Wt./Vol.	13. Waste Codes
	1.					
	2.					
	3.					
	4.					

14. Special Handling Instructions and Additional Information

15. **GENERATOR'S/OFFEROR'S CERTIFICATION:** I hereby declare that the contents of this consignment are fully and accurately described above by the proper shipping name, and are classified, packaged, marked and labeled/placarded, and are in all respects in proper condition for transport according to applicable international and national governmental regulations. If export shipment and I am the Primary Exporter, I certify that the contents of this consignment conform to the terms of the attached EPA Acknowledgment of Consent. I certify that the waste minimization statement identified in 40 CFR 262.27(a) (if I am a large quantity generator) or (b) (if I am a small quantity generator) is true.

Generator's/Offeror's Printed/Typed Name — Signature — Month Day Year

INT'L

16. International Shipments ☐ Import to U.S. ☐ Export from U.S. Port of entry/exit: ____

Transporter signature (for exports only): Date leaving U.S.:

TRANSPORTER

17. Transporter Acknowledgment of Receipt of Materials

Transporter 1 Printed/Typed Name — Signature — Month Day Year

Transporter 2 Printed/Typed Name — Signature — Month Day Year

DESIGNATED FACILITY

18. Discrepancy

18a. Discrepancy Indication Space ☐ Quantity ☐ Type ☐ Residue ☐ Partial Rejection ☐ Full Rejection

Manifest Reference Number:

18b. Alternate Facility (or Generator) — U.S. EPA ID Number

Facility's Phone:

18c. Signature of Alternate Facility (or Generator) — Month Day Year

19. Hazardous Waste Report Management Method Codes (i.e., codes for hazardous waste treatment, disposal, and recycling systems)

1.	2.	3.	4.

20. Designated Facility Owner or Operator: Certification of receipt of hazardous materials covered by the manifest except as noted in Item 18a

Printed/Typed Name — Signature — Month Day Year

EPA Form 8700-22 (Rev. 3-05) Previous editions are obsolete. DESIGNATED FACILITY TO DESTINATION STATE (IF REQUIRED)

VOID

Figure 13 Hazardous waste manifest.

handles mine products (such as portland cement, coal, gravel, and limestone) falls under MSHA regulations, not OSHA regulations.

Mine operators must comply with the safety and health standards enforced by MSHA, an agency within the Department of Labor (DOL). Like OSHA, MSHA may issue citations and propose penalties for violations. Unlike employees under OSHA, mine employees are subject to government sanctions for violating safety standards, such as smoking in or near mines or mining machinery. Similarly, employers and other supervisory personnel may be held personally liable for civil penalties or may be prosecuted criminally for violations of Mine Act standards.

> **NOTE**
> Mine Act standards are listed in 30 *CFR*, Parts 70, 71, 74, 75, 77, and 90. More information about MSHA regulations and procedures is available online at **www.msha.gov**.

4.4.2 ANSI and ASME Requirements

The American National Standards Institute (ANSI) and the American Society of Mechanical Engineers (ASME) are both private, non-profit organizations that promote uniformity in voluntary standards for engineering. Simply put, ANSI accredits standards developed by other organizations, while ASME actually develops codes and standards.

Although ANSI does not itself develop standards, it accredits standards that are developed by various standards organizations, government agencies, consumer groups, companies, etc. ANSI accreditation signifies that the procedures used by standards-developing organizations meet the Institute's requirements. Using accredited standards helps to ensure that the characteristics and performance of products are consistent, that people use the same definitions and terms, and that products are tested in the same way. ANSI also accredits organizations that perform product or personnel certification in accord with requirements defined in international standards. Additionally, ANSI designates specific standards as American National Standards, or ANS. There are approximately 9,500 American National Standards that carry the ANSI designation.

ASME focuses on developing codes and standards for mechanical devices. Its primary concern is delivering solutions to the practical challenges that engineering professionals encounter on the job. Committees of subject matter experts use an open, consensus-based process to develop ASME standards. ASME has produced approximately 600 codes and standards covering many technical areas, such as boilers and pressure vessels; elevators and escalators; piping and pipelines; flanges, fittings, and gaskets; and power plant systems and components. ASME's standards are used in more than 100 countries. ASME also has a large technical publishing operation, hosts numerous technical conferences and professional development courses every year, and sponsors many outreach and educational programs that promote engineering and allied sciences.

Government agencies cite many ASME standards as tools to meet their regulatory objectives. ASME standards are voluntary, however, unless the standards have been incorporated into a legally binding business contract or incorporated into regulations that are enforced by an authority that has jurisdiction, such as a federal, state, or local government agency.

> **NOTE**
> For more information about ANSI and ASME, visit the websites for these organizations at **www.ansi.org** and **www.asme.org**, respectively.

4.4.3 NFPA Requirements

The National Fire Protection Association (NFPA) is a non-profit US trade association that creates and maintains consensus-based standards and codes for preventing death, injury, and property losses due to fire, electrical, and related hazards. More than 300 ANSI-accredited NFPA codes and standards are available for adoption by local governments and are used throughout the United States, as well as in many other countries.

Some of the most widely used NFPA codes include the following:

- *NFPA 1 – Fire Code* (requirements for fire safety and property protection in new and existing buildings)
- *NFPA 70 – National Electrical Code®*
- *NFPA 70B – Recommended Practice for Electrical Equipment Maintenance*
- *NFPA 70E – Standard for Electrical Safety in the Workplace®*
- *NFPA 72 – National Fire Alarm and Signaling Code®*
- *NFPA 101 – Life Safety Code®* (requirements to protect occupants of new and existing buildings from fire, smoke, and toxic fumes)
- *NFPA 704 – Standard System for the Identification of the Hazards of Materials for Emergency Response* (the four-color diamond symbol for identifying the risks posed by hazardous materials, shown in *Figure 14*)

Figure 14 NFPA 704 diamond.

- NFPA 921 – *Guide for Fire and Explosion Investigations*
- NFPA 1670 – *Standard on Operations and Training for Technical Search and Rescue Incidents*
- NFPA 1901 – *Standard for Automotive Fire Apparatus*

> **NOTE**
> All NFPA codes and standards can be viewed (although not printed) free of charge at **www.nfpa.org**.

4.4.4 DHS and Coast Guard Requirements

The US Department of Homeland Security (DHS) is a cabinet department of the federal government. In contrast to the Department of Defense, which is charged with military actions abroad, the DHS operates in the civilian sector to protect the United States within, at, and outside its borders. DHS's mission is to prepare for, prevent, and respond to domestic emergencies, particularly terrorism.

Each component agency within DHS has a Designated Safety and Health Official (DSHO) who provides operational program management and oversight for safety and health programs, and develops policy, instructions, standards, requirements and metrics related to safety and health programs within the component.

The US Coast Guard (USCG) operates as a subcomponent of the DHS and is responsible for law enforcement, maritime security, national defense, maritime mobility, and the protection of natural resources. USCG health and safety requirements comply with or exceed applicable OSHA and other federal agency regulations, requirements, and standards. 29 *CFR* Part 1926, *Safety and Health Regulations for Construction*, applies to all USCG construction projects and sites.

USCG health and safety policies and requirements are covered in the *USCG Safety and Environmental Health Manual (USCG Commandant Instruction Manual 5100.47A)*. The manual covers a full range of standard health and safety topics (personal protective equipment, hazard communication, confined space entry, etc.) and outlines the notification, analysis, reporting, and recordkeeping requirements for a mishap analysis program.

> **NOTE**
> In USCG terminology, a mishap is essentially the same as an incident. Specifically, the USCG defines a mishap as any unplanned, unexpected, or undesirable event that causes injury, occupational illness, death, material loss, or damage. As explained in the *USCG Safety and Environmental Health Manual*, some mishaps are reportable; others are non-reportable. For more information, visit the Coast Guard website at **www.uscg.mil**.

4.4.5 USACE and NAVFAC Requirements

The US Army Corps of Engineers (USACE, sometimes shortened to CoE) is a federal agency within the Department of Defense. It is one of the world's largest public engineering, design, and construction management agencies. By providing security planning, force protection, research and development, disaster preparedness, and quick response to emergencies and disasters, the USACE supports both the US Department of Homeland Security and the Federal Emergency Management Agency (FEMA).

Safety and health requirements that apply to contractors working on USACE projects are published in the *Engineering Manual (EM) 385-1-1*. *EM 385-1-1* has been adopted by the US Army, Navy, and Air Force, as well as many independent contractors. Generally, EM regulations are in line with OSHA standards, although some EM regulations may be slightly more stringent than what is required by OSHA.

> **NOTE**
> A searchable, electronic version of the EM is available online through the USACE website at **www.usace.army.mil**.

In most cases, before beginning work on a USACE project, contractors must submit an Accident Prevention Plan (APP) that defines how they are going to manage their safety program under the contract. Other required elements within the APP, such as Activity Hazard Analyses (AHAs), are developed and added to the APP during the performance of the contract. AHAs are similar to JSAs/JHAs; they are based on the specific phases of work and the hazards involved and are developed by referencing the additional sections of the EM as needed. *EM 385-1-1* requires contractors to employ at least one competent person at each job site to manage, implement, and enforce the elements of their safety plans.

> **NOTE**
> An APP outline is provided in *EM 385-1-1, Appendix A*. Additional electronic tools, forms, templates, and checklists to assist contractors with implementing their health and safety programs are available through an HQ Safety Website link that can be found on the USACE website at www.usace.army.mil.

The Naval Facilities Engineering Command (NAVFAC) is the US Navy's engineering command. It manages the planning, design, and construction of shore facilities around the world for the US Navy and the Marine Corps. Safety and health requirements that apply to contractors working on NAVFAC projects are essentially the same as those for contractors working on USACE projects and are published in *EM 385-1-1*, as described above. Both USACE and NAVFAC require contractors to plan out each phase of a job before actually performing the work.

4.4.6 Training Requirements

Many regulations establish training requirements, which are listed in the regulations. Some programs have a set number of hours of training; for example the HAZWOPER (Hazardous Waste Operations) has a 40-hour initial course. Most training programs include annual or periodic refresher courses. It is important to keep accurate records of worker training. OSHA, insurance companies, and other government agents may inspect these training records.

There are many professional safety firms that offer training. These programs are available in a variety of formats, from standard lecture to online courses. Verify that the training meets the specific program requirements; some training may overlap.

There are different training needs, depending on job tasks. Most programs have several levels of training. The first level is for employees whose daily tasks are not directly related to any hazard. The second level is for those who may encounter a hazard. The third level is for those who work around the hazard every day. The final level is for those responsible for safety or emergency response.

The following are some of the training programs safety technicians in the construction industry may need to include in their site-specific safety programs:

- *Asbestos handlers* – 29 *CFR* 1926.1101
- *Confined space entry* – 29 *CFR* 1910.146
- *DOT HM-126* – Hazardous materials handling and transportation, 49 *CFR* 172
- *Hazard communication* – 29 *CFR* 1910.1200
- *HAZWOPER* – Hazardous Waste Operations and Emergency Response, 29 *CFR* 1910.120
- *Commercial pesticide applicators* – FIFRA laws managed by state agencies
- *Respiratory protection* – 29 *CFR* 1910.134

Additional Resources

29 *CFR*, Part 1926, *Safety and Health Regulations for Construction.*

30 *CFR*, Parts 48, 70, 71, 74, 75, 77, and 90, *Mine Act Standards.*

American National Standards Institute (ANSI), www.ansi.org

American Society of Mechanical Engineers (ASME), www.asme.org

American Society of Safety Engineers, www.asse.org

National Institute for Occupational Safety and Health (NIOSH), www.cdc.gov

Department of Homeland Security (DHS), www.dhs.gov

Department of Transportation (DOT), www.dot.gov

US Environmental Protection Agency (EPA), www.epa.gov

Federal Motor Carrier Safety Administration (FMCSA) www.fmcsa.dot.gov

Mine Safety and Health Administration (MSHA), www.msha.gov

US Naval Facilities Engineering Command (NAVFAC), www.navfac.navy.mil

National Fire Protection Association (NFPA), www.nfpa.org

Occupational Safety and Health Administration (OSHA), www.osha.gov

US Army Corps of Engineers (USACE), www.usace.army.mil

US Coast Guard (USCG), www.uscg.mil

4.0.0 Section Review

1. The OSHA Construction Standard for asbestos is _____.
   a. 29 *CFR* 1926.1101
   b. 29 *CFR* 1926.35B (1–6)
   c. 29 *CFR* 1926 Subpart AA
   d. 29 *CFR* Subpart F

2. The program requiring that hazardous wastes meet mandatory technology-based treatment standards before the wastes can be placed in a landfill is the _____.
   a. Contractor Pre-Qualification Program
   b. Brownfields Program
   c. Land Disposal Restrictions Program
   d. Site-Specific Safety Program

3. The Commercial Driver's License program is regulated by _____.
   a. ANSI
   b. DHS
   c. USACE
   d. DOT

4. The *Standard System for the Identification of the Hazards of Materials for Emergency Response* is NFPA code _____.
   a. *70*
   b. *101*
   c. *704*
   d. *1670*

SUMMARY

Site-specific safety plans must be customized for each project. These plans should be compiled from pre-bid job checklists and existing safety programs. The safety technician must analyze hazards for both probability and consequences, then tailor the site safety plan to minimize or eliminate these hazards through standard hazard control measures. It is important for an emergency action plan to be detailed and comprehensive; lives can depend on it.

In addition, certain aspects of site safety management, such as traffic control and emergency procedures, should be done cooperatively. Coordinate your site safety plan with the plans of other contractors who will be working on the site.

Review Questions

1. Two aspects of risk assessment are _____.
   a. scope and size
   b. consequences and probability
   c. hazards and controls
   d. plans and procedures

2. The site safety plan should be prepared using the information gathered in _____.
   a. safety audits
   b. the pre-bid checklist
   c. inspection reports
   d. site visits

3. For a site safety plan to be effective, the plan needs to include inspections, auditing, and _____.
   a. project costs
   b. prevailing wage information
   c. reporting
   d. annual revision

4. Training on the site emergency action plan should be provided to employees upon initial assignment and _____.
   a. weekly
   b. quarterly
   c. during random drills
   d. when the plan is changed

5. Workers can avoid injury from contact with toxic materials by using _____.
   a. a quarter mask respirator
   b. an SDS
   c. chemical-resistant suits
   d. leather work gloves

6. Before personnel are allowed to wear tight-fitting respiratory protection, they must have a medical evaluation and be _____.
   a. approved by the site supervisor
   b. fit tested to ensure the device seals properly
   c. certified to work in confined spaces
   d. certified by OSHA

7. When exposed to blood or other bodily fluids, personnel should follow _____.
   a. mandatory first aid treatment
   b. fluid collection protocols
   c. universal precautions
   d. sterilization procedures

8. According to OSHA fire protection requirements, the maximum travel distance to a Class A fire extinguisher is _____.
   a. 25 feet (8 meters)
   b. 50 feet (15 meters)
   c. 75 feet (23 meters)
   d. 100 feet (30 meters)

9. Which of the following is unacceptable when dealing with the media after an incident or emergency has occurred?
   a. Making off-the-record statements to friends and co-workers.
   b. Telling members of the media that an investigation is currently underway.
   c. Advising members of the media to speak to the company spokesperson.
   d. Expressing concern for the safety and well-being of any injured personnel.

10. The site safety plan must be coordinated with _____.
    a. the client, other contractors, and emergency services
    b. employees, safety managers, and foremen
    c. site security, doctors, and medical services
    d. the client, foremen, and emergency response personnel

11. A common tool for communicating safety information is a(n) _____.
    a. social media announcement
    b. note on employee pay stubs
    c. incident record
    d. bulletin board

12. Firms covered under OSHA PSM regulations are required to inform contractors of the _____.
    a. pre-bid checklist findings
    b. firm's experience modification rate
    c. site emergency response procedures
    d. safety auditing schedule

13. The Act requiring that federally listed species and habitat not be adversely affected during any activity that has federal involvement or is subject to federal oversight is _____.
    a. ESA
    b. TSCA
    c. SDWA
    d. RCRA

14. DOT rules for shipping hazardous materials are covered in _____.
    a. 29 *CFR* 1926
    b. 30 *CFR* 70
    c. 40 *CFR* 261
    d. 49 *CFR* 170

15. American National Standards are designated by _____.
    a. ASME
    b. ANSI
    c. NFPA
    d. ASSE

Trade Terms Introduced in This Module

Contractor pre-qualification: A process of screening contractors to allow them to bid on jobs for a specific company.

Defibrillator: An electronic device that administers an electric shock of preset voltage to the heart through the chest wall in an attempt to restore the normal rhythm of the heart during ventricular fibrillation.

Immediately dangerous to life and health (IDLH): A situation that poses a threat of exposure to airborne contaminants when that exposure is likely to cause death or immediate or delayed permanent adverse health effects or prevent escape from such an environment.

Intrinsically safe: An electric tool or device that is UL-rated to be explosion-proof under normal use. For example, a fuel pump must be intrinsically safe to prevent explosions sparked by static electricity.

Pre-bid checklist: A list of questions or issues that aids contractors in gathering the information needed to prepare a bid package.

Probability of occurrence: The likelihood that a specific event will occur, usually expressed as the ratio of the number of actual occurrences to the number of possible occurrences.

Process Safety Management (PSM): The Process Safety Management of Highly Hazardous Chemicals, 29 *CFR* 1910.119. An OSHA standard that covers certain chemical plants and refineries.

Risk assessment: The process of qualifying or ranking hazards given their probability and consequences.

Universal precautions: A set of precautions designed to prevent transmission of human immunodeficiency virus (HIV), hepatitis B virus (HBV), and other blood-borne pathogens when providing first aid or health care. Under universal precautions, blood and certain body fluids of all patients are considered potentially infectious for HIV, HBV, and other blood-borne pathogens.

Appendix A

CONTRACTOR PRE-QUALIFICATION FORM (PQF)

Contractor Prequalification Form (PQF)

GENERAL INFORMATION	
1. Company name:	Telephone Number:
	FAX Number:
Street Address:	Mailing Address:
Province: Postal Code:	Province: Postal Code:
2. Officers:	Years With Company
President:	
Vice-President:	
Treasurer:	
3. How many years has your organization been in business under your present firm name?	
4. Parent Company Name:	
City:	Province: Postal Code:
Subsidiaries:	
5. Under Current Management Since (Date):	
6. Contact for Requesting Bids:	
Title: Telephone: Fax:	
7. Contractor's PQF Completed By:	
Title: Telephone: Fax:	

ORGANIZATION
8. Form of Business   Sole Owner ☐   Partnership ☐   Corporation ☐
9. Percentage Owned:
10. Describe Services Performed: ☐ Construction ☐ Construction Design ☐ Maintenance ☐ Manpower and Resources ☐ Original Equipment Manufacturer and Installer ☐ Original Equipment Manufacturer and Maintenance ☐ Project Maintenance ☐ Service Work (e.g., Janitorial, Clerical, etc.) ☐ Other
11. Describe the Additional Services Performed:
12. List other types of work within the services you normally perform that you subcontract to others, including brokers:

Contractor Prequalification Form (PQF)

13.	Do you evaluate your subcontractor's health and safety program?
14.	Attach a list of the major equipment (e.g., cranes, forklifts, JLGs) your company has available for work at this facility, and the method of establishing the competencies to operate this equipment.
15.	Describe any affiliations with labour organizations.

16.	Annual Dollar Volume for the Past Three Years:	19 ___ $	19 ___ $	19 ___ $

17.	Largest Job During the Last Three Years: $

18.	Your Firm's Desired Project Size:	Maximum:	Minimum:
19.	Financial Rating:	D and B:	Net Worth:

COMPANY WORK HISTORY

20. Major jobs in Progress:

Customer/Location	Type of Work	Size $M	Customer Contact	Telephone	Fax

21. Major jobs Completed in the Past Three Years:

Customer/Location	Type of Work	Size $M	Customer Contact	Telephone	Fax

22.	Are there any judgments, claims or suits pending or outstanding against your company? If yes, please attach details.	Yes ☐ No ☐
23.	Are you now, or have you ever been, involved in any bankruptcy or reorganization proceedings? If yes to either of the above questions, please attach details.	Yes ☐ No ☐

HEALTH AND SAFETY PERFORMANCE

24.	From the last three years (including subcontractors):	19 ___	19 ___	19 ___
	• Number of fatalities?			
	• Number of lost time incidents?			
	• Number of medical aid injuries?			
	• Do you have a modified work program?			

25. Please list your past three years' recordable injury incidence rate (including subcontractors):
 _____, 19 ____   _____, 19 ____   _____, 19 ____

$$\frac{\text{Number of Lost Time Accidents} \times 200{,}000}{\text{Total Employee Hours (Yearly)}}$$

Contractor Prequalification Form (PQF)

26. Man hours (including those of the subcontractors) worked in the last three years:	Year	19____	19____	19____
	Hours	Field			
		Total			
27. Please list your overall Worker's Compensation Rating for the past three years. Please attach your company's WCB summary. _____, 19____ _____, 19____ _____, 19____					
28. Have you received an Alberta Labour OH&S stop work order, or equivalent, from another province in the last three years? Yes ☐ No ☐ Describe _____ _____					

HEALTH AND SAFETY MANAGEMENT
29. Highest ranking safety professional in your organization:
Title: Telephone: Fax:
30. Do you have, or provide: • A full-time health and safety representative?   Yes ☐   No ☐ • A full-time site health and safety representative?   Yes ☐   No ☐

HEALTH AND SAFETY PROGRAM AND PROCEDURES
31. Do you have a written Health and Safety Management Program?   Yes ☐   No ☐ Does the program address the following key elements: • Accountabilities and responsibilities for managers, supervisors, and employees?   Yes ☐   No ☐ • Employee participation?   Yes ☐   No ☐ • Hazard recognition and control?   Yes ☐   No ☐ • Management commitment and expectations?   Yes ☐   No ☐ • Periodic health and safety performance appraisals for all employees?   Yes ☐   No ☐ • Resources for meeting health and safety requirements?   Yes ☐   No ☐
32. Does the program include work practices and procedures such as: • Accident/Incident Reporting?   Yes ☐   No ☐ • Compressed Gas Cylinders?   Yes ☐   No ☐ • Confined Space Entry?   Yes ☐   No ☐ • Electrical Equipment Grounding Assurance?   Yes ☐   No ☐ • Emergency Preparedness, including an Evacuation Plan?   Yes ☐   No ☐ • Equipment Lockout and Tag Out (LOTO)?   Yes ☐   No ☐ • Fall Protection?   Yes ☐   No ☐ • Housekeeping?   Yes ☐   No ☐ • Injury and Illness Recording?   Yes ☐   No ☐ • Personal Protective Equipment (PPE)?   Yes ☐   No ☐ • Portable Electrical/Power Tools?   Yes ☐   No ☐ • Powered Industrial Vehicles (Cranes, Forklifts, JLGs, etc.)?   Yes ☐   No ☐ • Unsafe Condition Reporting?   Yes ☐   No ☐ • Vehicle Safety?   Yes ☐   No ☐ • Waste Disposal?   Yes ☐   No ☐

Contractor Prequalification Form (PQF)

33. Do you have written programs for the following:
• Hearing Conservation?
• Respiratory Protection?
   Where applicable, have employees been:
     ☐ Fit Tested?
     ☐ Medically Approved?
     ☐ Trained?
• WHMIS?
	Yes ☐ No ☐
	Yes ☐ No ☐
	
	Yes ☐ No ☐

34. Do you have a Substance Abuse Program? If yes, does it include the following: • Preemployment? • Random Testing? • Testing for Cause?	Yes ☐ No ☐ Yes ☐ No ☐ Yes ☐ No ☐ Yes ☐ No ☐
35. Medical: Do you conduct medical examinations for: • Preemployment? • Pulmonary? • Replacement Job Capability? • Respiratory? Describe how you will provide First Aid and other medical services for your employees while on site. Specify who will provide this service: _____ Do you have personnel trained to perform First Aid and CPR?	 Yes ☐ No ☐ Yes ☐ No ☐ Yes ☐ No ☐ Yes ☐ No ☐ Yes ☐ No ☐
36. Do you hold site safety meetings for: • Employees? Yes ☐ No ☐ Frequency _____ • Field Supervisors? Yes ☐ No ☐ Frequency _____ • New Hires? Yes ☐ No ☐ Frequency _____ • Subcontractors? Yes ☐ No ☐ Frequency _____ Are the health and safety meetings documented?	
37. Personal Protection Equipment (PPE): Is applicable PPE provided for employees? Do you have a program to ensure PPE is inspected and maintained?	 Yes ☐ No ☐ Yes ☐ No ☐
38. Do you have a corrective action process for addressing individual health and safety performance deficiencies?	Yes ☐ No ☐
39. Equipment and Materials: • Do you conduct inspections on operating equipment (e.g., cranes, forklifts, JLGs, etc.) in compliance with the regulatory requirements? • Do you have a system for establishing the applicable health, safety, and environmental specifications for the acquisition of materials and equipment? • Do you maintain operating equipment in compliance with the regulatory requirements? • Do you maintain the applicable inspection and maintenance certification records for operating equipment?	 Yes ☐ No ☐ Yes ☐ No ☐ Yes ☐ No ☐ Yes ☐ No ☐

Contractor Prequalification Form (PQF)

40. Subcontractors: Do you evaluate the ability of subcontractors to comply with applicable health and safety requirements as part of the selection process? Do you include your subcontractors in: • Audits? • Health and Safety Meetings? • Health and Safety Orientation? • Inspections? Do your subcontractors have a written Health and Safety Management Program? Do you use health and safety performance criteria in the selection of subcontractors?	Yes ☐ Yes ☐ Yes ☐ Yes ☐ Yes ☐ Yes ☐ Yes ☐	No ☐ No ☐ No ☐ No ☐ No ☐ No ☐ No ☐
41. Inspections and Audits: • Are corrections of the deficiencies documented? • Do you conduct health and safety inspections? • Do you conduct Health and Safety Management Program audits?	Yes ☐ Yes ☐ Yes ☐	No ☐ No ☐ No ☐
42. Craft Training: • Are employees' job skills certified, where required, by regulatory or industry consensus standards? • Have employees been trained in the appropriate job skills? • List crafts which have been certified: _____ _____ _____	Yes ☐ Yes ☐	No ☐ No ☐

HEALTH AND SAFETY TRAINING

	New Hires		Supervisors	

43. Safety Orientation Program:
• Do you have a Health and Safety Management Orientation Program for new hires and newly hired or promoted supervisors?
• Does this program provide instruction on the following:
  • Emergency Procedures?
  • Fire Protection and Prevention?
  • First Aid Procedures?
  • Incident Investigation?
  • New Worker Orientation?
  • Safe Work Practices?
  • Safety Intervention?
  • Safety Supervisors?
  • Toolbox Meetings?
  • WHMIS Training?
• How long is the orientation program?
	Yes ☐
	
	Yes ☐
	Yes ☐
	Yes ☐
	Yes ☐
	Yes ☐
	Yes ☐
	Yes ☐
	Yes ☐
	Yes ☐
	Yes ☐
	_____ Hours
		No ☐
		
		No ☐
		No ☐
		No ☐
		No ☐
		No ☐
		No ☐
		No ☐
		No ☐
		No ☐
		No ☐
			Yes ☐
			
			Yes ☐
			Yes ☐
			Yes ☐
			Yes ☐
			Yes ☐
			Yes ☐
			Yes ☐
			Yes ☐
			Yes ☐
			Yes ☐
			_____ Hours
				No ☐
				
				No ☐
				No ☐
				No ☐
				No ☐
				No ☐
				No ☐
				No ☐
				No ☐
				No ☐
				No ☐

44. Health and Safety Training Program: • Do you have a specific Health and Safety Training Program for supervisors? • Do you know the regulatory health and safety training requirements for your employees? • Have your employees received the required health and safety training and retraining?	Yes ☐ Yes ☐ Yes ☐	No ☐ No ☐ No ☐

Contractor Prequalification Form (PQF)

45. Training Records:
   - Do you have health and safety, and crafts training records for your employees? Yes ☐ No ☐
   - Do the training records include the following:
     - Date of Training? Yes ☐ No ☐
     - Employee Identification? Yes ☐ No ☐
     - Method Used to Verify Understanding? Yes ☐ No ☐
     - Name of Trainer? Yes ☐ No ☐
   - How do you verify understanding of the training? (Check all that apply)
     - ☐ Job Monitoring ☐ Written Test
     - ☐ Oral Test ☐ Other (List) _____
     - ☐ Performance Test _____

INFORMATION SUBMITTAL

Please provide copies of checked (√) items with the completed contractor's PQF:

_____ Accident/Incident Investigation Procedure.
_____ Example of Employee Health and Safety Training Records.
_____ Health and Safety Audit Procedure or Form.
_____ Health and Safety Incentive Program.
_____ Health and Safety Inspection Form.
_____ Health and Safety Orientation Outline.
_____ Health and Safety Program.
_____ Health and Safety Training for Supervisors (Outline).
_____ Health and Safety Training Program (Outline).
_____ Health and Safety Training Schedule (Sample).
_____ Housekeeping Policy.
_____ Respiratory Protection Program.
_____ Substance Abuse Program.
_____ Unsafe Conditions Reporting Procedure.
_____ WHMIS Program.

Note: Owner checks items to be provided with the contractor's PQF.

Individual to contact for clarification or additional information:

Name: _____ Telephone: _____

FAX: _____

OWNER'S USE ONLY

DO NOT FILL OUT - OWNER'S USE ONLY

Contractor is: ☐ Acceptable for Approved Contractors' List.

☐ Conditionally acceptable for Approved Contractors List.

Conditions: _____
_____
_____

Reviewed By: _____ Date: _____

Appendix B

PRE-JOB PLANNING SAFETY CHECKLIST

Pre-Job Planning Safety Checklist

1. **General Information:**

    Job Number: _____ Client: _____

    Location: _____

    Client Contact: _____ Phone: _____ Fax: _____

    Start Date: _____ Completion Date: _____

2. **Scope of Work:** Briefly describe the project and your scope of work. (Type and size of project, materials of construction, and construction methods)\

    Peak Employment: Company _____ Subcontractors _____

    Will the job involve any unusual or high risk work? If so, specify the nature of the potential problem(s) and the proposed solutions.
    _____
    _____
    _____
    _____

    Will the job involve any of the following? If so, describe:

    _____ Blasting
    _____ Pile Driving
    _____ Tunneling or major excavations
    _____ De-watering
    _____ Demolition
    _____ Work at extreme heights
    _____ Work over water
    _____ Underpinning
    _____ Handling or exposure to asbestos or hazardous wastes, lead, or OSHA regulated substances
    _____ Work in, on, or adjacent to equipment handling flammable, toxic, or otherwise hazardous chemicals
    _____
    _____

3. **Hazardous Processes, Materials, or Equipment:**

   Identify processes, materials, or equipment that may expose employees to hazardous conditions either in routine work or emergency situations. Obtain Material Safety Data Sheets or Hazardous Waste Sheets on all hazardous materials to which employees may reasonably be exposed. Find out how the client is complying with the OSHA Hazard Communications Standard and how this information will be conveyed to contractor employees. Find out if the proposed work falls under the OSHA Process Safety Management Standard 1910.119.

4. **Client Safety Rules & Procedures:**

   Obtain copies of all client safety rules, procedures, or manuals that contractors are expected to follow. Do not overlook emergency plans or procedures. List any unusual or special safety requirements.

5. **Medical Surveillance / Industrial Hygiene Monitoring Requirements:**

   List any medical surveillance or industrial hygiene monitoring requirements for contractor personnel. Determine who will be expected to provide these services. List any special requirements such as no beard policies, drug screening tests, showering facilities, and decontamination facilities.

6. **Regulatory Permit Requirements & Inspection Schedules**

   List all required permits and inspection schedules and who is responsible for obtaining these permits and coordinating the necessary inspections?

7. **Site Location & Layout:**

   A. Briefly describe site conditions and layout. (Rural, residential, commercial, industrial congested, etc.)

   B. Describe adjacent exposures. How might they be affected by the project? Is there any need to take photos of the adjacent structures or have a physical inspection made by an independent third party?

   C. Utilities: List all utilities available at or near the site. If possible, obtain the names and phone numbers of area utility representatives.

      1. Water:

      2. Electricity:

      3. Gas:

      4. Sewer:

      5. Other:

      6. Have all existing utilities been located and marked appropriately?

      7. Will an assured grounding program or ground fault circuit interrupters by used to protect temporary electrical circuits? Who is responsible for coordinating this activity?

   D. Field office, storage, and laydown areas:

      1. Any designated areas?

2. Accessible by trucks, forklifts, and cranes? Will vehicles be able to drive through as opposed to backing out?

3. Any restricted or congested roads / areas? If so, describe.

4. Parking facilities for employees, contractors, and visitors?

5. Any special requirements for temporary buildings with respect to location, size materials of construction, and etc.?

8. **Personnel Health & Safety:**

   A. Personal protective equipment: What equipment is required and who will provide it?

   _____ Hard Hats
   _____ Safety Shoes
   _____ Chemical Resistant Boots
   _____ Fire Retardant Clothing
   _____ Plano Safety Glasses, with or without side shields
   _____ Prescription Safety Glasses, with or without side shields
   _____ Gloves, specify: _____
   _____ Safety Harness and Lanyards, specify type: _____
   _____ Chemical Resistant Clothing, specify: _____
   _____ Life Lines
   _____ Personnel Nets
   _____ Debris Nets
   _____ Face Shields / Goggles
   _____ Respiratory Protection (Specify what type: _____)
   _____ Other, specify: _____

   B. What temporary toilet facilities will be needed? Will any showering facilities be required?

   C. Is a drinking water supply available on site?

   D. What method(s) will be used to protect wall and floor openings?

E. Any special scaffolding requirements? If so, specify:

F. Any materials or personnel hoists or elevators? If so, who will operate, inspect, and maintain them?

G. Overhead protection required at building entrances?

9. **Emergency Reporting & Response:**

    A. List all emergency communication systems and onsite phone numbers.

    B. How will onsite medical and First Aid be handled? Who will provide what type and level of care? What First Aid supplies and rescue equipment will be required / provided?

    C. Location and telephone number of the nearest off-site medical facilities and ambulance service companies.

    D. Name, address, and phone number of company physician(s):

    E. Obtain information on the client's emergency assembly areas and evacuation routes.

F. What onsite emergency notification and response procedures will be used? Obtain information on the client's emergency alarm system and signals. If possible, obtain lists of emergency alarm codes. Try to get a tape recording of the alarm signals.

G. What emergency escape equipment (if any) is needed? Who will provide it and where will it be located?

H. What offsite emergency reporting and response procedures will be used?

I. Name, address, and phone number of fire department.

J. Will contractor personnel be expected to serve on a Hazardous Materials Emergency Response Team? If so, specify the level and extend of contractor involvement.

10. **Fire Protection and Prevention:**

   A. Is there an adequate number of active fire hydrants on site? What provisions will be made to keep them unobstructed and accessible?

   B. List the number, type, and size of portable fire extinguishers to be provided and by whom:

   C. Will standpipes and / or sprinkler systems be required to follow the building up floor by floor? ____ Yes ____ No ____ N/A. Briefly describe type of systems, maintenance, and inspection requirements and responsibilities during the construction phase.

D. Is there a site Fire Brigade? Will contractors be expected to serve on this brigade?

E. Will temporary heating be used: _____ Yes _____ No _____ N/A. If so, what type?

F. Requirements for the storage and handling of flammable liquids. Any secondary containment required? Describe:

G. How, where, and at what frequency will trash be removed and disposed?

H. How will hazardous waste be handled?

11. **Administration:**

   A. Who will be responsible for collecting and disseminating Material Safety Data Sheets?

   B. Who will be responsible for hiring site personnel?

   C. What screening placement procedures will be used?

   _____ Written applications
   _____ Reference checks
   _____ DMV records check
   _____ Medical exams
   _____ Drug screening
   _____ Other, specify: _____

   D. Who will be responsible for New Employee / Contractor Orientation and what materials and / or handouts will be used?

   E. What types of safety inspections are required, at what frequency, and who will conduct them?

F. What types of safety meetings will be held, at what frequency, and with whom?

12. **Special Tools or Equipment Requirements:**

    Will any special tools or equipment be required If so, specify what types and brands of:

    A. Explosion proof lighting

    B. Non-sparking tools

    C. Air operated tools or equipment

    D. Ground fault circuit interrupters

    E. Extension cords

    F. Air moves or other gas freeing equipment

    G. Gas testing equipment

    H. Other (specify)

13. **Work Permit Requirements:**

    Obtain as much information as possible concerning the client's work permitting procedures. Are permits required for the following types of work?

    A. Hot / Hazardous work

    B. Cold / Safe work

    C. Vessel entry (confined space)

    D. Gas testing

    E. Equipment isolation including lockout / tag out

    F. Excavations

G. Vehicle entry

H. Critical lifts

I. Scaffolding

J. Other (specify)

14. **Training Requirements:** *

    List any special safety training requirements such as:

    A. Respirator

    B. Hazard communication

    C. Handling of certain hazardous materials (i.e., asbestos, lead). Specify the material(s).

    D. Hazardous waste site

    E. Response to spills or releases of hazardous materials. Specify what level of response will be required by contractor personnel.

    F. Certification for operators of Fork

    G. Other (specify)

    * **Important Note:** Find out who will perform and document this training, the client or the contractor. What training aids such as video tapes, films, slide / tape presentations are available from the client for contractor to use?

15. **Reporting Requirements:**

    List any special reporting requirements the client may have, such as:

    A. Accident, injuries, illnesses, or near misses

B. Safety meetings

C. Safety inspections

D. Regulatory agency visits

E. Other (specify)

16. **Site Security:**

    A. Describe the security measures required for the job during normal working hours and after hours.

    B. If controlled access is required, how will it be handled and by whom?

    C. Will any burglar alarm system, guard dog, or watchman services be used? _____ Yes _____ No _____ N/A. If so, describe:

    D. Is / will site lighting be adequate

17. **Public Liability:**

    A. Will any sidewalks or streets have to be blocked? _____ Yes _____ No _____ N/A. If so, describe:

    B. Any overhead protection and / or lighting required: _____ Yes _____ No _____ N/A. If so, describe:

    C. Briefly describe any barricading, lighting, signs, traffic control devices or flagmen that may be required:

18. **Environmental Hazards & Controls:**

    A. EPA, State, or local Right to Know Requirements: Describe:

    B. If asbestos or lead is suspected or anticipated, what special precautions, monitoring, and training will be required? Who will coordinate and pay for these activities?

    C. Will any wastes generated or leaving the site be classified as hazardous and require disposal as a hazardous waste? _____ Yes _____ No _____ N/A. If so, will it be necessary to obtain any types of generator numbers and or permits? Who will obtain the necessary permits?

    D. Will there be any open burning or on site landfills? _____ Yes _____ No _____ N/A. If so, describe:

    E. Will any site activities create noise, air, water, or ground pollution problems? _____ Yes _____ No _____ N/A. If so, describe potential problems and proposed solutions.

    F. Is secondary containment required for flammable / hazardous material storage?

19. **Cranes / Hoists and Lifting Devices:**

    A. Will there be any cranes, hoists, or lifting devices on site? _____ Yes _____ No _____ N/A. If so, describe:

    B. Who will operate, maintain, and inspect these devices? How will they be certified?

C. Any heavy, unusual, or critical lifts to be made? _____ Yes _____ No _____ N/A. If so, describe:

D. Any requirements for inspections, certifications, or load tests? _____ Yes _____ No _____ N/A. If so, describe:

E. Any requirements for critical lift plans? _____ Yes _____ No _____ N/A. If so, describe:

F. Any planned use or prohibition on use of crane suspended personnel baskets? _____ Yes _____ No _____ N/A. Explain:

20. **Signs, Posters, and Bulletin Boards:**

    A. List all signs, posters, notices required for the project:

    _____ OSHA Employee Rights poster
    _____ Worker's Compensation notice
    _____ Company / jobsite rules
    _____ Emergency phone numbers
    _____ Location of MSDS's
    _____ Employee access to medical & exposure records
    _____ Others, specify:

    B. Where will these notices and signs be posted?

21. **Contract Specifications:**

    A. Bonding requirements: (Specify)

A. Insurance requirements:

   1. Types of coverages & policy limits:

   2. Who will provide what coverages?

   3. Any special requirements needing insurance carrier approval?

   4. Have certificates of insurance been requested and received from?

      a. Client / owner
      b. Prime or general contractor
      c. Sub contractors
      d. Others

   5. Names, addresses, and phone numbers of key insurance company representatives:

      a. Agent / broker:

      b. Claims representative:

      c. Loss control or safety representative:

   6. Any hold harmless or indemnity agreements? If so, obtain copies.

   7. Any unusual safety and loss prevention requirements, such as:

      a. Site specific safety and health plan required?
      b. Site safety representative required?
      c. Drug, alcohol, and substance abuse policy required?
      d. Any pre-employment and or follow up medical exams required?
      e. Other, specify:

22. **Additional Information or Comments:**

Source of Information:

    Name: _____ Title _____ Date _____

    Name: _____ Title _____ Date _____

Report Prepared by:

    Name: _____ Title _____ Date _____

Report Reviewed by:

    Name: _____ Title _____ Date _____

Appendix C

Generic Safety Plan

Site Specific Outline	Include in Program	Comments
1. Introduction		
2. Purpose of Program		
3. Company Mission Statement		
4. Vision Statement		
5. Goals		
6. Plant Philosophy on Safety		
7. Security		
8. Conduct		
a. Conduct termed horseplay is forbidden		
9. Prohibited Items		
a. Drugs or alcohol prohibited on premises		
b. Firearms, explosives, pets, and televisions are not permitted on premises		
c. Cameras or other image capturing device prohibited without approval		
d. Finger rings prohibited in process units, process areas, maintenance shops, laboratories, electrical work, climbing ladders, mobile equipment, or where it constitutes a hand or finger hazard		
e. Neckties, scarves, or any dangling jewelry prohibited in process areas, maintenance ships and laboratories		
10. Substance Abuse Policy		
11. Plant Entry Procedures		
12. Plant Exit Procedures		
13. Maps		
14. PPE		
a. Contractor to provide all necessary PPE		
b. Contractor must train in use		
c. Flame resistant clothing		
i. Must wear in operating areas		
ii. FRC must meet NFPA standard and site requirements		
iii. FRC must be properly maintained and cleaned		
iv. Sleeves secured around wrists and shirttails tucked in		
v. Welding shirts not open pockets		
vi. FRC outermost garment		
vii. To protect against hazardous materials FRC worn under PPE		
viii. Raingear worn over FRC		
ix. Cold weather apparel worn under FRC		
d. Gloves		
i. Worn when hazard analyses deem necessary		
ii. Cloth gloves prohibited around revolving tools or machinery		
iii. Dispose of used gloves properly		
e. Safety glasses		
i. Comply with ANSI Z87.1 standard		
ii. With side shields		
iii. Clear, photogray, and tinted are permitted		
iv. Sun glasses not worn in confined spaces, buildings, or at night		
v. Metal frames safety glasses not permitted around electrical work		
f. Hard hats		
i. Meet ANSI Z89.1 and providing class E protection		
ii. Metal hats not allowed		
iii. Hats may not be modified		
iv. Goggles shall be affixed to hardhat		

Site Specific Outline	Include in Program	Comments
g. Hearing protection		
i. Required:		
1. When working in process units		
2. Where signs are posted		
3. In temporary high noise areas		
h. Safety shoes		
i. Meet ANSI Z41 standard		
ii. Use in process and maintenance areas		
iii. Other jobs may require safety toed shoes		
i. Additional PPE		
i. May be required for specific jobs:		
1. Respirators, acid suits, and gloves		
2. Plant will advise contractors of any required PPE		
15. Compressed Gas Cylinders		
16. Permits		
a. Define scope of work		
b. Identify Hazards/risk		
c. Establish control measures		
d. Links to other permits/operations		
e. Authorized		
f. Communicates work information		
17. Excavations		
18. Facial Hair Policies		
a. Contractors must comply with company/plant policy		
b. Cannot interfere with respiratory equipment sealing		
19. Respiratory Equipment		
20. Respiratory Protection		
21. Fire Equipment		
22. Fire Watch		
23. Solvents		
24. Cleaning Agents		
25. LOTO		
a. Procedural steps have been developed for:		
i. Shutting down		
ii. Isolating		
iii. Purging		
iv. Depressurizing		
v. Flushing		
vi. Neutralizing		
vii. Testing		
b. Color coded locks used and remain in place for service and maintenance		
c. Safe energy state is verified		
d. Contract employee apply personal lock before starting work		
e. Lock must have the contract employee's identity attached		
f. Individual personal locks remove at the end of each day or completes job		
26. Electrical Safety		
27. Housekeeping		
a. All areas kept orderly		
b. After job area cleaned up		
c. Useable materials returned to their proper location		
d. Scrap and surplus disposed of by contractor		

Site Specific Outline	Include in Program	Comments
28. Injury/Accident Reporting Procedures		
a. Contractors must provide first aid and medical services for employees		
b. Plant will provide assistance for severe injuries or special rescue		
c. Contractors required to report all incidents and near misses		
d. Documentation of incident and follow-up is required		
e. Anyone can report an emergency		
f. Dial emergency phone extensions on plant telephone		
g. Emergency button on plant radio		
h. Provide as much information as possible		
i. Do not hang up until your information is confirmed		
29. Safety Harness Requirements		
30. Vehicles		
a. Inspected and in safe work conditions		
b. Drivers trained and certified class of vehicle		
c. Seat belts are installed and worn		
d. Key left in vehicle at gates/process units		
e. Hand held devices not in use by driver (phones/radios)		
f. Management of change		
i. Risk assessment conducted		
ii. Development of a work plan		
iii. Control measures		
1. Equipment, facilities, and process		
2. Operations, maintenance, inspection procedures		
3. Training, personnel, and communication		
4. Documentation		
iv. Competent person authorization of the work plan through completion		
31. Traffic Regulations		
a. Compliance of traffic signs required		
b. Speed limits obeyed		
c. Pedestrians have the right of way		
d. Vehicles cannot block roadways and/or accesses		
e. Brief stops to load/unload passengers ok		
32. Signs and Barricades		
a. Must be approved such as		
i. Areas below overhead work		
ii. Hazardous leaks		
iii. Sand blasting		
iv. Asbestos		
v. Hydro blasting		
vi. Radiography		
vii. Open trenches		
viii. Roadways or walkways		
b. Erector of barricade must be tag it		
c. Tag must be identify the hazards inside barricade		
d. Tag must be in conspicuous location		
e. Do not enter barricaded areas or fire lanes, fire hydrants, or emergency equipment		
f. Use caution at railroad crossings		
33. Smoking Policies		
34. Smoking Areas		
a. Not permitted inside to plant controlled gates or company vehicles		
b. No lighters, matches, cigarettes, or smoking materials		

Site Specific Outline	Include in Program	Comments
35. Contractor Equipment		
a. Must furnish MSDS for hazardous materials		
b. Must have documentation and approval		
36. Contractor Areas		
37. Environmental Rules		
38. Labeling		
a. All storage containers must be labeled with identity and hazards		
b. Hazardous materials must be removed at the end of the job unless approved by job rep.		
39. MSDS Locations		
40. Confined Spaces		
a. Permit required		
b. Include areas without natural ventilation		
c. Include the following:		
i. Vessels		
ii. Vessel skirts		
iii. Storage tanks		
iv. Open floating roof tanks		
v. Railway tank cars		
vi. Tanker truck trailers		
vii. Tote bins		
viii. Excavations		
ix. Sewer systems		
d. Indentified with appropriate signage		
41. Heavy Equipment		
42. Trains/Crossings		
43. PSM		
44. Parking Lots		
45. Docks		
46. Lifesaving Equipment		
47. Vessels/Barges		
48. Ships		
49. Cranes		

Additional Resources

This module presents thorough resources for task training. The following resource material is suggested for further study.

California DOSH Incident and Illness Prevention Program, **www.dir.ca.gov/dosh/puborder.asp**

29 *CFR*, Part 1926, *Safety and Health Regulations for Construction.*

29 *CFR* 1926 Subpart AA, *Confined Spaces.*

29 *CFR* 1926 Subpart F, *Fire Protection and Prevention.*

30 *CFR*, Parts 48, 70, 71, 74, 75, 77, and 90, *Mine Act Standards.*

NIOSH, Publication Number 80-106, Criteria for a Recommended Standard: Working in Confined Spaces, September 1979.

NIOSH, Publication Number 87-113, A Guide to Safety in Confined Spaces. July 1987.

American National Standards Institute (ANSI), **www.ansi.org**

American Society of Mechanical Engineers (ASME), **www.asme.org**

American Society of Safety Engineers, **www.asse.org**

National Institute for Occupational Safety and Health (NIOSH), **www.cdc.gov**

Department of Homeland Security (DHS), **www.dhs.gov**

Department of Transportation (DOT), **www.dot.gov**

US Environmental Protection Agency (EPA), **www.epa.gov**

Federal Motor Carrier Safety Administration (FMCSA) **www.fmcsa.dot.gov**

Mine Safety and Health Administration (MSHA), **www.msha.gov**

US Naval Facilities Engineering Command (NAVFAC), **www.navfac.navy.mil**

National Fire Protection Association (NFPA), **www.nfpa.org**

Occupational Safety and Health Administration (OSHA), **www.osha.gov**

US Army Corps of Engineers (USACE), **www.usace.army.mil**

US Coast Guard (USCG), **www.uscg.mil**

Figure Credits

Miller and Long Co., Inc./elcosh.org, Module opener

Motorola Solutions, Inc., Figure 2

Courtesy of Amerex Corporation, Figure 3 (photos)

Workrite Uniform Company, Figure 5A

MSA The Safety Company, Figures 5B, 5G, 6A, 6B, 6D

Topaz Publications, Inc., Figures 5C, 5F

Photo courtesy of Bullard, Figure 5D

North Safety Products USA, Figure 5E, 6C

National Institute for Occupational Safety and Health (NIOSH): Respirator Selection Logic. DHHS (NIOSH) Publication No. 2005-100. Cincinnati, OH: NIOSH, 2004., Table 2

Physio-Control, Inc., SA01

© Niels Vos/Shutterstock.com, Figure 8

NIOSH/John Rekus/elcosh.org, Figure 9

Rhino Marking & Protection Systems, RhinoDome™ Test Station, TriView®, RhinoDome™, A-Tag™ Pavement Marker, Figure 10

Program Executive Office, Assembled Chemical Weapons Alternatives (PEO ACWA), U.S. Army, Figure 11

University of California, Riverside, Figure 12

U.S. Environmental Protection Agency, Figure 13

Construction Owners Association of Alberta, Appendix A

Ronald Sokol, Appendix C

Section Review Answer Key

Answer	Section Reference	Objective
Section One		
1. a	1.1.2	1a
2. c	1.2.0	1b
3. b	1.3.0	1c
Section Two		
1. c	2.1.4	2a
2. d	2.2.2	2b
3. c	2.3.0	2c
Section Three		
1. c	3.1.4	3a
2. a	3.2.0	3b
Section Four		
1. a	4.1.0	4a
2. c	4.2.0	4b
3. d	4.3.0	4c
4. c	4.4.3	4d

NCCER CURRICULA — USER UPDATE

NCCER makes every effort to keep its textbooks up-to-date and free of technical errors. We appreciate your help in this process. If you find an error, a typographical mistake, or an inaccuracy in NCCER's curricula, please fill out this form (or a photocopy), or complete the online form at **www.nccer.org/olf**. Be sure to include the exact module ID number, page number, a detailed description, and your recommended correction. Your input will be brought to the attention of the Authoring Team. Thank you for your assistance.

Instructors – If you have an idea for improving this textbook, or have found that additional materials were necessary to teach this module effectively, please let us know so that we may present your suggestions to the Authoring Team.

NCCER Product Development and Revision
13614 Progress Blvd., Alachua, FL 32615

Email: curriculum@nccer.org
Online: www.nccer.org/olf

❑ Trainee Guide ❑ Lesson Plans ❑ Exam ❑ PowerPoints Other _____

Craft / Level: _____ Copyright Date: _____

Module ID Number / Title: _____

Section Number(s): _____

Description:
_____
_____
_____
_____

Recommended Correction:
_____
_____
_____
_____

Your Name: _____

Address: _____

Email: _____ Phone: _____

Safety Orientation and Safety Meetings

OVERVIEW

This module describes how to prepare and deliver effective safety training using both formal safety meetings and toolbox/tailgate talks.

Module 75223

Trainees with successful module completions may be eligible for credentialing through NCCER's Registry. To learn more, go to **www.nccer.org** or contact us at 1.888.622.3720. Our website has information on the latest product releases and training, as well as online versions of our *Cornerstone* magazine and Pearson's product catalog.

Your feedback is welcome. You may email your comments to **curriculum@nccer.org**, send general comments and inquiries to **info@nccer.org**, or fill in the User Update form at the back of this module.

This information is general in nature and intended for training purposes only. Actual performance of activities described in this manual requires compliance with all applicable operating, service, maintenance, and safety procedures under the direction of qualified personnel. References in this manual to patented or proprietary devices do not constitute a recommendation of their use.

Copyright © 2018 by NCCER, Alachua, FL 32615, and published by Pearson, New York, NY 10013. All rights reserved. Printed in the United States of America. This publication is protected by Copyright, and permission should be obtained from NCCER prior to any prohibited reproduction, storage in a retrieval system, or transmission in any form or by any means, electronic, mechanical, photocopying, recording, or likewise. To obtain permission(s) to use material from this work, please submit a written request to NCCER Product Development, 13614 Progress Blvd., Alachua, FL 32615.

75223 V5

From *Safety Technology,* Trainee Guide. NCCER.
Copyright © 2018 by NCCER. Published by Pearson. All rights reserved.

75223
SAFETY ORIENTATION AND SAFETY MEETINGS

Objectives

When you have completed this module, you will be able to do the following:

1. Explain how to conduct a safety training program.
   a. List types of training.
   b. List training methods.
   c. Describe how to prepare and deliver safety training sessions.
2. Explain how to conduct a safety meeting.
   a. Describe how to deliver and document a formal safety meeting.
   b. Describe how to perform a toolbox/tailgate safety talk.

Performance Task

Under the supervision of your instructor, you should be able to do the following:

1. Deliver and document a formal safety meeting.

Trade Terms

A-B-C-D method
Audiovisual materials
Five Ps for Successful Safety Talks
Know-show-do
Open-ended questions

Prerequisite training
Recordable incident rate
Remedial training
Simulators
What-why-how

Industry Recognized Credentials

If you are training through an NCCER-accredited sponsor, you may be eligible for credentials from NCCER's Registry. The ID number for this module is 75223. Note that this module may have been used in other NCCER curricula and may apply to other level completions. Contact NCCER's Registry at 888.622.3720 or go to **www.nccer.org** for more information.

Contents

1.0.0 Conducting a Safety Training Program 1
   1.1.0 Types of Safety Training.. 1
      1.1.1 New Employee Orientation 1
      1.1.2 Site-Specific Orientation 1
      1.1.3 Craft-Specific Safety Training.. 2
      1.1.4 Supervisory Training 2
      1.1.5 Continuing Education / Recertification 2
   1.2.0 Methods of Training 3
      1.2.1 Sequence the Training Logically 3
      1.2.2 Know-Show-Do 3
      1.2.3 What-Why-How 3
      1.2.4 Hands-On Practice and Demonstration 3
      1.2.5 Mentoring Program 4
      1.2.6 Safety Training Scenario Workshops 5
      1.2.7 Audiovisual Training Aids 5
   1.3.0 Preparing and Delivering Safety Training... 6
      1.3.1 Coordinating Arrangements... 6
      1.3.2 Preparing to Train 7
      1.3.3 Delivering the Training 8
      1.3.4 Managing the Classroom 11
      1.3.5 Concluding the Training.. 12
2.0.0 Introduction to Safety Meetings 16
   2.1.0 Delivering and Documenting a Formal Safety Meeting 16
      2.1.1 Evaluating Safety Meetings 18
      2.1.2 Recordkeeping 20
   2.2.0 Toolbox/Tailgate Safety Talks 20
      2.2.1 Safety Talk Preparation 23
      2.2.2 Five Ps for Successful Safety Talks.. 23

Figures and Tables

Figure 1   Simulation and virtual reality. 4
Figure 2   Typical plant refinery. 5
Figure 3   Laptop and projector. 6
Figure 4   Portable screen. 6
Figure 5   U-shaped seating. 9
Figure 6   A-B-C-D method. 11
Figure 7A Participant evaluation form (1 of 2). 13
Figure 7B Participant evaluation form (2 of 2). 14
Figure 8   Typical safety meeting. 17
Figure 9A Typical speaker evaluation form (1 of 2). 18
Figure 9B Typical speaker evaluation form (2 of 2). 19
Figure 10 Cost-of-safety equation. 20
Figure 11 Self-evaluation checklist. 21
Figure 12 Recordable incident rate graph. 22
Figure 13 Attendance sheet. 22
Figure 14 The Five Ps for Successful Safety Talks. 23

Table 1   Common Uses of Visual Aids 7

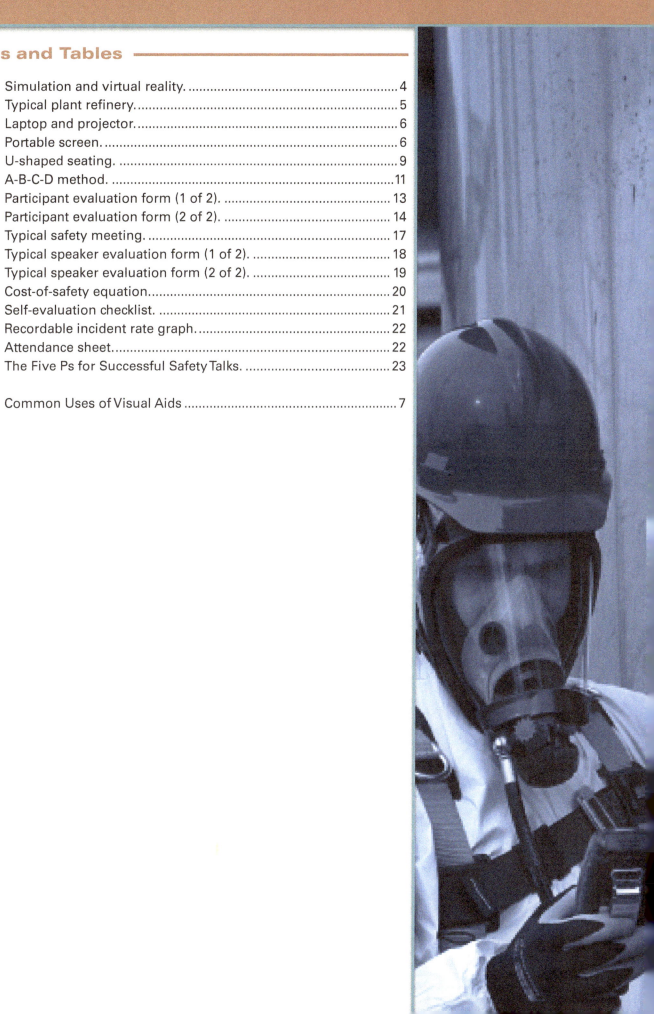

SECTION ONE

1.0.0 CONDUCTING A SAFETY TRAINING PROGRAM

Objective

Explain how to conduct a safety training program.
a. List types of training.
b. List training methods.
c. Describe how to prepare and deliver safety training sessions.

Trade Terms

A-B-C-D method: A device used to write training objectives that specifically cover the audience, behavior, conditions, and minimum degree of mastery.

Know-show-do: A method of teaching tasks or procedures in which the instructor first explains necessary background information, then demonstrates the task or procedure, and concludes by having the participants practice the task or procedure.

Prerequisite training: Training that provides the skills and knowledge required for a course that is to follow.

Remedial training: Training that is designed to address specific areas of difficulty, allowing participants to learn more about a larger course or topic.

Simulators: Teaching aids that are used to simulate a piece of equipment for training purposes rather than conducting training on the actual equipment.

What-why-how: A teaching method in which the instructor describes what an object, task, or procedure is, why it is used or done, and how to use or do it.

Safety training is a primary element in a safety program. The intent of a safety training program is to affect workers' behavior and actions by increasing their knowledge and/or improving their skills.

1.1.0 Types of Safety Training

There are two kinds of training that occur in the workplace: formal and informal. Formal training is a planned and organized activity, prearranged with a group of workers to teach them a specific group of skills or give them specific knowledge.

Informal training is not structured or planned, but occurs when an opportunity for training avails itself. These opportunities may arise in safety meetings, in one-on-one conversations with workers, or as you observe the workers performing various job duties and tasks.

While informal training is an important tool, formal training programs will be the focus of this module. Formal safety training can be broken down into the following major types:

- New employee orientation
- Site-specific orientation
- Craft-specific safety training
- Supervisory training
- Continuing education / recertification

> **NOTE:** Workers who have limited reading or English speaking skills should be assigned a competent translator. Trainee documentation should include the name of the translator.

1.1.1 New Employee Orientation

When new employees are hired, they must receive safety orientation. This training comes in two forms: new employee orientation and site-specific orientation.

New employees receive a lot of new information all at once, so you should plan on breaking their safety training into general and specific content. The new employee orientation should be a general overview of company safety policies and procedures, along with the employee's role and tasks.

1.1.2 Site-Specific Orientation

Site-specific orientation covers the high-risk areas or duties that exist on the job site, preventive measures to minimize the risk of incidents, and contingency plans in the case of an incident. In addition, you may want to cover subjects such as the following:

- The company chain of command
- Where to go for help in case of an emergency
- A walk-through tour of the job site showing safety equipment and escape routes
- The use of standard personal protective equipment (PPE)
- A description of any hazardous materials used on the site
- Emergency reporting and response procedures
- Where to find safety data sheets (SDSs)

After the safety orientation training has been completed, employees will be ready for more specific safety training on the job for which they were hired.

> **NOTE:** Follow-up training with new employees should be done within 30 to 60 days to ensure that workers have retained the information received during training and are doing the job safely.

1.1.3 Craft-Specific Safety Training

Craft-specific safety training should be offered to any new employees, anyone acquiring a new job or job duties, or anyone whose safety practices need improvement. This training must cover:

- Specific and correct work procedures
- Care, use, maintenance, and limitations of all tools and equipment, including required PPE
- Risks associated with any harmful or hazardous materials used, including warning properties, symptoms of overexposure, first aid, and cleanup procedures for spills

Job-specific safety training should include as much demonstration and hands-on practice as possible. JSAs, also referred to as *job hazard analyses (JHAs)*, are designed to inform workers about the specific hazards associated with a job. Use job safety analyses (JSAs) / task safety analyses (TSAs) and pre-task safety planning as training tools.

> **NOTE:** Emergency response personnel should be present when demonstrating the use of any hazardous material or action.

29 *CFR* 1926.21 states that "the employer shall instruct each employee in the recognition and avoidance of unsafe conditions and the regulations applicable to his work environment to control or eliminate any hazards or other exposures to illness or injury." This regulation requires employers to educate employees on how to recognize, avoid, and work safely around any and all hazards they will encounter on the job site. OSHA also requires job-specific safety training for a wide variety of areas.

1.1.4 Supervisory Training

Supervisors play a large role in ensuring a safe work place. To maximize their effectiveness, safety training should be provided for supervisors. Supervisory safety training should accomplish the following objectives:

- Reinforce the company's commitment to a safe work environment.
- Review in detail company safety policies and procedures.
- Explain how to analyze a job to look for potential health and safety risks.
- Explain how to conduct a safety meeting.
- Describe what to do if an unsafe condition exists.
- Describe the policy and procedure to follow after an incident.
- Explain how to coach, counsel, and discipline employees.
- Explain how to investigate and document an incident.

Supervisors have the ability to set the tone for their workers. A supervisor who looks out for workers' safety will help keep safety incidents and health risks to a minimum and reduce the impact of an incident by following procedures and plans. To do these things effectively, every supervisor should go through specific supervisor safety training.

1.1.5 Continuing Education / Recertification

A safety training program must include a continuing education / recertification component to ensure safety related knowledge and skills are kept current and active. Safety requirements can change with new technology, technical updates, expirations, new policies and procedures, or pertinent laws. A safety program should address new and reoccurring safety requirements by providing continuing education and recertification on a routine basis.

Periodic recertification ensures that employees maintain their ongoing commitment to safety and promotes the ongoing enhancement of safety knowledge and skills. Good recordkeeping is key to ensuring employees are being provided with the necessary continuing education / recertification safety related training.

A comprehensive approach to safety training requires that employees be given an opportunity for continual skills development. In some cases, government and safety regulations require that employees' skills be updated and recertified, which takes a significant amount of time. Because of these time demands, government and safety issues should be coupled with related skill development areas. For example, if government regulations require that training be given on a specific hazardous chemical, provide training that covers the chemical in a job-specific application. For maintenance personnel, this may be replacing a valve in the process piping that carries that hazardous chemical. Maintenance personnel would be given the opportunity to develop the skill of replacing the valve in addition to meeting the government regulations.

1.2.0 Methods of Training

Effective training is more than simply telling a room full of people what you know about the subject matter or talking about every slide of a presentation. Participants generally retain the following:

- 10 percent of what they read
- 20 percent of what they hear
- 30 percent of what they see
- 50 percent of what they see and hear
- 70 percent of what they see, hear, and respond to
- 90 percent of what they see, hear, and do

Therefore, your challenge as an instructor is to engage the participants in as many ways as possible. This section describes the following strategies for making safety training more engaging and effective:

- Sequencing the training logically
- Know-show-do method for teaching procedures or tasks
- What-why-how method for breaking down factual information
- Hands-on practice and demonstration
- Mentoring programs
- Audiovisual training aids

1.2.1 Sequence the Training Logically

Training sessions often present a lot of information over a relatively short period of time. The best way to prevent the information from becoming overwhelming is to teach the material in a logical sequence, starting with what the participants already know and using that to lead them into the new material.

For example, to begin a conversation about the Fire Triangle, start by asking what kinds of flammable materials the participants see in the room. After they have listed some flammable items, begin talking about how those items might start burning and what is needed to start, maintain, and put out a fire. Starting with a piece of information that is common knowledge allows the workers to put the new information into a useful context.

> **Did You Know?**
>
> **Construction Incidents**
>
> A recent Construction User's Round Table A-3 Report indicates that 25 percent of all construction incidents involve personnel who have been on the job less than 30 days.

1.2.2 Know-Show-Do

The know-show-do method is very effective for teaching procedures or tasks. To employ the know-show-do strategy, first provide the participants with the information that they need to know in order to perform the task. For example, if you are teaching them how to use a personal fall arrest system (PFAS), you will want the participants to know the requirements for fall protection, the purpose of the various harness straps and D-rings, demonstrate how to inspect a PFAS, and explain how a lanyard is attached to various anchor points. After the participants have the necessary background information, show them how to don a PFAS. Finally, have the participants practice donning a PFAS.

1.2.3 What-Why-How

The what-why-how method is a great way to break down factual information for a class. Use these three key questions consistently as you describe safety policies and procedures, equipment operation, and use of personal protective equipment. For example, you might use the what-why-how method to teach the use of hearing protection by first showing and discussing the different types of hearing protection devices. After thoroughly reviewing what equipment is used, discuss why it is important to use hearing protection equipment. Finally, teach the workers how to use hearing protective equipment. The how-to portion of the lesson would probably lead to a more detailed session using the know-show-do method to teach the participants when and how to select and use hearing protection.

1.2.4 Hands-On Practice and Demonstration

Hands-on practice and demonstrations are the best way to teach people how to operate equipment, perform tasks, or follow procedures. The most effective training course will dedicate a large portion of its time to allowing the participants to use the actual equipment or practice the desired procedure. At this time, you should assess the ability of each participant to accomplish the task.

Demonstrations are a good strategy for teaching skills because it gives learners something real to observe. A demonstration can be performed by an instructor, a member of the class, or an invited expert. Video or web-based training may also be used to present a demonstration. A review following a demonstration is important to give closure and to ensure trainees understand the procedure and the steps involved. Keep in mind that the majority of demonstrations are done to prepare trainees to practice the same task. Always give trainees a chance to practice the procedure as soon as possible following the demonstration while it is fresh in their mind. Demonstrate the step or tasks for the trainees, explaining why the steps are important and exactly what is being done, beginning with the simple and moving to the more complex. It may be necessary to demonstrate more complex tasks several times. Make sure that the trainees fully understand the process. Encourage questions and discussion to determine how quickly trainees are advancing and which areas are unclear.

There may be situations in which demonstrating or allowing hands-on practice on the actual equipment is not practical. For example, it may be too risky or costly to have new forklift drivers begin by operating a forklift. In this situation, the training should include other training strategies such as simulators, videos, or models. In some instances, a video of the procedure can be used to train employees on how to do something, although this is much less effective than hands-on practice.

Simulators are useful for teaching skills related to equipment or processes that may not be taken out of production or are impossible to access. They are especially useful when teaching procedures that may be hazardous if performed incorrectly. A crane simulator is a good example of this. There are a number of simulators available for the various crafts. Instructors can also create their own simulators, mock-ups, and models to demonstrate a process (*Figure 1*).

1.2.5 Mentoring Program

A formalized mentoring program can be an effective means for transferring the knowledge and skills needed to perform specific jobs tasks. The mentor approach provides an opportunity for trainees to observe and work with experienced co-workers. The mentoring program should have standards, master plans, or policies and procedures relating to the program. Establish policies and procedures in the following areas:

- Where and when mentoring takes place
- How mentors are selected, oriented, and trained
- How mentors and protégés are matched
- Who supervises mentoring pairs and how often that individual is in contact with each mentor/protégé pair
- Whom a mentor or a protégé should contact when problems arise
- How to handle complaints

Figure 1 Simulation and virtual reality.

- How to resolve problems in relationships or bring relationships to closure
- How to evaluate your success

A key component of the mentoring program involves selecting and training current members of the workforce as mentors. Mentors should be selected from the operations and maintenance workforce and trained on the basic requirements and goals of the program. Prospective mentors should be evaluated and selected for the following qualities:

- Thorough knowledge of the skills to be taught
- Safety awareness and positive work history
- Relevant work experience
- Effective communication and organization skills
- Commitment to the mentoring program

Once the mentors are selected and the program implementation is on a solid foundation, ensuring the program is well managed is crucial. A well-managed program promotes accuracy and efficiency, establishes credibility, and enables you to gauge progress effectively and identify areas that need improvement. Well-defined performance objectives for the individual mentoring pairs will allow you to evaluate the effectiveness of the program. Appropriate documentation and good recordkeeping is an important requirement for measuring the success of the program.

1.2.6 Safety Training Scenario Workshops

Safety training scenario workshops can be used to actively engage participants in discussions and activities focused on relevant safety topics. A safety scenario must be designed and developed that can be used to address both general and specific safety concerns. The training scenario should provide opportunity for individuals and groups to work through a realistic safety situation that they may encounter on the job. Once a scenario has been defined, the workshop activities are focused on capturing the participant's responses to the questions and safety concerns relative to the scenario. The workshop format allows the participants to apply and use their safety knowledge and skills in context with a realistic safety situation. The training scenario workshop also provides a means for the safety technician to evaluate and reinforce the safety knowledge and skills needed to promote a safe work site.

Training Scenario Example:

A pipe at a refinery (*Figure 2*) is leaking hot fluid from a pipe rack containing five lines 15' above ground. The pipe is wrapped in asbestos insulation. Company A is contracted to repair the leak. A tank located next to the pipe rack is having the hatch cover bolts replaced by Company B. Both jobs must continue at the same time to meet owner requirements. What is required to safely complete the projects on this multi-employer worksite? Who is responsible for maintaining control of this work at a multi-employee workplace? Who is required to verify containment of asbestos?

The following are guidelines for conducting the training scenario:

- *Go through the safety scenario together* – Project the scenario on a screen and as a group decide what option to take at each point.
- *Act it out* – Assign roles in the story and have participants read and act their lines once the group has decided what they should do.
- *Go through the scenario in small groups* – Divide participants into groups of four or so. Have them work through the scenario and then discuss the results with the larger group.
- *Require someone to defend each option* – To bring the discussion to a deeper level, assign each option to a participant: "Bob, you'll argue for option B every time, whether you agree with it or not. Give it your best shot!" You can do this in large or small groups. If you have four options at each decision point, you might create groups of four and, before they start the scenario, have each person choose an option to always defend.
- *Ask the group to improve the scenario* – Ideally, you tested an earlier version of the scenario on a sample of your audience and improved it based on their suggestions. Now do it again, but as a learning exercise. You could ask what options participants wanted to have but didn't, how the plot could be made more realistic, and how failure and success should be measured. Have groups design their own scenarios. After going through and discussing a scenario you wrote, form small groups and have them each design a branching scenario for their colleagues. You might offer a list of story ideas for them to choose from, each offering the opportunity to closely examine the complex decisions that happen on the job.

Figure 2 Typical plant refinery.

1.2.7 Audiovisual Training Aids

Proper use of audiovisual equipment can greatly improve the effectiveness of a class. When preparing to teach, you should plan on using some audiovisual aids to help the participants learn. These may include a computer presentation displayed by a projector, flipcharts, posters, whiteboards or chalkboards, videos, models, smart phone and tablet applications, learning management systems, and samples. Each of these can be used to great effect. However, overuse or misuse of audiovisuals can detract from the training by being distracting or overwhelming to the participants. *Figure 3* shows a laptop and projector. *Figure 4* shows an example of a portable screen.

Figure 3 Laptop and projector.

Table 1 lists some common audiovisual aids, their uses, and their advantages and disadvantages. This information will help you decide when and how to use some types of audiovisual equipment.

Audiovisuals can enhance your instruction, but they should not be relied upon to carry the training for you. As a rule of thumb, you should be able to conduct an effective course even if you lose or cannot use the audiovisuals you have prepared.

1.3.0 Preparing and Delivering Safety Training

Preparation and delivery of safety training includes the design, development, implementation and evaluation of the course materials and classes that support the training program. The implementation and evaluation phase focuses on class delivery and accessing the effectiveness of the individual classes and the safety program in general. Planning and preparing ahead of time is key to delivering successful safety training. This section includes the following topics related to preparing and delivering safety training:

- Coordinating arrangements
- Preparing to train
- Delivering the training
- Managing the classroom
- Concluding the training

1.3.1 Coordinating Arrangements

Coordinating the training arrangements means inviting the right people to be participants, getting a classroom that can accommodate all of the people, and making sure that the room, any necessary equipment, and all of the participants are available at the same time. The coordination process typically begins with a request or need for a safety related training session or meeting. Once the goals and objectives for the safety training are known, the appropriate people and training venue must be determined. Coordinating the appropriate resources involves the following activities:

- *Identifying and inviting the correct people* – Talk to the crew supervisor or foreman to determine who should be invited to the class. If there are a large number of participants, you may want to divide them up into more than one class. It is important to establish entrance requirements to ensure that everyone attending a class has received the necessary prerequisite training. If you are teaching a confined-space course, for example, you may want to require that all participants taking the course have received prior training in selecting and using respirators. If you have to stop the class to instruct one or two people on respirators, the rest of the class may not receive all of the training they need because the class will take longer than planned. The required prerequisite training should be included in your course announcement, and you should verify that all entrants

Figure 4 Portable screen.

Table 1 Common Uses of Visual Aids

Audiovisual Aid	Use	Advantages	Disadvantages	Requirements
Computer and projector	Used to show both static and moving (dynamic) text and graphics to a large group.	Presentation can provide a mixture of animations, graphics, text, and sound. This often makes training more effective, since it will appeal to the learners on many different levels.	Can be overused at the expense of other important teaching strategies, like hands-on training.	Programmed material created in advance. Electricity to run the projector and computer. A screen or blank wall in an area big enough to project the image.
Video/DVD	Used to show presentations incorporating both static and dynamic images to a group.	Excellent for showing procedural tasks being performed. Real reception.	Can be overused at the expense of other important teaching strategies. Unless specifically filmed locally in support of your instruction, videos may not accurately address the specific equipment being covered in the class.	Programmed video cassette or DVD. VCR or DVD player and television set. Electricity to power the player and television.
Flipcharts	Used to present or capture information so that it can be presented to a large group and preserved for later reference.	Flipcharts provide a useful means for capturing ideas from brainstorming sessions and group discussions. Flipchart pages can be posted in the classroom and saved for later reference.	Can be hard to read if writing is illegible or too small. It takes some practice to summarize and annotate information in real time.	Pads of flipchart paper, flipchart stands, large markers, and masking tape.
Whiteboard/ chalkboard	Used to present or capture information, illustrate a principle, or work out a problem with a large group.	Good for capturing ideas from brainstorming sessions or group discussions. Transparencies of schematics, drawings, or prints can be projected onto a whiteboard so that circuit paths can be traced or key points can be highlighted.	Whiteboards/ chalkboards can be hard to read if writing is illegible or too small. It takes some practice to summarize and annotate information in real time. Its use is not practical if it is desired to keep the recorded information for future reference.	Whiteboard/ chalkboard, dry erase markers/chalk, and erasers.
Tablet	Provides portable access to training content for individuals.	Can be used to support text-based, video-based, and e-learning content.	Small screens can limit the amount of viewable content being displayed to end user.	Tablet and network access to content management/ learning management portals
Simulation/virtual reality	Used when actual training on equipment or processes is difficult due to availability and accessibility.	Provides a way to train without subjecting people, equipment and processes to actual adverse consequences.	Cost can be a limiting factor depending on the complexity of the simulation and the supporting hardware and software.	Computer or specific simulator hardware and software.

have received the appropriate training. If participants need prerequisite training consider providing the required training before the full class starts. This is a better option than denying an applicant who may need the course.

- *Obtaining and securing a classroom* – Even if the majority of your class is going to be a hands-on course with equipment, you will still need a classroom area to conduct your class, especially if you plan to use slides, flipcharts, or a whiteboard. Work with the administrative staff, facilities crew, or supervisor to obtain and secure classroom space as close as possible to the equipment on which you will be training. Some facilities have classrooms or conference rooms available. Occasionally, training must be conducted in a lunchroom, break room, or office. It is important to make sure that the area you have for a classroom is large enough to accommodate all of the participants, and is as comfortable and free of distractions as possible.

- *Coordinating room and participant availability* – Perhaps the trickiest part of setting up a training class is trying to coordinate your schedule with the schedules of each of the participants, and then finding an available room for just that time. You may need to offer the class more than once to accommodate different schedules or shifts. While this increases the number of times you have to teach the class, it does make the classroom coordination much easier.

1.3.2 Preparing to Train

As the instructor, the class participants will see you as an expert. While this does not necessarily mean that you must know everything about the topic, you should be sufficiently prepared to either answer questions that arise or know where to look up the answers. Therefore, you must prepare yourself in advance of conducting the course, rather than relying on prepared materials and your general knowledge to carry you through. Don't be afraid to admit you don't know the answer. You can find the answer and follow-up at a later time.

When preparing to conduct a training course, make sure that you are well versed and familiar with the equipment, procedures, or tasks on which you will be training. Study any available documentation and regulations, practice the procedures and tasks yourself, and talk to the workers and their supervisors about their training needs and expectations. Documents to study may include as-built drawings of equipment, maintenance manuals, operation manuals, blueprints, company policies, OSHA regulations, or ANSI standards. This documentation will often form the foundation for the training, as well as provide valuable content.

Practice any procedures and tasks to be taught to make sure that you know how to safely and correctly perform them. For example, if you need to teach a group of workers how to set up and use a new band saw, practice with the saw beforehand to make sure that you can accurately demonstrate the correct procedure. You may wish to record your presentation beforehand and then review the video for potential areas of improvement.

While practicing, prepare your class notes on what you will say about the procedure and each step. Plan how you will teach key points such as setup, operation, troubleshooting, and repair/replace procedures. Make notes so that you will remember what you want to say when you are teaching the class.

Make sure that you talk to the workers and their supervisors about the training. You will want to know what kind of personnel will be in the class, what their roles are, and how much experience they have. This information will help you to tailor the information to meet the needs of the workers and the company.

Arrive at the training site well enough in advance to set up your classroom. You will need to set up an instructor area, arrange the classroom tables and chairs, set up your audiovisual equipment, and set out the participant materials. You may also wish to make arrangements for coffee, drinks, meals, and snacks.

Set up your training materials so that you can run the class smoothly and without a lot of pauses. You must set up the area from which you will do most of your speaking. You will need to keep your notes handy, along with any frequently used materials like manuals, markers, or a pointer. If space permits, you should set up an area to stage your supplies and materials. This may include any models or samples, extra participant materials, pens, pencils, pads, markers, and reference materials. These should be kept off to the side where you can get to them easily, but not where they will be in the way or distracting.

Arrange the participant tables and chairs the way you want them. Typically, if there is enough space, the tables are pulled together into a U-shape with the participants sitting around the outer perimeter. This arrangement allows for better interaction between the participants and the instructor. *Figure 5* shows a typical U-shaped seating strategy.

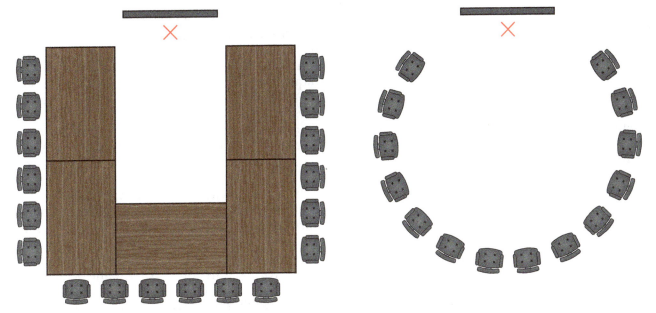

Figure 5 U-shaped seating.

If you are using any audiovisual equipment, you should set that up in advance as well. Any projectors should be set up, plugged in, and tested. If you are going to use a laptop to show a computer-driven slide presentation, make sure that you start up the computer and get the presentation running. It can be very embarrassing and time consuming to arrive in a classroom to find that a projector bulb is burned out or that your laptop isn't able to drive the multimedia projector.

If you are planning to use audio, make sure the speakers are loud enough for the classroom. Laptop speakers are generally audible only to people within a couple of feet of the laptop. If you plan on using flipcharts, you should make sure that you have enough flipchart stands and markers.

Finally, set up the participant materials. At a minimum, the participants should receive a notepad, pencil, and pen. Typically, participants receive copies of handouts and manuals. You may also want to distribute tent cards or name tags. These will be useful to you as you conduct the training so that you can address each participant by name. You may want to set each participant's area up with all of the class materials, or you may want to give them just notebooks, tent cards, pencils, and pens initially, and then distribute the other class materials as you get to them during the training.

Arriving at the training site early to set up the classroom, materials, and audiovisual equipment is essential to having a well-organized and successful training session. By doing this, you won't have to think about those details as you conduct the class, allowing you to focus on conducting the training.

1.3.3 Delivering the Training

After adequate preparation and planning, you are now ready to conduct the training session. Even though conducting a training session may seem like a daunting or frightening situation, the class can run smoothly with proper preparation. As you get more experience conducting training courses, you will become more confident and comfortable in front of a class. This section includes the following training delivery topics:

- Course and instructor introduction
- Class administration
- Icebreaker activities
- Reviewing the class objectives

Course and Instructor Introduction – It is important to set the proper tone from the very beginning of class. Practice handling the opening a few times before the class so that it runs as smoothly and efficiently as possible. This will help build your credibility with the participants from the outset. Your opening should include course and instructor introductions, class administration, a class icebreaker, and a review of the class objectives.

As the class starting time approaches, stand at the doorway to greet the participants as they arrive. If they are unfamiliar with the training site, show them where the restrooms are, where to keep their jackets, and where to get coffee, and then invite them to find their seats.

When you begin the class, introduce yourself by telling the class your name, company, job title, and background. Even if you know some or all of the class participants, it is important to reinforce your credibility and let them know why you are there.

After you have introduced yourself, introduce the course. Tell the participants the name of the course or the equipment that you will be talking about. Make sure that everyone in the room is in the correct class.

Class Administration – Some key points of information should be addressed to help the participants feel more comfortable in the classroom setting. These include the daily start and end times and frequency and length of breaks. The participants will appreciate knowing the course schedule. It is important to adhere to the schedule you present; otherwise, you will quickly lose credibility. You should also review the locations of restrooms, refreshments, and smoking areas. Explain the site emergency reporting and response procedures and where to go in the event of an emergency. Gauge how much information you give the students based on their familiarity with the training site.

At this point, explain the ground rules of the training and set the participants' expectations for the course. For example, if you would like to encourage questions and discussions throughout the delivery of the course, you should let the class know from the beginning. If you want to discourage cell phone or other interruptions, request that participants forward their calls or turn off their cell phones. Tell participants if you plan on spending most of the course time doing hands-on activities with equipment. It is important to anticipate and answer the questions that most participants have as they walk into a new class so that they can dedicate their focus to the course content.

Icebreaker Activities – A training class can feel like an artificial grouping of people who do not often work or associate with each other. In some cases, participants may not know others in the class. Other times, everyone in the class will be on the same crew or shift. Regardless, they are not typically used to being in a training class together. Icebreakers are activities that can help to overcome these barriers.

In its simplest form, the icebreaker can be as simple as going around the room and having everyone introduce themselves and give their titles and their years of experience. While this often suffices, some trainers like to enhance the icebreaker and speed up the process of familiarizing the class members with each other by adding other details into the introduction. An excellent question to ask everyone in the icebreaker is what they hope or expect to get out of the course. If you use this question, you may want to write the responses down on a sheet of flipchart paper and post it in the classroom after the icebreaker is completed.

One other technique that is often used in an ice breaker activity is pairing off the participants, having them ask each other the questions you prescribe, and then going around the room and having everyone take a turn introducing the person with whom they were just paired. This technique is useful in breaking down barriers such as office personnel/maintenance or operations personnel or first shift /second shift. These relationships are very real parts of the everyday workplace. Any steps you can take to remove those barriers from the classroom will make your job as an instructor much easier.

Be careful with icebreaker activities, as some participants may find them uncomfortable or embarrassing. Make sure that you know your audience well before asking them to do anything unusual or potentially embarrassing. There is no faster way to turn off a participant or an entire class than to force them to do something in an icebreaker that they would rather not do. If someone declines your request in an icebreaker, respect their choice and follow up with them one-on-one later to make sure that everything is all right.

Review the Class Objectives – Before jumping into the course content, it is important to review the course objectives with the class. A learning objective is a statement of what the learners will be expected to do once they have completed a specified course of instruction. It explains the knowledge, skills, and abilities (KSAs) that trainees are expected to learn. The US Office of Personnel Management defines KSAs as the attributes required to perform a job. These terms can be defined as follows:

- *Knowledge* – A body of information applied directly to the performance of a function. Having knowledge of pipe sizes and types would be an important knowledge attribute for pipefitters.
- *Skill* – An observable competence to perform a learned psychomotor act. Threading galvanized pipe is an example of a pipefitter skill.
- *Ability* – Competence to perform an observable behavior or a behavior that results in an observable product. The ability to interpret piping drawings is an example.

Objectives should be very simple sentences that state what the participants will be able to do once they have completed the training. This foreknowledge makes the training much more effective because it helps the participants to prepare themselves for what they are about to learn. An effective tool to use when writing objectives is the A-B-C-D method, illustrated in *Figure 6*. The A-B-C-D method helps you to make sure that each objective states the audience, behavior, conditions, and degree of acceptance.

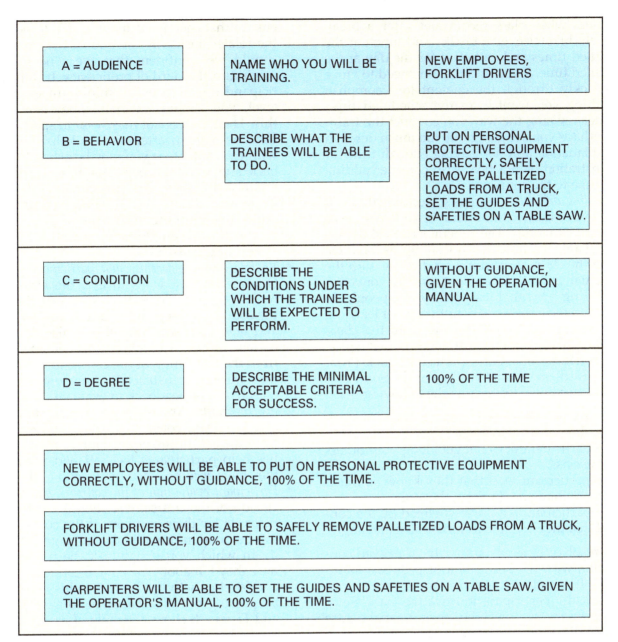

Figure 6 A-B-C-D method.

1.3.4 Managing the Classroom

As the instructor, you are responsible for making sure that the class runs smoothly, on schedule, and meets its objectives. This is known as *classroom management*, and it will often take care of itself if you have adequately prepared, organized, and outlined your class. Classroom management typically covers three elements:

- Pace of instruction
- Adherence to the schedule
- Handling difficult situations

Pace of instruction – It is the instructor's responsibility to make sure that the classroom pace is not too fast or slow. If content is covered too quickly, the participants will feel frustrated and they will not get as much value out of the training as they could. If content is covered too slowly, the participants will lose patience and their minds will begin to wander. Monitor the participants' comfort with the pacing by asking them questions. Use a mix of direct questions about the pacing of the course (e.g., *"Are we going too fast?"* or *"Did everyone get a chance to try that?"*) and questions that test for understanding (e.g., *"Who can tell me why we use a hearing protection?"* or *"Where did we say you could find that procedure in the manual?"*).

Adherence to schedule – It is very important to stick to the schedule you have set up. Begin the class at the designated time and do not wait for stragglers when class reconvenes after a break or lunch. Likewise, give breaks and lunch at the scheduled times and end class for the day at the scheduled time. If you find that you need to move too quickly through the content to accomplish everything you want to in the scheduled time, consider revising the way you conduct the training. Perhaps you are going into too much detail in the explanation and not leaving enough time for demonstration or practice. Also consider adding more time to the course.

Handling difficult situations – Occasionally, you may have to deal with difficult or rude people or awkward situations. The general rule is to defuse the situation as quickly and subtly as possible during the class, then address it directly with the individual or individuals in a private conversation during a break. Most of the time, people do not intend to be rude or disruptive. After learning that their behavior is detracting from the class, they are likely to improve their performance. Unfortunately, some participants will cause problems regardless of what you say or do. Some reasons for this behavior include the following:

- The participant was forced to attend a class they did not want to go to.
- The participant does not get along with others in the class.
- The participant feels that they know as much or more than you do about the topic.
- The participant is so overwhelmed by the content that they begin to feel frustrated and angry.

To resolve these cases, first try a private conversation with the participant or participants involved. If you do not see any improvement, you may ask the participant to leave the class. Since your primary responsibility to is to make sure that the class benefits each participant, you may be forced to ask someone to leave so that class may continue in a positive manner. If this does happen, make the request in private during a break so that the person may gather their materials and leave without further disrupting the class.

1.3.5 Concluding the Training

When you have finished delivering the course content and everyone has had enough time to practice the procedures and tasks covered during the course, there are a few things that must be done to conclude the course. These include the following:

- *Performance assessment* – Throughout the demonstrations and practice exercises of the course, you must monitor the participants' performance and offer coaching and feedback to ensure that each participant is able to perform the related tasks. By the end of the class, you should feel confident that everyone meets a minimum standard of proficiency. If some participants do not meet this minimum acceptable level, you should let their supervisor know this. Their supervisor may want to spend some time with them afterward to help them out with the content, or you may be asked to come back and tutor the participants to help them attain proficiency. The participants may need to attend **remedial training** and then re-attend the course in question. If none of these strategies works, the employee may need to be reassigned or dismissed.
- *Review course objectives and participant goals* – One of the best ways to end the class is to review the class objectives that you listed in the course opening and to briefly review when and how each was taught. It is also useful to review the participants' objectives and goals from the beginning of the course to verify that they were met. Instructors can use participant objectives and feedback to evaluate their training methods. Another way to evaluate course objectives and goals is by using the icebreaker from the beginning of the course. These activities help participants feel satisfied with what they have accomplished by the end of the class.
- *Participant evaluation of the course* – At the conclusion of the course, you should ask the participants for their opinions of the course. This can be done informally as a large group activity in which you simply ask the participants how they felt about the course content, pacing, and level of activity. Typically, this is handled by passing out an evaluation form. *Figure 7A* and *Figure 7B* show a sample of an evaluation form. The form is helpful because it allows each participant to anonymously voice his or her opinions on the course's content, flow, sequence, pacing, and level of activity. The evaluation is useful because it will provide valuable feedback so you can adjust and revise the training for the next class.
- *Course completion* – When conducting safety training, you will have to verify satisfactory course completion for each participant on the sign-in sheet or attendance list. It is important that the record of course completion be entered into their human resource records to prove compliance with OSHA and company policies/procedures. You may want to have personalized certificates of completion printed for each class participant. Handing out certificates symbolizes an accomplishment and is a great, positive way to conclude the class.

PARTICIPANT EVALUATION FORM – PAGE 1

Instructor _____   Name (Optional) _____

Date _____

	Poor 1	2	Average 3	4	Excellent 5	Not Applicable N/A
Introduction						
Title of lesson stated or displayed?						
Objectives stated?						
Motivation established?						
Lesson overviewed?						
Presentation						
Appropriate information level?						
Objectives covered?						
Presentation follows a logical sequence?						
Visual Aids						
Properly used?						
Illustrate the point?						
Visible by all?						
Out of sight when not in use?						
Used methods suggested in Guide?						
Questioning Techniques						
Different types asked?						
Sufficient number asked?						
Focused attention?						
Created discussion?						

Figure 7A Participant evaluation form (1 of 2).

PARTICIPANT EVALUATION FORM – PAGE 2

Instructor _____   Name (Optional) _____

Date _____

	Poor 1	2	Average 3	4	Excellent 5	Not Applicable N/A
Questioning *(continued)*						
Related information to applications?						
Adapted to level of participants?						
Summary						
Reviewed key points?						
Pointed out benefits to participants?						
Instructor Qualities						
Gestures & Mannerisms						
Eye contact						
Knowledge						
Voice						
Professional attitude						
Enthusiasm for subject						
Overall Comments						
Completed all material?						
Used Task Module properly?						
Followed all safety procedures?						
Additional Comments						
What did you like best in the way this lesson was presented?						
What one aspect of the presentation would you change and how would you present it differently?						

Figure 7B Participant evaluation form (2 of 2).

Additional Resources

Basic Safety Administration: A Handbook for the New Safety Specialist, Fred Fanning, CSP. 2003. Des Plaines, IL: The American Society of Safety Engineers (ASSE).

Construction Project Safety, John Schaufelberger and Ken-Yu Lin. 2014. Rockland, MA: RSMeans.

Construction Safety Planning, David V. MacCollum, P.E., CSP. 1995. Hoboken, NJ: John Wiley & Sons.

The Psychology of Safety Handbook, E. Scott Geller, Ph.D. 2016. Boca Raton, FL: CRC Press.

The Participation Factor—How to Increase Involvement in Occupational Safety, E. Scott Geller, Ph.D. 2002. Des Plaines, IL: The American Society of Safety Engineers (ASSE).

1.0.0 Section Review

1. Site-specific orientation, covers the high-risk areas or duties that exist on the job site, preventive measures to minimize the risk of incidents, and _____.
    a. material safety procedures for incidents
    b. OSHA inspection plans
    c. corrective action plans
    d. contingency plans in the case of an incident

2. Know-show-do is a very effective method for teaching _____.
    a. procedures or tasks
    b. company safety philosophy
    c. checklist items
    d. factual information

3. Entrance requirements are established to ensure that everyone attending a class has received the necessary _____.
    a. background checks
    b. training outlines
    c. prerequisite training
    d. safety data

Section Two

2.0.0 Introduction to Safety Meetings

Objective

Explain how to conduct a safety meeting.
a. Describe how to deliver and document a formal safety meeting.
b. Describe how to perform a toolbox/tailgate safety talk.

Performance Tasks

1. Deliver and document a formal safety meeting.

Trade Terms

Audiovisual materials: Materials such as photos, films, charts, and graphs that are designed to aid in learning or teaching by making use of both hearing and sight.

Five Ps for Successful Safety Talks: A technique for conducting effective safety talks involving five key elements: preparing, pinpointing, personalizing, picturing, and prescribing.

Open-ended questions: Questions that require more than a yes or no answer. These types of questions are used to encourage the audience to participate in the discussion.

Recordable incident rate: An equation that calculates the number of job-related injuries and illnesses, or lost workdays per 200,000 hours of exposure on a construction site.

Safety meetings are beneficial before the start of the day's tasks or after an incident occurs. Safety meetings can range from short, informal safety talks to long, formal meetings. Schedule safety meetings before the start of the days' tasks or after an incident occurs to provide the most benefit to the participants. Safety meetings can be used to exchange information regarding specific safety matters, defuse potential job disruptions by providing an outlet for problems, provide a written record of the actions taken to correct a problem, and establish an effective communications link between management and employees.

The length and location of a meeting depends on the safety issues to be discussed. Short, informal meetings cover specific jobs workers are doing. Formal safety meetings are longer and can feature a guest speaker, films, photos, or other audiovisual materials. Whether you are conducting a short, informal safety meeting or a formal safety meeting, it is important to provide workers with effective lessons that will help them do their jobs safely.

Depending on the employer and/or job site, formal safety meetings are held once a week for job-site personnel at a specific time and place. These meetings should last no more than an hour, and the audience should include all employees, including project managers, supervisors, and workers. On some jobs, subcontractors are included in the meetings. Many subcontractors have their own safety meetings as well.

Topics for a safety meeting can vary from hand-tool safety to emergency response. Safety meeting topics should always be practical and relevant to the work that is done on the site. Suggested safety meeting topics and resources for topics include the following:

- Recent incidents and near misses
- OSHA Fatal Facts
- NIOSH FACE reports
- Department of Energy Lessons Learned
- Safety data sheets (SDSs)
- Incident analysis
- Upcoming jobs or tasks
- National-interest topics such as National Fire Prevention Week
- OSHA training requirements
- Results of safety audits, inspections, and observations
- Examples of positive reinforcement (i.e., noticing and acknowledging employees who are doing things correctly)

2.1.0 Delivering and Documenting a Formal Safety Meeting

Safety meetings should be planned in advance. When they are planned in advance, the presented information will be organized and easy to understand. This helps to ensure that the message of the meeting, which will always be safety, gets through to the audience. As a safety technician, you will be responsible for facilitating these meetings. The following items should be considered when preparing for and conducting safety meetings:

- *Meeting site preparation* – Check to make sure all of the necessary equipment has been set up, if needed. Make sure equipment is working properly before the meeting starts. Contact your IT department if there are any issues with the equipment.

- *Time* – As a general rule, make sure that training is conducted during working hours. It should not be conducted during the employees' breaks or lunches or after the workday is complete.
- *Location* – Make certain that the meeting area is at a comfortable temperature and well lit. Check for distractions and background noise. Confirm that everyone can see and hear what is being presented.
- *Materials* – Provide handouts, visual aids, models, and examples to help your audience see as well hear the information being presented. This will reinforce what is being said in the presentation. Job site tools, equipment, and materials can also be used to reinforce concepts.
- *Presenter preparation* – Make sure you have a thorough understanding of the material you are about to present. This will help you to anticipate audience questions and needs. It will also increase your credibility as a presenter.
- *Group composition* – Ask all employees, supervisors, and upper-level managers to attend safety meetings. The presence of management demonstrates company interest in good safety practices.
- *Audience pet peeves* – Make sure you are aware of the things that will distract your audience and keep them from retaining the presented information. Some pet peeves include the following:
    - There is no real purpose or agenda for the meeting.
    - The objectives are unclear.
    - The topic doesn't apply to the audience.
    - The presenter is not prepared.
    - The meeting starts late or lasts too long.
    - There is no closure on the topic or a decision hasn't been made on the issue.
    - The meeting is used to complain about issues rather than solve them.
    - The presenter doesn't allow anyone in the audience to speak or participate in the discussion.
    - One person in the audience does all the talking, and everyone else is quiet.
- *Appropriate topic* – Make sure that the topic is related to work in progress, future activities, or recently completed tasks. Limit the number of topics to one or two per meeting.
- *Participant preparation* – Learn how to get the audience involved in the meeting (see the *Appendix*). State the objectives and purpose of the meeting at the beginning. This will help your audience know what they are expected to learn. Leave time at the end of the presentation for questions and answers.
- *Guest speakers / subject matter experts* – Invite manufacturers, industry experts, and contractors to discuss prior incidents or conduct demonstrations.

Good safety meetings incorporate all of these ideas. The key to making the presentation a success is knowing how to coordinate them. For example, before you can set up a time and place for a meeting, you must choose a relevant topic. Next, you need to make sure the meeting location has all of the equipment you may need, such as a chalk or drawing board, a TV and DVD player, or a projector and screen. Once you know that all of the equipment is in place, you can prepare any handouts or visual aids that will be used in the presentation.

Next, prior to the meeting date, you should practice what you are going to say. Practice until you are comfortable with the material. When the meeting begins, make sure the temperature of the room and the lighting are comfortable for the audience.

During your presentation, speak clearly and at a good pace. Watch the audience for signs of boredom or uneasiness (folded arms, leaning back in chairs, doodling, cell phone use, etc.). If you see participants losing interest, take a short break or involve them in the presentation when appropriate.

At the end of the presentation, leave room for a question and answer period to allow the participants to clarify anything that may not have been understood. Summarize the objectives in order to reinforce them. Incorporating all of these ideas ensures that the message of the meeting is understood and carried over into the audience's work. *Figure 8* shows a typical safety meeting.

Figure 8 Typical safety meeting.

2.1.1 Evaluating Safety Meetings

Conducting effective meetings can be a challenge. One way to improve the quality of meetings is to have the audience, including project managers and supervisors, provide feedback. The form shown in *Figure 9A* and *Figure 9B* is an example of a speaker evaluation form. This form provides the presenter with feedback about the effectiveness of the meeting. The feedback received from these forms should be taken seriously, especially if the overall score is average or below average.

This form can also be an effective tool in determining if the quality of safety meetings justifies their cost. This is important because coordinating and conducting safety meetings can

SPEAKER EVALUATION FORM – PAGE 1

Presenter _____ Name (Optional) _____

Date _____

	Poor 1	2	Average 3	4	Excellent 5	Not Applicable N/A
Introduction						
Title of lesson stated or displayed?						
Objectives stated?						
Motivation established?						
Lesson overviewed?						
Presentation						
Appropriate information level?						
Objectives covered?						
Presentation follows a logical sequence?						
Visual Aids						
Properly used?						
Illustrate the point?						
Visible by all?						
Out of sight when not in use?						
Used methods suggested in Guide?						
Questioning Techniques						
Different types asked?						
Sufficient number asked?						
Focused attention?						
Created discussion?						

Figure 9A Typical speaker evaluation form (1 of 2).

SPEAKER EVALUATION FORM – PAGE 2

Presenter _____   Name (Optional) _____

Date _____

	Poor 1	2	Average 3	4	Excellent 5	Not Applicable N/A
Questioning *(continued)*						
Related information to applications?						
Adapted to level of participants?						
Summary						
Reviewed key points?						
Pointed out benefits to participants?						
Instructor Qualities						
Gestures & Mannerisms						
Eye contact						
Knowledge						
Voice						
Professional attitude						
Enthusiasm for subject						
Overall Comments						
Completed all material?						
Followed all safety procedures?						
Additional Comments						
What did you like best in the way this lesson was presented?						
What one aspect of the presentation would you change and how would you present it differently?						

Figure 9B Typical speaker evaluation form (2 of 2).

be expensive. The equation in *Figure 10* shows exactly how expensive safety meetings can be. However, consider the cost of a meeting compared to the cost of an incident. If the effectiveness of meetings does not justify the cost, the number and length of meetings may be reduced. This would be detrimental to both the workers and the company itself. Using evaluation forms to make changes and improvements to presentations helps to document the effectiveness of meetings and therefore increase the likelihood that safety meetings will be done often and correctly.

Another way to improve the quality of meetings is to make their effectiveness a personal goal. *Figure 11* is an example of a form you can use to evaluate yourself as a presenter. By making sure you are doing the best possible job in presenting safety issues, you are ensuring that meetings continue to take place and that the quality of the meetings remains high. This can reduce the number of incidents and missed days of work.

2.1.2 Recordkeeping

Keeping records of both formal and informal safety meetings and toolbox/tailgate safety talks is an important part of any safety program. Recordkeeping provides documentation that proves that safety meetings are taking place, workers are attending, and actions are being taken to correct safety problems. Proving that safety meetings take place on a regular basis can help shield a company from legal liability, help control the recordable incident rate of the company, and justify adding safety meetings and training into the company's budget. For example, *Figure 12* shows the findings from a Construction Industry Institute report stating that the companies that included safety in their budgets had a lower recordable incident rate. This means less time and money spent on incidents. The Construction Industry Institute would not have been able to gather this information if accurate records of safety meetings had not been kept. The easiest way to document safety meetings is to record attendance by using a sign-up sheet like the one shown in *Figure 13*. In addition, the speaker evaluation forms can also be used as way to record safety meetings.

> **NOTE**
> Always review each safety meeting record to be sure that each attendee's name is clearly legible and that no fictitious names have been entered.

$E \times N \times R \times C$

WHERE:
- E = NUMBER OF EMPLOYEES
- N = NUMBER OF MEETINGS PER YEAR
- R = AVERAGE HOURLY WAGE RATE INCLUDING FRINGE BENEFITS
- C = AVERAGE LENGTH OF MEETING

EXAMPLE: 25 EMPLOYEES ATTEND 12 1-HOUR MEETINGS PER YEAR. THE AVERAGE HOURLY COST PER EMPLOYEE WITH BENEFITS = $27.00

$E \times N \times R \times C$
$25 \times 12 \times 1 \times 27 = \$8,100.00$

Figure 10 Cost-of-safety equation.

2.2.0 Toolbox/Tailgate Safety Talks

Toolbox/tailgate talks are short, informal safety meetings led by the crew supervisor for his or her crew. The safety technician's role in toolbox/tailgate talks is to act as a resource for the supervisor and attend the meetings when possible or needed. He or she should also maintain the records of toolbox talks and alert management if problems arise with the frequency/infrequency or quality of the meetings.

Toolbox/tailgate talks are designed to inform workers of specific hazards associated with a job. They can also act as a refresher to workers, reminding them about the hazards and safeguards of their job. Safety talks are usually held on site, near the location of the work.

Toolbox/tailgate safety talks can be held daily or weekly depending on company policy. They should always be conducted, however, when any of the following conditions exist:

- A certain job hasn't been performed for some time.
- New employees join the crew.
- A task and/or location poses specific hazards. For example:
  - Weather
  - Elevated work locations
  - Work over water
  - High-volume, high-speed traffic
  - Very limited sight distance in approach to a work zone
  - Areas known for high incident rates

SELF EVALUATION CHECKLIST

Ask yourself the following questions to determine your effectiveness as a presenter.

	YES	NO
TOPIC		
Are presented topics related to the work that is being done on site?		
Do you use other resources to look for relevant topics such as OSHA, NIOSH, eLCOSH, and the CDC?		
PREPAREDNESS		
Before each safety meeting do you:		
Inspect the job site for hazards related to your topic?		
Read over the material you plan to cover?		
Look up any terms or concepts you don't understand?		
Make sure you are familiar with any laws, regulations, and company rules related to the meeting's topic?		
Review reports of recent accidents on the site, including "near misses"?		
PARTICIPANT/CREW INVOLVEMENT		
Do you:		
Begin with a real-life example, or with information that will capture interest?		
Encourage full participation by the crew throughout the meeting?		
Invite the crew to ask questions and make suggestions related to the topic?		
Respond to questions that you can answer, and offer to find answers you don't know?		
Allow time at the end of the meeting for questions and suggestions on any safety issue?		
Ask the crew for feedback about the meeting?		
Involve the crew in preparing for and/or leading future safety meetings?		
FOLLOW UP		
Do you:		
Look into complaints, concerns, and suggestions that the crew brought up?		
Report back later to let the crew know what will be done?		
Keep good records of each tailgate meeting and other safety matters?		
CREDIBILITY/RESPECTABILITY		
Do you:		
Set an excellent safety example yourself?		
Invite crew members to come to you anytime with safety problems and suggestions?		
Encourage and reward safe work practices?		

Source: Oregon Occupational Safety and Health Administration (OR-OSHA)

Figure 11 Self-evaluation checklist.

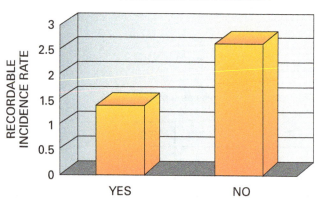

Figure 12 Recordable incident rate graph.

- Substantial changes in work conditions, weather, or procedures have occurred.
- A recent incident in this or another crew needs to be reviewed.
- A near miss has occurred.
- A supervisor feels employees are becoming lax about safety.

The best time to have a safety talk is before the workday begins. You can also review the accomplishments of the previous shift and set goals for the upcoming shift. This helps to remind workers about safety issues before they begin work rather than during or after work. Sometimes, however, if an incident occurs during the day, it is also appropriate to have a short safety talk after the event or at the end of the day to discuss it.

SIGN-IN FORM

TOPIC: ______________________________

Date Presented: ____________________ By: ____________________

Project Name/No.: ____________________ Location: ____________________

NAMES OF THOSE WHO ATTENDED THIS SAFETY MEETING

PRINTED NAME	SIGNATURE

Figure 13 Attendance sheet.

> **NOTE:** The Construction Industry Institute™ (CII) website (www.construction-institute.org) has more information about the effectiveness of safety meetings.

2.2.1 Safety Talk Preparation

Even though toolbox/tailgate talks are informal and take place on site, they should be planned in advance. The following are some helpful hints for organizing and conducting toolbox/tailgate safety talks.

- Start with a review of recent safe work and practices.
- Present short talks on topics related to current or upcoming work activities.
- Limit prepared talks to 10 to 15 minutes.
- Encourage employee participation.
- Use open-ended questions, ask for opinions, invite suggestions, and provide appropriate follow-up.
- Review recent near misses and workplace injuries.
- Discuss how near misses happened and how they could have been prevented.
- Review and discuss hazards encountered while working with other workers on site.
- Look ahead to potential safety hazards involved in upcoming work and remind workers of the proper use of safety equipment and the procedures to be followed in dealing with those hazards.

2.2.2 Five Ps for Successful Safety Talks

One technique that can help you give better safety talks is a technique called the Five Ps for Successful Safety Talks. You can easily apply this technique to safety talk topics that are related to work being done on a site.

The Five Ps for Successful Safety Talks are the following:

- *Prepare* – Think safety. Write ideas down for future use. Read safety materials thoroughly. Listen to others' ideas and attitudes. Organize and outline your talks. Practice what you are going to say. Know your audience and understand how to effectively communicate with them.
- *Pinpoint* – Concentrate on just one safety rule, first-aid hint, unsafe practice, or main idea. Too many topics can confuse your audience.

Figure 14 The Five Ps for Successful Safety Talks.

- *Personalize* – Establish common ground with your audience. Make the lesson personal and important. Use as many real-life examples as possible.
- *Picture* – Create clear mental pictures for your listeners. Appeal to both their ears and their eyes. Help them really see what you mean. Use visual aids.
- *Prescribe* – In closing your safety talk, tell workers what you expect them to do. Give them something to think about after the meeting.

The Five Ps technique is a good way to organize your thoughts and prepare a meeting that effectively communicates the purpose of the meeting: safety. *Figure 14* illustrates the Five Ps.

Additional Resources

Basic Safety Administration: A Handbook for the New Safety Specialist, Fred Fanning, CSP. 2003. Des Plaines, IL: The American Society of Safety Engineers (ASSE).

Construction Project Safety, John Schaufelberger and Ken-Yu Lin. 2014. Rockland, MA: RSMeans.

Construction Safety Planning, David V. MacCollum, P.E., CSP. 1995. Hoboken, NJ: John Wiley & Sons.

OSHA Fatal Facts, available at www.osha.gov/Publications/fatalfacts.html

The Psychology of Safety Handbook, E. Scott Geller, Ph.D. 2016. Boca Raton, FL: CRC Press.

The Participation Factor—How to Increase Involvement in Occupational Safety, E. Scott Geller, Ph.D. 2002. Des Plaines, IL: The American Society of Safety Engineers (ASSE).

The following websites offer resources for products and training:

Construction Industry Institute™ (CII), www.construction-institute.org

Occupational Safety and Health Administration (OSHA), www.osha.gov

2.0.0 Section Review

1. The easiest way to document safety meetings is to record attendance by using a _____.
   a. computer
   b. sign-up sheet
   c. feedback form
   d. checklist

2. Even though they are informal and take place on site, safety talks should be _____.
   a. vague and general
   b. spontaneous
   c. long and detailed
   d. planned in advanced

SUMMARY

Safety training is a fundamental requirement for ensuring that workers have the necessary safety skills and knowledge to perform their jobs safely. Proper preparation and planning ensures that the safety training program implementation and course delivery are done in an effective manner. A comprehensive safety training program needs to address the various types of training, including new employees, site-specific, craft-specific, supervisory, and continuing education training needs. Mentoring programs can also be used to support the safety program training initiatives.

Proper use of audiovisual training aids is another important aspect of the training program that should not be overlooked. Preparation and delivering of safety training includes coordinating arrangements, preparing to train, delivering the training, managing the classroom, and concluding the training. The overall goal is to ensure participants leave the safety training with a sense of confidence that they have the knowledge and skills to perform their jobs safely.

Safety meetings help to ensure workers are trained on how to work safely. They also provide an outlet and an opportunity for workers to share safety concerns, participate in the solutions that correct work place hazards, and provide feedback about the training they receive. Safety meetings provide safety technicians with opportunities to learn and improve on both the content and delivery of safety topics that address the specific needs of the workforce.

Safety technicians are typically responsible for coordinating and conducting all safety meetings, with the exception of toolbox/tailgate safety talks. These short, informal meetings are usually conducted by a crew's first-line supervisor. Safety technicians must know how to prepare, conduct, and record formal safety meetings. In addition, it is important to provide guidance to the supervisor on how to conduct effective and successful safety talks using appropriate resources.

Review Questions

1. Safety orientation training comes in two forms: new employee orientation and _____.
   a. craft-specific orientation
   b. procedural orientation
   c. site-specific orientation
   d. regulation orientation

2. The regulation that requires employers to instruct each employee to recognize, avoid, and work safely around any and all hazards they will encounter on the job site is _____.
   a. 29 CFR 1910.14
   b. 29 CFR 1926.11
   c. 29 CFR 1910.42
   d. 29 CFR 1926.21

3. A supervisor who looks out for workers' safety will help keep safety incidents and health risks to a minimum and reduce the _____.
   a. cost of training
   b. impact of an incident
   c. number of procedures
   d. number of complaints

4. The best way to prevent training information from becoming overwhelming is to teach the material _____.
   a. in a formal setting
   b. in an informal setting
   c. randomly
   d. in a logical sequence

5. Hands-on practice and demonstrations are the best way to teach people how to operate equipment, perform tasks, or _____.
   a. follow procedures
   b. justify plans
   c. submit comments
   d. evaluate policies and procedures

6. The key to delivering successful safety training is planning and _____.
   a. waiting for opportunities
   b. presenting old materials
   c. preparation ahead of time
   d. delaying delivery

7. Set up your training materials so that you can run the class smoothly and without a lot of _____.
   a. notes
   b. pauses
   c. details
   d. space

8. The length and location of a safety meeting depends on the _____.
   a. timing of the incident
   b. safety guidelines
   c. safety issues to be discussed
   d. time schedule for workers

9. One way to improve the quality of meetings is to have the audience, including project managers and supervisors, provide _____.
   a. feedback
   b. information
   c. refreshments
   d. statistics

10. Toolbox/tailgate talks are designed to inform workers of _____.
    a. specific orientation training
    b. upcoming events and notifications
    c. future regulations and laws
    d. specific hazards associated with a job

Trade Terms Introduced in This Module

A-B-C-D method: A device used to write training objectives that specifically cover the audience, behavior, conditions, and minimum degree of mastery.

Audiovisual materials: Materials such as photos, films, charts, and graphs that are designed to aid in learning or teaching by making use of both hearing and sight.

Five Ps for Successful Safety Talks: A technique for conducting effective safety talks involving five key elements: preparing, pinpointing, personalizing, picturing, and prescribing.

Know-show-do: A method of teaching tasks or procedures in which the instructor first explains necessary background information, then demonstrates the task or procedure, and concludes by having the participants practice the task or procedure.

Open-ended questions: Questions that require more than a yes or no answer. These types of questions are used to encourage the audience to participate in the discussion.

Prerequisite training: Training that provides the skills and knowledge required for a course that is to follow.

Recordable incident rate: An equation that calculates the number of job-related injuries and illnesses, or lost workdays per 200,000 hours of exposure on a construction site.

Remedial training: Training that is designed to address specific areas of difficulty, allowing participants to learn more about a larger course or topic.

Simulators: Teaching aids that are used to simulate a piece of equipment for training purposes rather than conducting training on the actual equipment.

What-why-how: A teaching method in which the instructor describes what an object, task, or procedure is, why it is used or done, and how to use or do it.

Appendix

MEETING TIPS: HOW TO GET THE CREW INVOLVED

Safety meetings work best if the whole crew actively participates. Here are some ways to encourage everyone to get involved.

Ask questions instead of lecturing – During the meeting, introduce each new point you want to make by asking the crew a question. After you ask each question, wait a short time to let people think. Then call on volunteers to answer. Use the answers as a springboard for discussion. Don't just read the answers. Avoid asking trainees to stand and read a section. Instead, engage them in an active discussion.

Ask about personal experience – If you ask a question and no one has an answer, rephrase the question. It may be too abstract. Try to make it more direct and personal. Ask if someone has had any personal experience that can help the group figure out an answer. For example, suppose no one can answer the question, "What are the health effects of breathing asbestos?" You could try to make the question more personal by asking, "Have you ever known anyone who got sick from working with asbestos? What kind of illness did they have?"

Limit the amount of time any one person can talk – If a crew member is talking too much, invite someone else to speak. Do it tactfully. For example, wait until the person pauses, quickly say "thank you," and then move along.

Role playing – Provide workers with a scenario involving a safety issue. Have the worker play the role of supervisor or safety technician. Ask them how they would resolve the issue in the scenario if they were the supervisor or safety technician.

Don't fake it – If someone has a question and you don't know the answer, don't guess or fake an answer. Write the question down. Promise that you will get back to the person, and then make sure you do.

Stick to the topic – If the crew's questions and comments move too far from the topic, tell them that their concerns can be addressed later; either in private conversation or in an upcoming safety meeting.

Competition – Add some activities that make workers compete against each other for prizes. For example, use the game show type format using questions about the topic and reward the winner. Winning prizes can range from a t-shirt to a gift certificate.

Additional Resources

This module presents thorough resources for task training. The following resource material is suggested for further study.

29 *CFR 1926. Safety and Health Regulations for Construction.* **www.ecfr.gov**
30 *CFR 48, 70, 71, 74, 75, 77, and 90. Mine Act Standards.* **www.ecfr.gov**
Basic Safety Administration: A Handbook for the New Safety Specialist, Fred Fanning, CSP. 2003. Des Plaines, IL: The American Society of Safety Engineers (ASSE).
Construction Project Safety, John Schaufelberger and Ken-Yu Lin. 2014. Rockland, MA: RSMeans.
Construction Safety Planning, David V. MacCollum, P.E., CSP. 1995. Hoboken, NJ: John Wiley & Sons.
OSHA Fatal Facts, available at **www.osha.gov/Publications/fatalfacts.html**
The Psychology of Safety Handbook, E. Scott Geller, Ph.D. 2016. Boca Raton, FL: CRC Press.
The Participation Factor—How to Increase Involvement in Occupational Safety, E. Scott Geller, Ph.D. 2002. Des Plaines, IL: The American Society of Safety Engineers (ASSE).
The following websites offer resources for products and training:
  Construction Industry Institute™ (CII), **www.construction-institute.org**
  Occupational Safety and Health Administration (OSHA), **www.osha.gov**

Figure Credits

CM Labs Simulations Inc. - www.cm-labs.com, Figure 1
U.S. Environmental Protection Agency, Figure 2
© zentilia/Shutterstock.com, Figure 3A
© iStockphoto.com/iShootPhotosLLC, Figures 3B
Elite Screens, Inc., Figure 4
Oregon Occupational Safety and Health Administration, Figure 11
Construction Industry Institute, Figure 12

Section Review Answer Key

Answer	Section Reference	Objective
Section One		
1. d	1.1.2	1a
2. a	1.2.2	1b
3. c	1.3.1	1c
Section Two		
1. b	2.1.2	2a
2. d	2.2.1	2b

NCCER CURRICULA — USER UPDATE

NCCER makes every effort to keep its textbooks up-to-date and free of technical errors. We appreciate your help in this process. If you find an error, a typographical mistake, or an inaccuracy in NCCER's curricula, please fill out this form (or a photocopy), or complete the online form at **www.nccer.org/olf**. Be sure to include the exact module ID number, page number, a detailed description, and your recommended correction. Your input will be brought to the attention of the Authoring Team. Thank you for your assistance.

Instructors – If you have an idea for improving this textbook, or have found that additional materials were necessary to teach this module effectively, please let us know so that we may present your suggestions to the Authoring Team.

NCCER Product Development and Revision
13614 Progress Blvd., Alachua, FL 32615

Email: curriculum@nccer.org
Online: www.nccer.org/olf

❏ Trainee Guide ❏ Lesson Plans ❏ Exam ❏ PowerPoints Other _____

Craft / Level: _____ Copyright Date: _____

Module ID Number / Title: _____

Section Number(s): _____

Description: _____

_____

_____

_____

Recommended Correction: _____

_____

_____

_____

Your Name: _____

Address: _____

_____

Email: _____ Phone: _____

Permits and Policies

Overview

This module provides an overview of the various work permits that are typically required on a construction site. It explains how to complete a hot work permit and confined space entry permits, and describes lockout/tagout procedures.

Module 75224

Trainees with successful module completions may be eligible for credentialing through NCCER's Registry. To learn more, go to **www.nccer.org** or contact us at 1.888.622.3720. Our website has information on the latest product releases and training, as well as online versions of our *Cornerstone* magazine and Pearson's product catalog.

Your feedback is welcome. You may email your comments to **curriculum@nccer.org**, send general comments and inquiries to **info@nccer.org**, or fill in the User Update form at the back of this module.

This information is general in nature and intended for training purposes only. Actual performance of activities described in this manual requires compliance with all applicable operating, service, maintenance, and safety procedures under the direction of qualified personnel. References in this manual to patented or proprietary devices do not constitute a recommendation of their use.

Copyright © 2018 by NCCER, Alachua, FL 32615, and published by Pearson, New York, NY 10013. All rights reserved. Printed in the United States of America. This publication is protected by Copyright, and permission should be obtained from NCCER prior to any prohibited reproduction, storage in a retrieval system, or transmission in any form or by any means, electronic, mechanical, photocopying, recording, or likewise. To obtain permission(s) to use material from this work, please submit a written request to NCCER Product Development, 13614 Progress Blvd., Alachua, FL 32615.

75224 V2

From *Safety Technology,* Trainee Guide. NCCER.
Copyright © 2018 by NCCER. Published by Pearson. All rights reserved.

75224
PERMITS AND POLICIES

Objectives

When you have completed this module, you will be able to do the following:

1. Explain how to issue various types of work permits.
    a. Describe how to complete a hot work permit.
    b. Describe how to use lockout and tagout devices.
    c. List the requirements for excavations.
    d. Identify types of chemical hazards.
    e. Identify other types of work permits.
2. Explain how to identify the atmospheric hazards in confined spaces.
    a. Identify the types of atmospheric hazards that may be present in a confined space.
    b. Describe how to perform atmospheric testing.
3. Explain how to prepare workers for confined space entry.
    a. List the types of information required on a confined space entry permit.
    b. Identify the roles and responsibilities of personnel in confined spaces.
    c. Identify and describe the types of emergency training required for workers in confined spaces.

Performance Tasks

Under the supervision of your instructor, you should be able to do the following:

1. Perform a pre-inspection.
2. Complete a hot work permit.
3. Use lockout and tagout devices.

Trade Terms

Affected employees
Asphyxiation
Atmosphere
Atmospheric contaminants
Authorized employees
Competent person
Confined space
Confined space entry permit system
Energy control procedures
Energy-isolating devices
Engulfment
Excavation
Fire watch
Hot work
Induced hazards
Inerting
Inherent hazards
Intrinsically safe
Lockout
Lower explosive limit (LEL) or lower flammable limit (LFL)
Non-permit required confined space
Other employees
Oxygen-deficient atmosphere
Oxygen-enriched atmosphere
Parts per million (ppm)
Permissible exposure limit (PEL)
Permit-required confined space (PRCS)
Safety data sheet (SDS)
Self-contained breathing apparatus (SCBA)
Specific gravity
Tagout
Trench
Upper explosive limit (UEL) or upper flammable limit (UFL)
Vapor density

Industry Recognized Credentials

If you are training through an NCCER-accredited sponsor, you may be eligible for credentials from NCCER's Registry. The ID number for this module is 75224. Note that this module may have been used in other NCCER curricula and may apply to other level completions. Contact NCCER's Registry at 888.622.3720 or go to www.nccer.org for more information.

Contents

1.0.0 Issuing Work Permits 1
   1.1.0 Hot Work Permits 3
   1.2.0 Lockout/Tagout Policies and Procedures 8
      1.2.1 LOTO Procedures Training 9
      1.2.2 LOTO Procedure 10
      1.2.3 LOTO Exceptions 12
      1.2.4 Additional LOTO Safety Requirements 12
   1.3.0 Excavation Policies and Procedures 13
   1.4.0 Chemical Hazards Policies and Procedures 16
   1.5.0 Additional Types of Work Permits 21
      1.5.1 Aerial Lifts 21
      1.5.2 Vehicles 21
      1.5.3 Electrical Work 21
      1.5.4 Facility Siting 21
2.0.0 Confined Spaces 23
   2.1.0 Atmospheric Hazards in Confined Spaces 24
      2.1.1 Oxygen 26
      2.1.2 Flammable Atmospheres 26
      2.1.3 Measuring Flammable Atmospheres 31
      2.1.4 Toxic Atmospheres 32
      2.1.5 Environmental Conditions 33
      2.1.6 Using Tools and Equipment in Confined Spaces 33
   2.2.0 Atmospheric Testing Procedures for Confined Spaces 34
      2.2.1 Testing Procedures for Oxygen 35
      2.2.2 Testing Procedures for Flammable Atmospheres 35
      2.2.3 Toxic Air Contamination Testing 36
      2.2.4 Ventilating Confined Spaces 37
3.0.0 Confined Space Entry Procedures 41
   3.1.0 Required Information on a Confined Space Entry Permit 42
   3.2.0 Roles and Duties of Personnel in Confined Spaces 42
      3.2.1 Entrant 42
      3.2.2 Attendant 42
      3.2.3 Supervisor 43
   3.3.0 Emergency Training 43
      3.3.1 Loss of Air Line 43
      3.3.2 Loss of Communications 43
      3.3.3 Inadequate Illumination 43
      3.3.4 Rescue Procedures 44

Figures and Tables

Figure 1 Take 5 for Safety. 4
Figure 2A Work permit (1 of 2). 5
Figure 2B Work permit (2 of 2). 6
Figure 3 Welding. 7
Figure 4 Safety sign. 8
Figure 5 Lockout/tagout devices. 9
Figure 6 Example of lockout tag. 10
Figure 7 Examples of lockout devices. 10
Figure 8 Group lockout. 13
Figure 9 Sloping the sides of an excavation site for different soil types. 14
Figure 10 Sloped trench and benched trench. 15
Figure 11 Trench shields. 15
Figure 12 GHS labels. 17
Figure 13A Safety data sheet (1 of 2). 18
Figure 13B Safety data sheet (2 of 2). 19
Figure 14 Hazardous material sign. 20
Figure 15 Confined space posting. 24
Figure 16 Different types of respirators. 25
Figure 17 Flammable vapor thresholds. 32
Figure 18 Gas detection meters. 34
Figure 19 Flammability detection meters. 35
Figure 20 Gas detector tubes. 36
Figure 21 Atmospheric air. 37
Figure 22 Air horn ventilation with bonding and grounding. 38
Figure 23 Positive ventilation system. 38
Figure 24 Fume extractor used for local exhaust. 39
Figure 25 Non-entry rescue. 44
Figure 26 Emergency response team responding to a confined space emergency. 45

Table 1 Sample Lockout/Tagout Procedure Checklist 11
Table 2 Symptoms of Oxygen Deficiency 26
Table 3A Summary of Hazardous Atmospheres 29
Table 3B Summary of Hazardous Atmospheres 30
Table 4 PELs of Common Toxic Gases 33

Section One

1.0.0 Issuing Work Permits

Objective

Explain how to issue various types of work permits.
a. Describe how to complete a hot work permit.
b. Describe how to use lockout and tagout devices.
c. List the requirements for excavations.
d. Identify types of chemical hazards.
e. Identify other types of work permits.

Performance Tasks

1. Perform a pre-inspection.
2. Complete a hot work permit.
3. Use lockout and tagout devices.

Trade Terms

Affected employees: Used in reference to LOTO policies and procedures. Affected employees are usually the equipment users or operators who must be able to recognize when a control procedure is in place. They also need to understand that they are not to remove the LOTO device or try to start the equipment during a control procedure.

Atmosphere: The air or climate inside a specific place.

Authorized employees: Used in reference to LOTO policies and procedures. Authorized employees perform the actual servicing or maintenance on the equipment. These workers need a high degree of training. They also need to know about the type and magnitude of the hazardous energy involved and how to safety isolate and control the energy source.

Competent person: According to OSHA, a person who is capable of identifying existing and predictable hazards in the surroundings, or working conditions which are unsanitary, hazardous, or dangerous to employees, and who has authorization to take prompt corrective measures to eliminate them.

Confined space: A workplace that has a configuration that hinders the activities of employees who must enter, work in, and exit the space.

Energy control procedures: Written documents containing information that authorized workers need to know in order to safely control energy while servicing or maintaining equipment or machinery; also called lockout/tagout procedures.

Energy-isolating devices: Devices that physically prevent the transmission or release of energy.

Excavation: Any man-made cut, cavity, trench, or depression in the earth's surface formed by earth removal.

Fire watch: Assigning trained personnel to watch for fires during welding and cutting operations, sound the alarm in the event of a fire, and extinguish any fires within the capacity of the available equipment. A fire watch must be maintained for at least one-half hour after completion of welding or cutting operations.

Hot work: Any work function that involves ignition or combustion; examples include welding, burning, cutting, and riveting.

Intrinsically safe: An electric tool or device that is UL-rated to be explosion-proof under normal use. For example, a fuel pump must be intrinsically safe to prevent explosions sparked by static electricity.

Lockout: The placement of lockout devices on an energy-isolating device in accordance with established procedures. The lockout device is intended to use a positive means, such as a lock, to secure the energy-isolating device in the Off or Safe position.

Other employees: All employees who are or may be in an area where energy control procedures may be utilized.

Permissible exposure limit (PEL): The limit set by OSHA as the maximum concentration of a substance that a worker can be exposed to in an 8-hour work shift. Most flammable gases have a PEL defined as 10 percent of the LEL/LFL, while toxic gases have individual PELs.

Permit-required confined space (PRCS): A confined space that has real or potential hazards including atmospheric, physical, electrical, or mechanical hazards as defined by OSHA.

Safety data sheet (SDS): A form that lists the hazards, safe handling practices, and emergency control measures for a specific chemical.

Tagout: The placement of tags or tagout devices on an energy-isolating device in accordance with established procedures. It is intended to indicate that the energy-isolating device and the equipment being controlled must not be operated until the tagout device is removed.

Trench: A narrow excavation made below the surface of the ground. The depth of a trench must be greater than the width. The width must not exceed 15 feet (4.6 m).

A number of work permit systems are used in the construction industry, each with safety as its primary focus. Permit systems are designed and developed to reduce the chances of incidents and injuries to workers by identifying potential hazards. Permit systems are also used to help prevent damage and loss of property. Work permits provide a means of communication between the host employer/controlling contractor and the personnel who are performing the work.

Site safety technicians are responsible for advising site management on all the safety issues for a given project. They must be aware of site requirements for work permits and have a good working knowledge of the hazards and safeguards associated with the various types of permits used on site.

There are a number of different types of work permits and related systems, including the following:

- Hot work permits
- Cold work permits
- Hazardous work permits
- Lockout/tagout devices
- Electrical work permits
- Excavation and trenching permits
- Line entry permits
- Confined space permits
- Crane/critical-lift work permits
- Ladder permits
- Aerial lift permits
- Elevated work permits
- Vehicle permits
- General/unit work permits (*e.g.*, in petrochemical facilities)

Not all of these permits are required by OSHA, but they may be required by the employer or job site. When work permits are issued, it is usually for a specific task. It is poor practice to issue one blanket work permit for the performance of a number of tasks. Furthermore, permits are usually issued for a short, specified period of time. Some permits may be valid for an entire workday or shift, while others may be valid only between the hours specified on the permit.

The host employer/controlling contractor usually issues permits. This person could be the owner of a site, but most likely it will be a contractor or construction manager. For the purposes of this module, this entity is referred to as the controlling employer.

The controlling employer is responsible for selecting qualified workers and subcontractors. Safety technicians may work directly for the controlling employer or for a subcontractor; either way, the job is the same. They need to make sure that the company follows the site's work permit policies and procedures. They also need to make sure that workers in their area(s) of responsibility are following the policies and procedures properly.

The controlling employer must explain site work permit requirements to all contractors and subcontractors. This person is also responsible for monitoring the work site, auditing the subcontractor's work, and ensuring that all work performed at the site is completed safely.

Safety policies may vary between the controlling employer, contractors, and subcontractors. Always follow the most stringent policies. It is critical that everyone on site agrees to the procedures being used.

Both the controlling employer and subcontractors must ensure that the workers they hire are trained to do the job, and that they understand the safety requirements associated with it. They are also responsible for ensuring that the equipment in use is safe and operating properly.

As a safety technician, your role can vary depending on the job site. On some sites, you may be working under the direction of the site safety manager; other times, you may be working on your own. It is important that your roles and responsibilities are clearly defined so that you can effectively ensure the safety of everyone on the job site.

Safety technicians may be involved in performing the following tasks:

- Training workers
- Pre-inspecting the work area
- Specifying job conditions and practices
- Completing any needed paperwork
- Observing permit-required work

For workers to be allowed to perform certain tasks, they may be required to complete additional training. In some cases, they will need to attend a class. Other times, the safety technician will conduct the training. It is important to maintain employee records that document any additional training that the workers receive. Include the type and date of the training and explain how the workers were able to demonstrate an understanding of their roles and responsibilities. Both the trainer and the worker should sign the training record.

The purpose of work permits is to reduce the risk of incidents and injuries to workers while specific tasks are being performed. They are also designed to reduce the chance of property damage and pollution. However, having a work permit doesn't guarantee that a job will be completed safely. A work permit is simply a tool, and like any tool it works best when it is used properly.

It is an excellent tool for making everyone on the job site aware of potential hazards and safeguards. It is also a useful tool for communicating to others on the site that a specific type of work is being performed.

For almost all work permits, a pre-inspection of the work area is needed. The safety technician may have a checklist or form to complete. The purpose of the pre-inspection is to locate and correct any conditions that could become a hazard while performing the work for which the permit is being requested. *Figure 1* shows an example of a pre-task planning form. Sometimes this type of paperwork needs to be shown to the controlling employer's workers before a work permit is issued.

Once the work area has been prepared and the necessary paperwork completed, a permit can be issued. Work permit policies vary from site to site. At some sites, you may receive a work permit from the controlling employer after you have completed a written pre-task checklist, inspection, or other item. Work permits must be written and documented per site policy (*Figure 2A* and *Figure 2B*).

Regardless of how permits are issued at the site, the worker and first-line supervisor are responsible for the safe completion of the task. They need to be sure that all of the necessary paperwork is completed and that the area and equipment are safe. The role of the safety technician during these operations is that of a site consultant.

The safety technician must observe the work area while the work is being performed and should also be available if the crew leader needs advice on assigning safety observer duties to other workers who have been trained for this role (such as a fire watch or confined space attendant). Safety observers are responsible for identifying and correcting any potential hazards on the spot. It may be necessary to continue to observe the work area after the task has been completed, such as when a fire watch is required at the site.

One of a safety technician's most important responsibilities is to serve as a resource to site management on safety, health, and in some cases, environmental regulations. This involves interacting with workers, subcontractors, and the public. In the role of safety technician, you will be a trainer, motivator, auditor, planner, and advisor; in short, a key player in the organization.

In addition, safety technicians also have the following responsibilities:

- In the absence of a site safety manager or supervisor, represent the company during visits by regulatory agencies
- Provide safety training for both new and experienced workers
- Participate in the development, review, or revision of Job Safety Analyses (JSAs), Task Safety Analyses (TSAs), work plans, incident reporting forms, and emergency action plans
- Audit and inspect the job site or work activities
- Anticipate, identify, and have management correct safety hazards
- Audit compliance with regulatory requirements
- Know how to complete required forms
- Observe work in progress to make sure safe work methods are used
- Use proper coaching techniques to correct unsafe behavior and reward safe behavior
- Conduct safety meetings
- Audit compliance with work permits
- Audit permit-required work areas
- Assist site management in conducting incident investigations
- Analyze data gathered during incident investigations
- Manage the site safety and health recordkeeping system
- Serve as a liaison between the job site and insurance company representatives
- Provide or coordinate first aid and access to follow-up medical care

Although there are a number of different types of work permits, this module primarily focuses on hot work permits, lockout/tagout procedures, confined space entry permits, and excavation and trenching permits. Confined space work permits are an important part of a permit safety program due to the potential hazards associated with confined spaces. 29 *CFR* 1910.146 covers the requirements for practices and procedures to protect employees in general industry from the hazards of working in a permit-required confined space (PRCS).

Safety technicians need to know when one or more work permits are required to perform a task. They must be able to spot and correct conditions that could become hazards while work requiring permits is being performed.

1.1.0 Hot Work Permits

Hot work permits are required because ordinary work conditions (such as maintenance) can become hazardous when hot work is performed in the area. A hot work permit is required whenever an open flame or other type of ignition source is required to complete a task. Common types of hot work include welding, grinding, burning, and

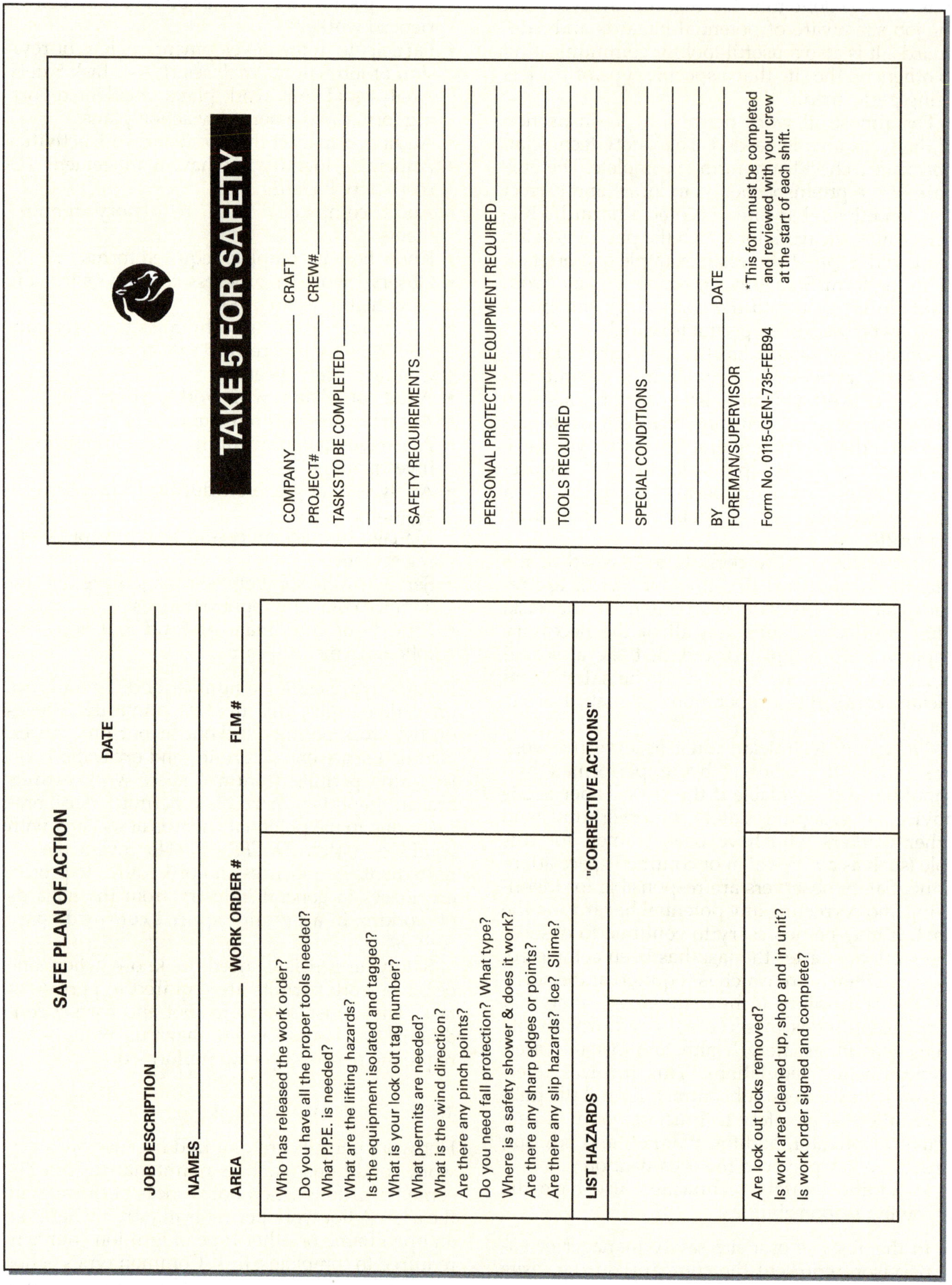

TAKE 5 FOR SAFETY

COMPANY _____ CRAFT _____
PROJECT# _____ CREW# _____
TASKS TO BE COMPLETED _____

SAFETY REQUIREMENTS _____

PERSONAL PROTECTIVE EQUIPMENT REQUIRED _____

TOOLS REQUIRED _____

SPECIAL CONDITIONS _____

BY _____ DATE _____
FOREMAN/SUPERVISOR

Form No. 0115-GEN-735-FEB94

*This form must be completed and reviewed with your crew at the start of each shift.

SAFE PLAN OF ACTION

JOB DESCRIPTION _____ DATE _____
NAMES _____
AREA _____ WORK ORDER # _____ FLM # _____

LIST HAZARDS
- Who has released the work order?
- Do you have all the proper tools needed?
- What P.P.E. is needed?
- What are the lifting hazards?
- Is the equipment isolated and tagged?
- What is your lock out tag number?
- What permits are needed?
- What is the wind direction?
- Are there any pinch points?
- Do you need fall protection? What type?
- Where is a safety shower & does it work?
- Are there any sharp edges or points?
- Are there any slip hazards? Ice? Slime?

"CORRECTIVE ACTIONS"

- Are lock out locks removed?
- Is work area cleaned up, shop and in unit?
- Is work order signed and complete?

Figure 1 Take 5 for Safety.

Burning - Welding - Hot Work Permit

Valid from _____ to _____ , _____   Master Card No._____
           (am/pm)   (am/pm)   DATE

1. Work Description
  Equipment Location or Area _____
  Work to be done:

2. Gas Test
  ☐ None Required

	Test Results	Other Tests	Test Results
☐ Instrument Check			
☐ Oxygen 20.8% Min			
☐ Combustilble % LFL			

Gas Tester Signature   Date   Time

3. Special Instructions   ☐ None   ☐ Check with issuer before beginning work

4. Hazardous Materials   ☐ None   What did the line / equipment last contain?

5. Personal Protection   ☐ Standard Equipment: welders hood with long sleeves; cutting goggles

  ☐ Goggles or Face Shield   ☐ Respirator   ☐ Forced Air Ventilation

  ☐ Standby Man   ☐ Other, specify: _____

6. Fire Protection   ☐ None Required   ☐ Portable Fire Extinguisher

  ☐ Fire Watch   ☐ Fire Blanket   ☐ Other, specify: _____

7. Condition of Area and Equipment

Required
Yes No   THESE KEY POINTS MUST BE CHECKED

Yes	No		
		a.	Lines disconnected & blanked or if disconnecting is not possible, blinds installed?
		b.	Lines steamed, purged, or otherwise properly cleared of combustibles?
		c.	Area and equipment satisfactorily clean of oil or combustibles?
		d.	Trenches, catch basins & sewer connections properly covered or sealed?
		e.	Immediate area and/or area under the work barricaded or roped off?
		f.	Adjoining equip. & operations checked to have any effect on the job?
		g.	Area fire suppression (fire water and sprinkler system) in service?

Comments

Figure 2A Work permit (1 of 2).

Burning - Welding - Hot Work Permit

8. Approval	Permit Authorization			Permit Acceptance		
	Area Supv.	Date	Time	Maint.Supv./Engineer Contractor Supv.	Date	Time
Issued by						
Endorsed by						
Endorsed by						

9. **Individual Review**

   I have been instructed in the proper Hot Work Procedures

   Signed   Signed

   Persons Authorized
   to Perform Hot Work

   _____   _____

   _____   _____

   _____   _____

   _____   _____

   _____   _____

   Fire Watch   _____   _____

10. **Job Completion**
    - ☐ Yes ☐ No   Is the work on the equipment completed?
    - ☐ Yes ☐ No   Has the worksite been cleaned and made safe?

    Workman answering above questions _____

    Issuer's Acceptance _____

    Forward to Production Superintendent within 7 days of job completion

Figure 2B Work permit (2 of 2).

soldering. A hot work permit must document that the fire prevention and protection requirements in 29 CFR 1910.252(a) have been implemented prior to beginning the hot work operation. It must also indicate the date(s) authorized for hot work and identify the object on which hot work is to be performed. The permit must be kept on file until completion of the hot work.

Hazards associated with hot work include the following:

- Gases
- Fumes
- Sparks
- Hot metal
- Heat
- Combustible dust or fibers
- Ignition of nearby construction materials

In these cases, the use of intrinsically safe (explosion-proof) equipment is required. For example, you cannot take a tablet or cell phone into certain hazardous areas, such as a Class I, Division 1 area (see 29 CFR 1926.407). All employees must be aware of the safety requirements of each area they enter.

On a typical site, the supervisor will notify the safety technician that a hot work permit is required. The safety technician and the crew leader should discuss the following questions:

- Is there another way to do the job?
- In what area does the job need to be performed?
- Can the area easily be made safe for hot work?
- Can the work be moved to another, safer area?

Once it is determined that the hot work permit is needed, the crew leader will complete a pretask planning form (similar to the one shown in *Figure 2A* and *Figure 2B*). This type of form may be used to apply for the hot work permit.

The crew leader and safety technician will inspect the work area for potential hazards. Flammable materials, such as wood shavings, must be removed or safeguarded. When an item needs to be moved, applicable codes should be reviewed to establish safe separation distances. The *NFPA 400, Hazardous Materials Code* consolidates fundamental safeguards for the storage, use, and handling of hazardous materials in all occupancies and facilities. *NFPA 400* and *NFPA 55* address fire and life safety requirements applicable to a wide range of substances.

Flammable items that cannot be moved must be covered with a fire-resistive material, such as specially designed fire-resistant blankets, curtains, or covers. In some cases, items may be dampened with water.

Assess the floors and nearby walls to determine whether cracks are present. Sparks could roll into a crack and smolder for hours before a fire becomes obvious. Ensure that all open sewers in the area are covered and sealed, since they can be a source of flammable methane gases as well as other gases and vapors.

While you are inspecting the area, check for adequate ventilation. This is important even if the area is not considered a confined area. Because welding (*Figure 3*), burning, and other hot work can form fumes and toxic gases, adequate air flow is important. The types of fumes generated by welding depend on the type of metal being welded and the type of welding rods used. Some welding fumes contain lead, cadmium, and chromium. Toxic gases produced by hot work include ozone and carbon monoxide.

Before any hot work can begin, a flammable gas test must be performed to ensure that no flammable gases or vapors are present. These gases and vapors not only represent a fire/explosion risk, but they are also potentially hazardous to workers who may be exposed to them through adsorption, ingestion, or inhalation.

Once the physical area has been made safe, the equipment needs to be inspected. During an equipment inspection, check the following:

- The tools/equipment that will be used during the operation
- Personal protective equipment that workers will wear
- Availability and proximity of safety equipment such as fire extinguishers, automated external defibrillators (AED) and eye wash stations

When the safety technician and crew leader are satisfied that the area is safe, safety policies and procedures may require that safety signs be hung in the area (*Figure 4*).

Figure 3 Welding.

Figure 4 Safety sign.

> **CAUTION**
> During hot work, the area temperature can increase rapidly. Observe workers for signs of heat-related problems such as heat stress, exhaustion, and dehydration.

After the hot work permit is issued but before the actual work begins, confirm that all required permits, inspection checklists, and other types of paperwork have been properly filled out. Also confirm that all hazards and safeguards have been reviewed with the work crew. If additional safety observers are required, the safety technician must establish a fire watch. During a fire watch, a trained, qualified person other than the welder or cutting operator must constantly scan the work area within a 35-foot (10.7 m) radius for fires or other events that could cause a hazard. Fire watch personnel must have ready access to fire extinguishers and alarms and know how to use them.

During hot work, potential hazards could include a change in conditions that cause the work area to become unsafe. For example, rain could leak in and puddle in an area where arc welding is taking place. Unsafe conditions could also include worker issues, such as unauthorized persons in the area. Air-sampling tests should be performed during long work periods to assess for health hazards such as vapors, fumes, gases, or areas with low oxygen levels.

After the hot work has been completed, continue to observe the area. Most fires begin after the work is finished. A spark that lands in an obscure location can smolder for hours before flames or smoke are detectable. The required observation time period will vary by site but must be maintained for a minimum of 30 minutes after the hot work is completed.

After the observation period, the crew leader must verify that all necessary information is properly recorded even if the work was done safely. The safety technician should be available to advise the crew leader as needed during this step.

> **NOTE**
> Work permits are not normally required for areas that have been specially designed for the work. For example, an area with concrete floor, fire-resistant walls, and special ventilation could be set aside for welding. This area may not need a work permit for welding.

1.2.0 Lockout/Tagout Policies and Procedures

Lockout/tagout (LOTO) policies and procedures are part of an energy control program. These programs are designed to prevent the unintended release of stored energy in machines or equipment. They consist of energy control procedures, a worker training program, and periodic inspections and audits. These procedures are written documents containing important information that authorized workers need to know in order to safely control energy while servicing or maintaining equipment or machinery.

Energy control procedures utilize energy-isolating devices that physically prevent the transmission or release of energy. There are many types of these devices, including manually operated circuit breakers, disconnect switches, valves, blocks, blinds, and blank flanges.

Lockout/tagout policies and procedures are designed to prevent the unplanned startup of machines or equipment during servicing or maintenance by using lockout or tagout devices (*Figure 5*) that isolate or identify energy sources. These energy sources could be any of the following:

- Electrical
- Mechanical
- Hydraulic
- Pneumatic
- Chemical
- Radiation (*e.g.*, x-ray equipment)
- Thermal
- Stored or residual
- Gravitational (when working with suspended or elevated equipment)

It is likely that your employer/contractor has already developed written LOTO procedures. If these procedures are not in place, you may need to assist with developing them. Include information that states exactly when the procedure must be implemented and specify the types of devices that should be used. The procedure needs to clearly state when a tagout device may be used in place of a lockout device.

Lockout and tagout devices must be provided by the employer. They must be standard in color, size, or shape and be easily identifiable and

Figure 5 Lockout/tagout devices.

1.2.1 LOTO Procedures Training

Safety technicians may be responsible for training workers on LOTO procedures. 29 *CFR* 1910.147, 1926.64, 1926.417, and 1926.702 define three categories of employees who may be involved in LOTO procedures. They are identified as follows:

- Authorized employees
- Affected employees
- Other employees

Authorized employees are those who actually perform the service or maintenance on the equipment. These workers need a high degree of training. They need to know about the type and extent of the hazardous energy involved. They also need to know how to safely isolate and control the energy source.

Affected employees are usually the equipment users or operators. These workers need to be able to identify when a control procedure is in place. They must understand that they are not to remove the LOTO device or try to start the equipment during the procedure.

Other employees are those workers who are in the area where the control procedure is taking place. All workers at the site must know the purpose and function of LOTO procedures and be able to identify a LOTO device. More importantly, all workers must understand that under no condition should they remove the device unless they applied it.

Lockout devices are better than tagout devices because they physically prevent the startup of the machine or equipment. In any situation where an energy source is lockable, a lockout device should be used. A tagout device may only be used if the employer can prove that it provides at least the same level of safety as a lockout device. The use of a tagout device when a lockout device can be applied is potentially dangerous, and site safety technicians should strongly discourage this practice.

During worker training, the uses and limits of tagout devices must be explained. Tagout devices are to be used when an energy source cannot be locked. Safety technicians should encourage their employers to modify or replace those types of energy sources so that they can be locked.

Tagout devices may be needed at your work site. If so, your workers will need to be trained on their use and limits. Keep in mind that tags may give workers a false sense of security. Remember that they are warning devices (*Figure 6*); they do not provide physical control of an energy source. *Figure 7* shows various lockout devices.

durable enough for the work environment. LOTO devices may be used only for their intended purpose. They must never be used to lock toolboxes, lockers, or other items. There must be an easy way to identify the person who applied the device. For example, the lock may be engraved with the worker's name or it may be written on the accompanying tag.

For safety reasons, lockout devices are always preferred over tagout devices. For example, when servicing a crane would you prefer a lockout device to prevent the operator from starting the crane, or a "Do Not Operate/Servicing in Progress" tag near the ignition switch? In which situation would you feel safer?

The LOTO procedure must include the following information:

- Steps for shutting down the equipment and releasing stored energy
- Steps to lock out or tag out the energy source
- Steps for testing that the LOTO device(s) are controlling the energy
- Requirements for releasing the LOTO device(s)

 NOTE The person who applies the LOTO device is usually the only person allowed to remove it.

In addition to these steps, the written procedures must include training requirements, periodic audit programs, and instructions on how to handle special situations. For example, the procedure should explain what to do when a LOTO period overlaps work shifts.

Figure 6 Example of lockout tag.

ELECTRICAL PLUG LOCKOUT

CIRCUIT BREAKER LOCK

BALL VALVE LOCKOUT

VALVE LOCK

Figure 7 Examples of lockout devices.

Tagout devices must be standardized and clearly identifiable. They must be securely attached to the energy source and must be durable enough to withstand the work area and/or the environment. The tag must carry the name of the person who applied it. Once a tag is attached, it may not be removed except by the person who attached it.

Tag attachments usually require at least 50 pounds of pressure to remove. This requirement is based on the strength of a standard $\frac{1}{4}$" (6 mm) nylon cable tie that has a rated unlocking strength of about 75 lb. (34 kg).

When tags are used, workers may need to use additional means to ensure safe conditions equal to that provided by lockout devices. This could include removing a circuit element, blocking a control switch, opening an extra disconnecting device, or removing a valve handle to reduce the chances that the equipment could be turned on.

Safety technicians are responsible for making sure that all workers receive periodic refresher training on LOTO policies and procedures.

1.2.2 LOTO Procedure

In order to safety implement a LOTO procedure, the authorized employee needs to be trained to perform the following steps:

Step 1 Before beginning, notify all affected and other workers of the upcoming maintenance.

Step 2 Identify and isolate all energy sources.

Step 3 Securely apply the LOTO device. Be certain to mark the device to identify who applied it.

Step 4 Test the energy source to determine that power has been isolated.

Step 5 Perform servicing or maintenance of the equipment.

Step 6 Release the LOTO device.

Step 7 Test the equipment to ensure that it is safe and operational.

Step 8 Complete the necessary paperwork.

Table 1 shows a sample checklist for lockout/tagout procedures. You can find this checklist and many others on the OSHA website (**www.osha.gov**).

> **NOTE**
> Always use a safety checklist to be sure that an important step will not be inadvertently overlooked.

Table 1 Sample Lockout/Tagout Procedure Checklist

Yes/No/NA	Lockout/Tagout Procedures
	Is all machinery or equipment capable of movement required to be de-energized or disengaged and locked out during cleaning, servicing, adjusting, or setting up operations, whenever required?
	Where the power disconnecting means for equipment does not also disconnect the electrical control circuit:
	Are the appropriate electrical enclosures identified?
	Are means provided to ensure the control circuit can also be disconnected and locked out?
	Is the locking out of control circuits in lieu of locking out main power disconnects prohibited?
	Are all equipment control valve handles provided with a means for locking out?
	Does the lockout procedure require that stored energy (mechanical, hydraulic, air, etc.) be released or blocked before equipment is locked out for repairs?
	Are appropriate employees provided with individually keyed personal safety locks?
	Are employees required to keep personal control of their key(s) while they have safety locks in use?
	Is it required that only the employee exposed to the hazard place or remove the safety lock?
	Is it required that employees check the safety of the lockout by attempting a startup after making sure no one is exposed?
	Are employees instructed to always push the control circuit stop button immediately after checking the safety of the lockout?
	Is there a means provided to identify any or all employees who are working on locked-out equipment by their locks or accompanying tags?
	Are a sufficient number of accident preventive signs or tags and safety padlocks provided for any reasonably foreseeable repair emergency?
	When machine operations, configuration, or size requires the operator to leave his or her control station to install tools or perform other operations, and when part of the machine could move if accidentally activated, is such element required to be separately locked or blocked out?
	In the event that equipment or lines cannot be shut down, locked out, and tagged, is a safe job procedure established and rigidly followed?

1.2.3 LOTO Exceptions

Under some conditions, OSHA requirements for LOTO policies and procedures will not apply. In these situations, workers must be provided with other protection that is just as effective as LOTO procedures. Always follow the most stringent requirements.

If the worker is not at risk for injury due to an unexpected release of energy, LOTO procedures are not needed. For example, when the equipment requires the use of a cord and plug, the authorized employee can control the plug during servicing.

LOTO policies and procedures may not apply during hot tap operations that involve transmission and distribution systems for gas, steam, water, or petroleum products; for example, in the following situations:

- When servicing must be performed on pressurized pipelines
- When continuation of the service is essential
- When shutdown of the system is impractical

> **CAUTION**
> Even though LOTO policies and procedures do not apply to hot tap operations, these are potentially hazardous activities that require special precautions and procedures.

There are times when equipment must be powered on during servicing; for example, when making fine adjustments such as centering a conveyor belt. Power is also needed when troubleshooting equipment to identify the cause of a problem. Although OSHA recognizes that these situations exist, the employer is still required to provide protection to workers performing these duties.

Employees performing minor tool changes, adjustments, and servicing during normal equipment operation may be exempt from LOTO policies. These types of servicing must be routine, repetitive, and integral to the use of the equipment. However, other safety precautions that provide effective protection, such as PPE and equipment de-energization, must still be used.

Even if the servicing is normally exempt from LOTO procedures, there are certain situations that require the use of LOTO policies and procedures, including the following:

- If machine guards or other safety devices must be removed or bypassed, exposing the worker to hazards at some time in the operation
- If the worker is required to place any part of his or her body in contact with the operational portion of the equipment
- If the worker needs to place any part of his or her body in the danger zone of the equipment

1.2.4 Additional LOTO Safety Requirements

There are a number of special conditions that can exist during LOTO procedures. These conditions include the following:

- *Temporary removal of LOTO devices* – OSHA permits LOTO devices to be temporarily removed under special conditions. For example, LOTO devices can be removed when power is needed to test or reposition equipment. In this case, the power may be applied under strictly controlled conditions. The following steps must be performed in this order:

Step 1 Clear the equipment of tools and materials.

Step 2 Ensure that all workers are out of the area.

Step 3 Remove all LOTO devices as specified by the site's policy.

Step 4 Energize the equipment and proceed with testing or repositioning.

Step 5 De-energize the equipment and isolate the energy source.

Step 6 Apply the LOTO device as specified by the site's policy.

Step 7 Test to confirm that the energy source has been isolated.

Step 8 Complete servicing or maintenance tasks.

- *Outside workers* – When outside workers such as subcontractors are working on your job site, you must ensure that the on-site workers and the outside workers are aware of each other's LOTO policies and procedures. Outside workers need to use the policies and procedures that at least meet the standards of the host work site.
- *Groups* – When service and maintenance need to be performed by a group of workers rather than a single worker, each authorized worker must be protected by their own LOTO device (*Figure 8*).
- *Complex lockouts* – In complex lockouts with large numbers of exposed personnel and/or large numbers of isolation points, it is impractical to have each worker affix a personal lock to each isolating device. In this case, the controlling employer will prepare an isolation log or

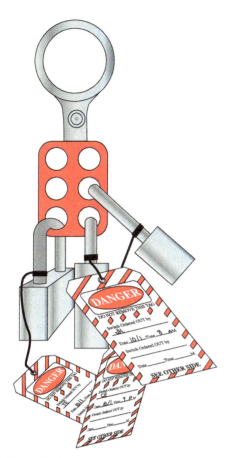

Figure 8 Group lockout.

master card listing all isolation points. The controlling employer's lock and accompanying tag will usually be attached to each isolation point. The keys to the locks will then be placed in a job lock box or cabinet that is fitted with one or more multiple lock adapters. Each exposed employee can then review the isolation log and affix their personal lock to the adapter on the box. In this situation, the controlling employer's lock will be the first on and last off. The contractor's lock will be the last on and first off. OSHA mandates that each exposed employee must have the right to inspect the application of the controlling employer's locks before affixing their lock to the job lock box.

> **NOTE**
> There are numerous variations to these types of systems. It is critically important for safety technicians to have an in-depth knowledge of their site-specific lockout/tagout and work permit systems.

- *Shift changes* – Specific LOTO policies and procedures must be in place to ensure that LOTO protection will continue during shift or personnel changes.

1.3.0 Excavation Policies and Procedures

OSHA regulations 29 *CFR* 1926.650, 1926.651, and 1926.652 recognize that excavation and trenching are very hazardous operations. A trench is a narrow excavation made below the surface of the ground. By definition, the depth of a trench must be greater than its width, and the width must not exceed 15 feet (4.6 m). An excavation is defined as any man-made cut, cavity, trench, or depression in the earth's surface formed by earth removal.

Although workers are exposed to many hazards during excavation and trenching, the primary hazard is cave-ins. Many incidents related to this type of work are the result of poor planning. Efforts to correct mistakes after the work has started increases the potential for cave-ins. It also slows down the work and adds to its cost.

Before the job begins, review your company's policies, procedures, and practices related to excavation and/or trenching. Verify that they are adequate to protect workers from potential hazards. As the safety technician, you are responsible for ensuring that all workers are aware of and adhere to these policies and procedures.

Work permits for excavations and trenching can be very complex. The site safety technician is responsible for advising the first-line supervisor when the plan is being verified before work begins. Using safety checklists can help ensure that information about the job site and all needed items are on hand.

When planning for excavation or trenching, the following factors should be considered:

- Traffic
- Surrounding buildings and their conditions
- Soil type
- Surface and ground water
- Overhead and underground utilities
- Weather

These and other conditions may require site studies and tests such as geological surveys and soil-type tests. OSHA categorizes soil and rock deposits into the following types (see *Figure 9*):[v]

- *Stable rock* – Natural solid mineral matter that can be excavated with vertical sides and remain intact while exposed. It is usually identified by a rock name, such as granite or sandstone.
- *Type A soil* – Cohesive soil that includes various types of clay and loam. No soil is Type A if it is fissured, subject to vibration, has previously been disturbed, has seeping water, or is part of a sloped, layered system with a slope of 4 horizontal to 1 vertical (4H:1V) or greater.

- *Type B soil* – Includes angular gravel, silt, silt loam, previously disturbed soils unless otherwise classified as Type C, soils that meet the requirements of Type A but are fissured or subject to vibration, dry unstable rock, and layered systems sloping into the trench at a slope less than 4H:1V.
- *Type C soil* – Includes granular soils such as gravel, sand and loamy sand, submerged soil, soil from which water is freely seeping, and submerged rock that is not stable. Type C also includes sloped, layered material with a slope of 4H:1V or greater.
- *Layered geological strata* – Where soils are configured in layers (*i.e.*, where a layered geologic structure exists), the soil must be classified on the basis of the soil classification of the weakest soil layer. Each layer may be classified individually if a more stable layer lies below a less stable layer (for example, where a Type C soil rests on top of stable rock).

Before any excavation actually begins, OSHA standards require the employer to estimate the location of utility lines that may be in the vicinity of the excavation. The following buried utilities may be encountered when trenching or excavating:

- Sewer lines
- Telephone cables
- Fuel lines, such as natural or propane gas
- Electric cables
- Water lines
- Other underground installations

In most states, your company will be required to contact local utility companies and the owners of the involved property before starting the excavation. Services such as 811 or One-Call can provide all of the information needed on buried utilities in the area. Contractors must call this service at least 72 hours in advance of doing any work. Adequate time must be built into the plan phase to allow for responses from these services.

When all the information about the job site has been collected, it's important to identify the equipment and materials that will be used. This includes personal protective equipment (see 29 CFR 1926.100 and 102) and other safety items, typically including the following:

- Safety shoes (steel toe or composite)
- Safety glasses

Figure 9 Sloping the sides of an excavation site for different soil types.

- Hardhats
- Reflective vests
- Respiratory protection
- Fall protection
- Atmospheric testing devices

Regardless of how often these types of jobs have been successful in the past, each new job must be carefully planned. Once the plan has been made and all needed local permits have been issued, the controlling employer will issue work permits at the site. Keep in mind that more than one type of work permit may be required.

Once excavation or trenching begins, OSHA standards require that a competent person inspect the site daily for potential cave-ins. They must also check for failure of protective systems and equipment, the presence of a hazardous atmosphere, or any other hazardous conditions. Re-inspections are required after natural events such as rain, or man-made events such as blasting that may increase the potential for hazards.

Since the primary hazard to excavation workers is the danger of cave-ins, OSHA requires that methods be used to prevent such incidents. Most of these incident-prevention methods need to be planned and approved by a registered professional engineer. Two methods used to prevent cave-ins include the sloped trench and benched trench (*Figure 10*).

The actual rise-to-run ratio of a slope depends on the soil type, water accumulation, and other factors. Another method to prevent cave-ins involves both sloping and shoring the side of the site. Trench boxes or shields made of wood, aluminum, or other suitable material may also be used (*Figure 11*).

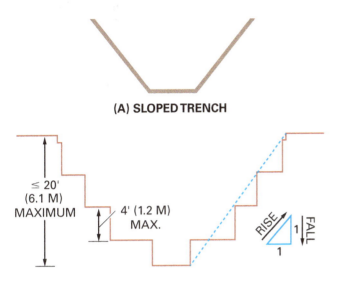

Figure 10 Sloped trench and benched trench.

Case History

An employee was installing a small-diameter pipe in a trench 3' (1 m) wide, 12' to 15' (4 to 5 m) deep, and 90' (27 m) long. The trench was not shored or sloped nor was there a box or shield to protect the employee. In addition, there was evidence of a previous cave-in. The employee re-entered the trench and a second cave-in occurred, burying him. He was found face-down in the bottom of the trench.

The Bottom Line: Following its inspection, OSHA issued a citation for serious violations of its construction standards:

- Employers must shore, slope, or otherwise support the sides of trenches to prevent their collapse per 29 *CFR* 1926.652(a)(1).
- Employers must protect employees with adequate personal protective equipment per 29 *CFR* 1926.95(a).
- Employers must provide an adequate means of exit from trenches per 29 *CFR* 1926.651(c)(2).
- Employees must be instructed to recognize and avoid unsafe conditions associated with their work per 29 *CFR* 1926.21(b)(2).

Had the required support been provided for the trench, it might not have collapsed.
Source: OSHA Fatal Facts Report #22

Figure 11 Trench shields.

Module 75224   Permits and Policies   15

1.4.0 Chemical Hazards Policies and Procedures

Chemical use is a common hazard at work sites. Some chemicals have a direct effect on the health and safety of workers, and some have long-term health consequences. The effect of exposure to some chemicals may not be detectable until years after the exposure.

OSHA estimates that annually more than 32 million workers are exposed to 650,000 hazardous chemical products in more than 3 million American workplaces. Many host employers list chemical hazards and precautions on their work permits. Some companies use a safe work permit; others use a hazardous work permit.

OSHA requires that all employers with hazardous chemicals at their sites prepare and use a written hazard communication program (29 CFR 1910.1200). The goal of the program is to be sure employers and employees know about work hazards and how to protect themselves in order to help reduce the number of chemical source incidents, illnesses, and injuries.

The current OSHA Hazard Communication Standard (HCS) is intended to align the United States with the Globally Harmonized System of Classification and Labeling of Chemicals (GHS). Currently, 72 countries participate in the GHS. The GHS is expected to prevent injuries and illnesses, save lives, and improve trade conditions for chemical manufacturers. The Hazard Communication Standard of 1983 gave workers the right to know, but the new Globally Harmonized System gives workers the right to understand, incorporating pictograms as a means to communicate.

The new hazard communication standard still requires chemical manufacturers and importers to evaluate the chemicals they produce or import and provide hazard information to employers and workers by labeling containers and supplying a safety data sheet (SDS) for each chemical. However, the old standard allowed chemical manufacturers and importers to convey this information on labels and safety data sheets in whatever format they chose. The modified standard provides a single set of harmonized criteria for classifying chemicals according to their health and physical hazards and specifies hazard communication elements for labeling and safety data sheets. The United States officially adopted the GHS on March 26, 2012. OSHA's adoption is actually a revision of the Hazard Communication Standard to align with the GHS. OSHA calls this revision HazCom 2012.

The most noticeable changes brought by GHS for most organizations are changes to safety labels, SDSs, and chemical classifications. *Figure 12* shows GHS labels for various materials. The GHS also standardizes the content and formatting of SDSs into 16 sections with a strict ordering. Labels also look quite different, with six standardized elements that include specific language depending upon the chemical classification.

The Hazard Communication Standard (HCS) requires chemical manufacturers, distributors, or importers to provide SDSs (formerly known as material safety data sheets or MSDSs) to communicate the hazards of chemical products. The HCS requires new SDSs to be in a uniform format to include the section numbers, the headings, and associated information as follows:

- *Section 1: Identification* – Product identifier; manufacturer or distributor name, address, and phone number; emergency phone number; recommended use; restrictions on use.
- *Section 2: Hazard(s) Identification* – All hazards regarding the chemical; required label elements.
- *Section 3: Composition/Information on Ingredients* – Information on chemical ingredients; trade secret claims.
- *Section 4: First Aid Measures* – Important symptoms/effects, acute, delayed; required treatment.
- *Section 5: Firefighting Measures* – Suitable extinguishing techniques, equipment; chemical hazards from fire.
- *Section 6: Accidental Release Measures* – Emergency procedures; protective equipment; proper methods of containment and cleanup.
- *Section 7: Handling and Storage* – Precautions for safe handling and storage, including incompatibilities.
- *Section 8: Exposure Controls/Personal Protection* – OSHA's Permissible Exposure Limit (PEL); ACGIH Threshold Limit Values (TLVs); and any other exposure limit used or recommended by the chemical manufacturer, importer, or employer preparing the SDS where available as well as appropriate engineering controls; personal protective equipment (PPE).
- *Section 9: Physical and Chemical Properties* – The chemical's characteristics.
- *Section 10: Stability and Reactivity* – Chemical stability and possibility of hazardous reactions.
- *Section 11: Toxicological Information* – Routes of exposure; related symptoms, acute and chronic effects; numerical measures of toxicity.

FLAME

- FLAMMABLES
- PYROPHORICS
- SELF-HEATING
- EMITS FLAMMABLE GAS
- SELF-REACTIVES
- ORGANIC PEROXIDES

FLAME OVER CIRCLE

- OXIDIZERS

EXPLODING BOMB

- EXPLOSIVES
- SELF-REACTIVES
- ORGANIC PEROXIDES

HEALTH HAZARD

- CARCINOGEN
- MUTAGENICITY
- REPRODUCTIVE TOXICITY
- RESPIRATORY SENSITIZER
- TARGET ORGAN TOXICITY
- ASPIRATION TOXICITY

EXCLAMATION MARK

- IRRITANT (SKIN AND EYE)
- SKIN SENSITIZER
- ACUTE TOXICITY (HARMFUL)
- NARCOTIC EFFECTS
- RESPIRATORY TRACT IRRITANT
- HAZARDOUS TO OZONE LAYER (NON-MANDATORY)

GAS CYLINDER

- GASES UNDER PRESSURE

CORROSION

- SKIN CORROSION/ BURNS
- EYE DAMAGE
- CORROSIVE TO METALS

ENVIRONMENT (NON-MANDATORY)

- AQUATIC TOXICITY

SKULL AND CROSSBONES

- ACUTE TOXICITY (FATAL OR TOXIC)

Figure 12 GHS labels.

- *Section 12: Ecological Information*
- *Section 13: Disposal Considerations*
- *Section 14: Transport Information*
- *Section 15: Regulatory Information*
- *Section 16: Other Information* – includes the date of preparation or last revision.

> **NOTE:** OSHA is not enforcing Sections 12 through 15 [29 *CFR* 1910.1200(g)(2)] since other agencies regulate this information.

Refer to *Appendix D* of 29 *CFR* 1910.1200 for a detailed description of SDS contents. Employers must ensure that SDSs (*Figure 13A* and *Figure 13B*) are readily accessible to employees.

To ensure chemical safety in the workplace, information about the identities and hazards of any chemical must be available and understandable to workers. OSHA's Hazard Communication Standard requires the development and dissemination of this information. Specifically, chemical manufacturers and importers are required to evaluate the hazards of the chemicals they produce or import and prepare labels and safety data sheets to convey the hazard information to their downstream customers. All employers with hazardous chemicals in their workplaces must have labels and safety data sheets available to their exposed workers and must train them to handle the chemicals appropriately.

Adoption of GHS brings major changes to many areas of the HCS, including the following:

- *Hazard classification* – Provides specific criteria for classification of health and physical hazards, as well as classification of mixtures.
- *Labels* – Chemical manufacturers and importers are required to provide a label that includes a harmonized signal word, pictogram, and hazard statement for each hazard class and category. Precautionary statements must also be provided.
- *Safety data sheets* – Now have a specified 16-section format.
- *Information and training* – To facilitate recognition and understanding, employers were required to train workers by December 1, 2013 on the new label elements and the format of the new safety data sheets.

OSHA anticipates the revised standard will prevent 43 fatalities and 585 injuries annually, with a net annualized savings of over $500 million a year.

Some of the health hazards of chemicals include the following:

- Irritation
- Sensitivity
- Carcinogenicity (ability to cause cancer)

Chemicals also have direct physical hazards, including the following:

- Flammability
- Corrosion
- Reactivity

The work site must have a written plan that describes how the OSHA standard will be used. The following are requirements for the written plan:

- The plan must label hazardous chemicals with the identity of the material and appropriate hazard warnings. See *Figure 14* for an example of a hazardous material posting.
- The plan must incorporate the use of an SDS for each hazardous chemical at the site. These may be obtained from the chemical manufacturers or distributors.
- The plan must include a training program that provides information to all employees who may be exposed to hazardous chemicals. This training must occur before the first work assignment involving hazardous chemicals, and additional training is needed whenever the hazard changes. Workers should be trained to handle chemicals safely and use approved policies and procedures, including the proper personal protective equipment. When an incident occurs during the use of hazardous chemicals, it must be properly documented. The incident report must include the following:

  – Name of the chemical
  – Circumstances under which the incident occurred
  – Names of all involved workers
  – Length of the exposure
  – Whether other chemicals were involved
  – Immediate action taken

Figure 14 Hazardous material sign.

GHS SAFETY DATA SHEET
WELD-ON® 705™ Low VOC Cements for PVC Plastic Pipe

Date Revised: DEC 2011
Supersedes: FEB 2010

SECTION 1 - PRODUCT AND COMPANY IDENTIFICATION

PRODUCT NAME: WELD-ON® 705™ Low VOC Cements for PVC Plastic Pipe
PRODUCT USE: Low VOC Solvent Cement for PVC Plastic Pipe
SUPPLIER:
MANUFACTURER: IPS Corporation
17109 South Main Street, Carson, CA 90248-3127
P.O. Box 379, Gardena, CA 90247-0379
Tel. 1-310-898-3300

EMERGENCY: Transportation: CHEMTEL Tel. 800.255-3924, 813-248-0585 (International)   **Medical:** Tel. 800.451.8346, 760.602.8703 3E Company (International)

SECTION 2 - HAZARDS IDENTIFICATION

GHS CLASSIFICATION:

Health		Environmental		Physical	
Acute Toxicity:	Category 4	Acute Toxicity:	None Known	Flammable Liquid	Category 2
Skin Irritation:	Category 3	Chronic Toxicity:	None Known		
Skin Sensitization:	NO				
Eye:	Category 2B				

GHS LABEL: OR   **Signal Word:** Danger   **WHMIS CLASSIFICATION:** CLASS B, DIVISION 2

Hazard Statements
H225: Highly flammable liquid and vapor
H319: Causes serious eye irritation
H332: Harmful if inhaled
H335: May cause respiratory irritation
H336: May cause drowsiness or dizziness
EUH019: May form explosive peroxides

Precautionary Statements
P210: Keep away from heat/sparks/open flames/hot surfaces – No smoking
P261: Avoid breathing dust/fume/gas/mist/vapors/spray
P280: Wear protective gloves/protective clothing/eye protection/face protection
P304+P340: IF INHALED: Remove victim to fresh air and keep at rest in a position comfortable for breathing
P403+P233: Store in a well ventilated place. Keep container tightly closed
P501: Dispose of contents/container in accordance with local regulation

SECTION 3 - COMPOSITION/INFORMATION ON INGREDIENTS

	CAS#	EINECS	REACH Pre-registration Number	CONCENTRATION % by Weight
Tetrahydrofuran (THF)	109-99-9		05-2116297729-22-0000	25 – 50
Methyl Ethyl Ketone (MEK)	78-93-3		05-2116297728-24-0000	5 – 36
Cyclohexanone	108-94-1		05-2116297718-25-0000	15 – 30

All of the constituents of this adhesive product are listed on the TSCA inventory of chemical substances maintained by the US EPA, or are exempt from that listing.
* Indicates this chemical is subject to the reporting requirements of Section 313 of the Emergency Planning and Community Right-to-Know Act of 1986 (40CFR372).
\# indicates that this chemical is found on Proposition 65's List of chemicals known to the State of California to cause cancer or reproductive toxicity.

SECTION 4 - FIRST AID MEASURES

Contact with eyes: Flush eyes immediately with plenty of water for 15 minutes and seek medical advice immediately.
Skin contact: Remove contaminated clothing and shoes. Wash skin thoroughly with soap and water. If irritation develops, seek medical advice.
Inhalation: Remove to fresh air. If breathing is stopped, give artificial respiration. If breathing is difficult, give oxygen. Seek medical advice.
Ingestion: Rinse mouth with water. Give 1 or 2 glasses of water or milk to dilute. Do not induce vomiting. Seek medical advice immediately.

SECTION 5 - FIREFIGHTING MEASURES

		HMIS	NFPA	
Suitable Extinguishing Media:	Dry chemical powder, carbon dioxide gas, foam, Halon, water fog.			0-Minimal
Unsuitable Extinguishing Media:	Water spray or stream.	Health 2	2	1-Slight
Exposure Hazards:	Inhalation and dermal contact.	Flammability 3	3	2-Moderate
Combustion Products:	Oxides of carbon, hydrogen chloride and smoke.	Reactivity 0	0	3-Serious
		PPE B		4-Severe
Protection for Firefighters:	Self-contained breathing apparatus or full-face positive pressure airline masks.			

SECTION 6 - ACCIDENTAL RELEASE MEASURES

Personal precautions: Keep away from heat, sparks and open flame.
Provide sufficient ventilation, use explosion-proof exhaust ventilation equipment or wear suitable respiratory protective equipment.
Prevent contact with skin or eyes (see section 8).
Environmental Precautions: Prevent product or liquids contaminated with product from entering sewers, drains, soil or open water course.
Methods for Cleaning up: Clean up with sand or other inert absorbent material. Transfer to a closable steel vessel.
Materials not to be used for clean up: Aluminum or plastic containers

SECTION 7 - HANDLING AND STORAGE

Handling: Avoid breathing of vapor, avoid contact with eyes, skin and clothing.
Keep away from ignition sources, use only electrically grounded handling equipment and ensure adequate ventilation/fume exhaust hoods.
Do not eat, drink or smoke while handling.
Storage: Store in ventilated room or shade below 44°C (110°F) and away from direct sunlight.
Keep away from ignition sources and incompatible materials: caustics, ammonia, inorganic acids, chlorinated compounds, strong oxidizers and isocyanates.
Follow all precautionary information on container label, product bulletins and solvent cementing literature.

SECTION 8 - PRECAUTIONS TO CONTROL EXPOSURE / PERSONAL PROTECTION

EXPOSURE LIMITS:

Component	ACGIH TLV	ACGIH STEL	OSHA PEL	OSHA STEL
Tetrahydrofuran (THF)	50 ppm	100 ppm	200 ppm	
Methyl Ethyl Ketone (MEK)	200 ppm	300 ppm	200 ppm	
Cyclohexanone	20 ppm	50 ppm	50 ppm	

Engineering Controls: Use local exhaust as needed.
Monitoring: Maintain breathing zone airborne concentrations below exposure limits.
Personal Protective Equipment (PPE):
Eye Protection: Avoid contact with eyes, wear splash-proof chemical goggles, face shield, safety glasses (spectacles) with brow guards and side shields, etc. as may be appropriate for the exposure.
Skin Protection: Prevent contact with the skin as much as possible. Butyl rubber gloves should be used for frequent immersion.
Use of solvent-resistant gloves or solvent-resistant barrier cream should provide adequate protection when normal adhesive application practices and procedures are used for making structural bonds.
Respiratory Protection: Prevent inhalation of the solvents. Use in a well-ventilated room. Open doors and/or windows to ensure airflow and air changes. Use local exhaust ventilation to remove airborne contaminants from employee breathing zone and to keep contaminants below levels listed above. With normal use, the Exposure Limit Value will not usually be reached. When limits approached, use respiratory protection equipment.

Figure 13A Safety data sheet (1 of 2).

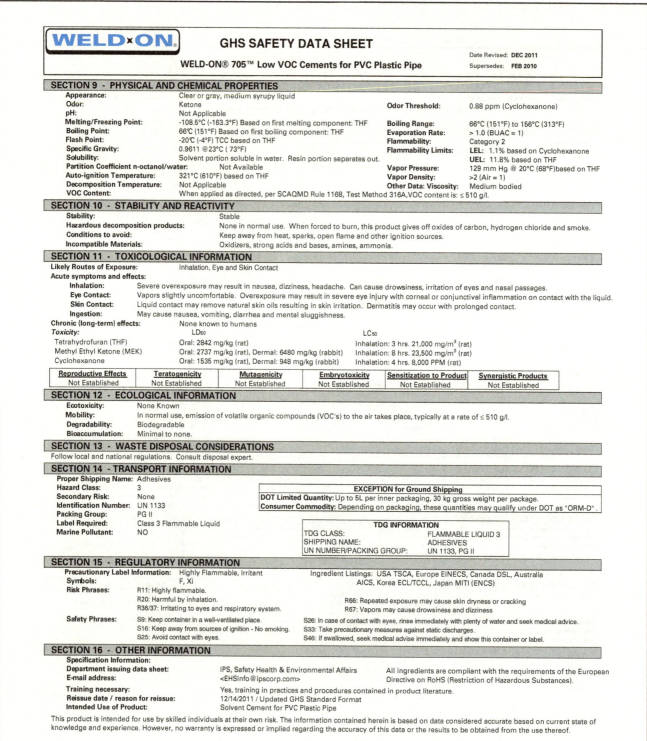

Figure 13B Safety data sheet (2 of 2).

1.5.0 Additional Types of Work Permits

Some types of jobs require additional work permits. The most common types of work that require special permits include working on aerial lifts or around vehicles, and some types of electrical work.

1.5.1 Aerial Lifts

Aerial lifts are vehicle-mounted devices that are used to raise workers above the ground. The term *aerial lift* typically applies to the following:

- Extensible boom platforms
- Aerial ladders
- Articulating boom platforms
- Vertical towers
- Any combination of the above

> **NOTE**
> There is no specific requirement for an aerial work permit. However, because they are so commonly used, safety technicians should review company safety policies and requirements for aerial lifts.

There are a number of safety precautions related to aerial lift operations, including the following:

- Test the lift each day prior to use.
- Allow only authorized persons to operate the lift.
- Ensure that while the lift is in operation, all involved workers follow approved safety policies and procedures, including the use of a fall arrest/restraint system.

1.5.2 Vehicles

A vehicle is defined as a car, bus, truck, trailer, or semi-trailer that is used to carry employees or move materials. It may be owned, leased, or rented by the employer. Vehicles must be maintained in a serviceable condition and must be used for their intended purpose. Operators of these vehicles need a valid permit or license in order to operate them. Be aware that vehicle entry may be restricted in certain areas because idling vehicles present a combustion hazard.

1.5.3 Electrical Work

Electrical work is safest when performed on de-energized circuits using lockout/tagout procedures. Special procedures are needed when working on energized electrical circuits. An electrical hot work permit is required when working on or near energized equipment with a potential of 50 volts or greater above ground. *NFPA 70E®, Standard for Electrical Safety in the Workplace*, helps owners, managers, and employees work together to ensure an electrically safe working area in compliance with 29 *CFR* 1910, Subpart S and 29 *CFR* 1926, Subpart K. For more information about *NFPA 70E®*, visit the NFPA website at **www.nfpa.org**.

Only qualified workers may perform hot electrical work. They must be thoroughly familiar with the equipment and use proper safety precautions to avoid incidents. Workers need to take special precautions, such as electrically isolating themselves from the equipment and ground by using electrical safety shoes and proper mats. They must de-energize all parts of the equipment that may be de-energized and use lockout/tagout procedures, if possible.

A competent person should observe the performance of electrical work. Safety technicians must ensure that correct protective devices and procedures are used and that all safety requirements are met. Depending on the state/employer requirements, the competent person observing hot electrical work must have yearly CPR certification/first aid/AED training. In addition, the observer needs to have access to a phone or radio to call for aid in case of an emergency, and must have the ability to immediately cut off all sources of electrical power.

1.5.4 Facility Siting

Facility siting is a way of identifying, evaluating, and managing risks related to fire, explosions, and toxins. Facility siting also determines distances for blast-resistant buildings and construction based on American Petroleum Institute (API) and NFPA standards. 29 *CFR* 1910.119, *Process Safety Management (PSM)* requires facility siting to be addressed in an organization's Process Hazard Analyses (PHAs).

The facility siting assessment may find potential safety hazards that require the use of permits. For example, a process-specific chemical hazard may require minimizing the presence of workers in a particular process area. You may also need to limit the use of portable buildings in this area. For this example, requiring permits for workers to access, occupy, and perform work in a given area or building within the facility will help to support the facility siting assessment findings. The types of known hazards at the facility determines other types of permits that may be needed to address any possible safety hazards.

Additional Resources

American Society of Safety Engineers, www.asse.org

Occupational Safety and Health Organization, www.osha.gov

Guidelines for Hot Work in Confined Spaces: Recommended Practices for Industrial Hygienists and Safety Professionals, Martin H. Finkel, CIH, CMC. 2000. Des Plaines, IL: The American Society of Safety Engineers (ASSE).

Complete Confined Spaces Handbook, John F. Rekus, MS, CIH, CSP. 1994. Boca Raton, FL: CRC/Lewis Publishers.

1.0.0 Section Review

1. Hot work permits are required because _____.
   a. state and local governments need to be able to control the types of work being performed in their jurisdiction
   b. ordinary work conditions can become safety hazards when hot work is being performed in the area
   c. workers need permission to be on a work site when welding is being performed
   d. a worker performing a job for the first time needs written instructions

2. Lockout and tagout devices are useful for _____.
   a. draining residual energy from components
   b. barricading dangerous work areas
   c. providing overcurrent protection to electrical devices
   d. isolating and identifying energy sources

3. At a minimum, excavation sites must be inspected _____.
   a. daily
   b. every other day
   c. weekly
   d. when first installed

4. First aid measures for exposure to a particular chemical can be found on the product's SDS in _____.
   a. Section 1
   b. Section 3
   c. Section 4
   d. Section 5

5. Electrical safety on the job site is covered by _____.
   a. *NFPA 70E*®
   b. *29 CFR 1910, Subpart B*
   c. *29 CFR 1926, Subpart M*
   d. *ISO 9000*

Section Two

2.0.0 Confined Spaces

Objective

Explain how to identify the atmospheric hazards in confined spaces.

a. Identify the types of atmospheric hazards that may be present in a confined space.
b. Describe how to perform atmospheric testing.

Trade Terms

Asphyxiation: Death due to lack of oxygen.

Atmospheric contaminants: Impurities in the air. Any natural or artificial matter capable of being airborne (other than water vapor or natural air) which in sufficient concentration can harm humans, animals, vegetation, or materials. Contaminants can exist in the form of liquids, solids, or gases.

Engulfment: To be in an environment in which the victim is covered with a material such as sand, gravel, or grain. The end result is often death due asphyxiation, strangulation, or crushing.

Inerting: The use of an inert gas such as nitrogen to supplant a flammable gas in a confined space.

Lower explosive limit (LEL) or lower flammable limit (LFL): The lowest concentration of air-fuel mixture at which a gas can ignite.

Non-permit required confined space: A workspace free of atmospheric, physical, electrical, and mechanical hazards that could cause death or injury.

Oxygen-deficient atmosphere: A body of air that does not have enough oxygen to sustain normal breathing. This is usually considered to be less than 19.5 percent by volume.

Oxygen-enriched atmosphere: A body of air that contains enough oxygen to be flammable or explosive. This is usually considered to be more than 23.5 percent by volume.

Parts per million (ppm): A measure of the concentration of a substance in a volume of air, water, or other medium. In a volume of air, it is the number of milliliters of a substance in a cubic meter of air: 10 ppm = 10 ml/m3. One part per million is roughly the equivalent of two soda cans full of gas in a house full of air.

Self-contained breathing apparatus (SCBA): A device allowing an individual to breath while in a toxic or oxygen-deficient atmosphere.

Specific gravity: The ratio of the density of a substance to the density of a reference substance (typically water).

Upper explosive limit (UEL) or upper flammable limit (UFL): The highest concentration of air-fuel mixture at which a gas can ignite.

Vapor density: The relative weight of gases and vapors as compared with some specific standard, which is usually hydrogen but may also be air.

A confined space is any space that is not designed for continuous occupancy and has a limited means of egress (entry and exit). These spaces are prone to the accumulation of toxic or flammable contaminants or may have an oxygen-deficient atmosphere. Confined or enclosed spaces include, but are not limited to, storage tanks, process vessels, bins, boilers, ventilation or exhaust ducts, sewers, underground utility vaults, tunnels, pipelines, and open-top spaces more than 4 feet (1.2 m) in depth, such as pits, tubs, vaults, and vessels.

Atmospheric hazards are the most common hazards in confined spaces. These hazards can take various forms. A lack of oxygen can lead to asphyxiation in a matter of seconds. Flammable gas can turn a confined space into a deadly explosion hazard. Toxic gases can accumulate with little or no warning. Confined spaces must be tested before workers are allowed to enter. When testing an environment, personnel should first test for oxygen, then flammability, and finally toxicity. Many types of detection equipment are available to test for oxygen depletion or flammable and toxic gases. Some meters provide several tests in a single multi-use unit. Depending on the test results, additional ventilation and/or respiratory equipment may be required for safe entry.

Confined spaces can be categorized by those that require work permits and those that do not. 29 *CFR* 1910.146 defines a permit-required confined space as a space that has one or more of the following characteristics:

- Contains or has a potential to contain a hazardous atmosphere
- Contains a material that has the potential for engulfing an entrant
- Has an internal configuration such that an entrant could be trapped or asphyxiated by inwardly converging walls or by a floor that slopes downward and tapers to a smaller cross-section
- Contains any other recognized serious safety or health hazard

A permit-required confined space program is the employer's overall program for controlling and protecting employees from permit-space hazards and for regulating employee entry into permit spaces. A permit system is the employer's written procedure for preparing and issuing permits for entry and for returning the permit-required space to service after entry work is completed.

Types of hazards typical of a confined space could include the following:

- Oxygen-deficient atmosphere
- Oxygen-enriched atmosphere
- Flammable atmosphere
- Toxic atmosphere
- Temperature extremes
- Engulfment hazards
- Environmental hazards such as noise, slick/wet surfaces, or falling objects
- Mechanical hazards

Briefly, the following safety procedures apply to confined spaces:

- All permit-required confined spaces must be clearly identified and posted.
- Some confined spaces need permits both to enter and to do work in the space.
- Workers must check the confined space before entering.
- No worker is allowed to enter a confined space unless authorized to enter.
- Hazards of the confined space and the work must be identified and eliminated or controlled to an acceptable level.
- The atmosphere in the confined space must be tested and deemed acceptable before entry and monitored for safety throughout the duration of an authorized entry.
- Proper personal protective equipment must be used in a confined space.
- The entry permit must have an SDS attached to it that provides information about any toxins the worker may encounter.
- Rescue plans must be developed in advance and communicated to all workers involved in the confined space operation.
- All confined space entrants must be thoroughly trained to perform their jobs before entering the confined space.

The site safety technician must work with the crew leader or first-line supervisor to complete the following tasks:

- Identify whether the space is a permit-required or non-permit required confined space.
- Ensure that all permit-required spaces are clearly posted (*Figure 15*).

Figure 15 Confined space posting.

- Help develop safety policies and procedures for entering and working in confined spaces.
- Ensure that all workers who need to enter and work in confined spaces are properly trained so that they know what to do if an emergency develops.
- Ensure that confined space policies and procedures include methods to retrieve workers if an emergency develops.
- Ensure that workers are trained and use the proper protective equipment for the confined space.
- Ensure that periodic safety audits are done to make sure that confined space policies and procedures are in place, in use, and provide workers with adequate protection from known hazards.

2.1.0 Atmospheric Hazards in Confined Spaces

Atmospheric hazards are the most common—and deadliest—hazards found in confined spaces. A hazardous atmosphere is identified by any of the following characteristics:

- Too little or too much oxygen
- Explosive or flammable gases, vapors, or dusts
- Toxic gases

No one is allowed to enter a confined space until atmospheric testing is done. If it is necessary to enter the space to perform the testing, the person entering the space must wear the appropriate respiratory protection. Respiratory protection must be used until the test results indicate that it is not needed.

Different hazards require the use of different types of respirators. Employees are responsible for wearing the appropriate respirator and complying with the respiratory protection program. Types of respirators include the following (*Figure 16*):

- Air-purifying respirator
- Powered air-purifying respirator (PAPR)
- Supplied-air respirator (SAR) or airline respirator
  - Demand mode
  - Continuous flow mode
  - Pressure-demand or another positive-pressure mode
- Self-contained breathing apparatus (SCBA)
  - Demand mode
  - Pressure-demand or other positive pressure mode (open/closed circuit)

Air-purifying respirators use filters or sorbents to remove harmful substances from the air. They range from simple disposable masks to sophisticated devices. They do not supply oxygen and must not be used in oxygen-deficient atmospheres or in other atmospheres that are immediately dangerous to life or health (IDLH).

Atmosphere-supplying respirators are designed to provide breathable air from a clean air source other than the surrounwding contaminated work atmosphere. They include supplied-air respirators (SARs) and self-contained breathing apparatus (SCBA) units.

(A) SELF-CONTAINED BREATHING APPARATUS (SCBA)

(B) SUPPLIED AIR MASK

(C) FULL FACEPIECE MASK

(D) HALF MASK

Figure 16 Different types of respirators.

Atmospheric contaminants are impurities that can take any of the following physical forms:

- *Dust* – Small particles generated by crushing solids
- *Fumes* – Small particles created by condensation from a vapor state, especially volatized metals
- *Mist* – Suspended liquid particles formed by condensation from gases or by liquid dispersion
- *Smoke* – An aerosol mixture resulting from incomplete combustion
- *Vapors* – Gaseous forms of materials that are liquids or solids at room temperature; many solvents generate vapors
- *Gases* – Materials that do not exist as solids or liquids at room temperature, such as carbon monoxide and ammonia

2.1.1 Oxygen

The percentage of oxygen in a confined space is a critical safety factor. In an oxygen-deficient atmosphere, the concentration of oxygen is too low, and this can quickly lead to asphyxiation. An oxygen-enriched atmosphere, in which the concentration of oxygen is too high, presents a flammability/explosion hazard. The normal range for oxygen concentration is between 19.5 and 23.5 percent. Below 19.5 percent is considered oxygen-deficient, above 23.5 percent it is oxygen-enriched. An ideal concentration is 21 percent oxygen.

An oxygen-deficient atmosphere in a confined space environment will have a rapid effect on workers, as shown in *Table 2*.

Even if the concentration of oxygen in a confined space is initially safe, many processes can cause oxygen deficiency. Hot work such as welding, cutting, or brazing consumes oxygen. Oxygen can also be depleted due to bacterial action such as fermentation, or by chemical reactions such as rust. Each additional person working in a confined space, coupled with the amount of physical activity, reduces the concentration of oxygen. The concentration of oxygen can also be reduced through displacement by other gases. Many welding activities require the use of argon or other inert gases that could unintentionally displace the oxygen in a confined space. Welding should not be performed in confined spaces without proper ventilation.

2.1.2 Flammable Atmospheres

A flammable atmosphere in a confined space is one that contains sufficient oxygen as well as flammable gases, vapors, or dust. Types of flammable gases commonly found in confined spaces include the following:

- Acetylene
- Butane
- Propane
- Hydrogen
- Methane
- Natural gas

Many flammable gases are heavier than air and will sink to the lower levels of storage tanks, pits, and vessels. Other gases are lighter than air and may rise and develop a flammable concentration near the top of a closed-top tank. It is critical to test for flammable gases at several depths in a confined space. To determine if the gas is lighter or heavier than air, check its **vapor density**. The vapor density is listed on the safety data sheet (SDS) in *Section 9: Physical and Chemical Properties*. Refer to the NIOSH Pocket Guide to Hazardous Chemicals for naturally occurring gases.

> **NOTE**
> Work in other areas can affect the atmosphere in a confined space. Gases created or released above or below a confined space can rise or sink and become trapped in the confined space. Be aware of the hazards around a confined space in addition to the hazards in the space itself.

Table 2 Symptoms of Oxygen Deficiency

Oxygen Volume Percentage	Symptoms
12% to 16% Oxygen	Deep breathing, fast heartbeat, poor attention, poor thinking, poor coordination
10% to 14% Oxygen	Faulty judgment, intermittent breathing, rapid fatigue (possibly causing heart damage), very poor coordination, lips turning blue
Less than 10% Oxygen	Nausea, vomiting, loss of movement, loss of consciousness, death
Less than 6% Oxygen	Spasmodic breathing, convulsive movement, and death in approximately eight minutes
4% to 6% Oxygen	Coma in 40 seconds, death

Case History

A construction foreman died from asphyxiation after entering a manhole with an uncontrolled hazardous atmosphere. Four construction workers were working in an inactive sewer system on a job site that was unoccupied for over a week. A few minutes after they started working, the crew noticed that the foreman was missing, and a manhole cover was removed. While one worker called emergency services, a second worker entered the manhole to assist the foreman, and found him unresponsive at the bottom of the manhole. When the second worker became disoriented inside the manhole, another worker used a fan to blow fresh air into the manhole and the worker was able to climb out. The foreman was retrieved by fire department personnel and was later pronounced dead due to asphyxiation.

Although the manhole was newly constructed and not yet connected to an active sewer system at the time of this incident, it contained a hazardous atmosphere that resulted in asphyxiation. The employer did not ensure that atmospheric hazards were identified and precautions for safe operations implemented before starting work at the site. It was determined that the following conditions contributed to this incident:

- Workers were not trained to recognize confined space hazards and take appropriate protective measures.
- The atmosphere in the manhole was not assessed to determine if conditions were acceptable before or during entry.
- Proper ventilation was not used to control atmospheric hazards in the manhole.
- Protective and emergency equipment was not provided at the work site.
- An attendant was not stationed outside the manhole to monitor the situation and call for emergency services.

The Bottom Line: Confined spaces present significant hazards and all appropriate safety procedures must be followed in order to prevent serious injury or death.

Source: OSHA Fatal Facts No. 12-2015.

Flammable or explosive conditions can also be created by the work in a confined space. For example, spray painting can create explosive gases or vapors. Transferring coal, grain products, nitrated fertilizers, or finely ground chemical products can result in a concentration of combustible dust.

Chemicals can combine to form flammable gases. For example, sulfuric acid reacts with iron to form hydrogen, and calcium carbide reacts with water to form acetylene. Both of these gases are highly flammable. Simply opening the door to a confined space and allowing the outside air to enter can also cause chemical reactions that form flammable gases.

Spaces and adjacent spaces that contain or have contained flammable liquids or gases must be tested for flammable atmospheres before entry. Such spaces can include the following:

- Fuel tanks
- Pump rooms
- Pipelines
- Recently painted or solvent-cleaned spaces
- Sewage tanks
- Any spaces adjacent to the spaces listed above
- Cargo tanks (for tank ships)

Hot work in a confined space is a major cause of explosions in areas that contain combustible gas. However, flammable atmospheres can just as easily be ignited through a single spark generated from static electricity, such as from clothing or carpeting when the air is dry. Even something as simple as flipping a light switch can create a tiny arc that could ignite an explosive atmosphere.

The main hazard of static electricity is the creation of sparks in an explosive or flammable atmosphere that could potentially set off an explosion or fire. If the potential for fire around flammables and combustibles exists, grounding and bonding may be required. Static electricity generated in a confined space by workers themselves is potentially hazardous. Use controls to prevent or reduce the buildup of electrical charges, including the following:

- Conductive flooring
- Conductive clothing and footwear (to allow the charge to be conducted away, these items must be free of dirt and other contaminants)
- Cotton or linen clothing instead of wool, silk, or synthetic materials

The transfer of liquids also has the potential for creating static electricity. To prevent static when transferring liquids, make sure that the nozzle is touching the bottom of the vessel so that the liquid discharges horizontally. Also, lower the rate of flow. These two measures will prevent the free fall and turbulence that generate static. Another control is the use of anti-static additives (as in fuels). The additive increases the conductivity or lowers the resistance of the liquid. It also reduces the time it takes for the static charge to leak through the wall of the container and to the ground.

Flammable atmosphere locations are categorized as Class I, Division 1 and 2; Class II, Division 1 and 2; and Class III, Division 1 and 2. Refer to *Table 3A* and *Table 3B*.

Class I locations are those in which flammable gases or vapors are or may be present in the air in quantities sufficient to produce explosive or ignitable mixtures. Class I locations include the following:

- *Class I, Division 1* – A Class I, Division 1 location is a location in which:
  - Ignitable concentrations of flammable gases or vapors may exist under normal operating conditions; or
  - Ignitable concentrations of such gases or vapors may exist frequently because of repair or maintenance operations or because of leakage; or
  - Breakdown or faulty operation of equipment or processes might release ignitable concentrations of flammable gases or vapors and might also cause simultaneous failure of electric equipment.

> **NOTE**
> This classification usually includes locations where volatile flammable liquids or liquefied flammable gases are transferred from one container to another; interiors of spray booths and areas in the vicinity of spraying and painting operations where volatile flammable solvents are used; locations containing open tanks or vats of volatile flammable liquids; drying rooms or compartments for the evaporation of flammable solvents; inadequately ventilated pump rooms for flammable gas or for volatile flammable liquids; and all other locations where ignitable concentrations of flammable vapors or gases may occur during normal operations.

- *Class I, Division 2* – A Class I, Division 2 location is a location in which:
  - Volatile flammable liquids or gases are handled, processed, or used, but in which the hazardous liquids, vapors, or gases will normally be confined within closed containers or closed systems from which they can escape only in case of the rupture or breakdown of such containers or systems, or in case of abnormal operation of equipment; or
  - Ignitable concentrations of gases or vapors are normally prevented by positive mechanical ventilation, and which might become hazardous through failure or abnormal operations of the ventilating equipment; or
  - Is adjacent to a Class I, Division 1 location, and to which ignitable concentrations of gases or vapors might occasionally be communicated unless such communication is prevented by adequate positive-pressure ventilation from a source of clean air, and effective safeguards against ventilation failure are provided.

> **NOTE**
> This classification usually includes locations where volatile flammable liquids or flammable gases or vapors are used, but which would become hazardous only in case of an incident or unusual operating condition. The quantity of flammable material that might escape in such cases, the adequacy of ventilating equipment, the total area involved, and the record of the industry or business with respect to explosions or fires are all factors that merit consideration in determining the classification and extent of each location.
>
> Piping without valves, checks, meters, and similar devices would not ordinarily introduce a hazardous condition even though used for flammable liquids or gases. Locations used for the storage of flammable liquids or of liquefied or compressed gases in sealed containers would not normally be considered hazardous unless also subject to other hazardous conditions.
>
> Electrical conduits and their associated enclosures separated from process fluids by a single seal or barrier are classed as a Division 2 location if the outside of the conduit and enclosures is a nonhazardous location.

Table 3A Summary of Hazardous Atmospheres

Hazardous	Class Subdivisions Class I Divisions	Groups Class I, Division Groups
Class I: Material present is a flammable gas or vapor	Division 1: Locations in which hazardous concentrations of flammable gases or vapors are present normally or frequently	Group A: Atmospheres containing acetylene
	Division 2: Locations in which hazardous concentrations of flammable gases or vapors are present as a result of infrequent failure of equipment or containers	Group B: Atmospheres containing hydrogen, manufactured gases containing more than 30% hydrogen by volume, or gases or vapors of equivalent hazard
		Group C: Atmospheres containing ethylene, cyclopropane, or gases or vapors of equivalent hazard
		Group D: Atmospheres containing propane, gasoline, or gases or vapors of equivalent hazard
	Class I Zones	**Class I, Zone Groups**
	Zone 0: Locations in which combustible material is present continuously or for long periods	Group IIC: Atmospheres containing acetylene or hydrogen or other gases or vapors meeting Group IIC criteria
	Zone 1: Locations in which combustible material is likely to be present normally or frequently because of repair or maintenance operations or leakage	Group IIB: Atmospheres containing acetaldehyde, ethylene, or other gases or vapors meeting Group IIB criteria
	Zone 2: Locations in which combustible material is not likely to occur in a normal operation and, if it does occur, will exist only for a short period	Group IIA: Atmospheres containing propane, gasoline, or other gases or vapors meeting Group IIA criteria

Table 3B Summary of Hazardous Atmospheres

Hazardous	Class Subdivisions Class II Divisions	Groups Class I, Division Groups
Class II: Material present is a combustible dust	Division 1: Locations in which hazardous concentrations of combustible dust are present normally or may exist because of equipment breakdown or where electrically conductive combustible dusts are present in hazardous quantities	Group E: Atmospheres containing combustible metal dusts including aluminum, magnesium, and other metals of similar hazards
	Division 2: Locations in which hazardous concentrations of combustible dust are not normally suspended in the air but may occur as a result of infrequent malfunction of equipment or where dust accumulation may interfere with safe dissipation of heat or may be ignitable by abnormal operation of electrical equipment	Group F: Atmospheres containing combustible carbonaceous dusts, including carbon black, charcoal, coals, or dusts that have been sensitized by other materials so that they present an explosion hazard
		Group G: Atmospheres containing combustible nonconductive dusts not included in Group E or F, including flour, grain, wood, and plastic
	Class III Divisions	**Groups**
Class III: Material present is an ignitable fiber or flying	Division 1: Locations in which easily ignitable fibers or materials producing combustible flyings are handled, manufactured, or used	No Groups
	Division 2: Locations in which easily ignitable fibers are stored or handled, except in the manufacturing process	

Class II locations are those that are hazardous because of the presence of combustible dust. Class II locations include the following:

- *Class II, Division 1* – A Class II, Division 1 location is a location in which:
  - Combustible dust is or may be in suspension in the air under normal operating conditions, in quantities sufficient to produce explosive or ignitable mixtures; or
  - Mechanical failure or abnormal operation of machinery or equipment might cause such explosive or ignitable mixtures to be produced, and might also provide a source of ignition through simultaneous failure of electric equipment, operation of protection devices, or from other causes; or
  - Combustible dusts of an electrically conductive nature may be present.

> **NOTE**
> Combustible dusts, which are electrically nonconductive, include dusts produced in the handling and processing of grain and grain products, pulverized sugar and cocoa, dried egg and milk powders, pulverized spices, starch and pastes, potato and wood flour, oil meal from beans and seed, dried hay, and other organic materials that may produce combustible dusts when processed or handled. Dusts containing magnesium or aluminum are particularly hazardous and the use of extreme caution is necessary to avoid ignition and explosion.

- *Class II, Division 2* – A Class II, Division 2 location is a location in which:
  - Combustible dust will not normally be in suspension in the air in quantities sufficient to produce explosive or ignitable mixtures, and dust accumulations are normally insufficient to interfere with the normal operation of electrical equipment or other apparatus; or
  - Dust may be in suspension in the air as a result of infrequent malfunctioning of handling or processing equipment, and dust accumulations resulting from there may be ignitable by abnormal operation or failure of electrical equipment or other apparatus.

> **NOTE**
> This classification includes locations where dangerous concentrations of suspended dust would not be likely but where dust accumulations might form on or in the vicinity of electric equipment. These areas may contain equipment from which appreciable quantities of dust would escape under abnormal operating conditions or be adjacent to a Class II Division 1 location, as described above, into which an explosive or ignitable concentration of dust may be put into suspension under abnormal operating conditions.

Class III locations are those that are hazardous because of the presence of easily ignitable fibers or flyings but in which such fibers or flyings are not likely to be in suspension in the air in quantities sufficient to produce ignitable mixtures. Class III locations include the following:

- *Class III, Division 1* – A Class III, Division 1 location is a location in which easily ignitable fibers or materials producing combustible flyings are handled, manufactured, or used.

> **NOTE**
> Easily ignitable fibers and flyings include rayon, cotton (including cotton linters and cotton waste), sisal or henequen, istle, jute, hemp, tow, cocoa fiber, oakum, baled waste kapok, Spanish moss, excelsior, sawdust, woodchips, and other material of similar nature.

- *Class III, Division 2* – A Class III, Division 2 location is a location in which easily ignitable fibers are stored or handled, except in process of manufacture.

2.1.3 Measuring Flammable Atmospheres

In order to check the flammability of an atmosphere, you'll need to know the limits associated with particular gases. These limits can be found in the NIOSH Pocket Guide to Chemical Hazards. The lower explosive limit (LEL) or lower flammable limit (LFL), is the lowest concentration of air-fuel mixture at which a gas can ignite. The upper explosive limit (UEL) or upper flammable limit (UFL) is the highest concentration that can be ignited. Explosion can occur at any point between the LFL and UFL. A confined space should not be considered safe unless the concentration of gas is well below the LFL.

The potential for a flammable atmosphere increases as the temperature rises (*Figure 17*). A rise in temperature, such as from morning to afternoon, requires additional testing. For example, assume a hazardous material is present and its flash point is 50°F (10°C). A test run with a combustible gas indicator in the morning when the temperature is 34°F (1°C) may not show a hazardous atmosphere. However, in the afternoon, when the temperature increases to 65°F (18°C), you would likely get a reading at or above the LFL. Confined spaces must be tested periodically to confirm that they are still safe to enter.

The permissible exposure limit (PEL) is set by OSHA standards at 10 percent of the LFL. If the concentration is greater than 10 percent of the LFL, you may not enter, according to OSHA. You must first ventilate the area to reduce the concentration of flammable vapors. Technically, the air would not be flammable at 90 percent of the LFL. However, the air in a confined space often has pockets of concentrated vapors. It is impossible to test every part of the air in a confined space, so OSHA has set 10 percent of the LFL as a safe limit for entry.

Technically, a concentration above the UFL would not be flammable. However, it is not safe to enter a space above the UFL. The air can easily become diluted when a worker enters the space, and the concentration will sink below the UFL into the flammable range. The atmosphere may also contain an oxygen-deficient or toxic atmosphere.

Flammable vapors may be diluted through ventilation, or through the process of **inerting**. Inerting is when a noncombustible gas such as nitrogen is released into the confined space so that the atmosphere is no longer combustible. Keep in mind that inerting will also drive out oxygen, creating an oxygen-deficient atmosphere. It is therefore important to understand that inert gases such as nitrogen, argon, and helium are a potential asphyxiation hazard.

2.1.4 Toxic Atmospheres

There is a wide range of substances in gas or particulate (dust) form that can cause or contribute to a toxic atmosphere in a confined space. Toxic atmospheres result from several sources, including manufacturing processes, storage procedures, and operations such as welding or paint application. Human error and mechanical malfunctions may also contribute to the production of toxic gases.

PELs for toxic substances can be found in *Section 8: Exposure Controls/Personal Protection* of the SDS. You can also find PELs in the NIOSH Pocket Guide to Chemical Hazards or other reference sources. The PEL for a particular substance is the maximum concentration of the substance that a worker may be exposed to in an 8-hour shift without any adverse effects. For gases, the PEL is listed in **parts per million (ppm)**. For particulates, PELs are listed in milligrams per cubic meter (mg/m^3), or in fibers per cubic meter for fibrous substances, such as asbestos. For solids, the PEL is listed as a percentage, such as % silica.

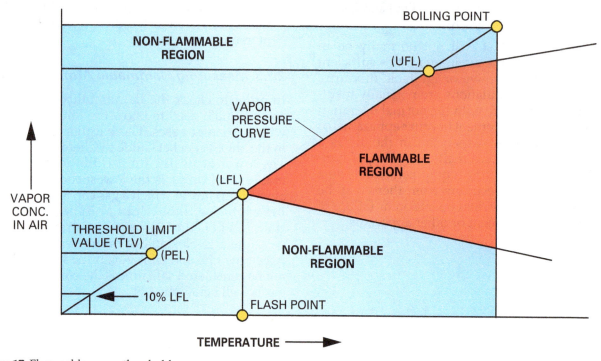

Figure 17 Flammable vapor thresholds.

> **Did You Know?**
> ## Hydrogen Sulfide Gas
> Hydrogen sulfide gas is a toxic gas that smells like rotten eggs. It is formed when waste containing sulfur is broken down by bacteria. Sewers, septic tanks, manholes, and well pits often contain hydrogen sulfide gas. You can smell hydrogen sulfide at a level of about 10 parts per billion (ppb) or 0.01 ppm. This is about the same as a thimbleful of hydrogen sulfide gas in a house full of air. However, if you are exposed for a while, you will no longer notice the odor. Always use an instrument to detect the presence of hydrogen sulfide gas. Do not rely on your sense of smell.

If applicable, this information is typically listed as an 8-hour time-weighted average (TWA) information on the product SDS. An SDS will also list short-term exposure limits (STEL), chemical-specific toxicity characteristics, health hazards, *specific gravity*, and reactivity hazards. PELs for several common toxic gases are listed in *Table 4*.

2.1.5 Environmental Conditions

The environmental conditions of a confined space need to be considered. Workers should not be subjected to extreme temperatures, noise, or physical hazards when working inside the space. Check for mechanical hazards, slippery surfaces, falling or dripping substances, and the general working conditions inside the confined space.

The environmental conditions may require the use of specialized equipment, such as protective suits for cooling if heat is an issue. Protective gear and clothing may also be needed to protect workers from frostbite or hyperthermia if the space is extremely cold. Noise levels should also be measured and addressed accordingly. For any given situation, equipment and clothing should be selected to provide an adequate level of protection. Over-protection as well as under-protection can be hazardous and should be avoided.

2.1.6 Using Tools and Equipment in Confined Spaces

Portable electric hand tools present a spark hazard and should not be used in a confined space. If these tools must be used, they must be double-insulated or protected by a ground fault circuit interrupter (GFCI). Portable electric tools, lights, and equipment used in confined spaces must meet the following minimum requirements:

- Industrial grade
- In good condition
- Operated at a maximum of 110 volts AC and connected to a GFCI or operated at no more than 12 volts DC

Additional care must be taken when using power tools if there is a potential for a flammable atmosphere. In this situation, the tools must be intrinsically safe. This means that they do not pose an explosion hazard under normal use. The tools must be certified by a nationally recognized testing laboratory, such as Underwriters Laboratory (UL), as being approved for use in hazardous locations.

Pneumatic tools and equipment used in confined spaces must be driven by air. Using any other compressed gas to drive tools can create fire or suffocation hazards. Using pneumatic tools does not eliminate the spark/fire hazard posed by electric tools. It only reduces the hazard by eliminating the electrical energy source.

> **Did You Know?**
> ## Carbon Monoxide
> Carbon monoxide (CO) is a hazardous gas that is colorless, odorless, and has roughly the same density as air. CO is a result of the incomplete combustion of materials such as wood, coal, gas, oil, and gasoline. It can also be formed by the decomposition of organic matter in sewers, silos, and fermentation tanks. It can be fatal at 1,000 ppm. It is dangerous even at 200 ppm because it quickly replaces oxygen in the bloodstream. Early symptoms of CO intoxication are nausea and headache.
>
> A safe reading on a combustible gas indicator is not an indication of a safe CO concentration. You must test specifically for carbon monoxide.

Table 4. PELs of Common Toxic Gases

Substance	PEL (PPM)
Carbon dioxide	5,000
Carbon monoxide	50
Hydrogen sulfide	20
Methane	1,000
Nitric oxide	25
Oxygen diflouride	0.05
Phosgene (carbonyl chloride)	0.1
Sulfur dioxide	5
Stoddard solvent	500

Compressed gas cylinders must not be brought into confined spaces unless they are part of a self-contained breathing apparatus (SCBA) or an emergency escape unit. The only gas cylinders that should be brought into a confined space are those used for respiratory protection.

2.2.0 Atmospheric Testing Procedures for Confined Spaces

The internal atmosphere of a confined space must be tested before anyone enters and again periodically thereafter if conditions change. Test for oxygen content, flammable gases and vapors, and toxic air contaminants, in that order. Anyone entering the space must be allowed to observe the pre-entry testing. Testing provides an evaluation of the hazards as well as verification of acceptable entry conditions.

Test equipment that measures both oxygen and a range of flammables is the best choice for initial testing. In a rapidly changeable atmosphere such as a sewer, the test equipment will be carried and used by the entrant to monitor the atmosphere. The entrant must test in the direction of movement before advancing. Tests must be repeated periodically to warn the entrant of any deterioration in atmospheric conditions. In some instances, the attendant may monitor the gas-testing instrument from the outside and alert the entrant(s) if there is a problem.

Gas detection meters are available in both single- and multiple-gas configurations (*Figure 18*).

Regardless of the type of meter, the oxygen concentration must be tested first. Test equipment may give misleading results in an oxygen-deficient atmosphere.

For example, an atmosphere might consist of 90 percent methane and 10 percent air. If you did not test for oxygen and went straight to a flammability test, the meter would show that this atmosphere is not explosive because there isn't enough oxygen for combustion. As outside air moves into the space, however, the atmosphere becomes highly explosive. Testing for and establishing a safe oxygen level (between 19.5 and 21.5 percent) must be the first step.

The best way to verify that a gas meter detects gas accurately and reliably is to test it with a known concentration of gas. This procedure, commonly referred to as a bump test, will verify whether the sensors in the instrument respond accurately and whether the alarms function properly. This is a qualitative function check in which a sample of the gas being detected is passed over the sensor(s) at a concentration and exposure time sufficient to activate all alarm settings.

A bump test or calibration check of portable gas monitors should be conducted before each day's use in accordance with the manufacturer's instructions. If an instrument fails a bump test or a calibration check, the operator should perform a full calibration before using it. If the instrument fails the full calibration, the employer should remove it from service and contact the manufacturer for assistance or service.

Figure 18 Gas detection meters.

2.2.1 Testing Procedures for Oxygen

Before beginning any atmospheric testing, the test equipment must be calibrated according to the manufacturer's instructions. Calibration ensures that the equipment provides accurate, dependable readings of oxygen concentrations. Oxygen meters must be adjusted to read 20.9 percent. Combustible gas detectors and toxic gas monitors should read zero.

Begin testing at the entrance to the space, moving the meter probe further into the space at 4-foot (1.2 m) intervals. Do not enter the space to test without appropriate respiratory protection. The initial testing for oxygen concentration in a confined space must be noted on the permit. Specify the time and the concentration reading. If ventilation is necessary, retest after ventilation. Be sure the oxygen levels are normal before entering the space.

The space must be periodically monitored for oxygen. There is no guarantee that it will still be safe the next day or even an hour later. Retesting must occur after any hot work in the space. Activities that create combustion consume oxygen. Retesting must be performed at regular intervals. This depends on several factors, including the size of the space, the number of entrants, and the work being performed. At a minimum, retesting must occur once each shift for the duration of the work permit.

Continual testing can be achieved by using single-gas disposable test meters. These units are similar to household smoke detectors and do not require calibration. They have an audible alarm when the oxygen concentration drops below normal levels. They may be placed in the confined space for the duration of the work permit.

The meters used for flammability testing are often also used for oxygen testing (*Figure 19*). Meters that test for flammability can be set to detect a combustible gas as a percentage of LEL.

2.2.2 Testing Procedures for Flammable Atmospheres

Some flammable gases such as propane and methane are heavier than air and will collect at the bottom of a confined space. Other gases such as hydrogen are lighter and will rise to the top. It is not sufficient to test the atmosphere only near the entrance of a confined space. The air may be very different at different levels inside the space. Tests should be done at a minimum of 4-foot intervals from top to bottom and bottom to top.

As with oxygen meters, the equipment must be calibrated according to the manufacturer's instructions before testing. Flammable gas detection equipment must sound an alarm before the

Figure 19 Flammability detection meters.

concentration reaches the LEL. OSHA standards define the highest permissible level of flammable gas in a confined space at 10 percent of the LEL for that particular gas. This does not mean that the atmosphere can be composed of 10 percent flammable gas.

Testing for concentrations of flammable dust is not as straightforward. A good rule of thumb is that the concentration of dust that meets or exceeds its LEL may be approximated as a condition in which the dust obscures vision at a distance of 5 feet (1.5 m) or less. Personnel should not attempt to work in an environment with such a high concentration of airborne dust or combustible fibers.

As with oxygen testing, conditions in the space may change. The initial test for flammable gases may not remain valid over time. Retest periodically to make sure the work area is safe. This is particularly important if the ambient temperatures vary widely throughout the day.

Certain work activities can contribute to the amount of flammable gas in the atmosphere. Operations within the space, such as purging welding lines, will contribute to the concentration of flammable gases. The application of protective coatings and paints can also create hazardous atmospheres. Operations outside the space can also change the air in a confined space. For example, a release of methane (a dense gas) on the ground may not immediately affect that particular floor but can quickly fill a connected below-ground space.

Testing must be redone after any operation that might increase gas concentrations. It should also be carried out at least once per shift for the duration of the work permit.

2.2.3 Toxic Air Contamination Testing

Confined spaces may also accumulate volatile organic compounds (VOCs). VOCs include solvents, paint thinners, and various fuels (gasoline, diesel, heating oil, kerosene, and jet fuel). Other VOCs are toluene, benzene, butadiene, hexane, and xylene, among others. It is important to test for the presence of all possible hazards in a confined space. Some types of toxic gases may be indicated on gas detection meters. However, there are many toxic gases that are not typically included on multi-use or single-use meters. These require the use of gas detector tubes (*Figure 20*).

Each tube tests for the presence of a different toxin. It indicates if the toxic agent is present, usually by a change in color. The tubes are a single-use product, so a new tube is needed for each test. Note that the tubes have a ±25 percent accuracy and only indicate the presence of the toxin; they do not effectively measure the concentration of the contaminant.

Before testing for toxic chemicals, make sure that oxygen and flammability tests have been completed and the atmosphere is safe. Calibrate the test meter (if it allows for toxic gas testing) and check for the presence of toxins. Note that toxins, like flammable gases, may be heavier or lighter than air. You must test at a minimum of 4-foot (1.2 m) intervals. All areas of the space will need to be tested.

If air toxins have been identified or are suspected, the specific concentration of the contaminant must be determined. This is done using a gas detector tube. First, select the appropriate tube for the suspected hazard. After calibrating the pump, open the tube and insert it in the pump or the sample line. The pump pulls the

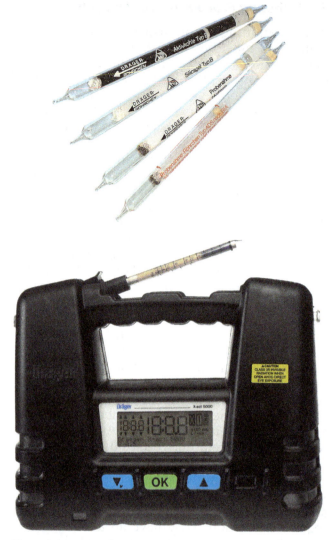

Figure 20 Gas detector tubes.

air sample through the tube and the toxin will change the color of the material in the tube. The concentration of toxic gas can be read directly from the markings on the tube. The tubes have limited accuracy due to the fact that it is often difficult to tell where the color changes on the tube.

Single-use gas detection tubes have the following limitations:

- Other gases may cause an inaccurate reading.
- Readings may vary among different tube manufacturers.
- You must only use tubes from the same pump manufacturer.

Always read and follow the instructions on the information sheet provided with each package of tubes. Be aware of the effects of cross-contamination from other gases and vapors in the environment.

After the initial testing of a confined space, you must purge or dilute the toxic levels of any gases present. You will need to periodically retest.

Retesting must be done after any process in or near the space that might produce toxic gases. For example, if scraping or blasting was done inside or near the confined space, retesting is required. At a minimum, it must be done at least daily for the duration of the work permit.

WARNING!
When wearing respiratory protection to protect against toxic vapors, make sure that the filter is rated for the specific toxin. Air-purifying respirators cannot be used if the amount of toxic vapor has not been determined. SCBA must be used until this determination is made, then the correct filter for an air-purifying respirator can be selected. Respirators may only be used if they reduce the toxin below the PEL and the contaminant has adequate warning properties.

CAUTION
Always check your test instrument for leaks. A leak in the sampling hose or pump will significantly change the readings. In addition to calibrating the instrument, you must perform a leak check as directed by the manufacturer. This should be done each time you use the equipment.

2.2.4 Ventilating Confined Spaces

To work safely in a confined space, there must be adequate ventilation. A ventilation system must be set up to circulate fresh air if supplied air is not used. In the event that the ventilation system stops working, a monitoring procedure must be in place to detect any increase in atmospheric hazards in time for the entrants to safely exit the space. Remember that most power tools create a spark hazard. If flammable vapors are present or suspected, special non-sparking or brass tools will be needed to prevent fires or explosions.

Ventilation is a critical aspect of confined space entry. It is important that fresh atmospheric air be used for ventilation air. Atmospheric air is approximately 21 percent oxygen and 78 percent nitrogen, with small amounts of carbon dioxide, hydrogen, helium, argon, and neon making up the remainder (see *Figure 21*).

WARNING!
When ventilating a space using supplied air, be aware that different gas cylinders have different connections. Proper labeling of hoses for different gases should be practiced in order to prevent potential safety hazards.

WARNING!
Pure oxygen should never be used to ventilate a space as it is explosive and creates a health hazard. Too much oxygen is as dangerous as too little.

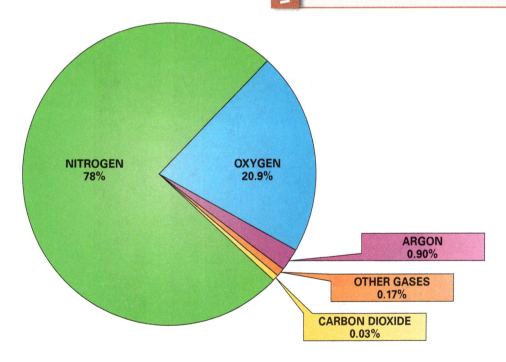

Figure 21 Atmospheric air.

Methods used to circulate fresh air into a confined space include the following:

- *Natural ventilation* – Natural ventilation is simply opening the hatches to allow air flow into the space. It is not usually enough to circulate fresh air into a confined space because of possible dead zones (areas where air can become trapped). These areas have the potential to contain a hazardous atmosphere even after the space is ventilated. Continuous air monitoring is recommended when using natural ventilation. Because working in the space will increase hazardous atmospheric conditions, a positive means of ventilation must be established and operated when work is in progress. The accepted practice and recommended level of ventilation is one complete air change every three minutes.
- *Air horns* – An air horn is often used to circulate air through a vessel, as shown in *Figure 22*. Air horns should be set to exhaust air out of the space. This will draw in clean air from other openings. Pneumatic air movers must use compressed air. Never use other gases such as nitrogen. Air movers or fans must be electrically bonded to the vessel. The vessel must also be grounded.
- *Air movers* – Air movers should be placed at the top of a vessel when vapors that are lighter than air may be present. This is the most

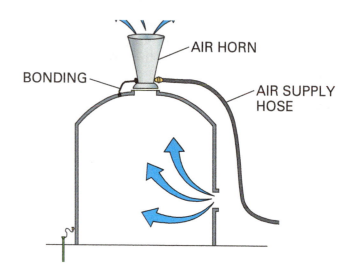

Figure 22 Air horn ventilation with bonding and grounding.

common ventilation configuration. More than one air mover and fresh-air inlet should be used whenever possible to ensure adequate ventilation of the entire space. Air movers should be placed at the bottom of a space when vapors that are heavier than air may be present. When this system is used, be certain the exhausted air does not create a hazard for area personnel. Air movers or fans must not be positioned to blow into a space. *Figure 23* depicts a common ventilation system for subsurface

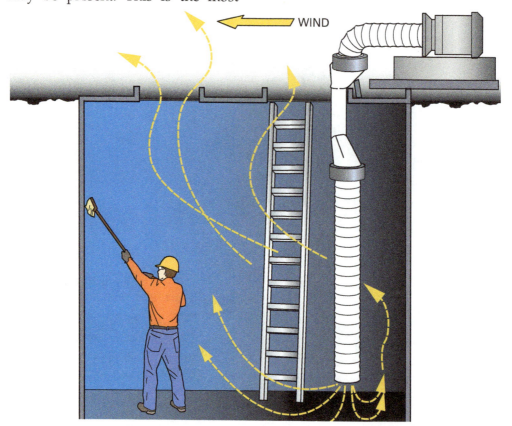

Figure 23 Positive ventilation system.

spaces. In this example, fans move clean air into the space. The system must be upwind so only clean, fresh air is drawn into the space. If it is downwind, the fans will blow the bad air back into the space. The system must be arranged to allow complete purging of the space. The exhausted air must be located so it will not be a hazard to other personnel or equipment. Additional air may be blown into a space in addition to purging ventilation. This provides clean air for general comfort and breathing.

> **WARNING!**
> Makeup air must not be blown into the space when there are flammable or toxic materials present or being generated by the work in progress. Only explosion-proof electric fans may be used to ventilate vessels or spaces that may contain flammable gases, vapors, or dust. Do not use pure oxygen as the makeup air.

– *Exhaust systems* – Local exhausting may also be required in addition to forced ventilation. A local exhaust system (*Figure 24*) is used when work such as welding generates fumes or vapors. The exhaust from this local exhaust system should be located so it does not present a hazard to other personnel or equipment.

When a worker must enter at the same opening that is used for ventilation, the equipment should minimize the restriction that may occur. For example, secure hoses so they do not present a tripping or entanglement hazard. Use flexible hoses when supplying air to spaces with smaller openings, such as manholes.

Retrieval equipment should be used whenever rescue may be difficult, there is a possibility of a fall, or self-contained or supplied breathing air is used.

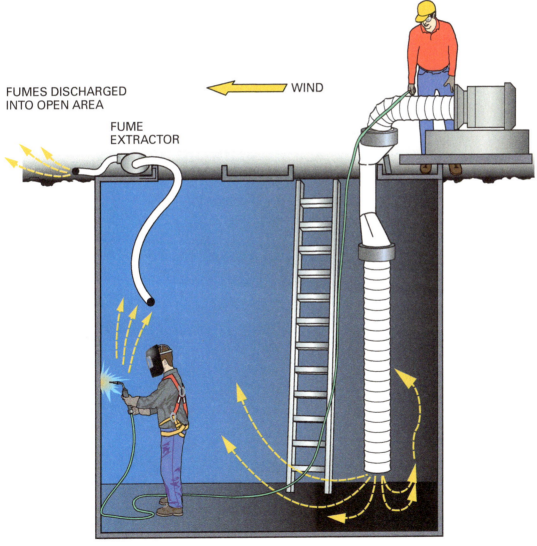

Figure 24 Fume extractor used for local exhaust.

Additional Resources

American Society of Safety Engineers, **www.asse.org**

Occupational Safety and Health Organization, **www.osha.gov**

Guidelines for Hot Work in Confined Spaces: Recommended Practices for Industrial Hygienists and Safety Professionals, Martin H. Finkel, CIH, CMC. 2000. Des Plaines, IL: The American Society of Safety Engineers (ASSE).

Complete Confined Spaces Handbook, John F. Rekus, MS, CIH, CSP. 1994. Boca Raton, FL: CRC/Lewis Publishers.

2.0.0 Section Review

1. An unprotected worker in an atmosphere of 5 percent oxygen can fall into a coma in _____.

    a. less than one minute
    b. 5 minutes
    c. 8 minutes
    d. 15 minutes

2. The concentration of dust in a confined space has probably exceeded its LEL if it obscures vision at a distance of _____.

    a. 5' (1.5 m) or less
    b. 7' (2.1 m)
    c. 9' (2.7 m)
    d. 15' (4.6 m) or more

SECTION THREE

3.0.0 CONFINED SPACE ENTRY PROCEDURES

Objective

Explain how to prepare workers for confined space entry.
  a. List the types of information required on a confined space entry permit.
  b. Identify the roles and responsibilities of personnel in confined spaces.
  c. Identify and describe the types of emergency training required for workers in confined spaces.

Trade Terms

Confined space entry permit system: The written procedure for preparing and issuing permits for entry, and for returning the permit-required space to service after entry work is completed.

Induced hazards: Hazards brought forward or created when people make incorrect decisions and/or take incorrect actions during the construction process (e.g., omission of protective features; physical arrangements that cause unintentional worker contact with electrical energy; oxygen deficient hazards created at the bottom of pits, shafts, or vaults; lack of safety factors in structural strength; and flammable atmospheres).

Inherent hazards: Hazards associated with specific types of equipment and the interactions of the equipment with the confined space. Includes hazards that cannot be eliminated without degrading the system or the equipment, or making them inoperative (e.g., high voltage, radiation generated by equipment, defective designs, omission of protective features, high or low temperatures, high noise, and high-pressure vessels and lines).

A confined space entry permit system is the written procedure for preparing and issuing permits for entry, and for returning the permit-required space to service after entry work is completed. Confined spaces requiring permits are those with a high likelihood of atmospheric or other hazards. OSHA's Confined Space in Construction Standard categorizes confined space hazards in construction into two basic classes: inherent hazards and induced hazards.

Before work can begin, the designated supervisor must approve the entry permit by signing it. The authorized permit will be posted at the entry point or made available to all entrants at the time of entry. The duration of the permit may not exceed the time required to complete the assigned task or job listed on the permit.

The supervisor will cancel the entry permit when operations covered by the permit have been completed. They will also cancel the permit if something occurs in or near the permit-required space that is not allowed by the permit. The supervisor will note any problems encountered during the entry operation on the permit. The employer must maintain a file of canceled entry permits for at least one year.

The construction confined space standard differs in several ways from the general industry confined space standard. The Confined Spaces in Construction standard includes the following requirements:

- Detailed coordination is required among multiple employers at the work site. For example, a gasoline-powered motor running near the entrance of a confined space could cause a potential buildup of carbon monoxide within the space.
- A competent person must evaluate the work site and identify confined spaces, including permit spaces.
- Continuous atmospheric monitoring is required whenever possible.
- Continuous monitoring of engulfment hazards is required.
- The standard allows for the suspension of a permit, instead of its cancellation. However, the space must be returned to the entry conditions listed on the original permit before re-entry.
- The standard makes the controlling contractor, rather than the host employer, the primary point of contact for information about permit spaces at the work site. The controlling contractor is also responsible for ensuring that employers outside a space know not to create hazards in the space, and that entry employers working in a space at the same time do not create hazards for one another's workers.

OSHA has added provisions to the construction confined space standard that clarify existing points and requirements in the general industry standard. These provisions include the following:

- Employers who direct workers to enter a space without using a complete permit system are required to prevent workers' exposure to

physical hazards through elimination of the hazard or isolation methods such as lockout/tagout.
- Employers who are relying on local emergency services for emergency services must arrange for responders to give the employer advance notice if they will be unable to respond for a period of time.
- Employers must provide training in a language and vocabulary that the worker understands.
- Several terms have been added to the definitions in the construction rule, such as *entry employer* to describe the employer who directs workers to enter a space, and *entry rescue* to clarify the differences in the types of rescue employers can use.

3.1.0 Required Information on a Confined Space Entry Permit

A confined space entry permit will include the following required information:

- Identification of the space to be entered
- Purpose of entry
- Date and authorized duration
- A means of tracking authorized entrants within the permit-required space; for example, a list of names, a roster, or other tracking system
- Names of all authorized attendants
- Name of the current entry supervisor, with a space for the signature or initials of the entry supervisor who originally authorized entry
- Known hazards of the confined space covered by the permit
- Measures used to isolate the space and to eliminate or control hazards before entry, including lockout or tagging of equipment, as well as procedures for purging, inerting, ventilating, and flushing spaces
- Acceptable entry conditions
- Test results (both initial and periodically recurring) accompanied by time, date, and the testers' names or initials
- Responsible rescue and emergency services and the means for calling for those services
- Communication procedures used by entrants and attendants to maintain contact
- Equipment to be provided, including PPE, test equipment, communications gear, alarm systems, and rescue equipment
- Other information needed in order to ensure employee safety
- Any additional permits that have been issued to authorize work in the space

3.2.0 Roles and Duties of Personnel in Confined Spaces

Working in permit-required spaces is a team effort involving entrants, attendants, and supervisors. All of these personnel must be trained to do the tasks they are expected to perform before any work in a confined space begins. Entrants, attendants, gas testers, and entry supervisors should have applicable skills training.

3.2.1 Entrant

Authorized entrants are those permitted by an employer to enter a permit-required space. Each entrant must be aware of the permit-required space hazards, including the symptoms and consequences of exposure. They must be trained on the proper use of all required test and work equipment.

Entrants must remain in regular communication with the authorized attendant. They must immediately notify the attendant if hazardous conditions or their warning signs arise. Entrants will leave the confined space immediately if hazardous conditions arise or if ordered to evacuate by the attendant.

3.2.2 Attendant

Authorized attendants are those who monitor the entrants' activities from outside the space. They must know all of the permit-required space hazards, including symptoms and consequences of exposure. The attendant is responsible for monitoring the entrants. This includes knowing at all times how many entrants are in the confined space and making sure that everyone involved is aware of the permit requirements. The attendant will stay out of the space during entry operations.

The attendant's duties are as follows:

- Keep unauthorized personnel away from the space.
- Maintain communication with the entrants.
- Keep track of the status of each entrant.
- Monitor activities inside and outside the confined space at all times.
- Determine if it is safe for entrants to remain in the space.

The attendant will order an evacuation if a hazardous condition arises. This includes detection of behavioral symptoms of hazard exposure or a problem outside the space that could harm the entrants.

If necessary, the attendant will activate approved rescue procedures. A non-entry rescue will be used if possible. Summoning rescue and other emergency services may be required if the attendant determines that the entrants may need assistance to evacuate. The attendant must not attempt an internal rescue, except as part of a trained rescue team. In that case, another person must assume the role of attendant.

3.2.3 Supervisor

The entry supervisor makes sure that the attendants and entrants follow entry permit procedures. The supervisor needs to know all of the permit-space hazards, including the symptoms and consequences of exposure. Supervisors are responsible for verifying the following:

- The entry permit is accurate and current.
- All permit tests have been conducted.
- All permit procedures and equipment are in place.
- Personnel comply with the provisions of the permit.
- Rescue personnel will be available in an emergency.
- Entry operations are consistent if another authorized person replaces an attendant or an entrant.

The supervisor must monitor the job and stop all entry operations and cancel the entry permit during a hazardous condition or when permit-required confined space work has been completed.

3.3.0 Emergency Training

In addition to hazard analysis training, all entrants, attendants, and supervisors must be trained in emergency escape procedures. Operating procedures for handling emergency situations must be reinforced with training and practice. Written procedures should not be accepted without testing them in simulated emergency scenarios. Effective procedures and practices can only be established through real-world testing. Training must include scenarios dealing with loss of air line, communications, and/or light.

3.3.1 Loss of Air Line

A supplied air respirator includes an air line connected to an outside air supply. Emergency training must include the possibility of an air line failure. SCBA can also fail without warning. If this occurs, the entrant must be prepared to evacuate the confined space as quickly as possible.

A limited air supply must be used efficiently. A personal egress bottle will provide a short supply of air for the entrant. Teach and practice air conservation, such as controlled breathing techniques, for situations in which the egress bottle is either too limited or not available.

It is usually wise to maintain an air line connection during escape procedures, even if the air line has failed. It is possible that the situation that caused the air line to fail has been corrected, providing a supply of air to the escaping entrant. If rescue efforts are needed, the air line connection may provide the best means of finding the entrant, especially in a complex space.

Entrants using SCBA and working as a team must be equipped with buddy breather hose attachments. If one entrant's SCBA fails, a fellow entrant may be able to supply them with breathable air until they both escape. Again, training and practice are important, and can mean the difference between a potentially fatal panic and an orderly, safe evacuation.

3.3.2 Loss of Communications

Communications equipment failure must be considered in any emergency training plan. Whatever the primary form of communication, alternatives and backup procedures must be established and practiced. A secondary communications device is one option. Another option is a personal distress device or some other signaling device. Emergency procedures could also be initiated by using a whistle, small air horn, tapping a tool, or a similar non-electronic means. Whatever tools are used must be used in training and tested regularly in practice. Each participant must understand the signals used.

Emergency procedure training must also establish the point at which the attendant should call in rescue team personnel if communication with the entrants is lost.

3.3.3 Inadequate Illumination

An emergency situation may involve a loss of power and loss of primary lighting or inadequate illumination. Entrants may need training to evacuate the space with little or no light. This includes accounting for team members, managing air supplies, maintaining communication, and operating rope systems.

Consider using secondary, intrinsically safe light sources. Each entrant can use a combination of flashlights, helmet lights, and chemical light sticks. In an emergency, the attendant should set up approved lighting at the entrance. This will mark the escape portal for entrants still in the space.

> **Think About It**
> ## Personal Distress Units
> The entrants on your team are planning to use wireless personal distress units, which provide an instant signal to the attendant if anything goes wrong. Will these devices provide an appropriate level of safety in a space lined with steel? What about reinforced concrete? Will line of sight affect either of these scenarios?

3.3.4 Rescue Procedures

Rescue procedures are necessary if one or more entrants are incapacitated or unable to exit without assistance. Safe and effective rescue operations require adequate planning and training. Rescue procedures must be specific to the space and set up in advance. Rescue personnel may be team members with specialized training, or a separate group dedicated to rescue operations.

Sixty percent of all confined space fatalities are would-be rescuers. Case histories of confined space incidents often mention one or more rescuers who died with the people they were trying to save. While time is certainly an important element in rescuing a co-worker, never attempt any rescue without proper training and equipment.

Non-Entry Rescue – Non-entry rescue (*Figure 25*) is the preferred method for extracting an entrant from a confined space. Personnel should not enter a confined space to respond to an emergency unless they have been properly trained and equipped.

If a confined space rescue is necessary, the attendant will summon the emergency rescue team. The attendant can also attempt to rescue entrants using only non-entry rescue equipment, unless it would increase the entrant's risk of injury. During an emergency, the attendant will monitor the situation until help arrives or the situation becomes unsafe for the attendant as well. The attendant must give the rescue team the following information:

- Number of victims
- Time exposed
- When the air was last monitored
- Victims' condition
- Hazards in the space

Retrieval systems should be used every time an entrant enters a confined space, unless it would increase the overall risk of entry or would not aid a rescue. Each entrant must wear a chest or full-body harness with an attached retrieval line. The line should be attached at the center of the entrant's back near shoulder level or at some other point that presents a profile small enough for the successful removal of the entrant. Wristlets are another option if the use of a harness is not feasible or might create a hazard to the entrant.

Attach the other end of the retrieval line to a mechanical device or fixed point outside the space so that rescue can begin as soon as possible. A mechanical device must be available to retrieve personnel from vertical spaces more than 5-feet (1.5 m) deep.

A typical rescue plan uses the following steps:

Step 1 As soon as a problem is discovered, the attendant gives the order to evacuate.

Step 2 If entrants cannot vacate the space under their own power, the attendant will contact the rescue team. The attendant must not enter the space.

> **WARNING!**
> An attendant who enters the space could also become incapacitated, which may result in the death of everyone in the space. The attendant's job is to notify the rescue team.

Step 3 If an entrant is incapacitated and is wearing a safety harness, the attendant and standby entrants must use the lifting device to remove the entrant from the space.

Figure 25 Non-entry rescue.

Entry Rescue – Only designated emergency responders can enter a confined space during an emergency. Each emergency responder must have full confined space entrant training in addition to emergency training. An emergency response team responding to a confined space emergency is shown in *Figure 26*.

Figure 26 Emergency response team responding to a confined space emergency.

The rescue team may be designated trained employees or an outside emergency response team. The supervisor must ensure that the team has the ability to perform rescue services and are able to reach a victim in time, depending on the hazards present.

Rescue team members must practice confined space rescue operations at least once annually. This practice should include the following:

- Hazard analysis
- Escape procedures
- Rescue team preparedness
- Use of all retrieval equipment and personal protective equipment

The practice of escape procedures must involve real challenges, including rapidly changing environments and problems with protective gear, especially failures in breathing apparatus. Address all potential rescue scenarios, including loss of air line, loss of communications, and loss of lighting. Practice should take place in simulated spaces similar to actual permit-required spaces or in the spaces themselves.

The safety of the entrants is the primary objective of any rescue operation in a hazardous environment. Before entering the confined space, rescuers must be familiar with the physical layout of the space and be aware of any hazards. All rescue personnel must be equipped with appropriate personal protective equipment and know how to use it. They must also be trained in basic first aid and cardiopulmonary resuscitation (CPR).

Additional Resources

American Society of Safety Engineers, www.asse.org

Occupational Safety and Health Organization, www.osha.gov

Guidelines for Hot Work in Confined Spaces: Recommended Practices for Industrial Hygienists and Safety Professionals, Martin H. Finkel, CIH, CMC. 2000. Des Plaines, IL: The American Society of Safety Engineers (ASSE).

Complete Confined Spaces Handbook, John F. Rekus, MS, CIH, CSP. 1994. Boca Raton, FL: CRC/Lewis Publishers.

3.0.0 Section Review

1. Required information on a confined space entry permit includes test results, both initial and _____.
   a. previously conducted
   b. future estimated
   c. periodically recurring
   d. anticipated periods

2. Entrants must remain in regular communication with the authorized _____.
   a. supervisor
   b. attendant
   c. employer
   d. assistant

3. Rescue team members must practice confined space rescue operations at least once _____.
   a. annually
   b. bi-annually
   c. every three years
   d. every five years

SUMMARY

Permits and policies used with a well-defined permit system help to support administrative requirements for preparing and issuing necessary permits. Permits and policies may need to be developed or modified to address conditional changes or for regulatory compliance. Safety technicians are responsible for auditing permit compliance to ensure that workers remain safe even when conditions change. Lockout/tagout procedures and work permit systems are the cornerstone of an effective safety program. Permits and policies that are properly managed and maintained support compliance and promote a safe working environment.

Working in confined spaces poses significant safety hazards for workers. Management of activities in confined spaces is one of the safety technician's most important duties. All personnel working in and around a confined space environment must take an active role in their own safety and the safety of their co-workers. Every member of the team needs to be aware of potential atmospheric hazards and the symptoms of exposure. Personnel need to know how the work processes may contribute to those hazards. Training is required before using test equipment to identify hazards.

Confined spaces in which atmospheric, engulfment, and entrapment hazards may be present are referred to as permit-required confined spaces. As the safety technician, you may be the primary person who manages and administers confined space work permits. The air in a confined space must be tested before entry and adequate ventilation must be established. Training must include emergency preparedness for evacuation and rescue procedures. Non-entry rescue should be used if possible. An entry rescue must only be done by personnel with the proper training and equipment. Training and practice will help to ensure that regulations are met and work is completed safely.

Review Questions

1. Work permits provide a method of communication between the host employer/controlling contractor and personnel that are performing the _____.
   a. audit
   b. inspection
   c. review
   d. work

2. Lockout/tagout policies and procedures are designed to _____.
   a. control the loss of property and equipment due to employee theft
   b. prevent unauthorized workers from using equipment
   c. keep unauthorized people away from certain parts of a work site
   d. prevent the unplanned startup of equipment during servicing

3. Energy control procedures are written documents that contain information that authorized workers need to know in order to safely control energy while servicing or maintaining _____.
   a. confined spaces
   b. equipment or machinery
   c. fire extinguishers
   d. facility siting

4. Although workers are exposed to many hazards during excavation and trenching, the primary hazard is _____.
   a. slippery conditions
   b. cave-ins
   c. dust particles
   d. static electricity

5. OSHA requires that all employers with hazardous chemicals at their sites prepare and use a written _____.
   a. list of authorized employees
   b. safety contract
   c. hazard communication program
   d. usage agreement

6. A company's Process Hazard Analyses are required to address _____.
   a. facility siting
   b. confined spaces
   c. safety audit shortfalls
   d. energized electrical work

7. Whenever possible, continuous monitoring of confined spaces may be required to ensure the atmosphere is safe for the entire _____.
   a. work shift
   b. work week
   c. duration of the entry
   d. tagout process

8. The most common type of hazard in a confined space is _____.
   a. atmospheric
   b. electric
   c. temperature
   d. noise

9. Do not allow anyone to enter a confined space before completing atmospheric _____.
   a. purging
   b. testing
   c. inerting
   d. release

10. An ideal oxygen concentration is _____.
    a. 2 percent oxygen
    b. 9 percent oxygen
    c. 15 percent oxygen
    d. 21 percent oxygen

11. Transferring coal, grain products, nitrated fertilizers, or finely ground chemical products can lead to a concentration of _____.
    a. combustible dust
    b. methane
    c. flammable liquids
    d. ignitable fibers

12. Pneumatic tools and equipment used in confined spaces must be driven by _____.
    a. oxygen
    b. argon
    c. neon
    d. air

13. Air is approximately 21 percent oxygen and 78 percent _____.
    a. argon
    b. hydrogen
    c. carbon dioxide
    d. nitrogen

14. Who can enter a confined space to rescue an entrant?
    a. The controlling employer
    b. Designated emergency responders
    c. The person who signed the entry permit
    d. Medical personnel

15. Sixty-percent of all confined space fatalities involve _____.
    a. would-be rescuers
    b. non-participants
    c. subcontractors
    d. untrained entrants

Trade Terms Introduced in This Module

Affected employees: Used in reference to LOTO policies and procedures. Affected employees are usually the equipment users or operators who must be able to recognize when a control procedure is in place. They also need to understand that they are not to remove the LOTO device or try to start the equipment during a control procedure.

Asphyxiation: Death due to lack of oxygen.

Atmosphere: The air or climate inside a specific place.

Authorized employees: Used in reference to LOTO policies and procedures. Authorized employees perform the actual servicing or maintenance on the equipment. These workers need a high degree of training. They also need to know about the type and magnitude of the hazardous energy involved and how to safety isolate and control the energy source.

Atmospheric contaminants: Impurities in the air. Any natural or artificial matter capable of being airborne (other than water vapor or natural air) which in sufficient concentration can harm humans, animals, vegetation, or materials. Contaminants can exist in the form of liquids, solids, or gases.

Competent person: According to OSHA, a competent person is one who is capable of identifying existing and predictable hazards in the surroundings, or working conditions which are unsanitary, hazardous, or dangerous to employees, and who has authorization to take prompt corrective measures to eliminate them.

Confined space: A workplace that has a configuration that hinders the activities of employees who must enter, work in, and exit the space.

Confined space entry permit system: The written procedure for preparing and issuing permits for entry, and for returning the permit-required space to service after entry work is completed.

Energy control procedures: Written documents containing information that authorized workers need to know in order to safely control energy while servicing or maintaining equipment or machinery; also called lockout/tagout procedures.

Energy-isolating devices: Devices that physically prevent the transmission or release of energy.

Engulfment: To be in an environment in which the victim is covered with a material such as sand, gravel, or grain. The end result is often death due asphyxiation, strangulation, or crushing.

Excavation: Any man-made cut, cavity, trench, or depression in the earth's surface formed by earth removal.

Fire watch: Assigning trained personnel to watch for fires during welding and cutting operations, sound the alarm in the event of a fire, and extinguish any fires within the capacity of the available equipment. A fire watch must be maintained for at least one-half hour after completion of welding or cutting operations.

Hot work: Any work function that involves ignition or combustion; examples include welding, burning, cutting, and riveting.

Induced hazards: Hazards brought forward or created when people make incorrect decisions and/or take incorrect actions during the construction process (e.g., omission of protective features; physical arrangements that cause unintentional worker contact with electrical energy; oxygen deficient hazards created at the bottom of pits, shafts, or vaults; lack of safety factors in structural strength; and flammable atmospheres).

Inerting: The use of an inert gas such as nitrogen to supplant a flammable gas in a confined space.

Inherent hazards: Hazards associated with specific types of equipment and the interactions of the equipment with the confined space. Includes hazards that cannot be eliminated without degrading the system or the equipment, or making them inoperative (e.g., high voltage, radiation generated by equipment, defective designs, omission of protective features, high or low temperatures, high noise, and high-pressure vessels and lines).

Intrinsically safe: An electric tool or device that is UL-rated to be explosion-proof under normal use. For example, a fuel pump must be intrinsically safe to prevent explosions sparked by static electricity.

Lockout: The placement of lockout devices on an energy-isolating device in accordance with established procedures. The lockout device is intended to use a positive means, such as a lock, to secure the energy-isolating device in the Off or Safe position.

Lower explosive limit (LEL) or lower flammable limit (LFL): The lowest concentration of air-fuel mixture at which a gas can ignite.

Non-permit required confined space: A workspace free of atmospheric, physical, electrical, and mechanical hazards that could cause death or injury.

Other employees: All employees who are or may be in an area where energy control procedures may be utilized.

Oxygen-deficient atmosphere: A body of air that does not have enough oxygen to sustain normal breathing. This is usually considered to be less than 19.5 percent by volume.

Oxygen-enriched atmosphere: A body of air that contains enough oxygen to be flammable or explosive. This is usually considered to be more than 23.5 percent by volume.

Parts per million (ppm): A measure of the concentration of a substance in a volume of air, water, or other medium. In a volume of air, it is the number of milliliters of a substance in a cubic meter of air: 10 ppm = 10 ml/m3. One part per million is roughly the equivalent of two soda cans full of gas in a house full of air.

Permissible exposure limit (PEL): The limit set by OSHA as the maximum concentration of a substance that a worker can be exposed to in an 8-hour work shift. Most flammable gases have a PEL defined as 10 percent of the LEL/LFL, while toxic gases have individual PELs.

Permit-required confined space (PRCS): A confined space that has real or potential hazards including atmospheric, physical, electrical, or mechanical hazards as defined by OSHA.

Safety data sheet (SDS): A form that lists the hazards, safe handling practices, and emergency control measures for a specific chemical.

Self-contained breathing apparatus (SCBA): A device allowing an individual to breath while in a toxic or oxygen-deficient atmosphere.

Specific gravity: The ratio of the density of a substance to the density of a reference substance (typically water).

Tagout: The placement of tags or tagout devices on an energy-isolating device in accordance with established procedures. It is intended to indicate that the energy-isolating device and the equipment being controlled must not be operated until the tagout device is removed.

Trench: A narrow excavation made below the surface of the ground. The depth of a trench must be greater than the width and the width must not exceed 15 feet (4.6 m).

Upper explosive limit (UEL) or upper flammable limit (UFL): The highest concentration of air-fuel mixture at which a gas can ignite.

Vapor density: The relative weight of gases and vapors as compared with some specific standard, which is usually hydrogen but may also be air.

Additional Resources

This module presents thorough resources for task training. The following resource material is suggested for further study.

American Society of Safety Engineers, **www.asse.org**

Occupational Safety and Health Organization, **www.osha.gov**

Guidelines for Hot Work in Confined Spaces: Recommended Practices for Industrial Hygienists and Safety Professionals, Martin H. Finkel, CIH, CMC. 2000. Des Plaines, IL: The American Society of Safety Engineers (ASSE).

Complete Confined Spaces Handbook, John F. Rekus, MS, CIH, CSP. 1994. Boca Raton, FL: CRC/Lewis Publishers.

Figure Credits

U.S. Coast Guard photo by Petty Officer 3rd Class Robert Brazzell, Module opener

Steven P. Pereira, CSP, Figure 2

Topaz Publications, Inc., Figures 3, 5

Honeywell Safety Products, Figure 7

Kundel Industries, Figure 11

U.S. Department of Labor, Figure 12

Weld-On Adhesives, Inc., a division of IPS Corporation, Figure 13

MSA The Safety Company, Figures 16A, 16B, 16D, 18, 19, 25

North Safety Products USA, Figure 16C

Draeger Safety, Inc., Pittsburgh, PA, Figure 20

Photo by Spc. Robert I. Havens, 105th Mobile Public Affairs Detachment, Figure 26

Section Review Answer Key

Answer	Section Reference	Objective
Section One		
1. b	1.1.0	1a
2. d	1.2.0	1b
3. a	1.3.0	1c
4. c	1.4.0	1d
5. a	1.5.3	1e
Section Two		
1. a	2.1.1	2a
2. a	2.2.2	2b
Section Three		
1. c	3.1.0	3a
2. b	3.2.1	3b
3. a	3.3.4	3c

NCCER CURRICULA — USER UPDATE

NCCER makes every effort to keep its textbooks up-to-date and free of technical errors. We appreciate your help in this process. If you find an error, a typographical mistake, or an inaccuracy in NCCER's curricula, please fill out this form (or a photocopy), or complete the online form at **www.nccer.org/olf**. Be sure to include the exact module ID number, page number, a detailed description, and your recommended correction. Your input will be brought to the attention of the Authoring Team. Thank you for your assistance.

Instructors – If you have an idea for improving this textbook, or have found that additional materials were necessary to teach this module effectively, please let us know so that we may present your suggestions to the Authoring Team.

NCCER Product Development and Revision
13614 Progress Blvd., Alachua, FL 32615

Email: curriculum@nccer.org
Online: www.nccer.org/olf

❏ Trainee Guide ❏ Lesson Plans ❏ Exam ❏ PowerPoints Other _____

Craft / Level: _____ Copyright Date: _____

Module ID Number / Title: _____

Section Number(s): _____

Description: _____

_____

_____

_____

_____

Recommended Correction: _____

_____

_____

_____

_____

Your Name: _____

Address: _____

_____

Email: _____ Phone: _____

Incident Investigations, Policies, and Analysis

Overview

This module describes how to conduct an incident investigation, including employee interviews and reporting requirements. It also explains how to analyze an incident to determine the root cause and prevent future incidents.

Module 75225

Trainees with successful module completions may be eligible for credentialing through NCCER's Registry. To learn more, go to **www.nccer.org** or contact us at 1.888.622.3720. Our website has information on the latest product releases and training, as well as online versions of our *Cornerstone* magazine and Pearson's product catalog.

Your feedback is welcome. You may email your comments to **curriculum@nccer.org**, send general comments and inquiries to **info@nccer.org**, or fill in the User Update form at the back of this module.

This information is general in nature and intended for training purposes only. Actual performance of activities described in this manual requires compliance with all applicable operating, service, maintenance, and safety procedures under the direction of qualified personnel. References in this manual to patented or proprietary devices do not constitute a recommendation of their use.

Copyright © 2018 by NCCER, Alachua, FL 32615, and published by Pearson, New York, NY 10013. All rights reserved. Printed in the United States of America. This publication is protected by Copyright, and permission should be obtained from NCCER prior to any prohibited reproduction, storage in a retrieval system, or transmission in any form or by any means, electronic, mechanical, photocopying, recording, or likewise. To obtain permission(s) to use material from this work, please submit a written request to NCCER Product Development, 13614 Progress Blvd., Alachua, FL 32615.

75223 V5

From *Safety Technology*, Trainee Guide. NCCER.
Copyright © 2018 by NCCER. Published by Pearson. All rights reserved.

75225
INCIDENT INVESTIGATIONS, POLICIES, AND ANALYSIS

Objectives

When you have completed this module, you will be able to do the following:

1. Explain how to conduct an incident investigation.
    a. Explain the difference between an accident and an incident.
    b. Describe how to conduct an incident investigation and interview.
    c. Describe how to complete an incident investigation form.
2. Explain how to analyze data to determine the cause of an incident.
    a. Describe the Three Levels of Incident Causation Model.
    b. Describe the Why Method.
    c. Identify and describe OSHA problem-solving techniques.
    d. Explain the importance of trend analysis.

Performance Tasks

Under the supervision of your instructor, you should be able to do the following:

1. Conduct an incident investigation interview.
2. Complete an incident investigation form.
3. Use one of the following methods to analyze data to determine the cause of an incident:
    - Three Levels of Incident Causation Model
    - Why Method
    - OSHA problem-solving techniques

Trade Terms

Accident
Interviewee
Incident
Pareto diagram

Reverse chronological order
Root cause analysis
Three Levels of Incident Causation Model
Why Method

Industry Recognized Credentials

If you are training through an NCCER-accredited sponsor, you may be eligible for credentials from NCCER's Registry. The ID number for this module is 75225. Note that this module may have been used in other NCCER curricula and may apply to other level completions. Contact NCCER's Registry at 888.622.3720 or go to www.nccer.org for more information.

Contents

1.0.0 Incident Investigations 1
   1.1.0 Accidents vs. Incidents 1
   1.2.0 Conducting an Incident Investigation 1
      1.2.1 Roles and Responsibilities 2
      1.2.2 Equipment Used in Incident Investigation 2
      1.2.3 Investigation Time Line 3
      1.2.4 Conducting Interviews 4
   1.3.0 Incident Investigation Reports 5
      1.3.1 Other Uses of Incident Investigation Reports 6
      1.3.2 Filling Out an Incident Investigation Report 6
2.0.0 Analyzing Data 13
   2.1.0 Three Levels of Incident Causation Model 13
      2.1.1 Level I (Direct Causes) 13
      2.1.2 Level II (Indirect Causes) 14
      2.1.3 Level III (Root Causes) 15
   2.2.0 The Why Method 16
      2.2.1 Sequence of Events: Why Method 16
   2.3.0 OSHA's Problem-Solving Techniques 17
      2.3.1 Change Analysis 17
      2.3.2 Job Safety Analysis 17
      2.3.3 Organizing and Reporting the Data 20
   2.4.0 Trend Analysis 21

Figures

Figure 1	Incident investigation equipment.	3
Figure 2	Incident scene sketch.	6
Figure 3	Sample incident investigation form.	7
Figure 4	OSHA Form 300.	9
Figure 5	OSHA Form 300A.	10
Figure 6	OSHA Form 301.	11
Figure 7	Three levels of causation.	14
Figure 8	Root causes.	15
Figure 9	Example cause-and-effect diagram for burns, strains, and cuts.	16
Figure 10A	JSA Form (Part 1 of 2).	18
Figure 10B	JSA Form (Part 2 of 2).	19
Figure 11	Example Pareto diagram.	21

Section One

1.0.0 Incident Investigations

Objective

Explain how to conduct an incident investigation.
a. Explain the difference between an accident and an incident.
b. Describe how to conduct an incident investigation and interview.
c. Describe how to complete an incident investigation form.

Performance Tasks

1. Conduct an incident investigation interview.
2. Complete an incident investigation form.

Trade Terms

Accident: An event that results in property damage and/or personal injury or death that could not have been prevented.

Incident: An unplanned, undesired event that may cause personal injury, illness, or property damage in various combinations or degrees from minor to catastrophic.

Interviewee: Person being interviewed.

Reverse chronological order: Events told in order from last to first.

Root cause analysis: A method of problem solving that seeks to identify the most basic cause(s) of an incident.

Incident investigations help to determine direct and indirect causes of incidents, prevent similar incidents from happening again, document facts, satisfy government regulations, detect trends, provide information on costs, and promote safety. The on-site supervisor is generally responsible for conducting incident investigations. The safety technician's role in incident investigations is to act as a resource for the supervisor. Once the investigation is complete, the safety technician is responsible for reviewing the report for completeness and accuracy. In a serious or potentially serious incident, the safety technician may assist in the investigation. Safety technicians are also responsible for following up with the site supervisor to ensure that the needed safety policies and procedures have been put in place and the control actions have been implemented. It is important, however, for safety technicians to know the entire incident investigation process, not just their role. Gaining a better understanding of the process will help make investigations more efficient and effective.

1.1.0 Accidents vs. Incidents

The major difference between an **accident** and an incident is distinguishing if the associated event was preventable. The term accident supports the mindset that an event was not preventable. Today, most safety professionals prefer to use the term incident because the fundamental concept of a good safety program is to find and remove hazards with the goal of preventing incidents. OSHA suggests using the term incident investigation to determine the root cause(s) of a reported event.

Near misses or close calls are a subset of incidents where the event did not result in injury, illness or property damage, but could have given different circumstances. Near miss/close call incidents can act as free warnings if they are reported and should be investigated accordingly. Consider a near miss/close call as a "near hit" and ensure appropriate corrective actions are taken so that a similar repeat event does not become an actual incident.

1.2.0 Conducting an Incident Investigation

The purpose of incident investigations is to determine the cause of the incident and recommend the corrective actions needed to prevent incidents from happening again. Incident investigations should be fact-finding rather than fault-finding activities. The emphasis of an investigation should never be on identifying who should be blamed for the incident, but on why the incident happened. Trying to focus the blame on one person can damage an investigator's credibility. It can also reduce the amount and accuracy of information received from workers. This does not mean that personal responsibility should not be determined. It means that the investigation should be concerned with only the facts.

Effective incident investigations are not emotional or judgmental. They are objective examinations that determine the reasons an incident happened. Incident investigations that focus on identifying and correcting root causes, not on finding fault or blame, also improve workplace morale and increase productivity, by demonstrating management's commitment to a safe work culture. This is known as root cause analysis.

When an incident results in personal injury or property damage, the incident must be investigated to find the cause or causes. The results of the incident investigation can pinpoint the corrective actions that will help prevent future incidents. In circumstances where an incident is deemed unpreventable a fact-based investigation will support the findings.

OSHA strongly encourages employers to investigate all incidents in which a worker was hurt, as well as near misses/close calls, in which a worker might have been hurt if the circumstances had been slightly different. Near misses need to be investigated because they also affect the work process. The potential consequences of some incidents are more serious than others. The amount of time and effort that goes into an investigation will depend on the seriousness of the incident or near miss. Always follow company policies and procedures when conducting an incident investigation. Note that owner-controlled insurance programs (OCIP) and contractor-controlled insurance programs (CCIP) requirements may supplement company policy.

1.2.1 Roles and Responsibilities

Investigating incidents requires cooperation from all levels of management and every worker on site. The foreman/supervisor's unique position gives him or her special priority and responsibility during an incident investigation. This person has certain qualifications and advantages other workers do not have. The following lists the qualities and advantages that make the supervisor the best person to investigate incidents.

- Supervisors know the most about the situation. They have daily contact and familiarity with the people, machines, and materials involved. They know the standard practices and circumstances in the area, as well as the hazards.
- Supervisors have a personal interest in identifying incident causes because they are often protective of their workers, machines, and materials. They can also give insight into how to correct the problem.
- Supervisors can take the most immediate action to prevent an incident from recurring because they are so familiar with the situation. Being in direct control of the people, procedures, and property in the area gives the supervisor the advantage of taking immediate corrective action and the greatest opportunity for effective follow-up.

> **Did You Know?**
> **Digital Evidence**
> Cameras are restricted in many work environments for a number of reasons—for example, to protect proprietary secrets, classified projects, or in high-profile projects. In some cases, companies have been known to confiscate and destroy personal cell phones that were brought into protected work environments or where employees were caught photographing incidents.

- Supervisors can communicate more effectively with workers. A worker may be employed by the company, but he or she works for and with the foreman/supervisor. Workers know the supervisor is interested in them and their safety. To a worker, the supervisor is a level of management that he or she should feel comfortable approaching. During an incident investigation, workers should feel free to be honest with their supervisor. This helps to ensure accurate information is obtained.

The disadvantage of a supervisor investigating his or her own incident is that it may be difficult to be unbiased. For example, the supervisor may have to admit some personal failure, such as not making the right work assignment. This is where the role of the safety technician becomes important. The safety technician will be able to provide an unbiased analysis of the incident; offer objective, corrective feedback; and provide quality control.

1.2.2 Equipment Used in Incident Investigation

Incident investigation requires specific equipment (*Figure 1*). The following is a list of equipment that is essential to an investigation.

- Camera, if permitted on site, to take pictures of the incident scene
- Tape measure to measure evidence and the distance between pieces of evidence
- Clipboard and writing pad to take notes
- Graph paper to draw diagrams of the incident scene
- Ruler to use as a scale reference in photos
- Pens and pencils to write notes and draw diagrams
- Incident investigation forms to properly document the incident

Figure 1 Incident investigation equipment.

- Flashlight and fresh batteries to see clearly if the incident area is dark

The following additional equipment can be helpful during an investigation.

- Incident investigator's checklists to make sure you are following your company's procedures
- Meters, detectors, and test equipment to gather any needed data about the incident
- Magnifying glass to see small details in evidence
- Sturdy gloves to protect your hands when handling materials
- High-visibility plastic tapes to mark off the incident area
- First aid kit to treat minor injuries
- Recording device to record information rather than writing it down
- Identification tags to identify and mark evidence
- Barricade markers/tape
- Masking tape to mark the incident area or seal containers that should not be opened
- Specimen containers to collect evidence
- GPS to document the location of the incident
- Chalk to identify the location of evidence
- Video camera to make a visual record of the incident scene
- Tarp to protect evidence from weather or contamination of the incident scene
- Blood borne pathogen kit
- Lockout/tagout set
- Personal protective equipment for the area, such as a hard hat, fall protection, safety glasses, etc.

1.2.3 Investigation Time Line

Gathering and preserving information during an investigation is critical to finding the cause or causes of an incident. The sooner the investigator is on the scene, the better. This way there is less chance of evidence tampering and workers are more likely to remember details. The following steps provide a sample time line for investigating incidents:

Step 1 The investigator should begin investigating the incident immediately after the injured person has been treated, before the scene can be changed, and before important evidence is removed, destroyed, or cleaned up. The investigator should look and listen for clues.

Step 2 The investigator should discuss the incident with the injured person if possible, after first aid or medical treatment has been administered. Care and tact are important when interviewing injured workers.

Step 3 The investigator should talk with witnesses who were at the scene of the incident, and those who arrived shortly thereafter. Talking to a witness should take place at the scene, if possible. If the situation is stressful for the witness, conduct the interview away from the incident scene. Interview witnesses separately, never as a group, because one person's story may influence others. Ask open-ended questions such as, "What did you see or hear?"

Step 4 The investigator should ask for all details that can give clues to the cause of the incident. Keep in mind that witnesses sometimes give more detailed information when they explain events in reverse chronological order. The investigator should encourage witnesses to share their ideas about what happened but only after the factual questions of who, what, where, when, why, and how have been answered. This may lead to more clues about the incident. If the incident involved equipment, the investigator must get details on the equipment such as the model number, recent maintenance, etc.

If the incident is fatal, these additional steps should be taken:

Step 1 Contact local authorities and cover the body, but do not move it.

Step 2 Take accurate measurements to define the physical relationship between the body and any equipment and materials involved.

Step 3 If allowed, photograph the incident scene and surroundings. Photos are important since the information will be lost once the scene has been cleaned up. Photos also allow you to revisit the scene. Do not take photos of the body unless there is a specific purpose for doing so, because they will generally be of limited value. Photos of the body can also cause needless heartache to the victim's family.

Step 4 Collect and identify any and all related material. Mark it in relation to the incident scene. This evidence may be used if the incident scene needs to be reconstructed.

Following these steps gives investigators the best chance at collecting accurate information and providing corrective feedback.

For complex comprehensive investigations, many companies use a team or a committee of workers to investigate incidents involving serious injury or property damage. This team or committee may replace the supervisor's investigation or may serve as a second-level investigation. Second-level investigations should be considered when:

- The incident has or could have had very serious results.
- The nature of the incident is very complex.
- The incident involved more than one supervisor's crew.
- The initial investigation did not clearly establish a full range of contributing factors or corrective and preventive actions.

> **NOTE:** When a team or committee investigates an incident, the team leader or chairperson must have enough authority and status in the organization to do whatever is needed to conduct a thorough investigation.

1.2.4 Conducting Interviews

Once you have documented the incident scene, it is important to start gathering data through the interview process. Conducting interviews is perhaps the most difficult part of an investigation.

The purpose of the incident investigation interview is to obtain an accurate and comprehensive picture of what happened. This is done by obtaining all pertinent facts, interpretations, and opinions. Your job is to construct a composite story using the various accounts of the incident and other evidence. In order to be effective, you must have a firm understanding of the techniques for interviewing.

Your first task when beginning the interviewing process is to determine who needs to be interviewed. Questions will need to be personalized for each person interviewed. Interviews should occur as soon as possible, but usually do not begin until things have settled down a bit. Some people you may want to consider for an interview include the following:

- *The injured worker* – Determine the specific events leading up to the incident.
- *Co-workers* – Find out if appropriate procedures were being used at the time of the incident.
- *Foreman or supervisor* – Get background information on the worker. He or she can also provide procedural information about the task that was being performed.
- *Jobsite superintendent* – This person can be the main source for information on related systems.
- *Maintenance workers* – Determine background on equipment and machinery maintenance.
- *Emergency responders* – Learn what they saw when they arrived and during the response.
- *Medical personnel* – Get medical information, as allowed by law.
- *Coroner* – This person can be a valuable source to determine the type and the extent of fatal injuries.
- *Police* – If a police report was filed, talk to the reporting officer.
- *The injured worker's spouse and family* – They may have insight into the worker's state of mind or other work issues.

Cooperation, not intimidation, is the key to a successful incident investigation interview. Interviewing injured workers and witnesses necessitates reducing their possible fear and anxiety, and developing a good rapport. It is counterproductive to give the impression that you are trying to establish blame. The purpose of the incident interview is to uncover additional information about the hazardous conditions, unsafe work practices, and related system weaknesses that contributed to the incident. Therefore, it is very important that the interviewer use effective techniques to establish a cooperative atmosphere.

It is also important to remember that you are conducting an incident investigation, not a criminal investigation, but your findings could be used in a legal setting. Your investigation must be accurate and professional. The following is a list of effective interviewing techniques that will help you find the facts, not assign fault.

- When interviewing, keep the purpose of the investigation in mind. You are attempting to determine the cause of the incident so that similar incidents will not recur. Make sure the interviewee understands this.
- Approach the investigation with an open mind. It will be obvious if you have preconceptions about the individuals or the facts.
- Go to the scene. Just because you are familiar with the location or the victim's job, don't assume that things are always the same. If you can't conduct a private interview at the location, find an office or meeting room that the interviewee considers a neutral location.
- Interview everyone involved including the injured workers, witnesses, and people involved with the process. Witnesses should be interviewed separately, never as a group.
- Put the person at ease. Explain the purpose of the investigation and your role. Sincerely express concern regarding the incident and a desire to prevent a similar occurrence.
- Express to the individual that the information given is important. Be friendly, understanding, and open minded. Be calm and unhurried.
- Let the individual talk. Ask background information such as name or job first. Ask the witness to tell you what happened but don't ask leading questions, interrupt answers, or make facial expressions.
- Ask open-ended questions to clarify particular areas. Try to avoid yes and no answer type questions. Don't ask "why," because these types of questions tend to make people respond defensively. For example: Do not ask: "Why did you drive the forklift with under-inflated tires?" Instead, ask: "What are the forklift inspection procedures? What are the forklift hazard reporting procedures?"
- Repeat the facts and sequence of events back to the person to avoid any misunderstandings.
- Take notes as carefully and accurately as possible. Let the individual read them if desired.
- Don't use a recording device unless you get permission because it may intimidate witnesses. If using a recording device, tell the interviewee that the purpose of the recording is to ensure accuracy.
- Ask for suggestions as to how the incident could have been avoided.
- Conclude the interview with a statement of appreciation for the interviewee's contribution. Ask him or her to contact you if he or she thinks of anything else. If possible, advise these people personally of the outcome of the investigation before it becomes public knowledge.

Understanding and applying this information during the interview process will help you establish a cooperative relationship so that you can obtain the facts. Intimidation and blaming will always result in an ineffective interview process.

1.3.0 Incident Investigation Reports

Incident investigation reports should document the full range of facts. The investigation should include thorough interviews of everyone with any knowledge of the event. Six key questions should be answered: who, what, when, where, how, and why. Facts should be presented carefully and clearly.

A good investigation is likely to reveal several contributing factors. It will also enable the investigator to recommend corrective actions that will help prevent future incidents. The report has to be thorough and accurate to justify the corrective actions recommended.

The incident investigator should avoid blaming the injured worker, even if the worker admits blame. Instead, the incident investigator should be objective and find all contributing causes. The error made by the worker is most probably not the root or basic cause. The worker who did not follow proper procedures may have been encouraged directly or indirectly by a supervisor or another worker to cut corners.

All supervisors and others who investigate incidents should be held accountable for describing causes carefully and clearly. When reviewing incident investigation reports, pay careful attention to phrases such as "the worker did not plan the job properly." While this statement may suggest an underlying problem with this worker, it does not identify all possible causes, preventions, and controls. It is the supervisor's responsibility to identify, anticipate, and report hazardous conditions to keep workers safe. The incident investigation report should list all the ways to correct the hazardous condition or unsafe activity. Some of these corrective actions can be accomplished quickly. Others may take time, planning, and money. Either way, the incident investigation report is the tool for identifying the cause of incidents and providing corrective actions that prevent more incidents. Each corrective action listed on the report should have an assigned person who is ultimately responsibility for the action, a completion date set and a place to mark completion of the item.

1.3.1 Other Uses of Incident Investigation Reports

The primary purpose of incident investigations is to prevent future occurrences. The information obtained through the investigation should also be used to update and revise the inventory of hazards and the company-established safety program. Reports should also be available to top management because they are ultimately responsible for safety on the site and must be aware of the results of investigations.

NOTE: In larger firms, the results should be shared with subsidiary/affiliated organizations.

1.3.2 Filling Out an Incident Investigation Report

Filling out an incident investigation report is an essential part of the investigation process. It helps to document incidents so that they can be properly analyzed and corrective actions can be made to the work process.

NOTE: Incident investigation reports can also be used to investigate work-related illnesses from single or multiple exposures to materials found on a job site. The illnesses can include conditions such as contact dermatitis or respiratory conditions caused by exposure to toxic gas.

When completing an incident investigation form, it is important to answer all questions on the form. Answers should be complete, specific, and factual. Do not include opinions or speculations in the report about why the incident happened. If the question does not apply to the incident, indicate this on the report with a written note that says N/A (not applicable). The report should also indicate if no answer was available for a question on the form. Any documentation supporting the investigation, such as witness statements, sketches of the scene (*Figure 2*), and photos, should be attached to the report.

The incident investigation report shown in *Figure 3* is an example of a report that is commonly used to report incidents. A complete form is shown in *Appendix A*. *Appendix B* lists common

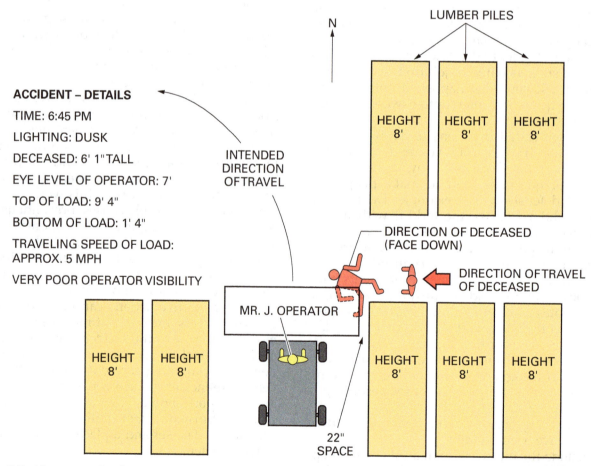

Figure 2 Incident scene sketch.

INJURY RELATED INCIDENT REPORT

USE THIS REPORT FOR ANY EVENT OTHER THAN AUTO THAT COULD HAVE RESULTED IN AN EMPLOYEE INJURY.

Employee Information

Employee Name		Employee Number	
Street Address		State & Zip Code	
Primary Language		Phone Number	
Position/Trade		Sex	☐ M ☐ F
Superintendent		PM Name	
Initial Indication	☐ Near Incident ☐ First Aid ☐ RX/Treatment ☐ Other: _____		

Jobsite Information

Job Name		Job Number	
Job Address		Job Phone	
Employee Began Work	☐ AM ☐ PM	Incident on Jobsite?	☐ Yes ☐ No
Incident Date		Time of Incident?	
Weather Conditions		Job in what County?	
Investigation Conducted by?		Investigator Cell #	
Private Owner/Describe Location			

Contacts

GC Name		Date / Time	/
Police		Date / Time	/
Fire Department		Date / Time	/
Ambulance		Date / Time	/
Helix Office		Date / Time	/
Family Member		Date / Time	/
OSHA / Cal-OSHA	☐	Date / Time	/
OSHA Reports or Citations? Describe:			

Initial Treatment & Forms

Injured Party Taken Where?		By Whom?	
Drug Tested?	☐ Yes ☐ No (if not, explain why)		
Worker Comp Form Completed?	☐ Yes ☐ No	Date	

DO NOT WRITE IN THIS AREA

Submitted to Carrier?	☐ Yes ☐ No	Claim Number	
Adjuster Info			
Notes			
The Event is:	☐ Recordable ☐ Lost Time ☐ Restricted Duty/Job Transfer ☐ 1st Aid		

Figure 3 Sample incident investigation form.

mistakes when completing OSHA forms. This report form meets the recordkeeping requirements specified by OSHA Form 301 (see *Figure 4*, *Figure 5*, and *Figure 6*). You will notice that the sample incident investigation report includes more information than the OSHA form. This is because OSHA does not provide a mandatory format for incident investigation reports but does require that specific information be documented and recorded. By including additional information on the report, companies can easily access information that would be helpful in reporting workers' compensation claims. The workers' compensation board determines what is needed on the forms. At the end of the year, all *OSHA Form 301* information is recorded on the year-end *OSHA Form 300*.

The following are additional requirements for reporting work-related incidents to OSHA:

- Within eight (8) hours after the death of any employee as a result of a work-related incident, you must report the fatality to the Occupational Safety and Health Administration (OSHA), US Department of Labor. Note that there may also be state reporting requirements.
- Within twenty-four (24) hours after the in-patient hospitalization of one or more employees or an employee's amputation or an employee's loss of an eye, as a result of a work-related incident, you must report the in-patient hospitalization, amputation, or loss of an eye to OSHA.
- You must report the fatality, inpatient hospitalization, amputation, or loss of an eye using one of the following methods:

    - By telephone or in person to the OSHA area office that is nearest to the site of the incident.
    - By telephone to the OSHA toll-free central telephone number, 1-800-321-OSHA (1-800-321-6742).
    - By electronic submission using the reporting application located on OSHA's public web site at **www.osha.gov**.

> **NOTE**
> MSHA has a 15-minute reporting requirement for all mining incidents. For more information, see **www.msha.gov**.

Figure 4 OSHA Form 300.

The figure shows OSHA Form 300, "...ted Injuries and Illnesses" (Log of Work-Related Injuries and Illnesses), /2004), containing:

Year ___
U.S. Department of Labor
Occupational Safety and Health Administration
Form approved OMB no. 1218-0176

Attention: This form contains information relating to employee health and must be used in a manner that protects the confidentiality of employees to the extent possible while the information is being used for occupational safety and health purposes.

...hat involves loss of consciousness, restricted work activity or job transfer, days away from work, or medical treatment beyond first ...s that are diagnosed by a physician or licensed health care professional. You must also record work-related injuries and illnesses ...through 1904.12. Feel free to use two lines for a single case if you need to. You must complete an injury and illness incident ...corded on this form. If you're not sure whether a case is recordable, call your local OSHA office for help.

Establishment name
City ___ State ___

Describe the case

(D) Date of injury or onset of illness (mo./day)

(E) Where the event occurred (e.g. Loading dock north end)

(F) Describe injury or illness, parts of body affected, and object/substance that directly injured or made person ill (e.g. Second degree burns on right forearm from acetylene torch)

Classify the case

CHECK ONLY ONE box for each case based on the most serious outcome for that case:

Remained at work

| (H) Days away from work | (I) Job transfer or restriction | (J) Other recordable cases |

Enter the number of days the injured or ill worker was:

| (K) Away From Work (days) | (L) On job transfer or restriction (days) |

Check the "injury" column or choose one type of illness:

(M) (1) Injury, (2) Skin Disorder, (3) Respiratory Condition, (4) Poisoning, (5) Hearing Loss, (6) All other illnesses

Page totals — 0 0 0 0 0 0 0 0 0 0

Be sure to transfer these totals to the Summary page (Form 300A) before you post it.

Page ___ 1 of 1

...average 14 minutes per response, including time to review ...review the collection of information. Persons are not required ...valid OMB control number. If you have any comments about ...partment of Labor, OSHA Office of Statistics, Room N-3644, ...ompleted forms to this office.

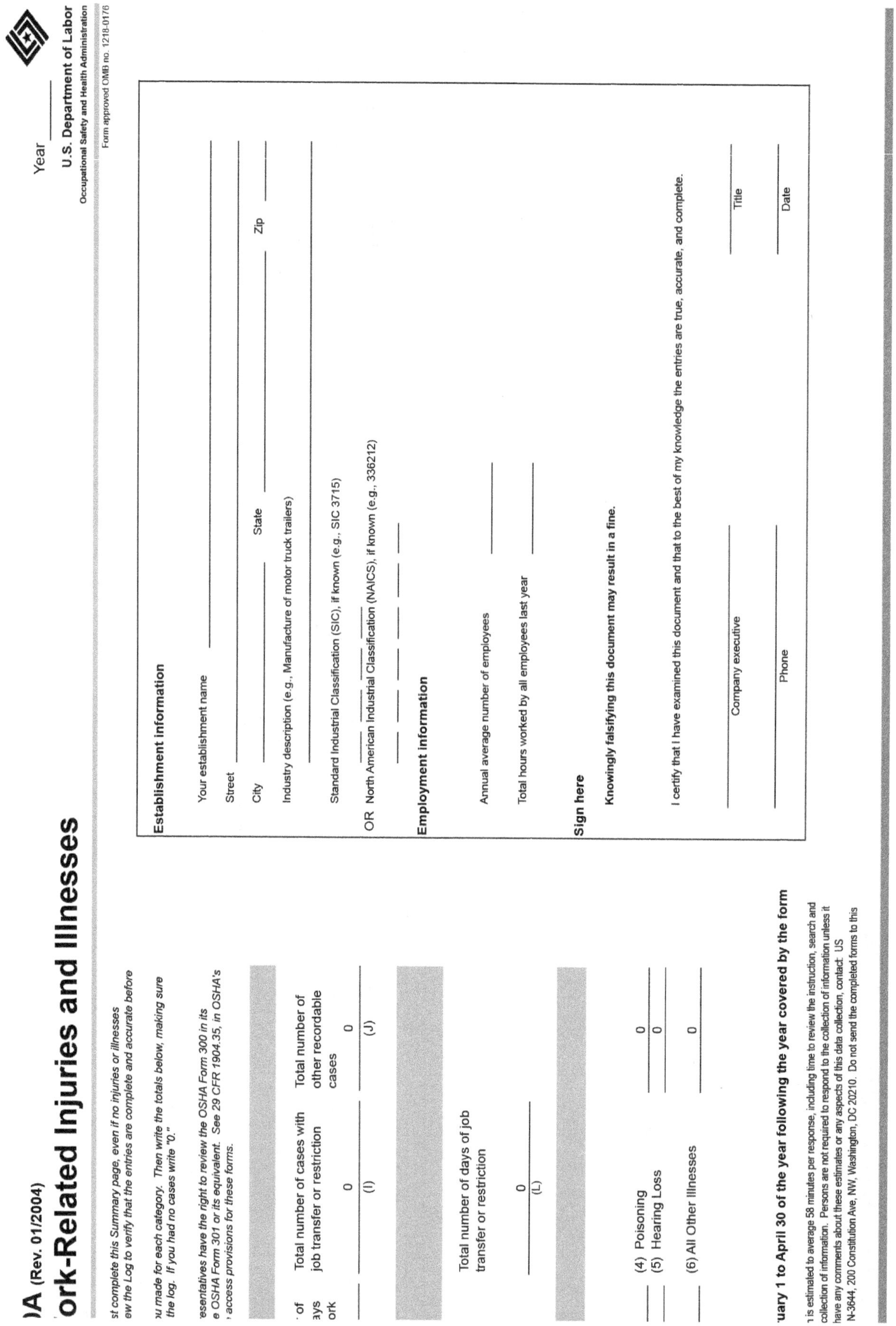
)A (Rev. 01/2004)

ork-Related Injuries and Illnesses

Year ______

U.S. Department of Labor
Occupational Safety and Health Administration

Form approved OMB no. 1218-0176

st complete this Summary page, even if no injuries or illnesses
ew the Log to verify that the entries are complete and accurate before

ɔu made for each category. Then write the totals below, making sure
the log. If you had no cases write "0."

esentatives have the right to review the OSHA Form 300 in its
e OSHA Form 301 or its equivalent. See 29 CFR 1904.35, in OSHA's
access provisions for these forms.

of ays ork

Total number of cases with job transfer or restriction 0 (I)

Total number of other recordable cases 0 (J)

Total number of days of job transfer or restriction 0 (L)

(4) Poisoning 0
(5) Hearing Loss 0
(6) All Other Illnesses 0

uary 1 to April 30 of the year following the year covered by the form

n is estimated to average 58 minutes per response, including time to review the instruction, search and collection of information. Persons are not required to respond to the collection of information unless it have any comments about these estimates or any aspects of this data collection, contact: US N-3644, 200 Constitution Ave, NW, Washington, DC 20210. Do not send the completed forms to this

Establishment information

Your establishment name ______

Street ______

City ______ State ______ Zip ______

Industry description (e.g., Manufacture of motor truck trailers) ______

Standard Industrial Classification (SIC), if known (e.g., SIC 3715) __ __ __ __

OR North American Industrial Classification (NAICS), if known (e.g., 336212) __ __ __ __ __ __

Employment information

Annual average number of employees ______

Total hours worked by all employees last year ______

Sign here

Knowingly falsifying this document may result in a fine.

I certify that I have examined this document and that to the best of my knowledge the entries are true, accurate, and complete.

Company executive ______ Title ______

Phone ______ Date ______

Figure 5 OSHA Form 300A.

U.S. Department of Labor
Occupational Safety and Health Administration

Form approved OMB no. 1218-0176

dent Report

Attention: This form contains information relating to employee health and must be used in a manner that protects the confidentiality of employees to the extent possible while the information is being used for occupational safety and health purposes.

Information about the employee

) Full Name _____

) Street _____

) City _____ State ____ Zip ____

) Date of birth _____

) Date hired _____

) ☐ Male
  ☐ Female

Information about the physician or other health care professional

) Name of physician or other health care professional _____

) If treatment was given away from the worksite, where was it given?

   Facility _____
   Street _____
   City _____ State ____ Zip ____

) Was employee treated in an emergency room?
   ☐ Yes
   ☐ No

) Was employee hospitalized overnight as an in-patient?
   ☐ Yes
   ☐ No

Information about the case

10) Case number from the Log _____ *(Transfer the case number from the Log after you record the case.)*

11) Date of injury or illness _____

12) Time employee began work _____ AM/PM

13) Time of event _____ AM/PM ☐ Check if time cannot be determined

14) What was the employee doing just before the incident occurred? Describe the activity, as well as the tools, equipment or material the employee was using. Be specific. Examples: "climbing a ladder while carrying roofing materials"; "spraying chlorine from hand sprayer"; "daily computer key-entry."

15) What happened? Tell us how the injury occurred. Examples: "When ladder slipped on wet floor, worker fell 20 feet"; "Worker was sprayed with chlorine when gasket broke during replacement"; "Worker developed soreness in wrist over time."

16) What was the injury or illness? Tell us the part of the body that was affected and how it was affected; be more specific than "hurt", "pain", or "sore." Examples: "strained back"; "chemical burn, hand"; "carpal tunnel syndrome."

17) What object or substance directly harmed the employee? Examples: "concrete floor"; "chlorine"; "radial arm saw." If this question does not apply to the incident, leave it blank.

18) If the employee died, when did death occur? Date of death _____

inutes per response, including time for reviewing instructions, searching existing data sources, gathering and maintaining the data needed, and completing and reviewing the collection of information. Persons are not required to respond to you have any comments about this estimate or any other aspects of this data collection, including suggestions for reducing this burden, contact: US Department of Labor, OSHA Office of Statistics, Room N-3644, 200 Constitution Ave, NW,

Figure 6 OSHA Form 301.

Additional Resources

The American Society of Safety Engineers (ASSE), **www.asse.org**

Occupational Safety and Health Administration (OSHA), **www.osha.gov**

Root Cause Analysis Handbook: A Guide to Effective Incident Investigation. 1999. Rockville, MD: Government Institutes.

The Psychology of Safety Handbook. E. Scott Geller, Ph.D. 2018. Boca Raton, FL: CRC Press.

1.0.0 Section Review

1. A fundamental concept of a good safety program is to focus on finding and fixing hazards, with an ultimate goal of _____.
   a. protecting property
   b. preventing incidents
   c. placing blame
   d. preventing cost

2. The disadvantage of a supervisor investigating his or her own incident is that it may be difficult to be _____.
   a. forthcoming
   b. unbiased
   c. proactive
   d. diligent

3. A good investigation is likely to reveal several _____.
   a. faulty parties
   b. missing elements
   c. contributing factors
   d. missing facts

Section Two

2.0.0 ANALYZING DATA

Objectives

Explain how to analyze data to determine the cause of an incident.

a. Describe the Three Levels of Incident Causation Model.
b. Describe the Why Method.
c. Identify and describe OSHA problem-solving techniques.
d. Explain the importance of trend analysis.

Performance Tasks

3. Use one of the following methods to analyze data to determine the cause of an incident:
   - Three Levels of Incident Causation Model
   - Why Method
   - OSHA problem-solving techniques

Trade Terms

Pareto diagram: A diagram used to perform a trend analysis by sorting data based on the frequency of occurrence.

Three Levels of Incident Causation Model: A method of incident investigation that seeks to identify three different factors: direct causes, indirect causes, and root causes.

Why Method: A method of incident investigation that is used to determine the root cause of an incident by asking the question "Why did this happen?" and investigating until the question "Why?" can no longer be asked.

It is important to analyze incident investigation data in order to discover why incidents happen. Once you know how an incident happened, you can start to correct the problem. The ultimate intent of the analysis is to scrutinize the data in an effort to prevent future incidents.

Data analysis is a special skill. As a safety technician, you will find that the ability to analyze data will be helpful in other areas of your work. This is because data analysis requires you to process information in a logical way. If you are able to do this, tasks such as conducting job safety analyses, planning safety meetings, and performing inspections and audits can be done quickly and effectively.

Because every incident is unique, there are different approaches and challenges with regards to analyzing incident investigation data. You may find that one approach is better or easier to use than another, or that a combination is best. The end objective of any analysis should be the same: finding out what caused the incident and what is necessary to fix the problem. The following are methods and models that can help you understand how and why incidents happen:

- Three Levels of Incident Causation Model
- Why Method
- OSHA's problem-solving techniques

2.1.0 Three Levels of Incident Causation Model

The causes of incidents can be classified by three different factors: direct causes, indirect causes, and root causes. These are also called the three levels of incident causation (*Figure 7*). Each level of causation represents one of these factors. It is important to be able to clearly recognize and understand the components of these three levels for analyzing incident data. If you understand what causes incidents, you will be better able to identify the causes when reviewing incident investigation forms, witness interviews, or physical evidence.

2.1.1 Level I (Direct Causes)

Level I represents direct causes of incidents. These are incidents resulting from the uncontrolled release of energy. These incidents may or may not cause injury or property damage, but they are still dangerous. When investigating an incident,

---

Did You Know?

Bureau of Labor Statistics

The Bureau of Labor Statistics (BLS) is the principal fact-finding agency for the federal government. The BLS operates an independent national statistical agency that collects, processes, analyzes, and disseminates essential statistical data to the American public, the US Congress, other federal agencies, state and local governments, business, and labor. They are experts in the field of data analysis. *Appendix C* of the BLS Data Integrity Guidelines, is an example of one procedure the BLS has in place to ensure the integrity of the data it collects, analyzes, and disperses.

Source: US Bureau of Labor Statistics

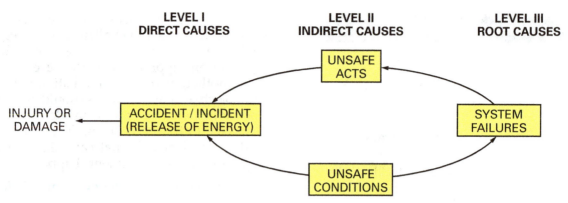

Figure 7 Three levels of causation.

be sure to look closely at energy sources that may have been released unintentionally. Common examples of the results of uncontrolled releases of energy include the following:

- Being struck by an object or equipment
- Being caught in between two objects
- Being struck by debris during the detonation of explosives
- Being cut or scraped by a jagged edge

These types of incidents are preventable if equipment is properly maintained and workers follow established safety guidelines.

2.1.2 Level II (Indirect Causes)

Level II represents indirect causes of incidents. Indirect causes are factors that contribute to an incident but are not the main cause. Indirect causes are also known as unsafe acts and conditions. In the past, investigators looked only for indirect causes of incidents, not the reasons behind the unsafe act or condition. This proved to be ineffective because indirect causes are often just the indicator of a greater problem.

It is estimated that 80 to 90 percent of incidents are caused by unsafe acts, and 10 to 20 percent are caused by unsafe conditions. It is important to be aware of both types of causes and use as many preventive measures as necessary, including design modification, proper training, appropriate equipment maintenance, and good worksite housekeeping.

Unsafe Acts – It is important to be able to recognize when a worker's behavior is unsafe. The following is a list of the most common unsafe acts found on a job site:

- Lack of proper and appropriate training
- Failing to use appropriate personal protective equipment
- Failing to communicate potentially hazardous conditions or unsafe behaviors to co-workers
- Failing to follow instructions or procedures
- Using incorrect or defective tools or equipment
- Lifting improperly
- Taking an improper working position
- Making safety devices inoperable
- Operating equipment at improper speeds
- Operating equipment without authority or proper training
- Servicing, repairing, or adjusting equipment while it is in motion or energized
- Loading or placing equipment or supplies improperly or dangerously
- Using equipment improperly
- Working under a suspended load or in an obviously dangerous area
- Working while impaired by alcohol or drugs (either legal or illegal)
- Engaging in horseplay

Unsafe Conditions – Unsafe conditions are physical conditions that are different from acceptable, normal, or correct conditions. The following is a list of the most common unsafe conditions:

- Poor housekeeping
- Poor illumination
- Poor ventilation
- Congested workplaces
- Defective tools, equipment, or supplies
- Extreme temperatures
- Excessive noise/pressure
- Fire and explosive hazards
- Hazardous atmospheric conditions
  - Gases
  - Dusts/fibers
  - Fumes
  - Vapors
- Inadequate supports or guards
- Inadequate warning systems
- Weather hazards

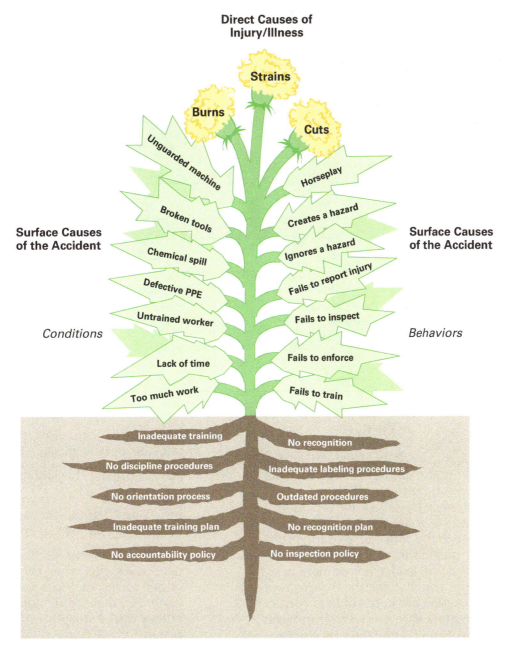

Figure 8 Root causes.

2.1.3 Level III (Root Causes)

Level III represents the root causes of incidents (*Figure 8*). The root or basic causes of incidents are the underlying reasons an incident happened. Root causes not only affect single incidents being investigated, they can also affect other future incidents and work problems. Once root causes are addressed and fixed, the occurrence of similar incidents should be minimized.

Contributing factors may involve equipment, process, people, material, environment, and management. A cause and effect diagram or fishbone diagram can be used to help identify the contributing factors and root causes associated with a specific incident or problem. *Figure 9* shows an example cause-and-effect diagram used to address burns, strains, and cuts.

Root causes should be corrected and fixed in a practical and timely manner. In some circumstances new procedures may need to be developed, new equipment may need to be procured, and formal training may be required. When appropriate, immediate temporary controls (ITCs) may need to be established until permanent fixes can be put in place.

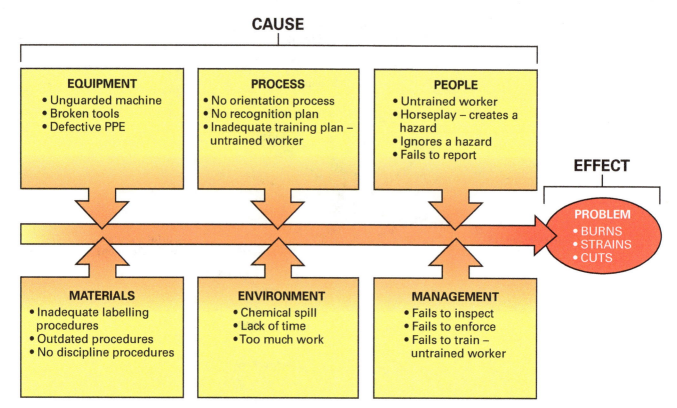

Figure 9 Example cause-and-effect diagram for burns, strains, and cuts.

2.2.0 The Why Method

The Why Method of incident investigation involves simply asking the question "Why did this happen?" and investigating until the question "Why?" can no longer be asked. By repeatedly asking why (five is a good rule), you can peel away layers to get to the root cause of the problem. The line of questioning may take several paths, each of which will lead to one or more basic or root causes. Once you discover the basic or root causes of incidents, you are better able to correct problems and create a safe work environment. The Why Method can be enhanced by combining it with the Sequence of Events Method. This combined methodology is called the Sequence of Events: Why Method.

2.2.1 Sequence of Events: Why Method

The Sequence of Events: Why Method is a systematic method of objectively identifying the root cause(s) of unsafe acts, conditions, and incidents. It involves identifying the sequence of events leading to the unsafe act or condition. This sequence-of-events model sees incidents as a chain of events that leads up to a failure.

The following steps are used in the Sequence of Events: Why Method.

Step 1 Identify the actual or potential injury, damage, or near miss.

Step 2 Ask, "Why did this situation occur?"

Step 3 Work in reverse chronological order, asking "Why?" until you can no longer ask "Why?" or you reach a dead end.

Step 4 Identify the events leading up to the incident. These preceding events can take one of two forms: they can be something that happened that should not have happened, or they can be something that did not happen but should have.

For example, a laborer on a construction site was told to go clean up the third floor of the building. He was specifically told to pick up all boards and scrap lumber. When the laborer got to the third floor, he found a pile of trash and scrap lumber covering a piece of plywood. He picked up all of the loose pieces of wood and trash and put them down the trash chute. He then proceeded to pick up the plywood. He picked up the piece of plywood and held it in front of him, which obstructed his view. When he stepped forward, he fell through a floor opening the plywood had been covering and suffered serious hip and back injuries.

The following sequence of events regarding this incident was determined by using the Sequence of Events: Why Method.

- *Incident event* – A laborer fell through a floor opening.
- *First preceding event* – Another contractor's temporary employee who covered the hole failed to secure the plywood in place and label the cover with the word HOLE as required by the OSHA fall protection standard.
- *Second preceding event* – The temporary contractor employee had been on the job only two days and had no knowledge of the OSHA requirements for covering floor openings and received no specific instructions from his supervisor other than to cover up that hole with a sheet of plywood.
- *Root cause* – This incident was caused by the lack of training and insufficient instructions to the temporary employee who covered the hole and the laborer who was not trained to identify the hazard. Find out why the worker(s) did not receive the proper training.

This method is effective because it goes beyond simply asking why. It allows for a deeper level of analysis because it requires you to record the chain of events leading up to the incident. When you are able to see all of the pieces of the incident, you can more easily determine the root cause.

2.3.0 OSHA's Problem-Solving Techniques

OSHA documents thousands of incidents that occur daily throughout the United States. These result from a failure of people, equipment, supplies, process, or surroundings to behave or function as expected. Incidents represent problems that must be solved through investigations. A successful incident investigation determines not only what happened, but also finds how and why the incident occurred. Formal problem-solving techniques can be used to solve problems of any degree of complexity. This section discusses two of the most commonly used procedures suggested by OSHA to analyze incident data:

- Change analysis
- Job safety analysis (JSA)/job hazard analysis (JHA)

2.3.1 Change Analysis

As its name implies, the change analysis technique emphasizes change. In terms of incident analysis problem solving, you must look for deviations from the norm. In order to do this, all situations that have resulted in an unanticipated change should be carefully identified and analyzed. Analyzing working conditions or behavior that were different from usual conditions or behavior can help identify the cause of the incident. Once you determine what was different or the source of the change, corrective actions should be taken.

The following steps are often used during the change analysis method:

Step 1 Define the problem. What happened?

Step 2 Establish the norm. What should have happened?

Step 3 Identify, locate, and describe the change.

Step 4 Specify what was and what was not affected.

Step 5 Identify the distinctive features of the change.

Step 6 List the possible causes.

Step 7 Select the most likely causes.

2.3.2 Job Safety Analysis

A job safety analysis (JSA), sometimes referred to as a job hazard analysis (JHA), is part of many existing incident prevention programs. A job safety analysis (e.g., welding a pipe) is a careful study of a job to find all of the associated hazards. A JSA also prescribes controls for each hazard. A JSA consists of a chart listing these steps, hazards, and controls (*Figure 10A* and *Figure 10B*). Review the JSA during the investigation if a JSA has been conducted for the job involved in an incident. Perform a JSA if one is not available. Perform a JSA as a part of the investigation to determine the events and conditions that led to the incident.

STARCON
a CIANBRO company
The Proactive Approach to Reducing Risk

JSA Grading — Score Using (2, 1, or 0) Maximum = 10   Total Score: _____
- Were Specific Hazards Identified Correctly?
- Were Detailed Job Steps Listed Correctly?   JSA Audited By _____
- Were Specific Hazards Mitigated Properly and How?   Date: _____
- Is JSA Filled Out Completely and Signed?   Time: _____
- Can TM's Tell Me The Location of the Eye Wash, Safety Shower and Evacutaion Area?

JSA Notes: _____

Insulation / Scaffold Job Safety Analysis (JSA)

Date: ___ Time: ___ T/M Name: ___ Permit #: ___
Supervisor: ___ Plant: ___
Plant Rep: ___ Unit Name: ___
Emergency Phone #: ___ Location: ___
Job Description: ___

STOP WORK OBLIGATION

THE BIG 5
- RESPONSIBILITY
- ACCOUNTABILITY
- INTERVENTION
- VALUE
- PRIDE

IF YOU SEE IT YOU OWN IT

Emergency Information
1. What Direction is the Wind Coming From?
2. Reviewed All Emergency Alarms / Phone Numbers with TM's: Yes___ No___
3. Escape Routes Reviewed Yes___ No___ List Escape Routes Below
   Primary: ___
   Secondary: ___
4. Location of Eye Wash Stations / Safety Showers: ___
5. My Evacuation Assembly Point Is: ___
6. Total Number of Team Members on Job: ___
   Supervisors Approval: ___
   Supervisors Review Date: ___ Time: ___

Team Member Signature and TM ID | Initial Changes
1. _____ ID# _____
2. _____ ID# _____
3. _____ ID# _____
4. _____ ID# _____
5. _____ ID# _____
6. _____ ID# _____
7. _____ ID# _____
8. _____ ID# _____
9. _____ ID# _____
10. _____ ID# _____
11. _____ ID# _____
12. _____ ID# _____

Job Scope Change / Modifications Discussion
Team Member: ___
Changes Made: ___

Supervisors Approval of Changes: ___
Date: ___ Time: ___
New JSA Required: Yes___ No___

PAUSE / WHAT IS DIFFERENT?
1. Scope   Yes___ No___
2. Unexpected Events Yes___ No___
3. Local Activities Yes___ No___
4. People   Yes___ No___
5. Pace   Yes___ No___
6. Fatigue   Yes___ No___
7. Area Conditions Yes___ No___
8. Other   Yes___ No___

HAZARD MANAGEMENT
- IDENTIFY
- ASSESS
- CONTROL
- RECOVER

Job Steps in Sequence	Hazards Identified	Hazard Control
1.		
2.		
3.		
4.		
5.		
6.		
7.		
8.		
9.		
10.		
11.		
12.		
13.		
14.		
15.		

Figure 10A JSA Form (Part 1 of 2).

Hazards Identified in Work Area	Hazard Control Methods	PPE Equipment Utilized
☐ Thermal Burns (Tracing/Hot Piping/Steam)	☐ 360 Degree Awareness	☐ Safety Glasses
☐ Asbestos	☐ Safety Nets Installed	☐ Goggles / Spoggles
☐ Overhead Power Lines	☐ Utilized Salt / De-Ice terials	☐ Side Shields (Prescription Glasses)
☐ Congested Area / Other Workers	☐ Abated / Encapsulated	☐ Safety Harness
☐ Insufficient Lighting	☐ Proper Distance Identified	☐ Double Retractable Device
☐ Heat Stress / Over Exertion	☐ Spotter Assigned	☐ Cheater Strap - Beam Strap
☐ Slips, Trips, Falls	☐ Twist and Release Method Used	☐ Radio (Battery Charged / Working Properly)
☐ Pinch Points	☐ Switch Covers Installed	☐ Hearing Protection
☐ Line of Fire	☐ Proper PPE Utilized for the Job	☐ Double Hearing Protection
☐ Snow, Ice, Rain, High Wind	☐ Lock Out / Tag Out in Place, Box#_____	☐ Leather Gloves
☐ Rotating Equipment	☐ Proper Barricades in Place / Signs Posted	☐ Cut Resistance Gloves (Kevlar)
☐ Rail Road Crossing / Tracks	☐ Continuous Monitor Utilized	☐ Nitrile Gloves (Alky Use)
☐ Leading Edge	☐ Proper Communication in Place	☐ Kevlar Sleeves (Tracing)
☐ Cuts, Punctures	☐ Proper Lifting Ergonomics Utilized	☐ H2S Monitor ☐ Tested
☐ Loud Noises	☐ Clean Work Area (House Keeping)	☐ Hard Hat
☐ Pump Switches / Electrical Equipment	☐ Tools Inspected and Safe for Use	☐ Steel Toe Boots
☐ Radiation Hazards	☐ Safety Equipment Inspected Prior to Use	☐ Rubber Boots
☐ Particles In Eyes	☐ Proper Equipment Grounding	☐ Face Shield (Z87 +)
☐ Dropping Materials	☐ Scaffold Inspected & Updated, Harness Inspected	☐ Fresh Air Equipment
☐ Chemical Exposure	☐ Respirator with Proper Cartridge	☐ (Vest, Horn, 5 Min Pack, 30 Min Pack, etc.)
☐ Sprains / Strains	☐ Proper donning / doffing of PPE	☐ Lock Out / Tag Out Locks in Place
☐ Elevated Work	☐ Other_____	☐ FR Clothing
☐ Sharp Objects	☐ Other_____	☐ Body Protection Other: (Bunker, Tyvek, Rain Suit)
☐ Other:_____	☐ Other_____	☐

Operations / Projects / Maintenance

1. Equipment Preparation and Transfer Complete   Yes _____ No _____ N/A _____
2. Job Scope Reviewed / Understood   Yes _____ No _____
3. Proper Safety Equip. On Job Site   Yes _____ No _____
4. Proper Permits Utilized:
   Cold Work _____ Confined Space _____ Alky _____ Hot Work _____
5. Special Permit _____   Yes _____ No _____
6. Joint Job Site Visit Conducted (JJSV)   Yes _____ No _____
7. Confined Space, Bottle Watch, Hole Watch, Fire Watch Attendant Assigned
   Yes _____ No _____ N/A _____
8. Proper Communications Set in Place   Yes _____ No _____
9. All Valves / Switches / Tubing / Blinds: Closed, De-Energized, and Verified:
   Yes _____ No _____ N/A _____
10. Atmospheric Testing Completed   Yes _____ No _____ N/A _____
11. Confined Space Procedure / Rescue Plan Reviewed With Certified Safety
    Attendant   Yes _____ No _____ N/A _____
12. Proper Tools and Equipment for Job Inspected   Yes _____ No _____
13. MSDS Reviewed and Understood   Yes _____ No _____ N/A _____
14. Material Transport Complete   Yes _____ No _____ N/A _____
15. Other: _____   Yes _____ No _____ N/A _____

System Breach Documentation / Variance

1. Physically Verified That No Hydrocarbon / Hazardous Material is Trapped
   Yes _____ No _____ N/A _____
2. Equipment in Place to Collect and Disperse any Spills
   Yes _____ No _____ N/A _____
3. Is There Hot Work Within a 50 foot Radius Horizontally or Vertically
   Yes _____ No _____ N/A _____
4. Safe Provisions for Egress From Work Site   Yes _____ No _____
5. Emergency Response Plan in Place Required   Yes _____ No _____
6. Safety Watch in Place   Yes _____ No _____ N/A _____
7. All Appropriate PPE in Place for Protection in Case of Leakage
   Yes _____ No _____ N/A _____
8. Process and Mechanical Witnessed the Sniff Testing
9. System Breach Checklist Completed   Yes _____ No _____ N/A _____
10. Field Verification Completed   Yes _____ No _____ N/A _____
11. All Signatures Acquired   Yes _____ No _____ N/A _____
12. System Breach Variance Required   Yes _____ No _____ N/A _____
13. Variance Documentation Complete and Approval Acquired
    Yes _____ No _____ N/A _____

Note: If Any Answer Above is a "No" Approval to Proceed Must be Obtained by Company and Site Leadership

Job Task Completion Review

1. Work Site Cleaned / Barricades Removed   Yes _____ No _____
2. Red Tags / LOTO Removed and Signed Off   Yes _____ No _____
3. Permits Pulled and Closed Out   Yes _____ No _____
4. Unit Sign-In / Sign-Out Logs Verified   Yes _____ No _____
5. Post JJSV Required / Completed   Yes _____ No _____
6. Open Manways / Holes Covered   Yes _____ No _____
7. Any Safety Involvement / Concerns   Yes _____ No _____
Concerns: _____

VISION for Safety at STARCON

Contractors and Customers / Owners by ensuring that our enabling and sustaining systems are in place and that our climate / culture supports our vision for safety as a core value. We will see all TM's Sub-Contractors, and Customer / Owners exhibiting positive behaviors that are reinforced and demonstrated through our commitment to Leading With Safety by:

- Focusing Resources Where the Higher Risk Occur
- Working with TM's Sub-Contractors, and Customer / Owners to Implement Lasting Improvements
- Identifying Hazards and Mitigating the Risk
- Intervening **(IF YOU SEE IT YOU OWN IT - TAKE ACTION)**
- Verifying Safety is Implemented Throughout All Work Phases
- Consistently Demonstrating and Supporting Safety Leadership at All Levels
  (Credibility, Action, Resolve, and Engagement)
- Continuously Investing in the Development of Safety Leaders

No One Gets Hurt!

SAFETY AND QUALITY ARE OUR CORE VALUES

Figure 10B JSA Form (Part 2 of 2).

2.3.3 Organizing and Reporting the Data

An incident investigation is not complete until a report is prepared and submitted to proper authorities. These reports help provide thorough documentation of an incident as well as create a structure for organizing the incident data. The information collected from change analyses and job safety analyses should be used for incident investigation reports. The following outline can be used as a tool to organize information gathered by the following methods:

1. Background Information
   - Where and when the incident occurred
   - Who and what were involved
   - Operating personnel and other witnesses

2. Account of the incident (what happened?)
   - Sequence of events
   - Extent of damage
   - Incident type
   - Agency or source, such as energy or hazardous material

3. Discussion (analysis of the how and why the incident happened)
   - Direct causes, such as energy sources or hazardous materials
   - Indirect causes, such as unsafe acts or conditions
   - Basic causes, such as management policies, personal factors, or environmental factors

4. Recommendations (to prevent a recurrence) for immediate and long-range action to remedy)
   - Basic causes
   - Indirect causes
   - Direct causes, such as reduced quantities, protective equipment, or protective structures

To get the most from your metrics, they must be simple and easily understood by every facet of the company. Good metrics will:

- Drive strategy and direction of the company
- Provide focus
- Assist in the decision making process
- Drive performance

Examples of different types of metrics include the following:

- Recordable Incident Rate (RIR)
- Days Away, Restricted, or Transferred (DART) Rate
- Lost Workday Case Rate (LWCR)
- Experience Modification Rate (EMR)
- Total Case Incident Rate (TCIR)
- Number of Days Without Lost Time Injury
- Project Safety Audit Scores

The importance of selecting appropriate metrics with measurable goals needs to be recognized. Collecting and analyzing data on safety related issues is critical from a compliance perspective and to ensure continuous improvement with a focus on a safe workplace. Selecting and analyzing the appropriate metrics is not easy, and requires periodic reassessment. Metrics must be selected with forethought on how they will be used to foster performance improvement and promote safety. After collecting and analyzing data, you should prioritize improvement opportunities based upon areas of concern from the data analysis report.

2.4.0 Trend Analysis

One of the many values of incident investigations is the trend analysis that can be studied after the investigation. By analyzing trends, one can get a much bigger and better picture of incident causation and take appropriate actions. By using completed incident reports, one can analyze the data by such factors as:

- *Time of day/hours worked in the previous 24 hours* – Is task or physical fatigue a factor? Can the workers be rotated between tasks to reduce fatigue? Should overtime be limited when performing certain tasks?
- *Length of employment/time on the job* – Are new workers improperly trained or experienced workers taking shortcuts?
- *Shift* – If accidents are occurring on the night shift, can dangerous tasks be moved to an earlier shift?
- *Supervisor/type of supervision* – Is the supervisor properly trained? Is closer supervision required for particular tasks?
- *Task being performed/body part injured* – If the same injuries occur repeatedly while performing certain tasks, can additional safeguards be installed or the equipment replaced? Have equipment guards been removed or safety devices defeated?
- *Type of injury* – If the trends show a spike in a specific type of injury (e.g., cuts on the hands or back injuries) consider targeted training for all employees on the importance of that safety topic.

Keeping track of incident investigation data and then using it to find trends will help improve the overall safety process. Make sure all data is forwarded to the corporate office for an even broader look. Trending analyses should include not only large incidents, but also near miss precursors and all other safety problems. The findings may surprise you.

A Pareto diagram (*Figure 11*) can be used to perform a trend analysis using data collected from a well-established incident investigation process. This type of diagram is typically used to sort data based on the frequency of occurrence and can help determine where resources should be allocated.

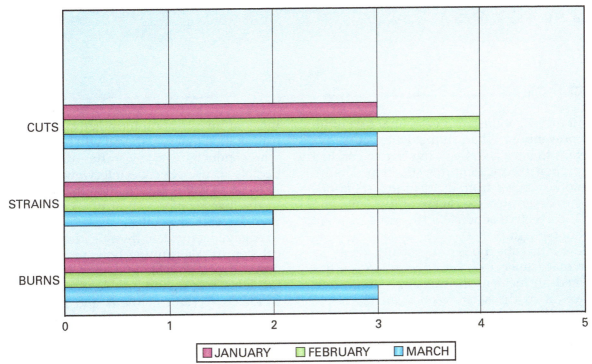

Figure 11 Example Pareto diagram.

Additional Resources

The American Society of Safety Engineers (ASSE), www.asse.org

Occupational Safety and Health Administration (OSHA), www.osha.gov

Gains from Getting Near Misses Reported. Bridges, William. 2012. AIChE 8th Global Congress on Process Safety.

Heinrich's Common Cause Hypothesis: A Tool for Creating Safety. Busch, Carsten. 2012. www.predictivesolutions.com.

Incident Investigation and Root Cause Analysis Course Materials. 2013. Process Improvement Institute, Inc.

2.0.0 Section Review

1. The causes of incidents can be classified by three different factors: direct causes, indirect causes, and _____.
   a. fundamental causes
   b. unrelated causes
   c. root causes
   d. stem causes

2. The Why Method can be enhanced by combining it with the _____.
   a. Sequence of Events Method
   b. Three Levels of Grief Model
   c. Cause and Effect Method
   d. Three Levels of Causation Model

3. A successful incident investigation determines not only what happened, but also finds how and why the _____.
   a. system failed
   b. incident occurred
   c. policies were not followed
   d. change occurred

4. Analyzing trends is most likely to result in _____.
   a. increased liability
   b. reduced incidents
   c. cost leveling
   d. reduced morale

SUMMARY

Incident investigations are an essential part of incident prevention. When an incident occurs, it is important to have a strategy in place to conduct a comprehensive incident investigation. The information collected from incident investigations is used to determine the causes of incidents and provide corrective actions so the same incident won't happen again.

Incident investigations are usually done by the foreman/supervisor. Once the supervisor has completed an incident investigation form, he or she gives it to the safety technician to review and submit to the appropriate personnel. The safety technician also acts as a resource for the supervisor during the investigation. It is important for both supervisors and safety technicians to know their role in incident investigations. It is also important for each to know what causes incidents, why investigations are conducted, how an incident investigation form is used, and how the information that is gathered will be used. Anyone who conducts interviews after an incident must be able to use effective interviewing techniques in order to gather information that will be useful in the investigation. All of this knowledge helps to make work sites safer.

Analyzing incident investigation data is a critical step in determining how and why incidents occur. If you are able to determine the causes of incidents, you should be able to minimize future incidents. Certain methods and models of data analysis may be more effective than others based on the type of incident. Well-defined policies and procedures are also needed to support the investigation process and for conducting data analysis. No matter which method is used, the ultimate goal is finding the cause of the incident and fixing the problem.

Review Questions

1. The on-site supervisor is generally responsible for conducting _____.
   a. incident investigations
   b. complex investigations
   c. police investigations
   d. criminal investigations

2. The major difference between an accident and an incident is distinguishing if the associated event outcome was _____.
   a. reckless
   b. preventable
   c. hazardous
   d. justified

3. The emphasis of an investigation should never be on identifying who should be blamed for the incident, but on _____.
   a. how much the incident will cost
   b. why the rules were broken
   c. how management is responsible
   d. why the incident happened

4. The greatest disadvantage of a foreman/supervisor conducting his or her own incident investigation is that he or she _____.
   a. is generally unfamiliar with the investigation process
   b. is usually too busy to conduct thorough investigations
   c. may find it difficult to remain unbiased
   d. cannot take immediate corrective actions

5. Gathering and preserving information during an investigation is critical to finding the cause or causes of _____.
   a. cost over-runs
   b. evidence tampering
   c. forgotten details
   d. an incident

6. An investigator should begin investigating an incident _____.
   a. before injured workers have been removed
   b. immediately after the injured worker has been treated
   c. before the next shift of workers arrives
   d. within 24 hours

7. Second-level investigations should be considered when the nature of the incident is very _____.
   a. simple
   b. complex
   c. costly
   d. strict

8. Emergency responders should be interviewed to learn _____.
   a. if they have seen similar incidents
   b. typical response times
   c. what best practices were used
   d. what they saw when they arrived at the scene

9. Incident investigation reports should document the full range of _____.
   a. motion
   b. facts
   c. liability
   d. blame

10. Because every incident is unique, there are different approaches and challenges with regards to analyzing incident investigation _____.
    a. regulations
    b. laws
    c. requirements
    d. data

11. Level I incidents involve _____.
    a. unsafe acts
    b. unsafe conditions
    c. uncontrolled releases of energy
    d. the failure to correct known on-site hazards

12. Unsafe conditions are physical conditions that are different from acceptable, normal, or _____.
    a. correct conditions
    b. correct situations
    c. unknown conditions
    d. unknown situations

13. Perform a JSA as a part of the investigation to determine the events and conditions that led to the _____.
    a. review
    b. filing
    c. incident
    d. listing

14. A job safety analysis (JSA) breaks a job into basic steps and identifies the _____.
    a. data required for reporting
    b. time it takes for each step
    c. cost of materials
    d. hazards associated with each step

15. The information collected from change analyses and job safety analyses should be used for _____.
    a. incident investigation reports
    b. determining who is at fault
    c. incident assignment review
    d. reporting fines to OSHA

Trade Terms Introduced in This Module

Accident: An event that results in property damage and/or personal injury or death that could not have been prevented.

Incident: An unplanned, undesired event that may cause personal injury, illness, or property damage in various combinations or degrees from minor to catastrophic.

Interviewee: Person being interviewed.

Pareto diagram: A diagram used to perform a trend analysis by sorting data based on the frequency of occurrence.

Reverse chronological order: Events told in order from last to first.

Root cause analysis: A method of problem solving that seeks to identify the most basic cause(s) of an incident.

Three Levels of Incident Causation Model: A method of incident investigation that seeks to identify three different factors: direct causes, indirect causes, and root causes.

Why Method: A method of incident investigation that is used to determine the root cause of an incident by asking the question "Why did this happen?" and investigating until the question "Why?" can no longer be asked

Appendix A

EXAMPLE INCIDENT INVESTIGATION REPORT FORM

INJURY RELATED INCIDENT REPORT

USE THIS REPORT FOR ANY EVENT OTHER THAN AUTO THAT COULD HAVE RESULTED IN AN EMPLOYEE INJURY.

Employee Information

Employee Name		Employee Number	
Street Address		State & Zip Code	
Primary Language		Phone Number	
Position/Trade		Sex	☐ M ☐ F
Superintendent		PM Name	
Initial Indication	☐ Near Incident ☐ First Aid ☐ RX/Treatment ☐ Other:_____		

Jobsite Information

Job Name		Job Number	
Job Address		Job Phone	
Employee Began Work	☐ AM ☐ PM	Incident on Jobsite?	☐ Yes ☐ No
Incident Date		Time of Incident?	
Weather Conditions		Job in what County?	
Investigation Conducted by?		Investigator Cell #	
Private Owner/Describe Location			

Contacts

GC Name		Date / Time	/
Police		Date / Time	/
Fire Department		Date / Time	/
Ambulance		Date / Time	/
Helix Office		Date / Time	/
Family Member		Date / Time	/
OSHA / Cal-OSHA	☐	Date / Time	/
OSHA Reports or Citations? Describe:			

Initial Treatment & Forms

Injured Party Taken Where?		By Whom?	
Drug Tested?	☐ Yes ☐ No (if not, explain why)		
Worker Comp Form Completed?	☐ Yes ☐ No	Date	

DO NOT WRITE IN THIS AREA

Submitted to Carrier?	☐ Yes ☐ No	Claim Number	
Adjuster Info			
Notes			
The Event is:	☐ Recordable ☐ Lost Time ☐ Restricted Duty/Job Transfer ☐ 1st Aid		

Equipment				
Equipment Involved	☐ Not Applicable	☐ Helix Vehicle	☐ Tool	☐ Other:
Model & Serial #			Helix ID #	
Equipment Owner	☐ Helix Electric	☐ Employee		☐ Other:

Describe results of the safety check performed and by whom:

Pictures Taken By		# of Pictures	
Camera Type		Who has Pictures	

Parties Involved (Attach Individual Statements)

Check box if Statement from Injured Party is Attached ☐   If not checked, explain why below:

List other Contractors/Subcontractors Involved:

1) Witness Name		Occupation	
Street Address		City & Zip Code	
Employed by		Phone Number	
Comments			

2) Witness Name		Occupation	
Street Address		City & Zip Code	
Employed by		Phone Number	
Comments			

3) Witness Name		Occupation	
Street Address		City & Zip Code	
Employed by		Phone Number	
Comments			

4) Witness Name		Occupation	
Street Address		City & Zip Code	
Employed by		Phone Number	
Comments			

Task & Incident Details

Describe **Task**:

Was the victim/involved employee **qualified** to perform the task? ☐ Yes ☐ No   If so, how do you know?

Had the victim/involved employee been observed doing this, or a similar task, correctly or incorrectly prior to the event? If incorrectly, what action was taken?

Describe **Incident**:

Nature and Extent of Injury:

Has the employee returned to work? ☐ Yes ☐ No	Anticipated Return to Work Date	
Do you have employee training documentation for associated activity? ☐ Yes, see attached		☐ No
Did the crew discuss the potential for this incident, and the safe work procedures to be followed to prevent it? ☐ Yes ☐ No *Please attach a copy of the meeting notes to support your findings.*		
List PPE worn at time of the incident:		

Corrective Action

If no corrective action was taken, explain:

Name		Action	☐ Verbal ☐ Written ☐ Suspension ☐ Termination
Reason			
Name		Action	☐ Verbal ☐ Written ☐ Suspension ☐ Termination
Reason			
Name		Action	☐ Verbal ☐ Written ☐ Suspension ☐ Termination
Reason			

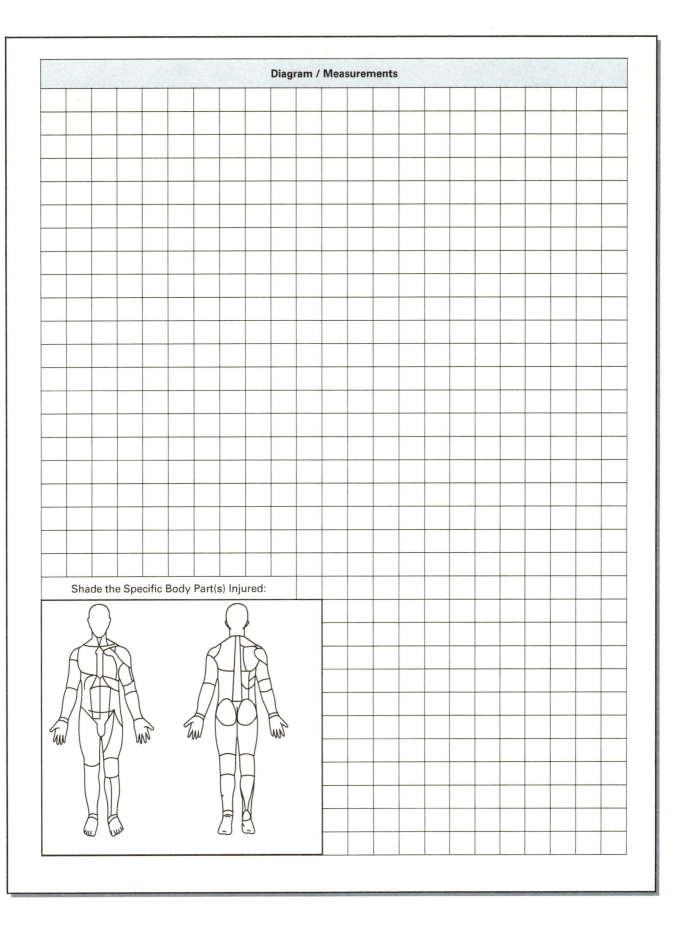

Post Incident Review
Causation Analysis

Specifically, what was the employee doing when injured?

What was the final factor?

Why did that happen? :

Why did that happen? :

Why did that happen? :

Why did that happen? :

Why did that happen? :

What steps were missed resulting in the injury/incident?

What does the employee **think** caused the incident?

☐ Using the information from this review, an activity hazard analysis (JHA) has been created identifying steps to complete the task safely and is attached.

Signatures			
Lead Investigator		Emp. #	
Superintendent		Emp. #	
Injured Employee		Emp. #	
Incident Review Team		Emp. #	
Incident Review Team		Emp. #	
Project Manager		Emp. #	
Date Completed:			

Statement of Injured/Involved Party

Employee Information			
Employee Name		**Employee Number**	
Street Address		**State & Zip Code**	
Position / Trade		**Phone Number**	
Job Name		**Sex**	☐ M ☐ F
Weather Conditions		**Incident on Jobsite?**	☐ Yes ☐ No
Date & Time of Incident	____/____/____ ____:____ ☐ AM ☐ PM		

What do you think caused the event when you were injured?

_____
_____
_____
_____
_____

What happened?

_____
_____
_____
_____
_____
_____
_____
_____
_____
_____
_____
_____
_____

Did site supervision ever observe you doing the same task, the same way, before this event? ☐ No ☐ Yes, describe below:

_____
_____
_____
_____

Signature: _____ Date: _____

Witness Statement

Witness Information				
Printed Name		**Employee Number**		*If Helix employee
Street Address		State & Zip Code		
Position / Trade		Phone Number		
Job Name		Sex	☐ M	☐ F
Weather Conditions		Incident on Jobsite?	☐ Yes	☐ No
Date & Time of Incident	___/___/___ ___:___ ☐ AM ☐ PM			

Statement:

_____
_____
_____
_____
_____
_____
_____
_____
_____
_____
_____
_____
_____
_____
_____
_____

Did you ever observe the employee doing the same task, the same way, before this event? ☐ No ☐ Yes, describe below:

_____
_____
_____

What prevented you from stopping the event?

_____
_____
_____

Signature: _____ Date: _____

Appendix B

TEN MOST COMMON MISTAKES FOUND ON OSHA INJURY AND ILLNESS RECORDKEEPING FORMS

Curtis Chambers, MS-OSH, CSP, of OSHA Training Services, Inc. recently posted the following OSHA Training blog discussing the Ten Most Common Mistakes Found on OSHA Injury and Illness Recordkeeping Forms.

1. **No case number appears on the Form 300 and Form 301.** See column A of the Form 300? This is where you have to enter a case number, which is a unique number (e.g., 01, 02, 03) that you make up and assign to each recordable incident. But be aware that this case number must ALSO appear on the corresponding OSHA Form 301 (Injury and Illness Incident Report) on line 10. And if you utilize an equivalent form in lieu of the OSHA 301 (such as a workers comp first report of injury form), this case number must appear somewhere on the alternate form as well.

2. **Checkmarks are entered in both Column H (days away from work) and Column I (job transfer or restriction) under the Classify the Case section of OSHA Form 300.** I see this happen often whenever an employee has to stay off work for a few days due to their injury, and then comes back to work on light duty for a couple of days. It seems logical that you would check both columns, right? However, the proper procedure in this case would be to check only one of the two boxes (the one with the most severe classification, which in this case is column H, days away from work). The reason you do not check both is that at the end of the year when you tally up the final numbers, you would actually show two incidents instead of one, and your incident rate would be artificially high.

3. **The employer does not enter both the number of days away from work (in Column K) AND number of days with job transfer or restriction (in Column L) of OSHA Form 300 when an employee has both.** This is the contrasting problem to the previous listed problem. Even though you would only check Column H under Classify the Case in the example above, you would still list the corresponding number of lost workdays in Column K AND days with job transfer or restrictions in Column L. So remember, in any incident involving both lost workdays and job restrictions or transfer, even though there would only be one checkmark (in column H), there would be days entered in both columns K and L.

4. **The employer mistakenly enters the number of scheduled workdays missed or with transfers/restricted duty instead of the number of Calendar days missed in Columns H and I of the OSHA Form 300.** This is often due to the record-keeper doing things the old way, when we used to fill out the OSHA 200 form. OSHA now requires you to enter the total number of calendar days missed or on light duty in these columns, not the number of scheduled workdays missed or on light duty. So even if an employee who normally works Monday through Friday with weekends off is injured on a Friday, and the doctor says he is not able to return to work or is on restricted duty until the following Monday, you would still have to count the Saturday and Sunday that the doctor said they could not work as lost or restricted workdays. [Reference *1904.7(b)(3)(iv) and (v)*].

5. **Incidents that should be recorded as an Illness are instead being classified as an Injury under Column M.** If an employee suffers flash-burn from welding, would you classify that as an injury or an illness? It should actually be classified as an illness (entered in line 6, titled "All Other Illnesses"). Incidents that are cumulative in nature (such as flash-burn, sunburn, noise-induced hearing loss, and cumulative trauma disorders) are generally considered as illnesses, whereas injury incidents typically result from a single exposure or event.

6. **Incidents affecting temporary service employees working under contract for an employer are not entered on that employer's OSHA injury and illness records.** The OSHA standards require you to enter all recordable incidents on your log suffered by any employee if you are directing their day-to-day activities. That is typically (although not always) the case where a company contracts with

a temporary help service or employee leasing company; the exception being when the temp service or leasing agency is also supervising the actual work being performed by their workers. [Reference 1904.31(b)(2)]

Recordable injuries or illnesses are not entered on the OSHA Form 300 within seven (7) calendar days of notification about the incident occurring, as required by OSHA. Typically, employers know about such incidents on the day they occur, and that is when the seven-day calendar day count begins to record the incident on your OSHA Form 300 (Log of Work-related Injuries and Illnesses). But in those cases where the employer does not learn about a recordable incident until a few days after it occurs, the seven days begins on the day they find out about it happening. So do not wait and update your logs once a month (or even once a quarter), as I so often see. Keep them up-to-date to within one week. [Reference 1904.29(b)(3)]

7. **Employers often under-report the "total hours worked by all employees last year" on the OSHA Form 300-A (Summary of Work-Related Injuries and Illnesses).** The total hours worked by all employees at the establishment during the corresponding year must be recorded on the form (see the right-hand side of form, under establishment information). This is important because under-reporting your hours worked will make your incident rates artificially high! Under-reporting is usually due to the employer forgetting to count the hours worked for ALL employees working at the establishment; this includes not just workers on the line, but also all salaried positions such as supervisors and managers, employees in support functions such as engineering, accounting, purchasing or human resources, and even the Big Kahuna himself (or herself). If they work at that site and are injured, that incident would go on your log, so you get to count their hours worked too. [Reference 1904.32(b)(2)(ii)]

8. **The signature certifying the accuracy of data on the OSHA Form 300-A is not that of a "Company Executive."** If you are the site safety manager, HR manager, or office manager, you most likely do not meet the definition of a company executive, and therefore should not be signing this particular form. OSHA defines an executive as the company owner (if the company is a sole proprietorship or partnership), a corporate officer, the highest ranking company official working at that site, or the immediate supervisor of the highest ranking person at the site. [Reference 1904.32(b)(4)]

9. **An employer does not maintain OSHA Injury and Illness Recordkeeping Forms for their company because they mistakenly claim the small employer exemption.** This is not actually addressing a mistake on a form; it addresses people who mistakenly think they don't have to fill out the forms. The OSHA standards say that an employer is generally exempt from maintaining these forms IF their company had ten (10) or fewer employees all of the previous year. There are two operative terms here; the first one is ALL, as in all of the previous year. You must look at your peak employment number throughout the entire year, and if that number was more than ten at any time during the previous year, then you must keep the forms the following year. Also, keep in mind that this is saying the total number of employees working for the COMPANY; you may employee ten or fewer employees at your particular site, but if there are other establishments within the same company (for example, a manufacturer with multiple facilities located throughout the state), you must count the total number of all COMPANY employees to see whether or not you qualify for that small employer exemption, not just the ones at your particular facility. [Reference 1904.1(b)(1) and (2)]

Source: Curtis Chambers, MS-OSH, CSP, of OSHA Training Services, Inc.

Appendix C

BUREAU OF LABOR STATISTICS DATA INTEGRITY GUIDELINES

The following guidelines must be followed by all Bureau of Labor Statistics (BLS) program offices and BLS employees to ensure the integrity of information maintained and disseminated by the BLS. Office of Management and Budget (OMB) information quality guidelines define integrity as the security of information—protection of the information from unauthorized access or revision to ensure that the information is not compromised through corruption or falsification.

Confidential Nature of BLS Records

Data collected or maintained by, or under the auspices of, the BLS under a pledge of confidentiality shall be treated in a manner that will ensure that individually identifiable data will be used only for statistical purposes and will be accessible only to authorized persons.

Pre-release economic series data prepared for release to the public will not be disclosed or used in an unauthorized manner before they have been cleared for release and will be accessible only to authorized persons.

Authorized persons include only those individuals who are responsible for collecting, processing, or using the data in furtherance of statistical purposes or for the other stated purposes for which the data were collected. Authorized persons are authorized access to only those data that are integral to the program on which they work and only to the extent required to perform their duties.

When non-BLS employees are granted access to confidential BLS data or Privacy Act data, they must be notified of their responsibility for taking specific actions to protect the data from unauthorized disclosure. The vehicle for providing this notification is the written contract or other agreement that authorizes them to receive the data. Accordingly, if a commercial contract, cooperative agreement, interagency agreement, letter of agreement, memorandum of understanding, or other agreement provides a non-BLS employee access to BLS confidential data or Privacy Act data, it must contain appropriate provisions to safeguard the data from unauthorized disclosure. The authorization document will state the purpose for which the data will be used and that all persons with access to the data will follow the BLS confidentiality policy, including signing the BLS non-disclosure affidavit. These provisions are required whether the data are accessed on or off BLS premises. They also are required when access to the data may be incidental to the work conducted under the contract or other agreement, such as in systems development projects, survey mail-out processing, etc.

Data Collection

The integrity of the BLS data collection process requires that all survey information be sound and complete. Data must be obtained from the appropriate company official or respondent, and the data entries must accurately report the data and responses they provided. The administrative aspects of the data collection process, such as work time reported and travel voucher entries, must be factually reported. Therefore, employees must not deliberately misrepresent the source of the data, the method of data collection, the data received from respondents, or entries on administrative reporting forms.

All BLS programs must follow the appropriate procedure for requesting authorization of processes for the electronic transmission of respondent-identifying data to or from respondents.

Procedures for Safeguarding Confidential Information

Program office managers are responsible for implementing procedural and physical safeguards to protect confidential information from disclosure or misuse within their offices, including:

- Preparing written procedures for the identification, labeling, handling, and disposal of confidential data. Ensuring that all employees within their organizations are familiar with and understand these procedures.

- Ensuring that new employees are informed about the different types of confidential data maintained in their work areas and the special precautions that are to be taken with their use, storage, and disposal.
- Developing data collection instruments and collection methodology in conformance with OMB guidelines on confidentiality.
- Ensuring that commercial contracts, cooperative and inter-agency agreements, letters of agreement, and memoranda of understanding, which give non-BLS employees access to confidential data, contain the proper confidentiality- and security-related clauses.

All BLS employees are responsible for following the rules of conduct in the handling of personal information contained in the records covered under the Privacy Act of 1974, which are in the custody of the BLS.

Dissemination of News and Data Releases

Public information documents require advance bureau-level clearance through the Associate Commissioner for Publications, who is responsible for seeing that each publication meets BLS publication standards and also the standards set by the Department of Labor, the Congressional Joint Committee on Printing, and OMB. BLS offices also are required to consult the Associate Commissioner for Publications before instituting an automated process to disseminate news releases or other products to the public.

No advance release of embargoed data shall be made unless directed by the Commissioner of Labor Statistics under the discretion granted under OMB Statistical Directive Number 3. BLS organizations shall strictly follow the Commissioner's specifications in making an advance release.

Data Security

The BLS has established appropriate computer security measures to safeguard the BLS' data processing environment against destruction or corruption of data or systems, unauthorized disclosure of data, and loss of service. These security measures are part of an overall management control process that includes program management, financial management, physical and personnel security, statistical data security, and information technology (IT) security. Associate, Assistant and Regional Commissioners, and Directors are assigned overall responsibility for directing the application of such controls to the Automated Information Systems and/or application systems, which they manage. The BLS Data Security Steering Committee provides overall direction to BLS security efforts.

Additional Resources

This module presents thorough resources for task training. The following resource material is suggested for further study.

The American Society of Safety Engineers (ASSE), **www.asse.org**

Occupational Safety and Health Administration (OSHA), **www.osha.gov**

Gains from Getting Near Misses Reported. Bridges, William. 2012. AIChE 8th Global Congress on Process Safety.

Heinrich's Common Cause Hypothesis: A Tool for Creating Safety. Busch, Carsten. 2012. **www.predictivesolutions.com**

Incident Investigation and Root Cause Analysis Course Materials. 2013. Process Improvement Institute, Inc.

Root Cause Analysis Handbook: A Guide to Effective Incident Investigation. 1999. Rockville, MD: Government Institutes.

The Psychology of Safety Handbook, E. Scott Geller, Ph.D. 2018. Boca Raton, FL: CRC Press.

Figure Credits

Wright-Patterson Air Force Base, 88th Air Base Wing Public Affairs, Module opener

Oregon Occupational Safety and Health Administration, Figure 2

Frank McDaniel, Figure 3, Appendix B

U.S. Department of Labor, Figures 4–6

Brett Richardson, Figure 10

Section Review Answer Key

Answer	Section Reference	Objective
Section One		
1. b	1.1.0	1a
2. b	1.2.1	1b
3. c	1.3.0	1c
Section Two		
1. c	2.1.0	2a
2. a	2.2.0	2b
3. b	2.3.0	2c
4. b	2.4.0	2d

NCCER CURRICULA — USER UPDATE

NCCER makes every effort to keep its textbooks up-to-date and free of technical errors. We appreciate your help in this process. If you find an error, a typographical mistake, or an inaccuracy in NCCER's curricula, please fill out this form (or a photocopy), or complete the online form at **www.nccer.org/olf**. Be sure to include the exact module ID number, page number, a detailed description, and your recommended correction. Your input will be brought to the attention of the Authoring Team. Thank you for your assistance.

Instructors – If you have an idea for improving this textbook, or have found that additional materials were necessary to teach this module effectively, please let us know so that we may present your suggestions to the Authoring Team.

NCCER Product Development and Revision
13614 Progress Blvd., Alachua, FL 32615

Email: curriculum@nccer.org
Online: www.nccer.org/olf

❏ Trainee Guide   ❏ Lesson Plans   ❏ Exam   ❏ PowerPoints   Other _____

Craft / Level: _____   Copyright Date: _____

Module ID Number / Title: _____

Section Number(s): _____

Description: _____

_____

_____

_____

_____

Recommended Correction: _____

_____

_____

_____

_____

Your Name: _____

Address: _____

_____

Email: _____   Phone: _____

OSHA Inspections and Recordkeeping

Overview

This module discusses the OSHA requirements for recordkeeping and explains how to manage the safety and health records for a job site. It also covers the two main types of OSHA inspections.

Module 75226

Trainees with successful module completions may be eligible for credentialing through NCCER's Registry. To learn more, go to **www.nccer.org** or contact us at 1.888.622.3720. Our website has information on the latest product releases and training, as well as online versions of our *Cornerstone* magazine and Pearson's product catalog.

Your feedback is welcome. You may email your comments to **curriculum@nccer.org**, send general comments and inquiries to **info@nccer.org**, or fill in the User Update form at the back of this module.

This information is general in nature and intended for training purposes only. Actual performance of activities described in this manual requires compliance with all applicable operating, service, maintenance, and safety procedures under the direction of qualified personnel. References in this manual to patented or proprietary devices do not constitute a recommendation of their use.

Copyright © 2018 by NCCER, Alachua, FL 32615, and published by Pearson, New York, NY 10013. All rights reserved. Printed in the United States of America. This publication is protected by Copyright, and permission should be obtained from NCCER prior to any prohibited reproduction, storage in a retrieval system, or transmission in any form or by any means, electronic, mechanical, photocopying, recording, or likewise. To obtain permission(s) to use material from this work, please submit a written request to NCCER Product Development, 13614 Progress Blvd., Alachua, FL 32615.

75226 V2

From *Safety Technology,* Trainee Guide. NCCER.
Copyright © 2018 by NCCER. Published by Pearson. All rights reserved.

75226
OSHA Inspections and Recordkeeping

Objectives

When you have completed this module, you will be able to do the following:

1. Explain how to follow OSHA and company requirements for recordkeeping.
    a. Identify OSHA requirements for recordkeeping.
    b. Classify illnesses and injuries per 29 *CFR* 1904.
    c. Explain how to document work-related illnesses and injuries using OSHA Forms 300, 300A, and 301.
    d. Describe how to manage safety and health records for a job site.
2. Explain how to comply with OSHA inspections.
    a. Describe the process for an on-site OSHA inspection.
    b. Define violations, citations, and penalties.
    c. Explain the difference between a focused inspection and a wall-to-wall inspection.
    d. Describe how to perform post-inspection follow-up.

Performance Tasks

This is a knowledge-based module; there are no performance tasks.

Trade Terms

Abatement
Abatement certification
Antigens
Calendar year
Chronic irreversible disease
Citation
Competent person
Compliance safety and health officers (CSHOs)
Contact dermatitis
Corrective action
Eczema

Good faith
Hypothermia
Infectious material
Imminent danger
Penalty
Recordable
Restricted work activity
Ulcers
Violation

Industry Recognized Credentials

If you are training through an NCCER-accredited sponsor, you may be eligible for credentials from NCCER's Registry. The ID number for this module is 75226. Note that this module may have been used in other NCCER curricula and may apply to other level completions. Contact NCCER's Registry at 888.622.3720 or go to **www.nccer.org** for more information.

Contents

1.0.0 Recordkeeping 1
  1.1.0 OSHA Recordkeeping Requirements 1
  1.2.0 Criteria for Classifying Illnesses and Injuries 2
    1.2.1 Work-Related Injuries or Illnesses 2
    1.2.2 Determining Whether an Illness or Injury Is Recordable 3
    1.2.3 Medical Treatment 4
    1.2.4 Cases Involving Restricted Work 4
    1.2.5 Determining Lost Work Days and Restricted Work Days 5
    1.2.6 Classifying Injuries 5
    1.2.7 Classifying Illnesses 6
  1.3.0 Documenting Illnesses and Injuries using OSHA Forms 7
    1.3.1 OSHA Form 300—The Log 7
    1.3.2 OSHA Form 300A—The Summary 9
    1.3.3 OSHA Form 301—Injury and Illness Incident Report 9
  1.4.0 Managing Safety and Health Records 13
    1.4.1 Recordkeeping Problem Areas 13
    1.4.2 Performing Recordkeeping Audits 13
2.0.0 OSHA Inspections 16
  2.1.0 The OSHA Inspection Process 18
    2.1.1 Presentation of the Inspector's Credentials 18
    2.1.2 Opening Conference 18
    2.1.3 Selection of Representatives 18
    2.1.4 Walk-Around Inspection 18
    2.1.5 Closing Conference 19
  2.2.0 Defining Violations, Citations, and Penalties 19
    2.2.1 Violations 19
    2.2.2 Citations 20
    2.2.3 Penalties 21
  2.3.0 Focused vs. Wall-to-Wall Inspections 23
    2.3.1 Program Qualifications 23
    2.3.2 Effective Written Safety Program 24
    2.3.3 Competent Person 26
    2.3.4 The "Fatal Four" Leading Hazards 26
    2.3.5 Modified Inspection Procedure 27
  2.4.0 Post-Inspection Follow-Up 27
    2.4.1 Reports to Management and Employees 28
    2.4.2 Abatement and Abatement Certification 28
    2.4.3 Corrective Actions 28
    2.4.4 Informal Conferences, Appeals, and Hearings 28
    2.4.5 Affirmative Defenses 29

Figures and Tables

Figure 1   Report incidents that result in a death or hospitalization 2
Figure 2   Recordable illness and injuries decision tree................................. 3
Figure 3   OSHA Form 300.. 8
Figure 4   OSHA Form 300A 10
Figure 5   Form 300A optional worksheet 11
Figure 6   OSHA Form 301 12
Figure 7   Records location list 14
Figure 8   The Construction Focused Inspection Guideline 25
Figure A01A   OSHA inspection checklist (page 1 of 2) 34
Figure A01B   OSHA inspection checklist (page 2 of 2) 35

Table 1   OSHA Inspection Statistics for 2011 thru 2016 17
Table 2   Federal OSHA Violations for 2011 thru 2016 20
Table 3   OSHA Violations and Penalties 22
Table 4   Significant State OSHA Violations and Penalties Issued in 2015 ... 23

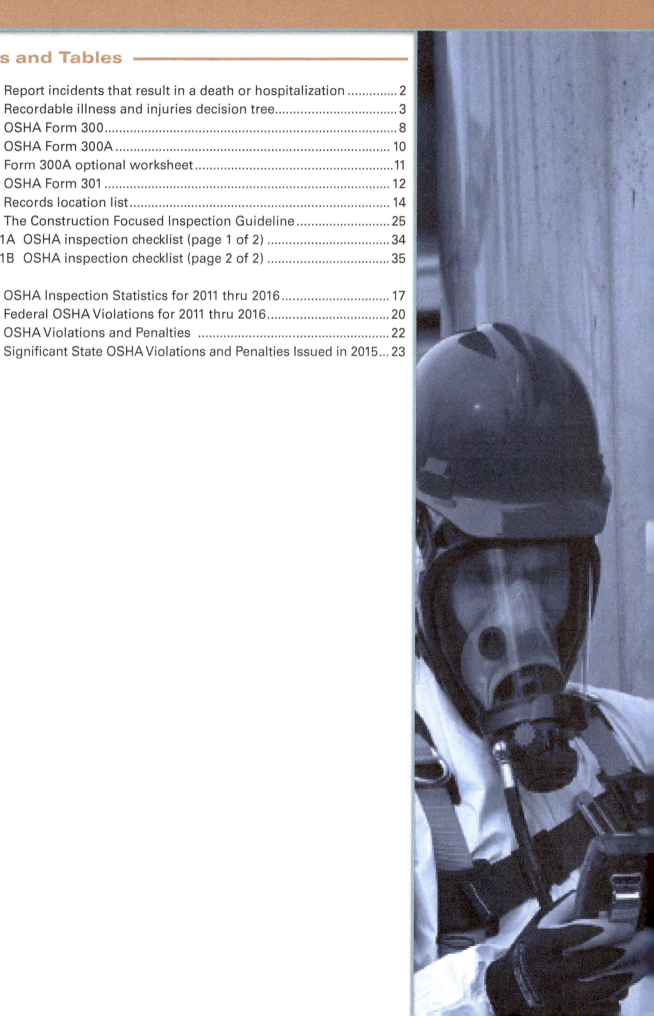

Section One

1.0.0 Recordkeeping

Objective

Explain how to follow OSHA and company requirements for recordkeeping.
a. Identify OSHA requirements for recordkeeping.
b. Classify illnesses and injuries per 29 *CFR* 1904.
c. Explain how to document work-related illnesses and injuries using OSHA Forms 300, 300A, and 301.
d. Describe how to manage safety and health records for a job site.

Trade Terms

Antigens: Toxins or enzymes capable of triggering an immune response similar to an allergy.

Calendar year: The period of a year beginning on January 1 and ending on December 31.

Chronic irreversible disease: A long-lasting, permanent disease.

Citation: A formal notice from OSHA that the company has allegedly violated a federal safety regulation and may be fined.

Contact dermatitis: An inflammation of the skin.

Eczema: An inflammatory condition of the skin characterized by redness, itching, and oozing sores that become scaly, crusted, or hardened.

Hypothermia: A life-threatening medical condition that can happen in cold weather when the body's temperature becomes too low.

Infectious material: Biological materials that carry contagious diseases and illnesses.

Penalty: A fine imposed on a company for violating a federal or state safety regulation. Penalties can also include prison time for criminal actions.

Recordable: An illness or injury is classified as recordable if it meets certain OSHA reporting guidelines.

Restricted work activity: Work activity assigned to those workers who have experienced a work-related injury or illness that prevents them from doing the routine functions of their jobs or from working a full workday.

Ulcers: Breaks in skin or mucous membrane that fester and erupt like open sores.

Violation: The act of violating federal or state safety regulations by failing to obey a safety standard or failing to keep the workplace free from hazards.

Recordkeeping is a critical part of a company's safety and health program. The data that is collected is used to help keep track of work-related injuries and illnesses. Once this data is gathered and analyzed, it can be used to help identify and correct problem areas to prevent future illnesses and injuries.

Recordkeeping not only provides information about illness and injury to management, but also informs workers about incidents that happen in their work area. When workers are aware of injuries, illnesses, and hazards in the workplace, they are more likely to follow safe work practices and report workplace hazards.

1.1.0 OSHA Recordkeeping Requirements

OSHA uses specific illness and injury information that is reported to them as part of the agency's site-specific inspection targeting program. The Bureau of Labor Statistics (BLS) also uses injury and illness records as the source data for the Annual Survey of Occupational Injuries and Illnesses. This report shows safety and health trends nationwide and industry wide.

As a safety technician, you are responsible for making sure recordkeeping on your site is done correctly. You must know all of the OSHA requirements for recording and classifying workplace illnesses and injuries. In addition, you may be responsible for coordinating and/or maintaining all job-site safety, health, and training records.

All employers covered by the Occupational Safety and Health Act (OSH Act) of 1970 are required to meet the recordkeeping regulations from 29 *CFR* 1904. If your company had more than 10 employees at any time during the last **calendar year**, you must keep OSHA injury and illness records. To determine if your company is exempt because of size, calculate your company's highest number of employees working at any given time during the last calendar year. If the number is greater than 10, you are required to keep OSHA illness and injury records. Keep in mind that a job-site safety technician is responsible for all workers on the site, including the subcontractors. All subcontractors are subject to the same requirements and normally share OSHA logs with the controlling contractor on a monthly basis.

If your company had 10 or fewer employees at all times during the last calendar year, you do not need to keep OSHA injury and illness records unless OSHA or the BLS informs you in writing that you must do so. Even if your employer is not required to keep injury and illness records, they are still required to report certain incidents to OSHA. Any fatality must be reported within eight hours and any inpatient hospitalization of one or more employees, amputation, or eye loss must be reported within 24 hours (*Figure 1*).

> **NOTE**
> Check 29 *CFR* 1904 if there is any question about whether your company is exempt from OSHA's recordkeeping rule.

OSHA also has specific requirements about the type of information that is recorded about job-related illnesses and injury, how it is collected, and how it is classified. OSHA Forms 300, 300A, and 301 are used to record this information. These forms are discussed in more detail later in this module.

1.2.0 Criteria for Classifying Illnesses and Injuries

OSHA has specific criteria for classifying illnesses and injuries. Each is based on what is considered recordable according to 29 *CFR* 1904. It is important to understand which types of injuries and illnesses are recordable so that they are recorded correctly.

Figure 1 Report incidents that result in a death or hospitalization.

Recordkeeping is a process that involves identifying and classifying specific information about an illness or injury to determine whether or not it is recordable. Consider the following four factors when gathering and analyzing data during the recordkeeping process:

- Is the illness or injury work related?
- Is the illness or injury recordable?
- What types of medical treatment needs to be recorded?
- How many days were missed or involved restricted work after the incident?

Record work-related injuries and illnesses that result in one of the following: death, amputation, eye loss, days away from work, restricted work or transfer to another job, medical treatment beyond first aid, loss of consciousness, or diagnosis of a significant occupation injury/illness that meet special recording criteria.

1.2.1 Work-Related Injuries or Illnesses

According to OSHA, in order to be considered work related, an injury or illness must be related to conditions or events in the work environment that caused or contributed to the illness or injury. For example, when a worker strains his or her back as a result of picking up equipment, the injury is considered work related. When an illness or injury is considered work related and meets certain criteria, it must be properly recorded. Before recording an illness or injury, make sure a thorough investigation has been completed. This will provide the needed documentation to support workers' compensation claims and will protect the company from legal liability.

> **NOTE**
> Illnesses and injuries that significantly aggravate a pre-existing condition, such as pre-existing asthma that is worsened during a job such as asbestos removal, are also considered work related if medical treatment is required or the injury results in lost workdays or restricted duty.

Symptoms that arise in workplace but are solely due to non-work-related event or exposure are not recordable. The following are examples of recordable and not recordable injuries:

- *Not recordable* – Joe trips over the family dog at home. His back begins hurting the next day at work and a doctor recommends physical therapy. The injury is not recordable.

- *Recordable* – A carpenter injures his hand at work. For three days he is reassigned to a desk job for the morning hours but works as a carpenter in the afternoon. The injury is a recordable case with three days of job transfer.

Healthcare Recommendations – You are not required to obtain the recommendation of a physician or licensed health care professional (PLHCP). However, if a recommendation is available, you must record the case accordingly. This is true even if the employee does not follow the recommendation. If a PLHCP changes the recommendation, record the most recent decision. Consider the following examples:

- *Example 1* – George is injured at work and is told by his physician not to lift anything over 30 lb. (13.6 kg) for five days, but he immediately resumes his full job duties. However, his job description requires him to lift more than 50 lb. (22.7 kg). The injury is a recordable case with five days of restriction.
- *Example 2* – Carl falls at work. After administering First Aid, the plant nurse instructs him to return to work, but he stays at home for two days. The injury is not recordable.
- *Example 3* – June pulls a muscle at work and her physician recommends she stay home for five days. The day before she is scheduled to return, she reports that her back is still hurting, and her doctor recommends taking an additional three days off. The case is recorded with eight days away from work.

1.2.2 Determining Whether an Illness or Injury Is Recordable

It is important to know when an illness or injury should be recorded. *Figure 2* shows a decision tree that can help you determine whether or not an illness or injury is recordable. In addition to the decision tree, OSHA has some specific criteria for judging the types of illnesses and injuries that must be recorded.

The following is a list of illnesses and injuries that must be recorded as required by OSHA:

- Work-related injuries and illnesses that result in death, loss of consciousness, days away from work, restricted work activity, job transfer, or medical treatment beyond first aid

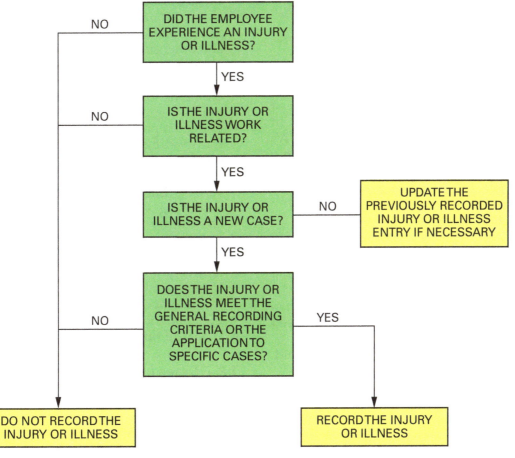

Figure 2 Recordable illness and injuries decision tree.

- Work-related injuries or illness that have been diagnosed by a physician or other licensed health care professional
- Work-related cases involving cancer, chronic irreversible disease, a fractured or cracked bone, or a punctured eardrum

OSHA has established additional criteria for cases requiring recordkeeping. These cases must be work related in order to be reported. Record the following conditions when you have verified that they are work related:

- Any needle-stick injury or cut from a sharp object that is contaminated with another person's blood or other potentially infectious material
- Any case requiring an employee to be removed from the site for medical treatment
- Any case in which a positive skin test or diagnosis has been made by physician or other licensed health care professional after exposure to a known case of active tuberculosis

Additional criteria such as the types of medical treatment and first aid should also be considered when recording illness and injuries.

> **NOTE:** The Mine Safety and Health Administration (MSHA) has different recording requirements. See www.msha.gov.

1.2.3 Medical Treatment

OSHA defines medical treatment as managing and caring for a patient for the purpose of fighting disease or disorder. Certain types of medical treatment are not applicable to OSHA's recordkeeping requirement. In your role as safety technician, you must know what treatments or procedures are considered medical treatment in terms of the OSHA requirement. The following is a list of the types of medical care that OSHA does not consider medical treatments and are not recordable:

- Visits to a doctor or health care professional solely for observation or counseling
- Diagnostic procedures, including administering prescription medications that are used solely for medical testing
- Any procedure that can be considered first aid

First aid is emergency care or treatment given to an ill or injured person before regular medical aid can be given. Since first aid is not considered recordable, make sure you know the type of treatment that is given to an ill or injured worker.

If a case is limited to first aid treatment and there is no lost time—days away from work (Column H) or job transfer or job restriction (Column I)—do not include the case on your OSHA 300 log. The case is not OSHA recordable, even if the first-aid treatment is administered at a health clinic, emergency room, hospital, or other medical treatment facility. First-aid treatment is defined in 29 *CFR* 1904.7(b)(5)(ii) and includes the following:

- Visits to a PLHCP solely for observation or counseling
- Diagnostic procedures, such as X-rays and blood tests, including the administration of prescription medications solely for diagnostic purposes (e.g., eye drops to dilate pupils)

> **NOTE:** An x-ray that comes up negative is not recordable.

- Using non-prescription medications at non-prescription strength (per box instructions)
- Administering tetanus immunizations
- Cleaning, flushing, or soaking wounds on the surface of the skin
- Using wound coverings such as bandages, gauze pads, butterfly bandages, Steri-Strips, and other similar coverings (wound closing devices are considered medical treatment)
- Using any non-rigid means of support such as elastic bandages, wraps, non-rigid back belts, etc.
- Using temporary immobilization devices while transporting a victim (e.g., splints, slings, neck collars, back boards, etc.)
- Drilling of fingernail or toenail to relieve pressure, or draining fluid from a blister
- Removing splinters or foreign material from areas other than the eyes by irrigation, tweezers, cotton swabs, or other simple means
- Removing foreign bodies from the eye using only irrigation or a cotton swab
- Using eye patches or finger guards
- Non-therapeutic massages (spa treatment)
- Using hot or cold therapy
- Drinking fluids for relief of heat disorder

1.2.4 Cases Involving Restricted Work

Workers can be placed on a restricted work schedule if they are ill or injured from a work-related incident and cannot perform the regular functions of their jobs. See 29 *CFR* 1904.7(b)(4)(i).

For recordkeeping purposes, an employee's routine functions are those work activities the employee regularly performs at least once per week.

Keep in mind, however, that not all illness and injury cases result in restricted work. According to OSHA's recordkeeping requirement, restricted work activity should be assigned to only those workers who have experienced a work-related injury or illness that prevents them from doing the routine functions of their jobs or from working a full day. Only a health care professional or the worker's employer can recommend a restricted work schedule. Make sure workers provide proper documentation, usually in the form of a doctor's note, to support the request for restricted work.

1.2.5 Determining Lost Work Days and Restricted Work Days

Determining lost workdays and restricted workdays is an important part of analyzing workplace illnesses and injuries. It is also an OSHA recordkeeping requirement.

The number of lost and restricted days can be calculated by counting calendar days from the day after the incident occurs. The day of the incident should not be included in this number. The totals should be separate if a single injury or illness involved both days away from work and days of restricted work activity. According to OSHA's recordkeeping rule, the number of lost or restricted workdays should stop being counted once the total of either, or the combination of both, reaches 180 days. See 29 *CFR* 1904.7(b)(3)(vii).

Job Restriction Examples – An injury or illness that prevents an employee from performing one or more routine functions of his or her job is recordable and can result in a restriction. A routine function is defined as a task performed at least once per week. Consider the following examples:

- *Example 1* – Every Friday, Cindy moves 15–40 lb. (6.8–18.1 kg) delivery boxes. A work-related injury prohibits her from lifting over 20 lb. for five days. The injury is a recordable case with five days of restriction.
- *Example 2* – Due to a work-related injury, Aaron's doctor tells him not to lift anything over 40 lb. (18.1 kg). His normal workweek does not require any lifting over 15 lb. (6.8 kg). The restriction is not recordable.
- *Example 3* – Todd is injured at work and is told to work only four-hour shifts for two days. The injury is a recordable case with two days of restriction.

1.2.6 Classifying Injuries

According to OSHA, an injury is any wound or damage to the body resulting from an event in the work environment. The following injuries are always classified as recordable:

- Fractures
- Chipped teeth
- Amputations
- Eye loss
- Loss of consciousness
- Electrocution

The following injuries are classified as recordable if they require treatment beyond first aid:

- Cuts
- Punctures
- Lacerations
- Abrasions
- Bruises
- Contusions
- Insect and animal bites
- Thermal, chemical, electrical, or radiation burns
- Sprains and strains resulting from a slip, trip, or fall

Any time any of these injuries happen on site, they must be recorded if they meet the appropriate criteria.

There are some instances in which an employee illness or injury does not need to be recorded as a work-related incident:

- Employee was present in the workplace as member of the public.
- Employee was engaged in a voluntary fitness program at work.
- The common cold and flu are not considered to be work-related contagious diseases.
- Mental illness, without the opinion of trained healthcare professional, is not automatically deemed work-related.
- Illnesses resulting from the employee's food brought in from an outside source (if food is provided by the employer on the job site, it is recordable).
- Illness or injury resulting from personal tasks completed during working hours.
- Illness or injury resulting from self-grooming, self-medication, or self-inflicted injuries (i.e., suicide attempts).

1.2.7 Classifying Illnesses

Illnesses can be defined as an unhealthy condition of body or mind. OSHA classifies recordable illness into the following four types:

- Skin diseases or disorders
- Respiratory conditions
- Poisoning
- Occupational illnesses

Knowing how to properly classify illnesses can speed up the recording process, allowing more time for other safety concerns, such as training or auditing. Make sure all illnesses can be verified as work related before they are recorded.

Skin Diseases or Disorders – Skin diseases or disorders are illnesses involving a worker's skin that are caused by exposure to chemicals, plants, weather, or other substances. The following are the types of skin diseases and disorders workers commonly get:

- Contact dermatitis
- Eczema
- Rashes caused by:
  - Irritants
  - Sensitizers
  - Poisonous plants
- Acne from oil
- Blisters
- Frostbite
- Sunburn
- Ulcers

Respiratory Conditions – Respiratory conditions are illnesses associated with breathing hazardous biological agents, chemicals, dust, gases, vapors, or fumes. Examples of the types of respiratory illnesses that workers commonly report include the following:

- *Silicosis* – A lung disease characterized by massive fibrosis of the lungs, caused by breathing in tiny bits of silica, a mineral that is part of sand, rock, and mineral ores such as quartz.
- *Asbestosis* – A lung disease caused by inhaling asbestos particles.
- *Pneumonitis* – An inflammation of the lungs that can be caused by many factors, including breathing in foreign particles, aspiration, viral infections, and inhalation of chemicals.
- *Pharyngitis* – An inflammation of the throat that may be caused by bacterial or viral infections.
- *Rhinitis* – An inflammation of the mucous membranes of the nose, a common condition among people with asthma.
- *Beryllium disease* – An inflammation of the lungs that can occur when beryllium dust or fumes are inhaled.
- *Tuberculosis (TB)* – An infectious lung disease caused by a type of bacterium called *mycobacterium tuberculosis* that spreads from person to person through microscopic droplets released into the air. TB screening tests may be required as part of a routine examination for new hires.
- *Occupational asthma* – A breathing condition triggered by inhaling airborne particles in the workplace.
- *Chronic obstructive pulmonary disease (COPD)* – A chronic lung disease, such as chronic bronchitis, emphysema, chronic asthma, and bronchiolitis. Smoking is the most common cause of COPD, with a number of other potential factors such as air pollution and genetics. In addition, second-hand smoke can also result in lung disease.
- *Hypersensitivity pneumonitis* – A lung disease that can occur when the lungs become inflamed from breathing in foreign substances (antigens), such as molds, dusts, and chemicals.
- *Chronic obstructive bronchitis* – A chronic cough lasting at least three months. Smoking or breathing in second-hand smoke or other fumes and dusts over a long period of time may cause chronic bronchitis.

Poisoning – Poisoning is caused when toxic or poisonous substances are ingested or absorbed into the body through skin or by breathing. The following are examples of poisons that can cause recordable illnesses:

- Metals such as lead, mercury, cadmium, or arsenic
- Gases such as carbon monoxide or hydrogen sulfide
- Organic solvents such as benzene, benzol, or carbon tetrachloride
- Insecticide sprays such as parathion or lead arsenate
- Chemical preservatives such as formaldehyde

Occupational Illnesses – Other illnesses are directly related to workplace activities and conditions. For example, some work sites can be very hot or very cold. Other sites may expose workers to radiation. The illnesses that result from these conditions are called *occupational illnesses*. The following are the types of illnesses that are commonly classified as occupational illnesses:

- Heatstroke
- Sunstroke
- Heat exhaustion

- Heat stress
- Hypothermia
- Frostbite
- Radiation sickness
- Tumors

1.3.0 Documenting Illnesses and Injuries using OSHA Forms

OSHA requires the use of the following three forms to record injuries and incidents at the work site:

- Form 300, *Log of Work-Related Injuries and Illnesses*: Used to record specific details about work-related illnesses and injury
- Form 300A, *Summary of Work-Related Injuries and Illnesses*: Shows the totals for the year in each category
- Form 301, *Injury and Illness Incident Report*: Used to record specific information about a work-related injury or illness

Employees, former employees, and their representatives have the right to review OSHA Forms 300 and 300A in their entirety. They also have limited access to OSHA Form 301 or another incident investigation reports. See 29 *CFR* 1904.35, OSHA's recordkeeping rule, for further details on the access provisions for Form 301 and other incident investigation reports.

1.3.1 OSHA Form 300—The Log

OSHA Form 300, sometimes called *the log*, is used to classify and record work-related injuries and illnesses and to note the extent and severity of each case. When an incident occurs, the log is used to record specific details about what happened and how it happened. Employers, especially larger companies with more than one site, must provide a separate form for each site that is expected to operate for one year or longer. OSHA Form 300 is shown in *Figure 3*.

> **NOTE**
> If the outcome or extent of an injury or illness changes after you have already recorded the case, simply draw a line through the original entry and write the new entry where it belongs. You must record the most serious outcome for each case.

There are circumstances in which the employee's name should not be entered on OSHA Form 300 (Log of Work-Related Injuries and Illnesses). Upon request, employers must provide workers, former workers, their personal representatives, and their authorized employee representative (union representative) access to injury and illness records. The names of employees must be left on OSHA Form 300 unless they are privacy concern cases. Employers may not use the Health Insurance Portability and Accountability Act (HIPAA) as a basis for removing employee names from the log before providing access. The recordkeeping rules require that employees, former employees, and employee representatives have access to the complete log, including names, except for privacy concern cases.

In accordance with HIPAA regulations, it is acceptable to leave the worker's name off the form when the following types of injuries or illnesses have occurred:

- An injury or illness to an intimate body part or to the reproductive system.
- An injury or illness resulting from a sexual assault, a mental illness, a case of HIV infection, hepatitis, or tuberculosis.
- Needle-stick injury or cut from a sharp object that is contaminated with blood or other infectious materials.
- Other illnesses, if the employee independently and voluntarily requests that his or her name not be entered on the log.

Enter the words "privacy case" in the space normally used for the employee's name when you cannot enter the worker's name on the OSHA 300 log. When privacy case is entered on the form, a separate, confidential list of the case numbers and workers' names must be kept so that you can update the cases and provide information to OSHA if asked to do so.

Details of an intimate or private nature do not need to be included on these forms but enough information must be entered to identify the cause and general severity of the injury or illness.

> **Did You Know?**
> In an OSHA standards interpretation letter dated August 2, 2004, OSHA held that the HIPAA privacy rule does not require employers to remove names of injured employees from the OSHA 300 log. This is due to the exception under HIPAA for records that are required by law. Since the OSHA 300 log is a required record, employers have no choice but to include all necessary information on it, including employee names and injury information.

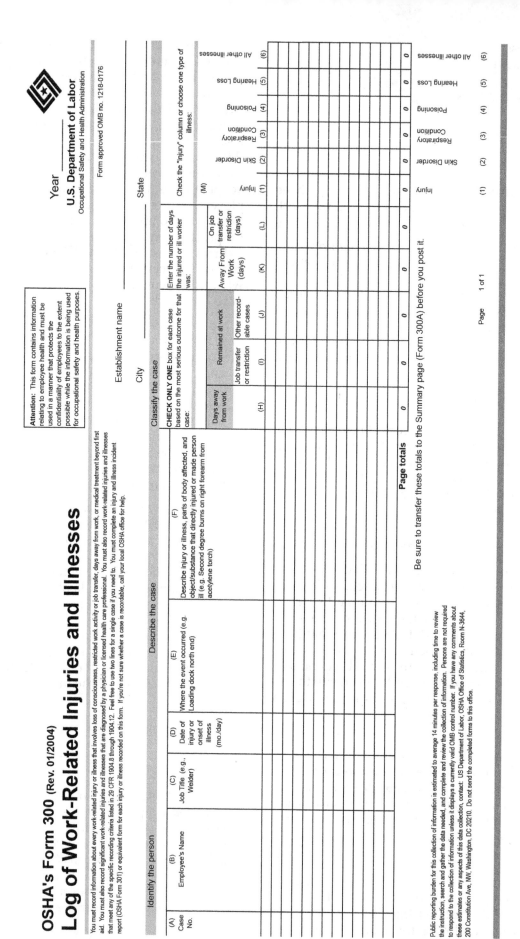

Figure 3 OSHA Form 300.

> **NOTE:** Use good judgment when describing the injury or illness on both the OSHA 300 and 301 forms when it is believed that information describing the case may identify the worker.

1.3.2 OSHA Form 300A—The Summary

Form 300A, also called *the summary*, shows the work-related injury and illness totals for the year in each category. All establishments covered by 29 CFR 1904 must complete this summary page, even if no work-related injuries or illnesses occurred during the year. *Figure 4* shows Form 300A. *Figure 5* shows the optional worksheet used to calculate the needed data for this form.

At the end of the year, the number of incidents in each category is counted, totaled, and transferred from the log (Form 300) to the summary (Form 300A). The summary is used to count individual entries for each category. The totals are then entered. If there were no cases on the log, a zero should be written in the total field.

After the summary has been completed, it is posted in a visible location so that workers can review and become aware of injuries and illnesses occurring in their workplace. Make sure all entries have been verified and are complete and accurate before posting the summary. The summary for the previous year must be posted from February 1st through the month of April. For example, calendar year 2015 must be posted February 1, 2016 through April 30, 2016.

The 300A summary must be posted at each job site in a conspicuous area where notices to employees are customarily placed. Even if there were zero incidents, the summary must still be posted. Copies of the form should be provided to any employees who may not see the posted summary because they do not regularly work onsite. The accuracy of the summary must be certified by the senior management executive, who must then sign the form.

> **NOTE:** Never post the log (Form 300). Only post the summary (Form 300A) at the end of the year.

1.3.3 OSHA Form 301—Injury and Illness Incident Report

Form 301, the injury and illness incident report (*Figure 6*), is one of the first forms you must fill out when a recordable work-related injury or illness has occurred. It is important to note that the injury and illness information needed to complete Forms 300 and 300A can also be collected from incident investigation reports. OSHA does not provide a mandatory format for incident investigation reports but does require that specific information be provided.

Form 301 is not required to be completed if an equivalent form is used that supplies all the information required by OSHA. Equivalent forms can include an incident investigation form, a state's first-report-of-injury form, a workers' compensation form, or an insurance report. To be considered an equivalent form, any substitute must contain all the information asked for on Form 300. Form 301 (or equivalent) should be one of the first forms completed after a work-related injury or illness occurs.

Form 301, or an equivalent incident investigation report, must be completed and submitted to authorized representatives within seven calendar days after you receive information that a recordable work-related injury or illness has occurred. Authorized persons include representatives from OSHA, Workers' Compensation, Public Health, law enforcement agencies, insurance auditors, and consultants. Before releasing this data, remove personally identifying information and provide only the information about the case section of the form.

Together, OSHA forms 300, 300A, and 301 help the employer and OSHA develop an understanding of the extent and severity of work-related incidents.

> **NOTE:** According to *Public Law 91-596* and 29 *CFR* 1904, OSHA's recordkeeping rule, you must keep Form 301 on file for five years following the year to which it pertains. Forms 300 and 300A must be updated and retained for three years.

Failure to maintain current and error-free workplace records can potentially result in OSHA violations. OSHA has raised the average cost of penalties, and frequently hands out willful and repeat citations. (A citation is a formal notice that a violation has taken place, and a penalty is a fine to discourage employers from repeat violations.) OSHA is also giving out more incident-by-incident (egregious) citations, rather than grouping incidents and giving one citation for all of them. Incident-by-incident citations were first employed in recordkeeping cases, and it's very easy to make a mistake on your records log and end up having to pay egregious violation fines for it.

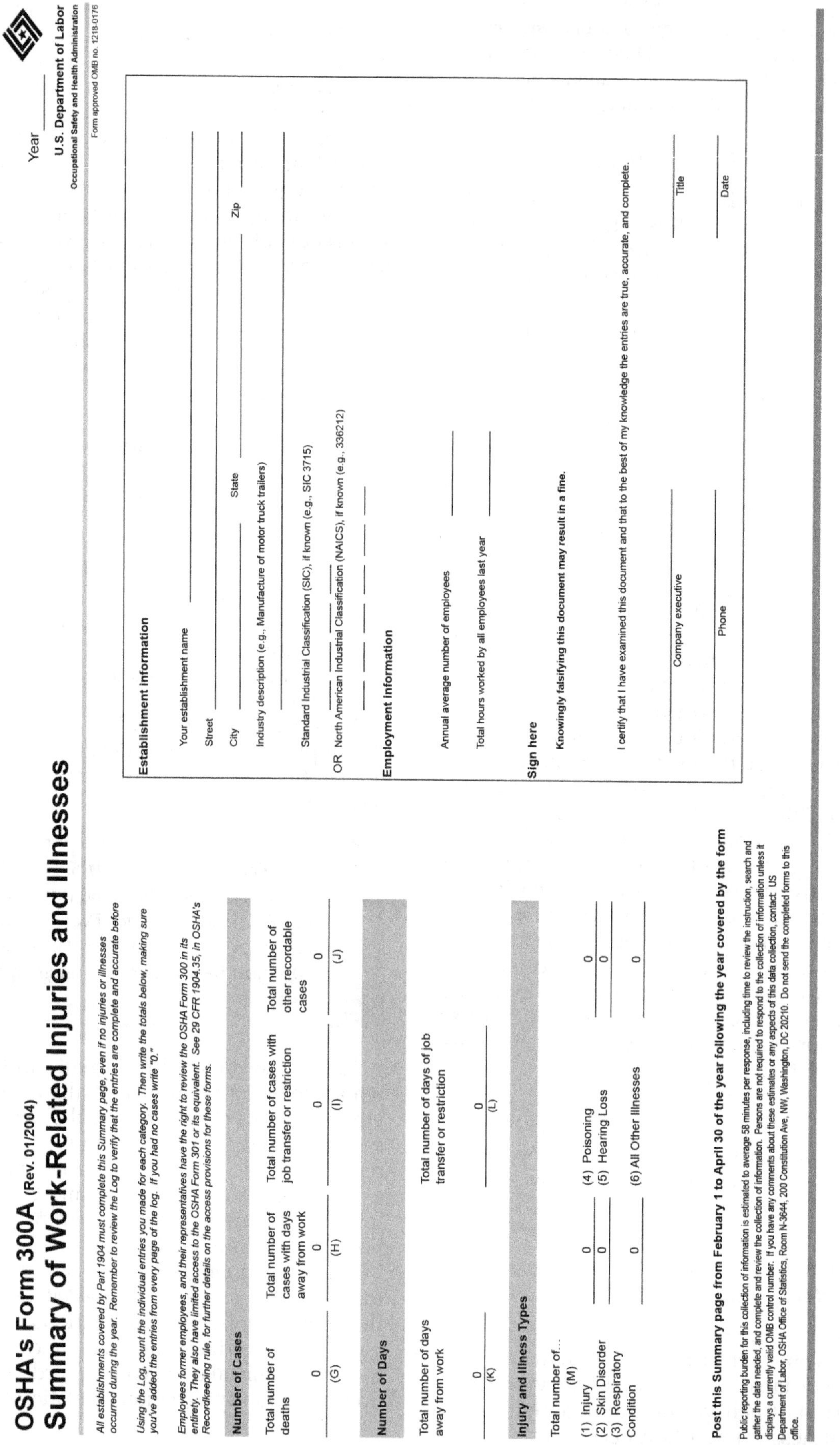
OSHA's Form 300A (Rev. 01/2004)

Summary of Work-Related Injuries and Illnesses

Year ____

U.S. Department of Labor

Occupational Safety and Health Administration

Form approved OMB no. 1218-0176

All establishments covered by Part 1904 must complete this Summary page, even if no injuries or illnesses occurred during the year. Remember to review the Log to verify that the entries are complete and accurate before

Using the Log, count the individual entries you made for each category. Then write the totals below, making sure you've added the entries from every page of the log. If you had no cases write "0."

Employees former employees, and their representatives have the right to review the OSHA Form 300 in its entirety. They also have limited access to the OSHA Form 301 or its equivalent. See 29 CFR 1904.35, in OSHA's Recordkeeping rule, for further details on the access provisions for these forms.

Number of Cases

Total number of deaths	Total number of cases with days away from work	Total number of cases with job transfer or restriction	Total number of other recordable cases
0	0	0	0
(G)	(H)	(I)	(J)

Number of Days

Total number of days away from work	Total number of days of job transfer or restriction
0	0
(K)	(L)

Injury and Illness Types

Total number of... (M)

(1) Injury	0	(4) Poisoning	0
(2) Skin Disorder	0	(5) Hearing Loss	0
(3) Respiratory Condition	0	(6) All Other Illnesses	0

Post this Summary page from February 1 to April 30 of the year following the year covered by the form

Public reporting burden for this collection of information is estimated to average 58 minutes per response, including time to review the instruction, search and gather the data needed, and complete and review the collection of information. Persons are not required to respond to the collection of information unless it displays a currently valid OMB control number. If you have any comments about these estimates or any aspects of this data collection, contact: US Department of Labor, OSHA Office of Statistics, Room N-3644, 200 Constitution Ave, NW, Washington, DC 20210. Do not send the completed forms to this office.

Establishment information

Your establishment name ____

Street ____

City ____ State ____ Zip ____

Industry description (e.g., Manufacture of motor truck trailers) ____

Standard Industrial Classification (SIC), if known (e.g., SIC 3715) ____

OR North American Industrial Classification (NAICS), if known (e.g., 336212) ____

Employment information

Annual average number of employees ____

Total hours worked by all employees last year ____

Sign here

Knowingly falsifying this document may result in a fine.

I certify that I have examined this document and that to the best of my knowledge the entries are true, accurate, and complete.

____ Company executive ____ Title

____ Phone ____ Date

Figure 4 OSHA Form 300A.

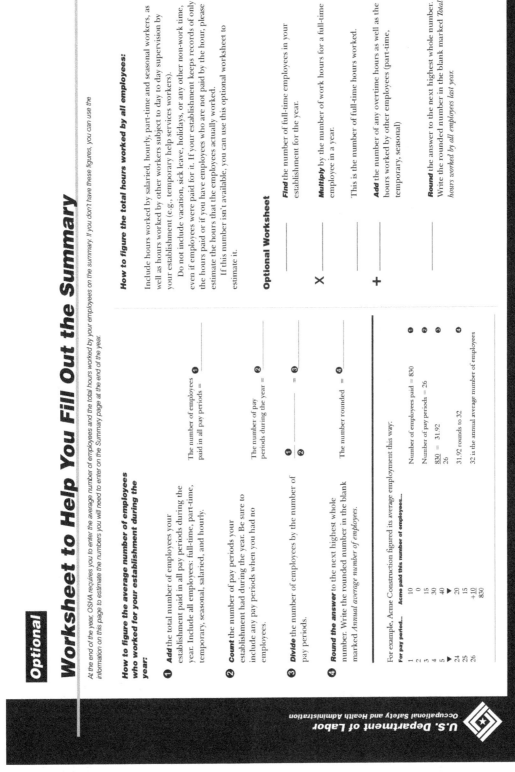

Figure 5 Form 300A optional worksheet.

OSHA's Form 301
Injuries and Illnesses Incident Report

U.S. Department of Labor
Occupational Safety and Health Administration

Form approved OMB no. 1218-0176

Attention: This form contains information relating to employee health and must be used in a manner that protects the confidentiality of employees to the extent possible while the information is being used for occupational safety and health purposes.

This *Injury and Illness Incident Report* is one of the first forms you must fill out when a recordable work-related injury or illness has occurred. Together with the *Log of Work-Related Injuries and Illnesses* and the accompanying *Summary*, these forms help the employer and OSHA develop a picture of the extent and severity of work-related incidents.

Within 7 calendar days after you receive information that a recordable work-related injury or illness has occurred, you must fill out this form or an equivalent. Some state workers' compensation, insurance, or other reports may be acceptable substitutes. To be considered an equivalent form, any substitute must contain all the information asked for on this form.

According to Public Law 91-596 and 29 CFR 1904, OSHA's recordkeeping rule, you must keep this form on file for 5 years following the year to which it pertains.

If you need additional copies of this form, you may photocopy and use as many as you need.

Completed by _____
Title _____
Phone _____ Date _____

Information about the employee

1) Full Name _____
2) Street _____
   City _____ State _____ Zip _____
3) Date of birth _____
4) Date hired _____
5) ☐ Male ☐ Female

Information about the physician or other health care professional

6) Name of physician or other health care professional _____
7) If treatment was given away from the worksite, where was it given?
   Facility _____
   Street _____
   City _____ State _____ Zip _____
8) Was employee treated in an emergency room?
   ☐ Yes ☐ No
9) Was employee hospitalized overnight as an in-patient?
   ☐ Yes ☐ No

Information about the case

10) Case number from the Log _____ *(Transfer the case number from the Log after you record the case.)*
11) Date of injury or illness _____
12) Time employee began work _____ AM/PM
13) Time of event _____ AM/PM ☐ Check if time cannot be determined
14) What was the employee doing just before the incident occurred? Describe the activity, as well as the tools, equipment or material the employee was using. Be specific. Examples: "climbing a ladder while carrying roofing materials"; "spraying chlorine from hand sprayer"; "daily computer key-entry."
15) What happened? Tell us how the injury occurred. Examples: "When ladder slipped on wet floor, worker fell 20 feet"; "Worker was sprayed with chlorine when gasket broke during replacement"; "Worker developed soreness in wrist over time."
16) What was the injury or illness? Tell us the part of the body that was affected and how it was affected; be more specific than "hurt", "pain", or "sore." Examples: "strained back"; "chemical burn, hand"; "carpal tunnel syndrome."
17) What object or substance directly harmed the employee? Examples: "concrete floor"; "chlorine"; "radial arm saw." If this question does not apply to the incident, leave it blank.
18) If the employee died, when did death occur? Date of death _____

Public reporting burden for this collection of information is estimated to average 22 minutes per response, including time for reviewing instructions, searching existing data sources, gathering and maintaining the data needed, and completing and reviewing the collection of information. Persons are not required to respond to the collection of information unless it displays a current valid OMB control number. If you have any comments about this estimate or any other aspects of this data collection, including suggestions for reducing this burden, contact: US Department of Labor, OSHA Office of Statistics, Room N-3644, 200 Constitution Ave, NW, Washington, DC 20210. Do not send the completed forms to this office.

Figure 6 OSHA Form 301.

1.4.0 Managing Safety and Health Records

In addition to injury and illness statistics, other records must be maintained. Some records are required by OSHA, and others are kept as a matter of good business practice. The safety technician is often required to maintain or coordinate the maintenance of these records. Records typically required by OSHA include the following:

- Employee job-related medical records
- Employee job-related exposure records
- Job- or task-specific training records
- Crane and hoisting equipment inspection and maintenance records
- Respiratory protection inspection records
- Fire extinguisher inspection records

Other commonly kept (but not required by OSHA) safety, health, and training records include the following:

- Job applications
- Alcohol and substance abuse policy acknowledgement forms
- Internal jobsite inspections, audits, and observation reports
- Disciplinary action reports

How and by whom these records are kept will vary depending on company policies and procedures. As a safety technician, you may have responsibility for some or all such records. Even if you are not responsible for keeping the records, you should know who does and where the records are kept. *Figure 7* is a sample records location list that can be used to document where records are kept and by the respective company representative.

OSHA has specific requirements for the maintenance of employee job-related medical and exposure records that are outlined in 29 *CFR* 1910.1020. As a safety technician, you should be well versed on these requirements, including employee access, which means the right to review and copy.

Employee job-related medical and exposure records should be kept confidential and under lock and key. Only personnel authorized by senior management may review medical and exposure records. The purpose of such reviews should be limited to that which is necessary to manage the safety and health process and protect the health and well-being of the employees. Critical, confidential employee medical and exposure records may only be released to members of management on a need-to-know basis. Such information should not include any physical conditions or medical problems. The information should only relate to the employee's fitness for duty. For example, the employee may not lift more than 15 pounds, or may use a respirator for escape purposes only. Such information must be kept confidential with regards to individual medical records.

Employees have a right to see and copy their personal medical and exposure records. Employee exposure records by job title or description should also be available to employees and their designated representatives. Employees should be informed of their right to see their job-related medical and exposure records and your procedures for granting access. This should be done at least annually.

1.4.1 Recordkeeping Problem Areas

There are several recordkeeping issues that can become potential problems for many employers. Common issues include the following:

- Improperly completed OSHA 300 Forms and related material.
- Unnecessarily recorded incidents. Never make your records look like you have more accidents than you actually do.
- No coordination between workers' compensation recordkeeping and OSHA recordkeeping. OSHA injury and illness records and workers' compensation records are independent of one another but should be kept in coordination. Sometimes OSHA inspectors will ask to see your workers' compensation log, and if it doesn't match up with your OSHA log, they will ask why.
- Employers fail to perform audits or make corrections to documents. It is a best practice to audit records on a yearly basis. Make sure they are accurate before you complete your summary.

1.4.2 Performing Recordkeeping Audits

The first step when auditing your records is to identify everyone in your organization who is responsible for maintaining injury and illness records. You should also determine each individual's business reasons for maintaining and accessing records. To the extent possible, try to consolidate injury and illness recordkeeping into one system that can be easily maintained and tracked. If multiple systems are being used, ensure that all records are complete and recorded in a consistent manner.

RECORDS LOCATION LIST

Type of Record	Responsible Party	Record Location
Employee Applications		
Employee Work-Related Medical Records		
Employee Work-Related Exposure Records		
Employee Training Records & Acknowledgement Forms		
OSHA 300 Logs		
Accident Report Forms & Investigating Report Documentation		
Inspection Report Forms & Inspection Documentation		
Safety Meeting Report Forms & Meeting Minutes		
Motor Vehicle and Mobile Equipment Maintenance & Inspection Report Forms		
Crane Repair & Annual Inspection Records		
Fire Extinguisher Inspection Records		
Job-Specific Work Permits (Confined Space Entry, LOTO, Hot Work)		

Figure 7 Records location list.

Perform self-audits of your records and logs on an annual basis, making corrections as needed when you find errors. Be sure to check your state OSHA rules to maintain state compliance. Some states have the exact same rules as federal OSHA rules, but others have added supplementary regulations.

Occupational injury and illness records have several distinct functions or uses. One use is to provide information to employers whose employees are being injured or made ill by hazards in their workplace. The information in OSHA records makes employers more aware of the kinds of injuries and illnesses occurring in the workplace and the hazards that cause or contribute to them. When employers analyze and review the information in their records, they can identify and correct hazardous workplace conditions on their own. Injury and illness records are also an essential tool to help employers manage their company safety and health programs effectively.

Additional Resources

OSHA 2002 Recordkeeping Simplified, James E. Roughton, CRSP, CHMM, CSP. 2002. Burlington MA: Butterworth-Heinemann.

The following websites offer resources for products and training:

American Society of Safety Engineers (ASSE), **www.asse.org**

Mine Safety and Health Administration (MSHA), **www.msha.gov**

Occupational Safety and Health Administration (OSHA), **www.osha.gov**

1.0.0 Section Review

1. All employers covered by the Occupational Safety and Health Act of 1970 (OSHA) are required to meet the recordkeeping regulations from 29 CFR _____.
   a. *Part 1904*
   b. *Paragraph 4c*
   c. *Section 1049*
   d. *Subparagraph 19a*

2. Recordkeeping is a process that involves identifying and classifying specific information about an illness or injury to determine whether or not it is _____.
   a. curable
   b. life threatening
   c. preventable
   d. recordable

3. To record injuries and incidents at the work site, OSHA requires the use of Forms _____.
   a. 101, 102, and 103
   b. 300, 300A, and 301
   c. 1926A, B, and C
   d. 3000A, B, and C

4. OSHA's specific requirements for the maintenance of employee job-related medical and exposure records are outlined in _____.
   a. *29 CFR 1910.1020*
   b. *26 CFR 1911.1022*
   c. *25 CFR 1810.1024*
   d. *24 CFR 1710.1026*

Section Two

2.0.0 OSHA Inspections

Objective

Explain how to comply with OSHA inspections.
a. Describe the process for an on-site OSHA inspection.
b. Define violations, citations, and penalties.
c. Explain the difference between a focused inspection and a wall-to-wall inspection.
d. Describe how to perform post-inspection follow-up.

Trade Terms

Abatement: The correction of the safety hazards or violations resulting from an OSHA citation.

Abatement certification: A written record that shows that the safety hazard or standard violation noted in the citation has been corrected.

Competent person: A person who is capable of identifying existing and predictable hazards, and who has authority to take prompt corrective measures to eliminate them.

Compliance safety and health officers (CSHOs): OSHA officials who conduct work-site inspections to enforce safety regulations.

Corrective action: Actions the company must take to correct safety violations.

Good faith: Good faith means that the employer makes a serious effort to comply with relevant OSHA standards to the best of his or her ability.

Imminent danger: A hazard that could cause death or serious physical harm immediately, or before the danger could be eliminated through normal enforcement procedures.

OSHA periodically inspects work sites to make sure safety standards are met. They will also schedule an inspection after a serious incident. OSHA inspections should never be taken lightly because OSHA can levy large fines or shut down operations if violations are discovered.

A good safety management program is your best defense against fines. You must make sure all applicable OSHA rules are enforced on your job site. OSHA has developed a focused inspection program for construction sites. If you have a written program and a **competent person** on site, your company may qualify for a more limited inspection.

Twenty-six states, along with Puerto Rico and the US Virgin Islands, currently have their own OSHA-approved health and safety programs. In order for states to have their own OSHA programs, their rules must be at least as stringent as the federal standards. State plans must set workplace safety and health standards that are at least as effective as OSHA standards. Many state plans adopt standards identical to federal OSHA. State plans have the option to promulgate standards covering hazards not addressed by federal OSHA standards. A state plan must conduct inspections to enforce its standards, cover state and local government workers, and operate occupational safety and health training and education programs. The state rules and procedures may differ from the federal standards explained in this module. You can find out which states have their own programs on OSHA's web site at **www.osha.gov**. It is your responsibility to be aware of the local regulations covering your job site.

OSHA representatives called **compliance safety and health officers (CSHOs)** inspect job sites. Their role is to enforce OSHA regulations. They are health and safety professionals who help employers reduce job hazards and enforce health and safety regulations. Before an inspection, CSHOs review the history of the business, the operations, and appropriate standards. They may also bring testing equipment to check compliance. Always be polite and do not attempt to mislead or lie to an official. If you do, any fines you receive will be multiplied.

OSHA will schedule an inspection when they think it is necessary. There are over seven million workplaces covered by federal and state OSHA standards. This makes it difficult to inspect every site. As a result, OSHA must prioritize inspections to deal with the most hazardous conditions first. The following is a general OSHA triage policy for prioritizing and scheduling inspections (from highest to lowest priority):

1. **Imminent danger** – Hazards that could cause death or serious physical harm receive top priority. Compliance officers will ask employers to correct these hazards immediately or remove endangered employees.

2. *Severe injuries and illnesses* – Employers must report:
   - All work-related fatalities within 8 hours
   - All work-related inpatient hospitalizations, amputations, or losses of an eye within 24 hours

3. *Worker complaints* – Allegations of hazards or violations also receive a high priority. Employees may request anonymity when they file complaints.
4. *Referrals* – Hazards noted by other federal, state, or local agencies, individuals, organizations or the media receive consideration for inspection.
5. *Targeted inspections* – Inspections aimed at specific high-hazard industries or individual work-places that have experienced high rates of injuries and illnesses also receive priority.
6. *Follow-up inspections* – Checks for abatement of violations cited during previous inspections are also conducted by the agency in certain circumstances.

Another reason for an inspection is if an OSHA inspector happens to see a potential violation as they are driving by or if they witness imminent danger.

In 2017, there were over 32,000 OSHA inspections. *Table 1* shows the inspection statistics for the years 2011 through 2016.

OSHA usually conducts inspections without giving any notice. If they do give notice, it will typically be less than 24 hours. In some cases, notice is required to ensure that key people will be at the site. An inspection can be delayed for good cause if it would produce a more effective inspection. In some cases, an inspection can be delayed until after business hours so that inspectors have access to all work areas.

> **NOTE**
> If OSHA sends a notice of inspection, it must be clearly posted in advance so that everyone on site is aware of the inspection.

Ensure that you are prepared to respond if an OSHA inspection does occur. This includes knowing individual duties and responsibilities, knowledge of established safety rules and procedures, and understanding how an OSHA inspection is conducted. For starters, before a CSHO ever sets foot on the job site, you should:

- Review the OSHA and state standards applicable to your location. Remember the more stringent standard always applies.
- Be aware of any national and local programs that are currently being emphasized by OSHA.
- Develop procedures for a visit and train employees on those procedures.
- Have records (300 logs, training records, etc.) readily available and up to date.
- Prepare a bag with essential items needed to address the CSHO, including a camera, checklist (see *Appendix*), company safety standards, and the OSHA standards.
- Make sure workers are properly trained on the safety requirements of the job site.

An OSHA inspector is required to present their credentials prior to beginning any inspection. Upon presentation of the credentials, the inspector will conduct an opening conference before inspecting the workplace. The decision to either consent to an inspection or demand a warrant should be made with legal counsel. During the opening conference, the CSHO will explain the purpose and scope of the inspection and obtain background information on the employer and the work being conducted.

You should have a company representative accompany the CSHO and take notes and photographs of the matters addressed during the inspection. An OSHA investigator should never be left unattended. Employers have the right to

Table 1 OSHA Inspection Statistics for 2011 thru 2016

OSHA Inspection Statistics	FY 2011	FY 2012	FY 2013	FY 2014	FY 2015	FY 2016
Total Inspections	40,614	40,961	39,228	36,163	35,820	31,948
Total Programmed Inspections	23,329	23,078	22,170	19,222	16,527	12,731
Total Unprogrammed Inspections	17,285	17,883	17,058	16,941	19,293	19,217
– Fatality/Catastrophe Investigations	851	900	826	850	912	890
– Complaints	8,765	9,573	9,505	9,570	9,037	8,870
– Referrals	4,776	4,864	4,024	3,829	4,705	4,641
– Referrals (Employer Reported)	NA	NA	NA	NA	1,864	2,050
– Other Unprogrammed Inspections	2,893	2,546	2,703	2,525	3,686	2,766

Notes:
Other unprogrammed inspections include Fatality/Catastrophe, Monitoring, Follow-Up, Unprogrammed Related, and Unprogrammed Other Inspections.
FY 2011 includes OSHA Information System (OIS) data and OSHA's Integrated Management Information System (IMIS) data that is of limited comparability to previous years that only include IMIS data.

ensure that everyone present on the job site, including OSHA, adheres to safety policies and procedures. This includes ensuring that the location, manner, and time of OSHA inspections do not interfere with safety and that appropriate personal protective equipment is utilized.

Representatives should ensure that they are accurate, truthful, and professional when communicating with OSHA. If a question is unclear, clarification should be sought. Do not offer opinions, make assumptions, or volunteer information.

2.1.0 The OSHA Inspection Process

An OSHA inspection is a formal process with five predefined steps. Recording each step of the inspection process is important to ensure all aspects of the inspection are recognized by the impacted stakeholders. The primary steps of an OSHA inspection should take place in the following order:

Step 1   Presentation of the inspector's credentials

Step 2   Opening conference

Step 3   Selection of representatives

Step 4   Walk-around inspection

Step 5   Closing conference

The following sections provide additional details for each of the steps in the OSHA inspection process. Here are several key points:

- Confirm the identity of the OSHA compliance officer and find out the scope of the inspection.
- The inspection will mostly entail a compliance officer taking photographs and notes, as well as asking questions of the employer and employees.
- The employer will learn from the compliance officer what hazards were found during the inspection and should fix those hazards as soon as possible.

2.1.1 Presentation of the Inspector's Credentials

When an OSHA inspector arrives, ask to see their credentials. These credentials will show the inspector's photo and serial number. Write down the inspector's name, serial number, and supervisor's name. Notify your office when the inspector arrives. The company legal counsel will determine if a search warrant is required.

Even though OSHA has the authority to enter and inspect a work site, the US Supreme Court said in the 1978 Barlow Decision that employers could invoke the Fourth Amendment of the US Constitution to prevent unwanted entry onto a site. When this is done, OSHA must get a search warrant before conducting an inspection. Few employers actually feel the need to exercise this right. Learning your company's policy on this issue will save valuable time for both your company and the inspector.

2.1.2 Opening Conference

After the inspector's credentials have been presented, the CSHO will explain the reason for the inspection and the applicable standards. The inspector will indicate whether it is a focused inspection or a wall-to-wall inspection. If it is a focused inspection or a complaint, take the CSHO directly to the affected area. Follow company policy for procedures. Always make detailed notes during the opening conference and request a copy of an employee complaint, if there is one. The complaint may be edited to remove the employee's name, if the employee requests it. Be sure that your OSHA posters are up, assured grounding materials are available, and hazard communication and safety programs are available for review.

2.1.3 Selection of Representatives

Certain people can go with the CSHO during an inspection. One person represents the employer. This can be the safety technician or manager. An employee representative also has the right to go along. The employee representative can be a member of the employees' union, trade association, or employee safety committee. On large construction sites, the CSHO may request representatives from all sub-contractors in addition to the general contractor. The best practice is to choose representatives before you are inspected. This will limit confusion, save time, and help you make sure that all parties are properly represented.

2.1.4 Walk-Around Inspection

After the opening conference, the CSHO and the representatives go through the workplace, inspecting for workplace hazards. The CSHO may talk with several employees. It must be a reasonable number, and work interruptions should be limited. Stop work on the job site if practical and the job allows—focus attention on the inspection. The potential for citations is if the workers are not working.

Workers are not required to discuss anything with the inspector. However, if they do, they cannot be punished because of anything they say or show the CSHO during the inspection.

The compliance officer will discuss any violations noted during the walk-around. Always make a note of every violation that the inspector points out. During this time, ask the CSHO for technical advice on how to correct the hazard. If possible, correct all violations on the spot. Be sure that the inspector notes your correction. You may still be cited for a violation even if it is corrected, but immediate correction shows **good faith** and can will help minimize fines.

The following tips should help to ensure that the inspection is done efficiently and effectively.

- Do not ask if something is or is not in compliance.
- Be polite.
- Emphasize the safety precautions taken on the project.
- If you see a violation, ask workers to correct it immediately.
- Take photos of the same items photographed by the inspector. Never allow the CSHO to stage a photo. Ask inspectors to be specific about what they are photographing and make sure photos are taken from multiple angles.

2.1.5 Closing Conference

The closing conference will review the hazards found on the site and the CSHO will review any apparent violations with the employer. Check each problem against the standard and ask for an explanation if it is not absolutely clear. Be sure to ask the CSHO how to correct the problem. If you disagree with the CSHO, be polite but explain your opinion. Discuss a possible time period to fix the problem and set realistic goals for doing so. If the problem is not fixed on time, the fines are increased.

The inspector's role is to enforce the rules, and it is important that you understand that perspective. The Directorate of Enforcement Programs (DEP) supports OSHA's mission of enforcing standards. Through policy vehicles such as Directives and Interpretations, DEP provides guidance on how CSHOs enforce OSHA standards and how employers are expected to comply with OSHA standards.

The CSHO must explain that violations may result in citations and penalties. He or she must also answer all questions about the inspection. Do not discuss proposed fines with the CSHO. He or she will not tell you how much the fines will be because the OSHA Area Director sets the fines after the CSHO has submitted a report. The CSHO should, however, inform you of your rights and responsibilities. Your rights and responsibilities are listed in the official OSHA booklet, Employer Rights and Responsibilities Following a Federal OSHA Inspection, *OSHA 3000-11R 2016*. The information in this booklet should be used as a discussion guide during your closing conference with the CSHO. For each apparent violation found during the inspection, the compliance officer has discussed or will discuss the following with you:

- Nature of the violation
- Possible abatement measures you may take to correct the condition
- Possible abatement dates you may be required to meet
- Any penalties that the area director may issue

2.2.0 Defining Violations, Citations, and Penalties

A *violation* is the act of breaking or not following any applicable OSHA standard. This includes the general duty clause for employers to provide a safe workplace. A *citation* is a formal notice that a violation has taken place. A *penalty* is a fine to discourage employers from repeat violations.

> **NOTE**
> You can be cited for having an unsafe workplace even if the recommended safety measures are not expressly written in the rules. Employers have a general duty to provide a safe workplace and can be cited for violating that duty. Use best practices to make sure you comply with the general duty clause to provide a safe workplace.

To avoid violations, be proactive and focus on prevention. Adopt a positive safety culture with a focus on safety and health management that will help prevent or reduce workplace injuries and illnesses.

2.2.1 Violations

You can be cited for violating any OSHA standard that applies to your job site. *Table 2* shows violation statistics for the years 2011 through 2016. The violations are categorized depending on their gravity and the willingness of the employer to correct them. The penalties increase for each group. These groups include the following:

- *Other-than-serious violations* – These are violations affecting safety and health, but they are unlikely to cause death or serious physical injury.
- *Serious violations* – These violations are likely to cause death or serious physical injury. The employer should or does know about the hazards causing the violation.

Think About It

OSHA's Top 10 List

According to OSHA's website (**www.osha.gov**), the following were the top 10 most frequently cited standards by Federal OSHA in fiscal year 2017:

1. Fall protection, construction (*29 CFR 1926.501*)
2. Hazard communication standard, general industry (*29 CFR 1910.1200*)
3. Scaffolding, general requirements, construction (*29 CFR 1926.451*)
4. Respiratory protection, general industry (*29 CFR 1910.134*)
5. Control of hazardous energy (lockout/tagout), general industry (*29 CFR 1910.147*)
6. Ladders, construction (*29 CFR 1926.1053*)
7. Powered industrial trucks, general industry (*29 CFR 1910.178*)
8. Machinery and Machine Guarding, general requirements (*29 CFR 1910.212*)
9. Fall Protection – Training Requirements (*29 CFR 1926.503*)
10. Electrical, wiring methods, components and equipment, general industry (*29 CFR 1910.305*)

How many of these standards apply to your work site?

- *Willful violations* – These violations involve those situations in which the employer intentionally and knowingly commits a violation of the standards, or is aware of the hazard and makes no reasonable effort to fix it. Willful violations are divided into repeat, failure to abate, and egregious violations. If you acknowledge that there is a violation, then it will be categorized as willful—seek legal advice and/or follow company policies before speaking to any violations. Be cautious when speaking or answering questions from the OSHA inspector to avoid any misunderstandings.
- *Repeated violations* – A repeat violation is a violation of any standard, regulation, rule, or order where, upon re-inspection, a similar violation is found. To be the basis of a repeat citation, the original citation must be final. A citation that is being contested may not serve as the basis for a subsequent repeat citation. Repeat violations mean the same rules are broken by one company. These problems could be on different job sites and at different times.
- *Failure-to-abate violations* – Failure-to-abate violations is a violation for failure to correct a previous violation OSHA has specifically cited on a particular job site.
- *Egregious violations* – When the hazard is so great it seems intentional, it is considered egregious. When the egregious policy is applied, the fines will be multiplied by the number of exposed workers. This is the most serious category.

2.2.2 Citations

A CSHO is required by law to issue citations for violations of safety and health standards. They are not allowed to issue warnings. In most cases, the citations are prepared by the OSHA Area Director and mailed to the employer. Citations include the following:

- A description of the violation and the applicable OSHA standard allegedly violated
- The proposed penalty
- The date by which the hazard must be corrected

Table 2 Federal OSHA Violations for 2011 thru 2016

OSHA Violation Statistics	FY 2011	FY 2012	FY 2013	FY 2014	FY 2015	FY 2016
Total Violations	85,514	78,723	78,186	67,680	64,763	58,702
Total Serious Violations	62,115	57,112	58,316	49,616	47,934	42,984
Total Willful Violations	594	423	319	439	527	524
Total Repeat Violations	3,229	3,034	3,139	2,966	3,088	3,146
Total Other-Than-Serious Violations	19,306	18,054	16,290	14,503	13,016	11,895

A copy of the citation must be posted at or near the place the violation occurred. The notice must be posted for at least three working days or until the problem is corrected—whichever is longer. You must correct the problems listed in the citation. Inspectors may come back to check that they have been corrected. If the problems are not corrected on time, stronger penalties apply.

An OSHA citation must be taken care of right away. The employer may request an informal conference with the OSHA Area Director. The purpose of this conference is to resolve any issues arising from the inspection. The employer may provide additional information at the conference. At the informal conference, the parties can enter into a settlement agreement that can revise the citations and penalties.

If the issues are not resolved at the informal conference, the employer may appeal the citation. Employers have 15 working days to appeal a citation. They must file an intention to contest the OSHA citation before the independent OSHA Review Commission. Employers should hire an attorney or other legal counsel if they decide to fight an OSHA citation.

> **NOTE**
> Always notify your main office of an OSHA inspection and any citations. The Safety Manager or Director in conjunction with senior management will make a decision about contesting citations. They will also notify other job sites about any citations. Similar conditions on other job sites can be considered repeat violations.

2.2.3 Penalties

OSHA will impose fines on companies to discourage them from breaking the rules. More than $140 million in OSHA fines are collected each year. A good safety program can help your company avoid these penalties.

OSHA regulations set specific penalties for violations, as shown in *Table 3*. Note that the Bipartisan Budget Act of 2015 required the Department of Labor to adjust its civil monetary penalties to account for inflation. These amounts may be adjusted by the Area Director. Some penalties may be lowered for good faith compliance. Good faith compliance means that the employer made a serious effort to follow the rules to the best of their ability (i.e., made a good faith effort to comply).

To show good faith, you must have a written safety and health program. The program must be an active part of your daily operations. OSHA has voluntary guidelines that include required programs, such as Hazard Communication, Lockout/Tagout, and other programs for construction. Other trade associations and safety organizations have sample programs that you can use. You must adjust these programs to fit the safety needs of your company.

If you are cited for OSHA violations following an inspection, penalties may vary depending on the type of citation. Civil penalties for OSHA violations are based on the gravity of the violation. The gravity of a violation is the primary consideration when calculating penalties and is established by assessing the severity of the injury/illness that could result from a hazard and the probability of an injury or illness occurring. Additional violations for which citations and proposed penalties may be issued upon conviction include the following:

- Falsifying records, reports, or applications can bring a fine of $10,000 or up to six months in jail, or both.
- Violations of posting requirements can bring a civil penalty of up to $7,000.
- Assaulting a compliance officer, or otherwise resisting, opposing, intimidating, or interfering with compliance officers while they are engaged in the performance of their duties is a criminal offense, subject to a fine of not more than $5,000 and imprisonment for not more than three years.

In an extreme case, you can go to jail. Willful violations can bring criminal charges. If an employee died and the employer is convicted of a willful violation, he or she faces a fine or imprisonment for up to six months, or both. Criminal convictions have stiff penalties. They can be up to $250,000 for an individual, or $500,000 for a corporation.

OSHA will refer a criminal case to the Criminal Division of the Department of Justice (DOJ). The DOJ will review the case and determine if prosecution is reasonable. The case will be referred to the appropriate US Attorney, who will bring criminal charges. The case will be heard in federal court, not by a review board or an administrative law judge. To obtain a criminal conviction, the Department of Justice must prove the following beyond a reasonable doubt:

- Violation of a specific OSHA standard (not a general duty violation)
- Committed by employer
- Violation conduct was direct cause of employee's death
- A willful violation

Table 3 OSHA Violations and Penalties

Violation Type	Description	Penalties
Willful	A willful violation exists under the OSH Act where an employer has demonstrated either an intentional disregard for the requirements of the act or a plain indifference to employee safety and health.	Penalties range from $5,000 to $70,000 per willful violation. If an employer is convicted of a willful violation of a standard that has resulted in the death of an employee, the offense is punishable by a court-imposed fine or by imprisonment for up to 6 months, or both. A fine of up to $250,000 for an individual, or $500,000 for a corporation, may be imposed for a criminal conviction.
Serious	Section 17(k) of the OSH Act provides that "a serious violation shall be deemed to exist in a place of employment if there is a substantial probability that death or serious physical harm could result from a condition which exists, or from one or more practices, means, methods, operations, or processes which have been adopted or are in use, in such place of employment unless the employer did not, and could not with the exercise of reasonable diligence, know of the presence of the violation."	OSHA may propose a penalty of up to $7,000 for each violation.
Other-Than-Serious	This type of violation is cited in situations where the accident/incident or illness that would be most likely to result from a hazardous condition would probably not cause death or serious physical harm, but would have a direct and immediate relationship to the safety and health of employees.	OSHA may propose a penalty of up to $7,000 for each violation.
De Minimis	De minimis conditions are those where an employer has implemented a measure different from one specified in a standard, that has no direct or immediate relationship to safety or health.	These conditions do not result in citations or penalties.
Failure to Abate	A failure to abate violation exists when a previously cited hazardous condition, practice or non-complying equipment has not been brought into compliance since the prior inspection (i.e., the violation remains continuously uncorrected) and is discovered at a later inspection. If, however, the violation was corrected, but later reoccurs, the subsequent occurrence is a repeated violation.	OSHA may propose a penalty of up to $7,000 for each violation.
Repeated	An employer may be cited for a repeated violation if that employer has been cited previously, within the last five years, for the same or a substantially similar condition or hazard and the citation has become a final order of the Occupational Safety and Health Review Commission (OSHRC). A citation may become a final order by operation of law when an employer does not contest the citation, or pursuant to court decision or settlement.	Repeated violations can bring a civil penalty of up to $70,000 for each violation.

Criminal enforcement under the OSH Act does occur, with 84 cases having been prosecuted since 1970, with some defendants serving prison sentences. In 2012, there were 13 cases referred for possible criminal prosecution.

Twenty-four states and Guam are covered under the federal OSHA program. Twenty-six states, along with Puerto Rico and the US Virgin Islands, have their own OSHA programs.

> **NOTE**
> For the construction industry federal OSHA reported 31,955 citations, 12,766 inspections, and penalties of $65,210,254 during the period October 2014 through September 2015. Penalties reflect current rather than initial amounts.

The violations and penalties assessed by state OSHA programs can be a significant source of large fines. *Table 4* highlights six states in 2015 that have levied six-figure penalties totaling about $2.9 million against 10 companies for alleged safety and health violations.

It is important to recognize that many states have as much authority as federal OSHA. Make sure to monitor both federal and state occupational safety and health enforcement practices. Always adhere to the most stringent requirements.

For both federal and state OSHA programs, willful violations can be significantly larger than serious violations. In some cases, a penalty for a willful violation may be more than 30 times higher than a serious violation. For example, the average state penalty issued in 2012 for a serious violation was $974 compared to an average of $35,744 for a willful violation. The difference between a willful and a serious violation is the employer's willingness to make a reasonable effort to fix hazardous conditions. A frequent violator is much more likely to be cited for a willful violation. A good safety program can prevent expensive willful violation citations.

2.3.0 Focused vs. Wall-to-Wall Inspections

OSHA launched the Focused Inspections Initiative to recognize responsible contractors who have effective safety and health programs. The inspections that take place under this program are very different from regular OSHA inspections. The new policy aims for an overall improvement in construction job-site safety.

Previously, construction inspections were all-inclusive (wall-to-wall). The CSHO would try to find all the violations on a construction site. As a result, CSHOs spent too much time and effort on a few projects. The contractor was likely to be cited for many violations, but these violations were not the cause of construction fatalities. The Focused Inspections Initiative directs CSHOs to spend more time looking for hazards that are most likely to cause fatalities and serious injuries to workers. OSHA has determined that this approach is effective in protecting overall worker safety.

2.3.1 Program Qualifications

A CSHO will decide if a project qualifies for a focused inspection by reviewing the project's safety and health plan. In order to qualify, the project must have a written safety and health program or plan, and a designated person responsible for and capable of implementing the program or plan.
The CSHO must note in each case file why a focused inspection was or was not conducted. The CSHO may use a written guideline to determine if a project qualifies for the program (*Figure 8*). Ask for a copy of the completed form at the time of the inspection. All contractors and employee representatives must be informed why a focused or a comprehensive (wall-to-wall) inspection is being conducted. This may be accomplished either by personal contact or by posting the guidelines. A request for a search warrant should not affect the determination as to whether a project will receive a focused inspection.

Table 4 Significant State OSHA Violations and Penalties Issued in 2015

State	General Description	Penalties
California	A meat byproducts processing company, a door manufacturer, a refinery, and two construction firms with violations	$1.6 million
Michigan	A foundry	$638,450
Washington	A bulk shipping firm	$218,450
New Mexico	A water utility	$144,000
North Carolina	A company that regulates and administers transportation programs	$140,000
Indiana	A shipyard	$112,500

Case History

Two cases from the second half of 2012 are recent examples of criminal prosecutions under the OSH Act:

- In July 2012, a federal grand jury in Texas indicted an environmental services company and its former president on conspiracy charges for illegally transporting hazardous materials that resulted in the death of two employees. The 13-count indictment alleged Port Arthur Chemical and Environmental Services and its former president willfully failed to provide protective measures to limit employees' exposure to hydrogen sulfide.
- Two months later, four former managers from Atlantic States Cast Iron Pipe Company lost their appeal of prison sentences and fines in connection with several safety and environmental violations and the resulting cover-up. Atlantic was placed on four years' probation and required to pay an $8 million fine. Jail sentences ranging from six to 70 months (almost six years) were upheld.

Since penalties are larger for EPA violations than those involving the OSH Act, prosecutors often use the environmental infractions to increase penalties.

If the OSHA inspector is investigating a complaint or incident, the CSHO may choose to conduct a focused inspection. If the CSHO observes no coordination by the general contractor, prime contractor, or other such entity to ensure that all employers provide adequate protection for their employees, a comprehensive inspection will be conducted.

2.3.2 Effective Written Safety Program

An effective safety and health plan must meet the requirements of 29 *CFR* 1926 Subpart C, *General Safety and Health Provisions*. Subpart C contains the basic requirements of a safety and health program on construction sites. It also contains the definitions for key terms used in the construction standards, such as *competent person*, *qualified person*, *approved*, and *suitable*. The plan must include provisions for the following:

- Project safety analysis
- Evaluation of subcontractors to conform to the plan
- Supervision and training
- Control of hazardous operations
- Documentation
- Employee involvement
- Emergency response

The plan must also be assessed based on the size and complexity of the project. In addition, the employer must have a written and implemented safety and health program such as OSHA's voluntary Safety and Health Management Guidelines. This includes programs required under the OSHA standards, such as Hazard Communication, Lockout/Tagout, and other programs for construction.

Some employers and employees may also be asked to participate in OSHA's Voluntary Protection Programs (VPP). In the VPP, management, labor, and OSHA establish cooperative relationships at workplaces that have implemented a comprehensive safety and health management system. Approval into the VPP is OSHA's official recognition of the outstanding efforts of employers and employees with regard to occupational safety and health. The VPP is designed to do the following:

- Recognize outstanding achievement of those who have successfully incorporated comprehensive safety and health programs into their total management system
- Motivate others to achieve excellent safety and health results in the same way
- Establish a relationship between employers, employees, and OSHA that is based on cooperation rather than coercion

Statistical evidence for the success of the VPP is noteworthy. The average VPP worksite has a Days Away, Restricted, or Transferred (DART) case rate of 52 percent below the industry average. These sites typically do not start out with such low rates. Reductions in injuries and illnesses begin when the site commits to the VPP approach to safety and health management and the challenging VPP application process. Fewer injuries and illnesses mean greater profits as workers' compensation premiums and other costs plummet. Entire industries benefit as VPP sites evolve into models of excellence and influence practices industry-wide. VPP participation can also lead to lower employee turnover and increased productivity and cost savings.

CONSTRUCTION FOCUSED INSPECTION GUIDELINE

This guideline is to assist the professional judgment of the compliance officer to determine if there is an effective project plan, to qualify for a Focused Inspection.

	Yes/No
PROJECT SAFETY AND HEALTH COORDINATION; are there procedures in place by the general contractor, prime contractor or other such entity to ensure that all employers provide adequate protection for their employees?	
Is there a **DESIGNATED COMPETENT PERSON** responsible for the implementation and monitoring of the project safety and health plan who is capable of identifying existing and predicable hazards and has authority to take prompt corrective measures?	
PROJECT SAFETY AND HEALTH PROGRAM/PLAN* that complies with 1926 Subpart C and addresses, based upon the size and complexity of the project, the following:	
Project Safety Analysis at initiation and at critical stages that describes the sequence, procedures, and responsible individuals for safe construction.	
Identification of work/activities requiring planning, design, inspection or supervision by an engineer, competent person or other professional.	
Evaluation/monitoring of subcontractors to determine conformance with the project Plan. (The Project Plan may include, or be utilized by subcontractors.)	
Supervisor and employee training according to the Project Plan including recognition, reporting and avoidance of hazards, and applicable standards.	
Procedures for controlling hazardous operations such as : cranes, scaffolding, trenches, confined spaces, hot work explosives, hazardous materials, leading edges, etc.	
Documentation of: training, permits, hazard reports, inspections, uncorrected hazards, incidents and near misses.	
Employee involvement in hazard: analysis, prevention, avoidance, correction and reporting.	
Project emergency response plan.	
*FOR EXAMPLES, SEE OWNER AND CONTRACTOR ASSOCIATION MODEL PROGRAMS, ANSI A10.33, A10.38, ETC.	
The walk-around and interviews confirmed that the Plan has been implemented, including:	
The four leading hazards are addressed: falls, struck by, caught in/between, electrical.	
Hazards are identified and corrected with preventative measures instituted in a timely manner.	
Employees and supervisors are knowledgeable of the project safety and health plan, avoidance of hazards, applicable standards, and their rights and responsibilities.	
THE PROJECT QUALIFIED FOR A FOCUSED INSPECTION	

Figure 8 The Construction Focused Inspection Guideline.

The OSHA Special Government Employees (SGE) program offers private and public sector safety and health professionals and other qualified participants the opportunity to exchange ideas, gain new perspectives, and grow professionally while serving on OSHA's VPP onsite evaluation teams. If you are employed at a VPP site, see how you can help OSHA and VPP while gaining valuable experience.

2.3.3 Competent Person

The focused inspection program requires that a competent person be responsible for implementation and monitoring of the safety plan. According to OSHA, a competent person is one who is capable of identifying existing and predictable hazards in the surroundings or working conditions which are unsanitary, hazardous, or dangerous to employees, and who has authorization to take prompt corrective measures to eliminate them. The on-site safety technician may fulfill this role with the support of additional safety managers as needed.

2.3.4 The "Fatal Four" Leading Hazards

The inspection will focus on the four leading hazards that cause the majority of deaths on construction job sites. The four major causes of death are:

- Falling from elevations
- Being struck by falling objects or vehicles
- Being caught in/between machinery, equipment, or soil
- Receiving an electrical shock

The inspector will focus attention on hazardous conditions that cause these types of incidents. For example, the CSHO will check possible sources of electrical shock, overhead power lines, power tools and cords, outlets, and temporary wiring.

Your safety program should cover all of these hazards or determine that they do not apply to your work site. You must be able to show the CSHO that you have training programs in place to deal with these hazards. The workers must be following the job safety programs.

VPP Star Mobile Workforce Participant Brandenburg Completes Wrigley Field Demolition Project

Major League Baseball's Chicago Cubs have started an extensive renovation project at the Cubs' historic Wrigley Field, a designated Chicago landmark. The number one goal expressed by both Wrigley Field Management and the general contractor was safety. Brandenburg Industrial Service Company was selected to demolish and remove the outfield wall and bleachers based on its commitment to providing a safe work environment. Brandenburg's history of participation in OSHA's cooperative programs, the OSHA Challenge, and subsequently the Voluntary Protection Program, was integral to the project and achieving the desired level of safety.

Brandenburg held daily safety meetings and Pepper's Safety Manager addressed potential work hazards, as well as the adjacent contractor activities, all of which made for excellent communication for the workers and the overall project. Additionally, Brandenburg's Safety Manager conducted weekly safety meetings addressing the upcoming work activities, as well as several weekly safety audits. As part of Brandenburg's VPP Best Practices, each Brandenburg worker completes a Daily Safety Task Analysis Card, which is used as a reminder for the associated hazards, preventive actions, and PPE requirements for the individual task being performed. This commitment to safety resulted in zero injuries or illnesses during the project, which lasted approximately 1,400 total man-hours.

> ### Construction's Fatal Four
>
> Out of 4,693 worker fatalities in private industry in calendar year (CY) 2016, 991 or 21.1 percent were in construction—that is, one in five worker deaths last year were in construction. The leading causes of worker deaths on construction sites were falls, followed by struck by object, electrocution, and caught-in/between. These hazards, known as the *Fatal Four*, were responsible for more than half (63.7 percent) the construction worker deaths in 2016, BLS reports. Eliminating the Fatal Four would save 631 workers' lives in America every year.
>
> - Falls – 384 out of 991 total deaths in construction in CY 2016 (38.7 percent)
> - Struck by object – 93 (9.4 percent)
> - Electrocutions – 82 (8.3 percent)
> - Caught-in/between – 72 (7.3 percent)

2.3.5 Modified Inspection Procedure

If your company qualifies for the Focused Inspection Program, the inspection process will be modified. The inspection will include the same five steps that are used in a regular inspection (as follows):

Step 1 Presentation of the inspector's credentials

Step 2 Opening conference

Step 3 Selection of representatives

Step 4 Walk-around inspection

Step 5 Closing conference

The difference between the regular inspection and the modified inspection is that the opening conference will include a broader review of safety documentation, and the walk-around inspection is more limited.

Review of the site safety and health plan and supporting documentation is a critical step in the inspection process. The inspector reviews all appropriate documentation, including the log of injuries and illness, which is required when jobs are scheduled for more than 12 months or have more than 10 employees. A good safety program will help you avoid violations and citations. Always follow company policy for OSHA inspections.

The walk-around inspection for a modified procedure is limited. Ask what the inspector wants to see and take them directly to that area of the job site. The CSHO will check that the safety and health plan is in force through interviews and observations. They will inspect for the four leading hazards and other serious hazards. Workers must be following standards or best practices listed in the safety plan. Inspectors may interview workers to see if they know the safety rules. The CSHO is not required to inspect the entire project. Inspectors should be escorted directly to the part of the site they request to see and a job site representative is allowed to be present during employee interviews.

If the CSHO sees that workers are not following the safety program, they will stop the focused inspection. The company then must face the usual wall-to-wall inspection. A serious violation does not automatically trigger a wall-to-wall inspection. It is up to the CSHO to use professional judgment and decide.

Although the walk-around inspection will focus on the four leading hazards, citations will be issued for any serious violations. Citations will also be issued for other-than-serious violations that are not fixed immediately. Other-than-serious violations that are corrected will not usually be cited or documented.

After the walk-around, there is a post-inspection conference. This is similar to the regular inspection process just described. The CSHO will inform you of any violations. Ask for the OSHA booklet that explains your rights and responsibilities.

2.4.0 Post-Inspection Follow-Up

If a violation is found by the CSHO, you must record and correct the problem. If the problem is not fixed right away, the CSHO will likely schedule another inspection. Always correct violations immediately whenever possible. Normal follow-up to an inspection includes the following:

- A report to management
- A plan to correct any problems
- A date for reinspection

The best way to prevent repeat violations is to do a self-inspection before OSHA returns.

> **NOTE**
>
> All violations must be corrected by the date set in the citation. If corrections are not made, the fines will be increased according to the delay.

2.4.1 Reports to Management and Employees

A written report must be prepared as soon as possible so that nothing is overlooked. The report should give management a record of what the inspection covered. It should mirror what the CSHO has on file, including the following:

- Documents reviewed by the CSHO
- Photographs
- A summary of employee statements
- Noted violations
- Violations corrected on the spot
- Any outstanding violations noted by the CSHO

Just as with an all-inclusive inspection, violations stemming from focused inspections must be corrected or appealed within 15 working days from receipt of the citation. If there are items that were not fixed, a plan must be made to correct them. Safety training or more equipment may be needed.

When a citation is received, it must be posted until the violation is corrected, or for three working days—whichever is longer. If the employer chooses to contest, the notice to employees of contest must be posted at the same location for three days. Any settlement agreements must also be posted.

2.4.2 Abatement and Abatement Certification

Abatement is the correction of the safety hazards or violations that led to an OSHA citation. Abatement certification is a written record that states that the problem noted in the citation has been fixed. If the hazard is corrected during an OSHA inspection, an abatement certification does not need to be filed. Abatement certification can be a simple one-page letter or a more complex plan. In either case, each of the violations must be listed, along with the corrective action.

Employers must submit an abatement certification to OSHA within 10 working days after the abatement date. The abatement date is set in the citation. This date is the deadline to correct the violation.

> **NOTE:** The date on the citation can only be changed by OSHA or the court.

2.4.3 Corrective Actions

Corrective actions are steps taken to fix the problem. Corrective actions can include training or retraining workers on safety procedures, or establishing toolbox talks. They may also require additional safety equipment, signs, or personal protective equipment. Corrective actions must be recorded. Keeping records of these actions helps provide proof that the needed changes have been made. This makes the OSHA's re-inspection process faster and more efficient.

Employees have a right to know of any corrective action or abatement plans. The documents must be posted with the other notices. Involve employees in correcting safety problems. This will help ensure that every worker becomes responsible for job-site safety. Some employers place notices in paycheck envelopes.

> **NOTE:** Failure to correct a violation has a penalty of up to $7,000 for every day the violation continues beyond the abatement date.

Make sure violations are corrected in a timely manner to prevent the initial fine from increasing. Daily accrual of penalties can add up quickly and become very costly if not addressed. For example, missing an abatement day by only five days can increase the initial $7,000 penalty to $42,000.

2.4.4 Informal Conferences, Appeals, and Hearings

If the employer believes that the citation is unreasonable, they may ask for an informal conference. This must happen within 15 working days of receipt of the citation. The informal conference is held to work out any disagreements. OSHA and the employer may make a settlement agreement to resolve the dispute. The primary goal is to protect workers and eliminate hazards. The settlement agreement must be posted.

Employers who are not satisfied after the informal hearing may request a formal hearing or appeal. This must be done by filing a Notice of Contest within 15 working days from receipt of the citation. The matter is then assigned to an administrative law judge who will formally review the case. At the appeal, the employer must be able to show why the citation was unreasonable.

2.4.5 Affirmative Defenses

At the appeal hearing, the employer can present an affirmative defense. An affirmative defense is an explanation or reason for the condition cited as a violation. An affirmative defense excuses the employer from a violation that has otherwise been proven by the CSHO. The employer must prove all aspects of the affirmative defense at the appeal hearing. Common affirmative defenses include the following:

- *Unpreventable employee misconduct* – To show unpreventable employee misconduct, the employer must show that the hazard was unknown to the employer and in violation of a safety policy that was effectively communicated and uniformly enforced.
- *Infeasibility of compliance* – Infeasible means that compliance with the regulation is functionally impossible. The employer must show that compliance with the standard is infeasible or would prevent performance of required work. The employer must also prove that there is no alternative means of employee protection.
- *Compliance poses a greater hazard* – Another affirmative defense is that following the standard would be a greater hazard to employees than noncompliance. The employer must prove that compliance would be more dangerous and that there are no alternative means of employee protection. The employer must also prove that the application of a variance from the standard is inappropriate.
- *Multi-employer work site* – A multi-employer work site is a job site where several employers are working. This is common in large construction projects. During an inspection, several employers can be cited for violations. These include the following:
  - The employer whose employees are exposed to the hazard
  - The employer who actually created the hazard
  - The employer who is responsible, by contract or actual practice, for safety and health conditions at the work site
  - The employer who is responsible for correcting the hazard
- To prove an affirmative defense of a multi-employer work site, the employer must prove all five of the following elements:
  - The employer did not create the hazard.
  - The employer did not have the responsibility or authority to remove the hazard.
  - The employer did not have the ability to correct or remove the hazard.
  - The employer can demonstrate that those responsible for controlling or correcting the hazard have been specifically notified of the hazard to which the employees are/were exposed.
  - The employer has instructed the employees to recognize the hazard and, where necessary, informed them how to avoid the dangers associated with it.
    ~ Where feasible, an exposing employer must have taken appropriate alternative means of protecting employees from the hazard.
    ~ When extreme circumstances justify it, the exposing employer shall have removed the employees from the job to avoid citation.

Did You Know?

Voluntary Consultation Service

OSHA offers a voluntary compliance assistance program to small and medium sized businesses. Any firm with 250 or fewer employees may receive free assistance from OSHA to identify workplace hazards. A full-service consultation is available to help employers establish effective workplace health and safety programs. It can cover the entire work site or be limited to a few specific problems. The compliance assistance program is similar to an inspection in that it can include an opening conference, a walk-around, and a closing conference. The benefit is that no citations are issued. It will help you to recognize the hazards in your workplace. The Construction Health and Safety Officer (CHSO) will suggest options for improving workplace safety. The CHSO will also suggest possible sources for further assistance or training. Under certain circumstance, your firm can be granted a one-year exemption from general enforcement inspections.

Additional Resources

Employer Rights and Responsibilities Following a Federal OSHA Inspection (OSHA 3000-11R 2016), US Department of Labor. Available at www.osha.gov/Publications/osha3000.pdf

The following websites offer resources for products and training:

American Society of Safety Engineers (ASSE), **www.asse.org**

Occupational Safety and Health Administration (OSHA), **www.osha.gov**

2.0.0 Section Review

1. When an OSHA inspector arrives, ask to see their _____.
   a. passport
   b. warrant
   c. credentials
   d. citation list

2. A violation is the act of breaking or not following any applicable _____.
   a. OSHA standard
   b. best practice
   c. safety recommendation
   d. OPM directive

3. An effective safety and health plan must meet the requirements of _____.
   a. *26 CFR 1910 Subpart A*
   b. *27 CFR 1926 Subpart B*
   c. *29 CFR 1910 Subpart C*
   d. *29 CFR 1926 Subpart C*

4. If the CSHO finds a violation, it is best to _____.
   a. wait until management decides if it must be corrected
   b. correct it immediately whenever possible
   c. wait for the official citation notice before making any changes
   d. appeal the citation before making any corrections

SUMMARY

Good recordkeeping is essential to an effective safety and health program. Without good records, it is difficult to analyze safety and health data and take the corrective actions to eliminate hazards. 29 *CFR* 1904, the recordkeeping rule, gives the requirements and criteria necessary for recording illness and injury information promptly and accurately. As a safety technician, you should have a good working knowledge of regulatory and company recordkeeping requirements.

An OSHA inspection is a formal process. Recording each step of the inspection process is important. Both management and employees must be notified at certain points in the process, and an employer representative must be present. Any violations found in an inspection should be corrected immediately to reduce potential fines. Your best defense against hazards, citations, and fines is a good safety program.

Many different circumstances can prompt an OSHA inspection, ranging from a workplace death to mere chance. Establishing a predefined plan for addressing an OSHA inspection will ensure you are ready before the OSHA inspector arrives. OSHA offers many resources online and in written booklets to help you comply with safety regulations. They also offer a free consultation service that can help you avoid expensive fines and increase the safety of your job site.

If you have a written safety program and a safety person on site, you could qualify for a focused inspection, which allows inspectors to focus on the most hazardous conditions in the construction industry rather than smaller, less important issues.

Review Questions

1. OSHA uses specific illness and injury information that is reported to them as part of the agency's site-specific _____.
   a. violation prevention program
   b. fine reduction process
   c. inspection targeting program
   d. citation review process

2. When classifying illnesses and injuries, it is important to understand which types of injuries and illnesses are recordable so that they are _____.
   a. recognized and avoided
   b. recorded correctly
   c. prevented and minimized
   d. penalized accordingly

3. OSHA defines medical treatment as managing and caring for a patient for the purpose of _____.
   a. eliminating liability
   b. preventing re-occurrence
   c. aggravating preconditions
   d. fighting disease or disorder

4. First aid is emergency care or treatment given to an ill or injured person _____.
   a. before regular medical aid can be given
   b. after the wound has healed
   c. after medical treatment has been given
   d. during the ambulance ride

5. The number of lost and restricted work days for an injured employee is calculated by counting calendar days from the day _____.
   a. before the incident occurs
   b. after the employee hire date
   c. after the incident occurs
   d. after the doctor visit

6. A breathing condition triggered by inhaling airborne particles in the workplace is occupational _____.
   a. sclerosis
   b. asthma
   c. lupus
   d. pneumonia

7. Heatstroke and sunstroke are illnesses that are commonly classified as _____.
   a. occupational hazards
   b. weather diseases
   c. conditional issues
   d. occupational illnesses

8. Skin diseases or disorders are illnesses involving a worker's skin that are caused by _____.
   a. poor personal hygiene
   b. excessive use of protective equipment
   c. exposure to chemicals, plants, or other substances
   d. improper use of safety clothes

9. OSHA Form 300, the Log of Work-Related Injuries and Illnesses, is used to record specific details about work-related _____.
   a. conditions and problems
   b. violations and penalties
   c. illnesses and injury
   d. citations and events

10. OSHA inspections should never be taken lightly because OSHA can levy large fines if _____.
    a. work is difficult
    b. competitors complain
    c. funding is withheld
    d. violations are discovered

11. Imminent danger situations include hazards that could cause death or serious physical harm and receive _____.
    a. top OSHA priority
    b. very little consideration
    c. the same priority as other situations
    d. low OSHA priority

12. Employers have a general duty to provide a safe workplace and can be cited for _____.
    a. random reasons
    b. declaring exceptions
    c. violating that duty
    d. redefining rules

13. The Focused Inspections Initiative directs CSHOs to spend more time looking for hazards that are most likely to _____.
    a. cause repeat violations
    b. reduce the cost of penalties
    c. increase the cost of penalties
    d. cause fatalities and serious injuries to workers

14. If your company qualifies for the Focused Inspection Program, the inspection process will be _____.
    a. modified
    b. eliminated
    c. enhanced
    d. waived

15. All violations stemming from a focused inspection must be corrected or appealed _____.
    a. by the first Monday of the month
    b. within 15 working days from receipt of the citation
    c. before the date selected by the onsite safety manager
    d. before the end of the first full week following the inspection

Trade Terms Introduced in This Module

Abatement: The correction of the safety hazards or violations resulting from an OSHA citation.

Abatement certification: A written record that shows that the safety hazard or standard violation noted in the citation has been corrected.

Antigens: Toxins or enzymes capable of triggering an immune response similar to an allergy.

Calendar year: The period of a year beginning on January 1 and ending on December 31.

Chronic irreversible disease: A long-lasting, permanent disease.

Citation: A formal notice from OSHA that the company has allegedly violated a federal safety regulation and may be fined.

Competent person: A person who is capable of identifying existing and predictable hazards, and who has authority to take prompt corrective measures to eliminate them.

Compliance safety and health officers (CSHOs): OSHA officials who conduct work-site inspections to enforce safety regulations.

Contact dermatitis: An inflammation of the skin.

Corrective action: Actions the company must take to correct safety violations.

Eczema: An inflammatory condition of the skin characterized by redness, itching, and oozing sores that become scaly, crusted, or hardened.

Good faith: Good faith means that the employer makes a serious effort to comply with relevant OSHA standards to the best of his or her ability.

Hypothermia: A life-threatening medical condition that can happen in cold weather when the body's temperature becomes too low.

Infectious material: Biological materials that carry contagious diseases and illnesses.

Imminent danger: A hazard that could cause death or serious physical harm immediately, or before the danger could be eliminated through normal enforcement procedures.

Penalty: A fine imposed on a company for violating a federal or state safety regulation. Penalties can also include prison time for criminal actions.

Recordable: An illness or injury is classified as recordable if it meets certain OSHA reporting guidelines.

Restricted work activity: Work activity assigned to those workers who have experienced a work-related injury or illness that prevents them from doing the routine functions of their jobs or from working a full workday.

Ulcers: Breaks in skin or mucous membrane that fester and erupt like open sores.

Violation: The act of violating federal or state safety regulations by failing to obey a safety standard or failing to keep the workplace free from hazards.

Appendix

EXAMPLE OSHA INSPECTION CHECKLIST PROVIDED AS A GENERAL GUIDE

Checklist for OSHA Compliance Inspections

Company Name: _____ Date: _____ Time: _____

Name of person using checklist: _____

Inspectors have statutory authority to: ○ Arrive unannounced ○ Enter without delay and at reasonable times ○ Inspect and investigate the workplace: 　■ during regular working hours 　■ at other reasonable times 　■ within reasonable limits and in a reasonable manner	○ Question privately any employee or employer ○ Other provisions ● Confidentiality—Names of complaints can be kept confidential ● Participation in inspection

ARRIVAL / OPENING CONFERENCE

Credentials: A person states their intention to conduct an occupational safety inspection of your farm. Ask this person for their credentials.

● Federal Credentials　　　　　　　　　　　　　　　　　　　　　Yes ☐　　　No ☐

Name of Compliance Safety and Health Officer (CSHO) _____

If credentials are acceptable, proceed to next item. To verify credentials, call area OSHA Director. See OSHA contacts on next page.

Purpose and Scope of Inspection

What is the impetus for the inspection?　☐ Employee complaint　☐ Program Inspection　☐ Referral
　　　　　　　　　　　　　　　　　　　　☐ Other (describe) _____

NOTE: Ask to see employee complaint or referral. Attach photocopy to your final notes. Inspector's failure to provide details of employee complaint (other than identification of employee) may be cause for appeal.

Contact phone number (s) of additional farm management team or individual responsible for safety program to be involved in the inspection process.

Ask the inspector what is the purpose and intended scope of the inspection (provide summary).

Employee Participation

With above information on purpose and scope of inspection, consult with the CSHO as to appropriate employee representation. If necessary, contact employee representative to attend the inspection.
Summarize the agreement regarding employee participation in the inspection.

Miscellaneous Items

● Plan and state your proposed route of inspection that will cover the purpose and scope of inspection.
● Gather up notebook, checklist, camera, two-way radio or cell-phone, and list of farm management team.

Figure A01A　OSHA inspection checklist (page 1 of 2).

ON-SITE INSPECTION	
Records and written programs: Examples of items you should be prepared to show.	
• OSHA 300 logs • HazCom program, MSDS records • Employee training records	• Confined space programs • Lockout/tagout • Respiratory protection program
Notes, photos, and measurements	
• <u>Notes</u>–Names of people participation in on-site inspection, times, places visited, CSHO's comments, names of people spoken to, your observations, etc. • <u>Corrections</u>–Where possible, immediately correct violations pointed out by the CSHO. Make a note and take a photo of your actions. • <u>Photos</u>–If the CSHO takes a photo, you take the same photo. Ask CSHO why the photo was taken. • <u>Measurements</u>–Take any measurement taken by CSHO, or ask for copy or reading.	
CONCLUDING THE INSPECTION	
Closing conference: At the conclusion of the on-site inspection, ask for a closing conference.	
At the closing conference, allow the CSHO to address their findings. Take careful notes on their statements at the closing conference. If you are less than completely clear about their findings, restate your understanding of their findings to the CSHO for agreement. If they have not addressed the following issues, be sure to ask for answers • What are the alleged violations? • What are the CSHO's next steps in the process? • Will there be further on-site inspection prior to issuance of any citations or 'decision not to issue'? • When can your farm expect to receive any 'decision not to issue' or citations?	
After the CSHO departs	
• Formalize your notes, photos, and measurements.	

Figure A01B OSHA inspection checklist (page 2 of 2).

Additional Resources

This module presents thorough resources for task training. The following resource material is suggested for further study.

Employer Rights and Responsibilities Following a Federal OSHA Inspection (OSHA 3000-11R 2016), US Department of Labor. Available at **www.osha.gov/Publications/osha3000.pdf**

OSHA 2002 Recordkeeping Simplified, James E. Roughton, CRSP, CHMM, CSP. 2002. Burlington MA: Butterworth-Heinemann.

The following websites offer resources for products and training:

American Society of Safety Engineers (ASSE), **www.asse.org**

Mine Safety and Health Administration (MSHA), **www.msha.gov**

Occupational Safety and Health Administration (OSHA), **www.osha.gov**

Figure Credits

MSA The Safety Company, Module opener

Wright-Patterson Air Force Base, 88th Air Base Wing Public Affairs, Figure 1

U.S. Department of Labor, Figures 3–6, 8, Tables 1–5, Appendix

Steven P. Pereira, CSP, Figure 7

Brandenburg Industrial Service Company, SA01

Section Review Answer Key

Answer	Section Reference	Objective
Section One		
1. a	1.1.0	1a
2. d	1.2.0	1b
3. b	1.3.0	1c
4. a	1.4.0	1d
Section Two		
1. c	2.1.1	2a
2. a	2.2.0	2b
3. d	2.3.2	2c
4. b	2.4.0	2d

NCCER CURRICULA — USER UPDATE

NCCER makes every effort to keep its textbooks up-to-date and free of technical errors. We appreciate your help in this process. If you find an error, a typographical mistake, or an inaccuracy in NCCER's curricula, please fill out this form (or a photocopy), or complete the online form at **www.nccer.org/olf**. Be sure to include the exact module ID number, page number, a detailed description, and your recommended correction. Your input will be brought to the attention of the Authoring Team. Thank you for your assistance.

Instructors – If you have an idea for improving this textbook, or have found that additional materials were necessary to teach this module effectively, please let us know so that we may present your suggestions to the Authoring Team.

NCCER Product Development and Revision
13614 Progress Blvd., Alachua, FL 32615

Email: curriculum@nccer.org
Online: www.nccer.org/olf

❏ Trainee Guide ❏ Lesson Plans ❏ Exam ❏ PowerPoints Other _____

Craft / Level: _____ Copyright Date: _____

Module ID Number / Title: _____

Section Number(s): _____

Description: _____

_____

_____

_____

Recommended Correction: _____

_____

_____

_____

Your Name: _____

Address: _____

Email: _____ Phone: _____

Glossary

A-B-C-D method: A device used to write training objectives that specifically cover the audience, behavior, conditions, and minimum degree of mastery.

Abatement: The correction of the safety hazards or violations resulting from an OSHA citation.

Abatement certification: A written record that shows that the safety hazard or standard violation noted in the citation has been corrected.

Acceptable level of risk: The level of risk that is reasonable when working in hazardous conditions.

Accident: An event that results in property damage and/or personal injury or death that could not have been prevented.

Affected employees: Used in reference to LOTO policies and procedures. Affected employees are usually the equipment users or operators. These workers need to recognize when a control procedure is in place. They also need to understand that they are not to remove the LOTO device or try to start the equipment during a control procedure.

Ambient noise levels: Background noise that is related to the jobs done on a work site.

Antigens: Toxins or enzymes capable of stimulating an immune response, similar to an allergy.

ARCS Model of Motivation: A strategy for motivating people developed by Dr. John Keller. The strategy states that to motivate individuals you must gain their attention, make the issue relevant to them, help them to feel confident that they can be successful, and provide them with a sense of satisfaction once they have achieved their goal.

Asbestos: A natural mineral that forms long crystal fibers, used in the past as a fire retardant. It is a known carcinogen.

Asbestos containing material (ACM): An object that is comprised of asbestos and other compounds.

Asbestosis: Scarring of the lung tissue caused by inhaled asbestos fibers. This terminal condition is caused by asbestos exposure.

Asphyxiation: Death due to lack of oxygen.

Atmosphere: The air or climate inside a specific place.

Atmospheric contaminants: Impurities in the air. Any natural or artificial matter capable of being airborne, other than water vapor or natural air, which, if in high enough concentration, harm man, other animals, vegetation, or material. They include many physical forms, liquid, solids, and gases as listed below.

Audible: When a noise or sound is heard or capable of being heard.

Audiovisual materials: Materials such as photos, films, charts, and graphs that are designed to aid in learning or teaching by making use of both hearing and sight.

Audit: To review safety policies and procedures to see if they are adequate and being used.

Authorized employees: Used in reference to LOTO policies and procedures. It refers to those employees who actually perform the servicing or maintenance on the involved equipment. These workers need a high degree of training. They also need to know about the type and magnitude of the hazardous energy and how to safety isolate and control the energy source.

Baseline: In medical surveillance programs, this refers to the initial health status of the person. Subsequent medical reports are compared to the baseline.

Behavioral-based safety (BBS): A proactive method of safety management based on psychology. It requires systematic workplace observation and analysis of unsafe behaviors, resolution of problems, and is coupled with training and incentives for behavior modification.

Bioaccumulate: The natural process by which chemicals become concentrated in higher levels of the food chain as larger animals consume many smaller contaminated animals or plants.

Calendar year: The period of a year beginning on January 1 and ending on December 31.

Causal links: A relationship between two events or occurrences in which one is the cause of the other.

Chronic irreversible disease: A long-lasting, permanent disease.

Citation: A formal notice from OSHA that the company has allegedly violated a federal safety regulation and may be fined.

Communication: A process by which information is exchanged between individuals through a common system of symbols, signs, or behavior.

Competent person: A person who is capable of identifying existing and predictable hazards, and who has authority to take prompt corrective measures to eliminate them; according to OSHA, a competent person is one who is capable of identifying existing and predictable hazards in the surroundings or working conditions which are unsanitary, hazardous, or dangerous to employees, and who has authorization to take prompt corrective measures to eliminate them.

Compliance safety and health officers (CSHOs): OSHA officials who conduct work-site inspections to enforce safety regulations.

Confined space: A workplace that has a configuration that hinders the activities of employees who must enter, work in, and exit the space.

Confined-space entry permit system: The written procedure for preparing and issuing permits for entry, and for returning the permit-required space to service after entry work is completed.

Consensus standards: Standards developed through the cooperation of all parties who have an interest in the use of the standard (*e.g.,* the *National Electrical Code®*). Consensus standards rely on the expertise of manufacturers, inspectors, craft professionals, maintenance personnel, and safety professionals.

Consequences: The final outcome of actions and behaviors; something that happens as a result of a set of conditions or actions.

Construction User's Round Table (CURT): CURT describes itself as an autonomous organization that provides a forum for the exchange of information, views, practices and policies of various owners at the national level. Similar groups, called Local User Councils, function at the local level and seek to address problems of cost, quality, safety and overall cost effectiveness in their respective areas.

Contact dermatitis: An inflammation of the skin.

Contractor pre-qualification: A process of screening contractors to allow them to bid on jobs for a specific company.

Corrective action: Actions the company must take to correct safety violations.

Cross-training: Training workers to do multiple jobs.

Culm: Coal refuse.

Defibrillator: An electronic device that administers an electric shock of preset voltage to the heart through the chest wall in an attempt to restore the normal rhythm of the heart during ventricular fibrillation.

Dielectric: A nonconductor of electricity, especially a substance with electrical conductivity of less than one-millionth of a siemens.

Direct cause: The immediate cause of an injury or illness, not accounting for any underlying unsafe behaviors or conditions.

Direct labor costs: Costs that can be directly related to an incident such as medical costs, workers' compensation insurance, benefits, and liability and property damage insurance payments.

Diversity: Differences between individuals, particularly with regard to race, religion, ethnicity, and gender.

Eczema: An inflammatory condition of the skin characterized by redness, itching, and oozing sores that become scaly, crusted, or hardened.

Energy control procedures: Written documents containing information that authorized workers need to know to safely control energy while servicing or maintaining equipment or machinery. These procedures are also called lockout/tagout procedures.

Energy-isolating devices: Devices that physically prevent the transmission or release of energy.

Engulfment: To be in an environment in which you are covered with a material such as sand, gravel, or grain. The end result is often death due asphyxiation, strangulation, or crushing.

Ergonomics: The applied science of equipment design, as for the workplace, intended to maximize productivity by reducing operator fatigue and discomfort.

Ethnic groups: Large groups of people classed according to common racial, national, tribal, religious, linguistic, or cultural origin or background.

Excavation: Any man-made cut, cavity, trench, or depression in the earth's surface formed by earth removal.

Experience modification rate (EMR): A lagging indicator of illness and injury rates based on insurance claims and predicted claims over a three-year period. It is most often applied by the insurance industry.

Feedback: The communication that occurs after a message has been sent and received. This communication enables the sender to determine whether his or her message has been accurately received.

Fire watch: Assigning trained personnel to watch for fires during welding and cutting operations, sound the alarm in the event of a fire, and extinguish any fires within the capacity of the equipment available. A fire watch must be maintained for at least a half hour after completion of welding or cutting operations.

Five Ps for Successful Safety Talks: A technique for conducting effective safety talks involving five key elements: preparing, pinpointing, personalizing, picturing, and prescribing.

Flaw: A part of the design of equipment, parts, or a process that creates a hazard or operational or maintenance difficulties.

Generators: Firms that create hazardous waste.

Good faith: Good faith means that the employer makes a serious effort to comply with relevant OSHA standards to the best of his or her ability.

Gross income: Income before deductions (*i.e.*, taxes).

Gross unloaded vehicle weight (GUVW): The weight of a liquid cargo trailer without liquids; also known as *dry weight*.

Gross vehicle weight rating (GVWR): The maximum allowed gross vehicle weight for a vehicle.

Hazardous conditions: Circumstances or objects that cause injury or illness. Most hazardous conditions arise as a result of unsafe (at-risk) behaviors.

Hazardous waste: A discarded material that has dangerous properties; it may be ignitable, corrosive, toxic, and/or reactive.

Hazardous waste manifest: A manifest is similar to a bill of lading. It is a shipping document that must be used for shipping waste that is considered hazardous by DOT and EPA standards.

Hot work: Any work function that involves ignition or combustion. Examples include welding, burning, cutting, and riveting.

Hypothermia: A life-threatening medical condition that can happen in cold weather when the body's temperature becomes too low.

Immediately dangerous to life and health (IDLH): A situation that poses a threat of exposure to airborne contaminants when that exposure is likely to cause death or immediate or delayed permanent adverse health effects or prevent escape from such an environment.

Imminent danger: A hazard that could cause death or serious physical harm immediately, or before the danger could be eliminated through normal enforcement procedures.

Incidence rate: A lagging indicator of illness and injury rates based on a Bureau of Labor Statistics (BLS) formula. It is measured in annual incidents per 100 workers.

Incident: An unplanned, undesired event that may cause personal injury, illness, or property damage in various combinations or degrees from minor to catastrophic.

Indirect cause: The underlying cause of an injury or illness. Categories include hazardous conditions and unsafe (at-risk) behaviors.

Induced hazards: Hazards brought forward or created when people make incorrect decisions and/or take incorrect actions during the construction process (*e.g.*, omission of protective features; physical arrangements that cause unintentional worker contact with electrical energy; oxygen deficient hazards created at the bottom of pits, shafts, or vaults; lack of safety factors in structural strength; and flammable atmospheres).

Inerting: The use of an inert gas, such as nitrogen, to supplant a flammable gas in a confined space.

Infectious material: Biological materials that carry contagious diseases and illnesses.

Inherent hazards: Hazards associated with specific types of equipment and the interactions of the equipment with the confined space. Include hazards that cannot be eliminated without degrading the system or the equipment, or making them inoperative (*e.g.*, high voltage, radiation generated by equipment, defective designs, omission of protective features, high or low temperatures, high noise, and high-pressure vessels and lines).

Inspection: The act of checking an area to identify, report, and correct hazards to workers, materials, and equipment.

Interviewee: Person being interviewed.

Intrinsically safe: An electric tool or device that is UL-rated to be explosion-proof under normal use. For example, a fuel pump must be intrinsically safe to prevent explosions sparked by static electricity.

Jargon: Technical terminology known only by people who work directly with the technology being discussed.

Job: A regular activity performed to achieve some end.

Job safety analysis (JSA): A careful study of a job or task to find all of the associated hazards and identify methods of safeguarding workers against each hazard; a method for studying a job to identify hazards and potential incidents associated with each step and developing solutions that will eliminate, minimize, and prevent hazards and incidents. Also called a job hazard analysis (JHA).

Know-show-do: A method of teaching tasks or procedures in which the instructor first explains necessary background information, then demonstrates the task or procedure, and concludes by having the participants practice the task or procedure.

Lagging indicator: A measure of performance based only on historical reporting.

Leading indicator: A measure of performance based on predictive observations and analysis.

Light ballasts: The parts of fluorescent lights that contain electric capacitors, which may contain PCBs.

Lockout: The placement of lockout devices on an energy-isolating device, in accordance with established procedures. The lockout device is intended to use a positive means, such as a lock, to secure the energy-isolating device in the off or safe position.

Lower explosive limit (LEL) or lower flammable limit (LFL): The lowest concentration of air-fuel mixture at which a gas can ignite.

Mesothelioma: An aggressive and terminal form of cancer that typically develops in the outer lining of the lungs. It is almost exclusively caused by exposure to asbestos.

Message: The information that the sender is attempting to communicate to the receiver.

Near miss: An unplanned event that did not result in injury, illness, or damage but had the potential to do so. Also called a *near hit* or a *close call*.

Net income: Income after deductions (*i.e.*, taxes and expenses).

Non-permit required confined space: A workspace free of atmospheric, physical, electrical, and mechanical hazards that could cause death or injury.

Non-verbal communication: Communication achieved through non-spoken means such as body language, facial expressions, hand gestures, and eye contact.

Observation: Watching a worker during the performance of his or her job for the purpose of determining whether the worker is working safely or committing an unsafe act.

Open-ended questions: Questions that require more than a yes or no answer. These types of questions are used to encourage the audience to participate in the discussion.

Other employees: All employees who are or may be in an area where energy control procedures may be utilized.

Oxygen-deficient atmosphere: A body of air that does not have enough oxygen to sustain normal breathing. This is usually considered less than 19.5 percent by volume.

Oxygen-enriched atmosphere: A body of air that contains enough oxygen to be flammable or explosive. This is usually considered more than 23.5 percent by volume.

Paraphrase: A restatement of a text, passage, conversation, or work process that is explained without changing its meaning.

Pareto diagram: A diagram used to perform a trend analysis by sorting data based on the frequency of occurrence.

Parts per million (ppm): A measure of concentration of a substance in a solution like air or water. In a volume of air, it is the number of milliliters of a substance in a cubic meter of air: 10 ppm = 10 ml/m^3. One part per million is about the same as two soda cans full of gas in a house full of air.

Pattern: An indication of how predictable the reoccurrence of an event is.

Penalty: A fine imposed on a company for violating a federal or state safety regulation. Penalties can also include prison time for criminal actions.

Permissible exposure limit (PEL): The limit set by OSHA as the maximum concentration of a substance that a worker can be exposed to in an 8-hour work shift. Most flammable gases have a PEL defined as 10 percent of the LEL/LFL, while toxic gases have individual PELs.

Permit-required confined space (PRCS): A confined space that has real or potential hazards including atmospheric, physical, electrical, or mechanical hazards defined by OSHA.

Polychlorinated biphenyls (PCBs): A group of man-made chemicals, which were widely used as dielectric fluids or additives.

Potentially responsible party/parties (PRP): An individual or firm who may be liable for paying the costs of a Superfund cleanup; a defendant in a Superfund lawsuit.

Pre-bid checklist: A list of questions or issues that aids contractors in gathering the information needed to prepare a bid package.

Pre-task plan: A process for identifying and evaluating potential hazards associated with a given task or work assignment.

Prerequisite training: Training that provides the skills and knowledge required for a course that is to follow.

Probability: The chance that something will happen.

Probability of occurrence: The likelihood that a specific event will occur, usually expressed as the ratio of the number of actual occurrences to the number of possible occurrences.

Process Safety Management (PSM): *The Process Safety Management of Highly Hazardous Chemicals, 29 CFR 1910.119.* An OSHA standard that covers certain chemical plants and refineries.

Profit: Net income over a given period of time.

Receiver: The person to whom the sender is communicating a message.

Recognition: Something that is given or returned verbally, through body language, or in writing for good or bad behavior.

Recordable: An illness or injury is classified as recordable if it meets certain OSHA reporting guidelines.

Recordable incident rate: An equation that calculates the number of job-related injuries and illnesses, or lost workdays per 200,000 hours of exposure on a construction site.

Reinforcement: Supporting or strengthening of a behavior or action.

Release of energy: Events or conditions that release energy from systems, machines, or pieces of equipment.

Remedial training: Training that is designed to address specific areas of difficulty, allowing participants to learn more about a larger course or topic.

Reportable quantity (RQ): The amount of a chemical that, when spilled, must be reported to the National Response Center.

Restricted work activity: Work activity assigned to those workers who have experienced a work-related injury or illness that prevents them from doing the routine functions of their jobs or from working a full workday.

Reverse chronological order: Events told in order from last to first.

Risk assessment: The process of qualifying or ranking hazards given their probability and consequences.

Root cause: The deepest level, system-related cause of an injury or illness. Requires in-depth analysis to discover. Also referred to as a *basic cause*.

Root cause analysis: A method of problem solving that seeks to identify the most basic cause(s) of an incident.

Safety audit: A predictive method of safety management requiring observation and reporting by an unbiased individual.

Safety data sheet (SDS): A form that lists the hazards, safe handling practices, and emergency control measures for a specific chemical.

Safety inspection: A predictive method of safety management requiring observation and reporting by a workforce supervisor.

Secondary containment: A barrier that collects chemical overflow or spills from their original containers.

Self-contained breathing apparatus (SCBA): A device allowing an individual to breath in a toxic or oxygen-deficient atmosphere.

Sender: The person who creates the message to be communicated.

Silica: A common mineral often referred to as *quartz*. It is present in soil, sand, granite, and other types of rocks.

Silicosis: A form of lung disease caused by breathing in crystalline silica dust.

Simulators: Teaching aids that are used to simulate a piece of equipment for training purposes rather than conducting training on the actual equipment.

Specific gravity: The ratio of the density of a substance to the density of a reference substance (typically water).

Storm water run-off: Rain that is not absorbed by the soil. Uncontrolled rainwater flows over land and picks up dirt and other contaminants and carries these to the nearest water body.

Structured on-the-job training (SOJT): Training that takes place during the performance of the job with the guidance of documentation or a facilitator.

Superfund: The common name for the Comprehensive Environmental Response, Compensation, and Liability Act.

Surface mines: Mines in which the mining operation is done near the surface of the ground.

Swales: Shallow trough-like depressions that carry water mainly during rainstorms or snow melts.

Tagout: The placement of tags or tagout devices on an energy-isolating device, in accordance with established procedures. It is intended to indicate that the energy-isolating device and the equipment being controlled must not be operated until the tagout device is removed.

Target organ: A specific organ in the human body most affected by a particular chemical.

Task: A discrete step or portion of a job.

Three Levels of Incident Causation Model: A method of incident investigation that seeks to identify three different factors: direct causes, indirect causes, and root causes.

Trench: A narrow excavation made below the surface of the ground. The depth of a trench must be greater than the width. The width must not exceed 15' (4.6 m).

Trend: The tendency to take a particular direction.

Ulcer: A break in skin or mucous membrane that festers and erupts like an open sore.

Uncontrolled release of energy: Energy that is released as result of an energy source that is uncontrolled. Energy sources can include tools, equipment, machinery, temperature, pressure, gravity, or radiation.

Universal precautions: A set of precautions designed to prevent transmission of human immunodeficiency virus (HIV), hepatitis B virus (HBV), and other blood-borne pathogens when providing first aid or health care. Under universal precautions, blood and certain body fluids of all patients are considered potentially infectious for HIV, HBV, and other blood-borne pathogens.

Unsafe behavior: Action taken or not taken that increases risk of injury or illness. Also called *at-risk behavior*.

Unwarrantable failures: An unwarrantable failure violation (second within 90 days) refers to a situation in which the mine operator knew or should have known that a violation existed and yet failed to take corrective action.

Upper flammable limit (UFL) or upper explosive limit (UEL): The highest concentration of air-fuel mixture at which a gas can ignite.

Vapor density: The relative weight of gases and vapors as compared with some specific standard, usually hydrogen, but sometimes air.

Verbal communication: Transfer of information through spoken word. This process involves a sender, receiver, message, and feedback.

Violation: The act of violating federal or state safety regulations by failing to obey a safety standard or failing to keep the workplace free from hazards.

Visual communication: Communication through visual aids such as signs, postings, and hand signals.

Wetlands: Lowland areas, such as marshes or swamps, which are saturated with moisture, especially when regarded as the natural habitat of wildlife.

What-why-how: A teaching method in which the instructor describes what an object, task, or procedure is, why it is used or done, and how to use or do it.

Why Method: A method of incident investigation that is used to determine the root cause of an incident by asking the question "Why did this happen?" and investigating until the question "Why?" can no longer be asked.

Written communication: Transfer of information through the written word.

Index

A

Abatement, (75226):16, 17, 33
Abatement certification, (75226):16, 28, 33
ABC model, (75220):7–8
Absenteeism, (75201):1, (75205):13
Acceptable level of risk, (75219):12, 50
Accidents. *See also* Incidents
   defined, (75225):1, 25
   incident vs., (75201):2, (75225):1
   statistics, (75201):2
ACIG. *See* American Contractors Insurance Group (ACIG)
ACM. *See* Asbestos containing material (ACM)
Activator, ABC model, (75220):8
Administrative hazard control methods, (75219):17
AEDs, (75222):16
Aerial lifts permits, (75224):21
Affected employees, (75224):1, 9, 49
Affirmative defenses, OSHA inspections, (75226):28
AHERA. *See* Asbestos Hazard Emergency Response Act (AHERA)
Air horns, (75224):38
Air line loss, (75224):43
Air movers, (75224):38–39
Alarms, (75222):8
Alcohol abuse policies and programs, (75201):9–10, 17
Alcohol use, on-the-job, (75220):7
Ambient noise levels, (75219):12, 17, 34, 50
American Contractors Insurance Group (ACIG), (75219):20
American National Standards (ANS), (75201):20, (75222):30
American National Standards Institute (ANSI)
   PPE standards, (75222):10
   purpose, (75222):30
   regulatory requirements, (75201):19–20
   site safety plan standard, (75222):20
American Society of Mechanical Engineers (ASME), (75201):19–20, (75222):30
Amosite, (75219):24
Annual Survey of Occupational Injuries and Illnesses (BLS), (75226):1
ANS. *See* American National Standards (ANS)
ANSI. *See* American National Standards Institute (ANSI)
Antigens, (75226):1, 6, 33
ARCS Model of Motivation, (75205):1, 7, 17
Asbestos
   defined, (75219):23, 50
   materials containing, (75219):56
   safety programs, (75219):24–25, 26
Asbestos containing material (ACM), (75219):23, 24–25, 26, 50, 56
Asbestos Hazard Emergency Response Act (AHERA), (75219):24, (75222):25
Asbestosis, (75219):23, 24, 50, (75226):6
Asbestos School Hazard Abatement Act (ASHAA), (75219):24, (75222):25
ASHAA. *See* Asbestos School Hazard Abatement Act (ASHAA)
ASME. *See* American Society of Mechanical Engineers (ASME)
Asphyxiation, (75224):23, 49
Asthma, occupational, (75226):6

Atmospheres
   confined spaces
      contaminants, (75224):26
      flammable, (75224):26–35
      oxygen- deficient/enriched, (75222):14, (75224):26
      toxic, (75222):14, (75224):32–33
   contaminants, (75224):23, 26, 49
   defined, (75224):1, 49
   explosive, (75224):27–28
   flammable, (75224):26–32
   hazardous, inspection for, (75224):14
   oxygen-deficient, (75224):23, 26, 50
   oxygen-enriched, (75224):24, 26, 50
   testing
      equipment, (75224):34, 35, 36
      for flammability, (75224):35–36
      oxygen, (75224):34, 35
      pre-entry, (75224):34
      repeating, (75224):36, 37
      toxic air, (75224):36–37
   toxic, (75222):14, (75224):32–33, 36–37
At-risk behavior, (75219):8, 20, (75220):1–2, 4–7
Attendants, confined space roles and duties, (75224):42–43
Attention, ARCS Model of gaining, (75205):7
Attitude, poor employee, (75205):12
Audible, (75219):12, 50
Audiovisual materials
   defined, (75223):16, 26
   to engage learners, (75223):5–6, 7
Audits
   defined, (75221):1, 28
   highlighted in text, (75221):2
   recordkeeping, (75226):13, 15
   safety, (75201):12, (75221):1, 8, 17–18, 28
Authorized employees, (75224):1, 9, 49

B

B12 vitamins, (75220):7
Ballasts, PCBs in, (75219):23, 28–29, 50
Barricades, (75219):18
Barriers, performance, (75220):2–3
Baseline, (75219):23, 44, 50
BBS. *See* Behavioral-based safety (BBS)
A-B-C-D method, (75223):1, 10–11, 26
Behavior
   ABC model of, (75220):8
   at-risk, (75219):8, 20, (75220):1–2, 4–7
   distracted, (75220):6
   fatigued, (75220):6
   impairment factors, (75220):3–4
   inattentive, (75220):6
   misconduct, (75220):4
   observation in reinforcing, (75220):2
   performance barriers, (75220):2–3
   rewarding vs. punishing, (75221):10
   unsafe, (75219):20–22, (75221):1, 8, 28
Behavioral-based safety (BBS), (75205):1, 9–12, 17, (75221):12, 28
Behavioral-based safety (BBS) programs, (75219):20–21
Behavioral Law of Effect, (75220):1–2

Behavior modification, (75205):9–10
Beryllium disease, (75226):6
Bioaccumulate, (75219):23, 28, 50
Blood alcohol content standards, (75201):17
BLS. *See* Bureau of Labor Statistics (BLS)
Bronchitis, chronic obstructive, (75226):6
Brownfield Program (EPA), (75219):33
Bureau of Labor Statistics (BLS)
   Annual Survey of Occupational Injuries and Illnesses, (75226):1
   Data Integrity Guidelines, (75225):35–36
   function, (75225):13
   incident rate reporting, (75221):15
   incident rate tracking, (75221):13–14
Buy Quiet initiative (NIOSH), (75219):16, 34

C

CAA. *See* Clean Air Act (CAA)
Caffeine, (75220):7
Calendar year, (75226):1, 33
Cameras, restrictions on, (75225):2
Carbon monoxide (CO), (75224):33
Causal links, (75220):10, 12, 21
CDC. *See* Centers for Disease Control and Prevention (CDC)
CDL. *See* Commercial Driver's License (CDL) Program
Cell phones, restrictions on, (75225):2
Centers for Disease Control and Prevention (CDC), (75201):15, (75222):25
CERCLA. *See* Comprehensive Environmental Response, Compensation, and Liability Act (CERCLA)
CESQGs. *See* Conditionally exempt small quantity generators (CESQGs)
Change analysis, (75225):17
Chemical hazards, (75224):16–20
Chemical management, (75219):39
Chemical-specific safety programs
   asbestos, (75219):24–25, 26
   lead, (75219):25, 27–28
   PCBs, (75219):28–30
   silica, (75219):30
Chronic irreversible disease, (75226):1, 4, 33
Chronic obstructive pulmonary disease (COPD), (75219):30, (75226):6
Chrysotile, (75219):24
Citation, (75226):1, 9, 33
Citations
   defined, (75226):1, 9, 33
   OSHA inspections, (75226):20–21, 28–29
Clean Air Act (CAA), (75201):15, (75222):26
Clean Water Act (CWA), (75201):15, (75219):31, (75222):26
Close call incidents, (75225):1
CMV. *See* Commercial motor vehicle (CMV) drivers; Commercial Motor Vehicle (CMV) Safety Act
CO. *See* Carbon monoxide (CO)
Coaching, (75220):7
CoE. *See* US Army Corps of Engineers (USACE/CoE)
Commercial Driver's License (CDL) Program, (75201):16–18, (75222):28
Commercial motor vehicle (CMV) drivers, (75201):16–18
Commercial Motor Vehicle (CMV) Safety Act, (75201):16–17, (75222):28
Communication
   defined, (75205):1–2, 17
   distractors, (75205):2
   in a diverse workforce, (75205):5–6
   interference, (75219):37
   non-verbal, (75205):1, 2, 17
   stress and, (75205):6
   verbal, (75205):1, 2–3, 17
   visual, (75205):1, 3–5, 17
   written, (75205):1, 3–5, 17
Communications loss, (75224):43
Competent person
   defined, (75224):1, 49, (75226):16, 26, 33
   for electrical work, (75224):21
   OSHA requirements, trenching, (75224):14
Compliance
   infeasibility of, (75226):29
   posing a greater hazard, (75226):29
Compliance safety and health officers (CSHOs)
   accompanying, (75226):17–18
   citations, issuing, (75226):20
   communicating with, (75226):18
   defined, (75226):16, 33
   focused vs. wall-to-wall inspections, (75226):23–24, 27
   inspection process
      closing conference, (75226):19
      opening conference, (75226):18
      pre-inspection requirements, (75226):17–19
      presentation of credentials, (75226):18
      selection of representatives, (75226):18
      walk-around inspection, (75226):18–19
   modified inspection procedure, (75226):27
Comprehensive Environmental Response, Compensation, and Liability Act (CERCLA), (75201):15, (75219):32, (75222):26
Conditionally exempt small quantity generators (CESQGs), (75219):39
Conferences, OSHA inspections
   closing, (75226):19
   informal, (75226):28
   opening, (75226):18
Confidence, ARCS Model of instilling, (75205):8
Confined space entry permit
   authorization, (75224):41
   information required, (75224):42
   standards, (75224):41–42
Confined space entry permit system, (75224):41, 49
Confined spaces
   atmospheres
      contaminants, (75224):26
      flammable, (75224):26–35
      oxygen- deficient/enriched, (75222):14, (75224):26
      toxic, (75224):32–33
   atmospheric testing
      equipment, (75224):34, 35, 36
      for flammability, (75224):35–36
      oxygen, (75224):34, 35
      pre-entry, (75224):34
      repeating, (75224):36
      toxic air contamination, (75224):36–37
   categories, (75224):23
   defined, (75224):1, 23, 49
   emergency entry, (75222):14
   emergency training
      inadequate illumination, (75224):43
      loss of air line, (75224):43
      loss of communications, (75224):43
      rescue procedures, (75224):44–45
   environmental conditions, (75224):32–33
   hazards
      atmospheric, (75222):14, (75224):23–35
      explosions, (75224):27–28
      spark, (75224):33
   highlighted in text, (75224):2
   hot work in, (75224):27

non-permit required, (75224):23, 24, 50
permit-required (PRCS), (75224):1, 3, 23–24, 50
roles and duties
   attendants, (75224):42–43
   entrants, (75224):42
   supervisors, (75224):43
safety procedures, (75224):24
tools and equipment in, (75224):33–34
types of, (75224):23
ventilating, (75224):37–39
Confined Spaces in Construction standard, (75224):41
Consensus standards, (75201):14, 26
Consequences
  ABC model, (75220):8
  defined, (75219):12, 50, (75220):1, 21
Construction, lean, (75201):5
Construction User's Round Table (CURT), (75201):4, 8, 26
Consumer Product Safety Commission (CPSC), (75221):10
Contact dermatitis, (75226):1, 6, 33
Containment, secondary, (75219):23, 39, 50
Contaminants, atmospheric, (75224):23, 26, 49
Continuing education, (75223):2
Contractor pre-qualification, (75222):1, 36
Contractor pre-qualification form (PQF), (75222):37–42
COPD. *See* Chronic obstructive pulmonary disease (COPD)
Corrective action, (75226):16, 28, 33
Counseling, (75220):7
CPSC. *See* Consumer Product Safety Commission (CPSC)
Craft-specific training , (75223):2
Crane operations, (75219):37
Crocidolite, (75219):24
Cross-training, (75219):12, 17, 50
CSHOs. *See* Compliance safety and health officers (CSHOs)
Culm, (75201):14, 19, 26
CURT. *See* Construction User's Round Table (CURT)
CWA. *See* Clean Water Act (CWA)

D
Danger
  immediately dangerous to life and health (IDLH), (75222):4, 36
  imminent, (75226):16, 33
DART. *See* Days Away, Restricted, or Transferred (DART)
Data
  analysis
    organizing for, (75225):20
    OSHA's problem-solving techniques, (75225):17–19
    purpose of, (75225):13
    Three Levels of Incident Causation Model, (75225):13–16
    Why Method, (75225):16–17
  BLS integrity guidelines, (75225):35–36
  organizing the, (75225):20
  reporting the, (75225):20
Days Away, Restricted, or Transferred (DART), (75221):12, (75226):24
Defibrillator, (75222):4, 36
Demonstration to engage learners, (75223):3–4
DEP. *See* Directorate of Enforcement Programs (DEP)
Department of Health and Human Services (DHHS), (75201):15, (75222):25
Department of Homeland Security (DHS), (75201):20–21, (75222):31
Department of Justice (DOJ), (75226):21
Department of Labor, civil monetary penalties, (75226):21
Department of Transportation (DOT)
  CDL Program, (75222):28
  Commercial Motor Vehicle (CMV) Safety Act, (75222):28

hazardous materials shipping, (75219):41–42, (75222):28, 29
hazardous waste shipping regulation, (75219):41–42
regulatory requirements, (75201):16–18
Designated Safety and Health Official (DSHO), (75201):20
Design for hazard control, (75219):16–17
DHHS. *See* Department of Health and Human Services (DHHS)
DHS. *See* Department of Homeland Security (DHS)
Dielectrics, (75219):23, 28, 50
Direct cause, (75201):4, (75221):12, 20, 21, 28, (75225):13–14
Direct labor costs, (75201):4, 8, 26
Directorate of Enforcement Programs (DEP), (75226):19
Disciplinary action, unsafe behavior, (75219):21–22
Disease. *See also specific diseases*
  beryllium, (75226):6
  chronic irreversible, (75226):1, 4, 33
  chronic obstructive pulmonary (COPD), (75219):30, (75226):6
  skin, (75226):1, 6
Distraction, (75220):6
Diversity, (75205):1, 17
DOT. *See* Department of Transportation (DOT)
Drug abuse policies and programs, (75201):9–10
Drug-Free Workplace Act, (75201):10
DSHO. *See* Designated Safety and Health Official (DSHO)

E
Eczema, (75226):1, 6, 33
Electrical work permits, (75224):21
Electromagnetic interference (EMI), (75219):37
Electromagnetic radiation, (75219):37
Emergency action plans
  benefits of, (75201):11, (75222):4
  chain of command, establishing a, (75222):5
  communications equipment, (75222):6
  effective, (75222):4
  elements of, (75222):5
  emergency response teams, (75222):6, 8
  media, dealing with the, (75201):11, (75222):18
  medical assistance, (75222):15
  personal protection, (75222):10–14
  personnel, accounting for, (75222):6
  pre-planning for
    external threats, (75222):16
    fire, (75222):16–17
    severe weather, (75222):16
    trapped workers, (75222):15–16
  requirements, (75222):4
  security, (75222):15
  in the site safety plan, (75222):22
  training, (75222):8, 10
Emergency response teams, (75222):6, 8
Emergency training, confined spaces
  inadequate illumination, (75224):43
  loss of air line, (75224):43
  loss of communications, (75224):43
  rescue procedures, (75224):44–45
EMI. *See* Electromagnetic interference (EMI)
Employees
  absenteeism, (75205):13
  affected, (75224):1, 9, 49
  alcohol use, on-the-job, (75220):7
  attitudes, poor, (75205):12
  authorized, (75224):1, 9, 49
  disciplinary action, unsafe behavior, (75219):21–22
  distracted, (75220):6
  fatigued, (75220):6

Employees (*continued*)
   impaired, (75220):3–4
   inability to work with others, (75205):12–13
   inattentive, (75220):6
   misconduct, (75220):4
   other, (75224):1, 9, 50
   performance barriers, (75220):2
   turnover, (75205):13–14
   unpreventable misconduct, (75226):28
Employer Rights and Responsibilities Following a Federal OSHA Inspection (OSHA), (75226):19
EMR. *See* Experience modification rate (EMR); Experience Modification Rate (EMR)
Endangered Species Act (ESA), (75219):33, (75222):27
Energy
   control procedures, (75224):1, 8, 49
   release of, (75201):1, 2, 26
   uncontrolled release of, (75219):1, 50
Energy drinks, (75220):7
Energy-isolating devices, (75224):1, 8, 49
Engineering hazard controls, (75219):16–17
Engulfment, (75224):23, 24, 49
Entrants, confined space roles and duties, (75224):42
Environmental Protection Agency (EPA)
   Brownfield Program, (75219):33
   compliance, (75221):10
   hazardous materials
      disposal regulation, (75219):40–41
      shipping regulation, (75219):41–42, (75222):28, 29
   Identification and Listing of Hazardous Waste, (75219):40
   Land Disposal Restrictions (LDR) Program, (75219):32–33, (75222):27
   "Managing Your Hazardous Wastes," (75219):40
   National Pollutant Discharge Elimination System (NPDES), (75222):21
   National Pollutant Discharge Elimination System (NPDES) permits, (75222):27
   pesticide registration, (75219):32
   regulatory requirements, (75201):15–16
   underground storage tank regulations, (75222):26
Environmental regulations
   Asbestos Hazard Emergency Response Act (AHERA), (75219):24
   Asbestos School Hazard Abatement Act (ASHAA), (75219):24
   Clean Air Act (CAA), (75222):26
   Clean Water Act (CWA), (75219):31, (75222):26
   Comprehensive Environmental Response, Compensation, and Liability Act (CERCLA), (75219):32, (75222):26
   Endangered Species Act (ESA), (75219):33, (75222):27
   Federal Insecticide, Fungicide, and Rodenticide Act (FIFRA), (75219):32, (75222):27
   Land Disposal Restrictions (LDR), (75219):32–33
   Migratory Bird Treaty Act, (75219):33, (75222):27–28
   National Historic Preservation Act (NHPA), (75219):33
   National Pollutant Discharge Elimination System (NPDES), (75219):31, (75222):21
   Oil Pollution Act, (75222):27
   PCB regulation, (75219):28
   Residential Lead-Based Paint Reduction Act, (75219):27
   Resource Conservation and Recovery Act (RCRA), (75219):39, (75222):26
   Safe Drinking Water Act (SDWA), (75222):27
   Toxic Substances Control Act (TSCA), (75219):28, (75222):27
Environment-related job-site concerns
   affecting construction, (75219):23–24
   ambient noise levels, (75219):34
   chemical management, (75219):39
   chemical-specific safety programs
      asbestos, (75219):24–25, 26
      lead, (75219):25, 27–28
      PCBs, (75219):28–30
      silica, (75219):30
   electromagnetic radiation, (75219):37
   hazardous waste management, (75219):39–43
   medical surveillance programs, (75219):42, 44–46
   mold, (75219):35–37
   nanotechnology, (75219):37–39
   training requirements, (75219):46
   weather hazards, (75219):34–35
EPA. *See* Environmental Protection Agency (EPA)
Equipment
   confined spaces, (75224):33–34
   incident investigation, (75225):2–3
   used in incident investigations, (75225):2–3
Ergonomics, (75219):45, (75220):10, 15, 21
ESA. *See* Endangered Species Act (ESA)
Ethnic groups, (75205):1, 17
Excavations, (75224):1, 2, 13–15, 49
Execution barriers, (75220):3
Exhaust systems, (75224):38–39
Experience Modification Rate (EMR), (75201):4, 7, 26, (75221):12, 17, 28
Explosions, confined space, (75224):27
Explosive limit
   lower (LEL), (75224):23, 31, 50
   upper (UEL), (75224):23, 31, 50
Exposure, defined, (75219):12
Exposure records, (75226):13

F

FACE. *See* Fatality Assessment and Control Evaluation (FACE) program
Facility citing permits, (75224):21
Failures, unwarrantable, (75201):14, 19, 26
Fatal Four hazards, (75221):1, (75226):26–27
Fatalities
   Fatal Four hazards, (75221):1, (75226):26–27
   heat stress, (75219):35
   OSHA reporting requirements, (75226):2
   statistics, (75201):1, (75221):1, 12
Fatality Assessment and Control Evaluation (FACE) program, (75220):1
Fatigue, (75220):6
FCC. *See* Federal Communications Commission (FCC)
Federal Communications Commission (FCC), (75219):37
Federal Highway Administration (FHWA), (75201):16–17
Federal Insecticide, Fungicide, and Rodenticide Act (FIFRA), (75219):32, (75222):27
Feedback, (75205):1, 2, 17, (75220):3, 7
FHWA. *See* Federal Highway Administration (FHWA)
FIFRA. *See* Federal Insecticide, Fungicide, and Rodenticide Pollution Act (FIFRA)
Fines, OSHA violations, (75226):19, 21–23
Fire emergencies, (75222):16–17
Fire extinguishers, (75222):7
Firefighting, (75222):8, 9
Fire watch, (75224):1, 3, 8, 49
Fish and Wildlife Service (USFWS), (75219):33, (75222):27–28
Five Ps for Successful Safety Talks, (75223):16, 23, 26
Flammable gas testing, (75224):7
Flammable limit
   lower (LFL), (75224):23, 31, 50
   upper (UFL), (75224):23, 31, 50
Flaw, (75219):12, 16, 50

Focused Inspections Initiative (OSHA), (75226):23
Focused vs. wall-to-wall inspections, (75226):23–27

G

Gas detection meters, (75224):34
Gases
  flammable, (75224):26–28
  testing for
    in confined spaces, (75224):23
    in hot work, (75224):7, 8
General Safety and Health Provisions (OSHA), (75226):24
Generators, (75219):23, 50
GHS. *See* Globally Harmonized System of Classification and Labeling of Chemicals (GHS)
Globally Harmonized System of Classification and Labeling of Chemicals (GHS), (75224):16–18
  Safety Data Sheet example, (75224):18–19
Good faith, (75226):16, 19, 21, 33
Good faith compliance, (75226):21
Gravity, specific, (75224):23, 33, 50
Gross income, (75201):4, 6, 26
Gross unloaded vehicle weight (GUVW), (75201):14, 17, 26
Gross vehicle weight rating (GVWR), (75201):14, 17, 26
GUVW. *See* Gross unloaded vehicle weight (GUVW)
GVWR. *See* Gross vehicle weight rating (GVWR)

H

Hands-on practice to engage learners, (75223):3–4
Hand switches, (75219):19
Hazard analysis
  classifying and prioritizing
    risk assessment formula, (75219):12–13
    risk assessment matrix, (75219):13–14
  flow charts, (75219):16, 54–55
Hazard Communication Standard (HCS) (OSHA), (75224):16, 18
Hazard identification and controls
  hierarchy of controls, (75219):15, 17
  methods
    administrative controls, (75219):17
    engineering/design, (75219):16–17
    job site interventions, (75219):20–23
    personal protective equipment, (75219):17
    safety devices, (75219):17–19
  process, (75221):4
  safety planning, pre-bid, (75222):1, 2
Hazardous conditions, (75221):1, 8, 28
Hazardous materials
  shipping regulations, (75219):41–42, (75222):28, 29
  storage, use, handling standards, (75224):7
  training programs, (75219):46
Hazardous Materials Code (NFPA), (75224):7
Hazardous situations, identifying, (75219):1
Hazardous waste
  conditionally exempt small quantity generators (CESQGs), (75219):39
  defined, (75219):23, 50
  identifying, (75219):40–41
  medical surveillance programs, (75219):42, 44–46
  shipping, (75219):41–42
  small quantity generators (SQGs), (75219):39
Hazardous waste management
  asbestos, (75219):25
  environment-related job-site concerns, (75219):39–43
  fines for improper disposal, (75219):27, 30
  Land Disposal Restrictions (LDR), (75219):32–33
  lead contaminated materials, (75219):27
  PCBs, (75219):28, 30

Resource Conservation and Recovery Act (RCRA) requirements, (75219):39–40
Hazardous waste manifest, (75219):23, 41, 50
Hazardous Waste Operation (HAZWOPER), (75219):46
Hazard recognition techniques
  job observations, (75219):8, 11
  job safety analysis (JSA), (75219):1–7
  pre-job planning checklists, (75219):8
  pre-task planning, (75219):7, 9
  safety inspections, (75219):8, 10
Hazards. *See also* Health risks; Incidents
  adjacent to the site, (75222):21
  atmospheric
    characteristics, (75224):24
    flammable gases, (75224):26–32
    oxygen, (75224):26
    respirators for, (75224):24–25
  chemical, (75224):16–20
  confined space
    atmospheric, (75224):23–34
    explosions, (75224):27–28
    spark hazards, (75224):33
  correcting
    Immediate Temporary Control (ITC), (75219):14
    permanent fix, (75219):14–15
    post-correction actions, (75219):14–15
    quick fix, (75219):14
  defined, (75201):2
  excavations and trenching, (75224):13
  hot work, (75224):7, 8
  induced, (75224):41, 49
  inherent, (75224):41, 49
  weather-related, (75219):35
HAZWOPER. *See* Hazardous Waste Operation (HAZWOPER)
HCS. *See* Hazard Communication Standard (HCS) (OSHA)
Health Insurance Portability and Accountability Act (HIPAA), (75226):7
Health risks. *See also* Hazards
  asbestos exposure, (75219):24–25
  electromagnetic radiation, (75219):37
  ergonomics to alleviate, (75219):45
  heat stress, (75219):34–35
  hot work, (75224):7
  lead exposure, (75219):25
  medical surveillance programs, (75219):42, 44–45
  mold, (75219):35–36
  oxygen deficiency, symptoms of, (75224):26
  PCB exposure, (75219):28
  personal protective equipment, (75219):34–35
  silica dust exposure, (75219):30
  weather hazards, (75219):34–35
Hearing protection, (75219):34
Hearings, OSHA inspections, (75226):28
Heat stress, (75219):34–35
Heinrich, H. W., (75221):2
Heinrich's triangle, (75221):2, 3
HIPAA. *See* Health Insurance Portability and Accountability Act (HIPAA)
Hot work
  before beginning, (75224):8
  in confined spaces, (75224):27
  defined, (75224):1, 49
  equipment inspections, (75224):7
  fire watch, (75224):8
  flammable gas testing prior to, (75224):7
  hazards associated with, (75224):7, 8
  highlighted in text, (75224):2

SAFETY TECHNOLOGY INDEX I.5

Hot work (*continued*)
   post-work observation, (75224):8
   ventilation, ensuring adequate, (75224):7
Hot work permits
   Burning - Welding example, (75224):5–6
   common types, (75224):3, 6
   inspections prior to issuing, (75224):7
   requirement for, (75224):3
   signage, (75224):8
Hurricane-related hazards, (75219):35
Hydrogen sulfide gas, (75224):33
Hypothermia, (75226):1, 7, 33

I

Identification and Listing of Hazardous Waste (EPA), (75219):40
IDLH. *See* Immediately dangerous to life and health (IDLH)
Illnesses and injuries
   classifying
      illnesses, (75226):6
      injuries, (75226):5
      OSHA criteria for, (75226):2
   documentation, OSHA, (75226):7–12
   healthcare recommendations, (75226):3
   lost work days, determining, (75226):5
   medical treatment, defined, (75226):4
   occupational illnesses, (75226):6–7
   recordable vs. not recordable, (75226):2–6
   recordkeeping requirements, OSHA, (75226):4
   restricted work, (75226):4–5
   work-related, (75226):2–3
Illnesses and injury forms (OSHA)
   Injury and Illness Incident Report (Form 301), (75226):7, 9, 12
   Log of Work-Related Injuries and Illnesses (Form 300), (75226):7–8
   Optional Worksheet to Help You Fill Out the Summary (Form 300A optional), (75226):11
   privacy concerns, (75226):7
   records, keeping, (75226):9
   Summary of Work-Related Injuries and Illnesses (Form 300A), (75226):7, 9, 10
Illumination, inadequate, (75224):43
Immediately dangerous to life and health (IDLH), (75222):4, 36
Immediate Temporary Control (ITC), (75219):14
Imminent danger, (75226):16, 33
Impairment factors, (75220):3–4
Imprisonment, OSHA violations, (75226):21, 23
Inattention, (75220):6
Incentive barrier, (75220):3
Incidence rates
   calculating, (75221):14
   Days Away, Restricted, or Transferred (DART), (75221):12
   defined, (75221):12, 28
   Lost Time (LTR), (75221):12
   recordable, (75221):13–15
   recording, OSHA requirements, (75221):13–14
   statistics, evaluating, (75221):14–15
   Total Recordable (TRIR), (75221):12
Incident investigation
   conducting
      equipment used, (75225):2–3
      interviews, (75225):4–5
      roles and responsibilities, (75225):2, 4
      time line, (75225):3–4
   data analysis
      organizing for, (75225):20
      OSHA's problem-solving techniques, (75225):17–19

      purpose of, (75225):13
      Three Levels of Incident Causation Model, (75225):13–16, 25
      Why Method, (75225):16–17
   data reporting, (75225):20
   effective, (75225):1
   purpose of, (75225):1–2, 6
   reporting requirements
      MSHA, (75225):8
      OSHA, (75225):8
   responsibility for, (75225):1
   second-level, (75225):4
   term usage, (75225):1
   trend analysis, (75225):21
Incident investigation reports
   blame, avoiding, (75225):5
   example, (75225):6–7, 26–32
   forms
      completing, (75225):5
      mistakes in, common, (75225):8, 33–34
   key questions, (75225):5
   requirements
      MSHA, (75225):8
      OSHA, (75225):8
Incident rate, recordable, (75223):16, 20, 26
Incident Report (OSHA), (75225):11
Incidents. *See also* Hazards
   accident vs., (75225):1
   analysis, (75221):22
   catastrophic, (75221):3
   causes of
      direct, (75201):4, (75221):20, 21, (75225):13–14
      human factors, (75220):1–2
      indirect, (75201):4–5, (75221):12, 20, 21, 28, (75225):14
      at-risk behavior, (75219):8
      root, (75201):5, (75219):14–16, (75221):20, 21, (75225):15–16
      Swiss Cheese Model, (75221):3
      transportation, (75221):12
   close call, (75225):1
   costs
      calculating, (75201):6
      community, (75221):2
      company, (75221):2
      direct, (75201):6
      financial, (75221):1, 2
      future insurance, (75201):7
      indirect, (75201):6
      to individuals, (75221):2
      insured, (75201):6
      total incurred, (75201):6
      uninsured, (75201):6
   defined, (75201):2, 19, (75225):1, 25
   energy sources released, (75219):1
   event analysis, (75221):22
   Experience Modification Rate (EMR), (75221):17
   fatalities
      causes of, (75221):1–2, 16
      Fatal Four hazards, (75221):1, (75226):26–27
      heat stress, (75219):35
      OSHA reporting requirements, (75226):2
      statistics, (75201):1, (75221):1, 12
   injury analysis, (75221):16, 22
   injury-causing, (75219):1
   investigation and analysis, (75201):12, (75225):1
   methods to reduce
      administrative controls, (75219):17
      engineering/design, (75219):16–17

hierarchy of controls, (75219):15, 17
job site interventions, (75219):20–23
personal protective equipment, (75219):17
safety devices, (75219):17–19
near miss, (75221):1, 2, (75225):1
near miss/close call, (75225):1
OSHA and
   investigation recommendations, (75225):1
   reporting requirements, (75226):2
preventable, (75225):14
preventing, (75201):4, (75219):11, 12, 20
reducing, (75220):10
reporting, (75201):19, (75226):2
statistics, (75201):1, 2, 10, 16, (75221):1, 12
term usage, (75201):2, (75225):1
Workers' Compensation Experience Modification Rate (EMR), (75221):17
Income
   gross, (75201):4, 6, 26
   net, (75201):4, 6, 26
Indirect cause, (75201):4–5, (75221):12, 20, 28, (75225):14
Indirect costs, (75201):6
Induced hazards, (75224):41, 49
Inerting, (75224):23, 49
Infectious material, (75226):1, 4, 33
Inherent hazards, (75224):41, 49
Injury analysis, (75221):16, 22
Injury and Illness Incident Report (Form 301) (OSHA), (75226):7, 9, 12
Injury and Illness Recordkeeping Forms (OSHA), (75225):33–34
Injury-causing incidents, (75219):1
Injury statistics, (75201):1
Inspections. *See also* Occupational Safety and Health Administration (OSHA) inspections
   defined, (75221):1, 28
   highlighted in text, (75221):2
   OSHA requirements, (75221):10
   safety, (75201):12, (75221):1, 3, 4–7, 17–18, 28
Interlocks on safety gates, (75219):19
Interviewee, (75225):1, 5, 25
Interviews in incident investigations, (75225):4–5
Intrinsically safe, (75222):4, 6, 36, (75224):1, 7, 50
ITC. *See* Immediate Temporary Control (ITC)

J
Jargon, (75205):1, 5, 17
JHA. *See* Job hazard analysis (JHA)
Job, (75220):1, 21
Job hazard analysis (JHA)
   defined, (75220):10, (75221):22
   example, (75220):22–23
   function, (75223):2
   problem-solving techniques, (75225):17–19
Job observations, (75219):8
Job safety analysis (JSA)
   completing a, (75221):22
   defined, (75201):1, 26, (75221):12, 22, 28
   example, (75219):2, 7, (75220):13, (75221):23, 29–30
   function, (75223):2
   pre-task planning, (75220):15–17
   problem-solving techniques, (75225):17–19
   process, (75219):1–2
   reasons for, (75220):10–11
   sample forms, (75219):3–7, 51–53
   steps in a
      common errors, (75220):15
      data collection, (75220):12, 14

hazard identification, (75220):14, 15
preparation, (75220):12
risk factor identification, (75220):14
select the job for analysis, (75220):11–12
solution development, (75220):14–15
Job site interventions, (75219):20–23
JSA. *See* Job safety analysis (JSA)

K
"Keeping Workers Safe during Hurricane Cleanup and Recovery" (OSHA), (75219):35
Knowledge barriers, (75220):2–3
Know-show-do, (75223):1, 3, 26

L
Labor costs, direct, (75201):4, 8, 26
Lagging indicator, (75221):12, 25, 28
Land Disposal Restrictions (LDR) Program (EPA), (75219):32–33, (75222):27
LDR. *See* Land Disposal Restrictions (LDR) Program (EPA)
Leading indicator, (75221):12, 28
Lead safety programs, (75219):25, 27–28
Lean construction, (75201):5
LEL. *See* Lower explosive limit (LEL)
LFL. *See* Lower flammable limit (LFL)
Light ballasts, (75219):23, 28–29, 50
Lightning, (75219):35
Limited inspections, (75226):16
Liquids transfer, static electricity with, (75224):28
Listening, barriers to effective, (75205):3
Lockout, (75224):1, 8, 50
Lockout devices, (75224):9–10
Lockout/tagout (LOTO), (75224):8–9
   exceptions, (75224):12
   policies, purpose of, (75224):8
   procedure
      checklist, (75221):6, (75224):10–11
      employees involved, (75224):9
      purpose of, (75224):8
      requirements, (75224):9
      steps in the, (75224):10
      training requirements, (75224):9
   special conditions
      complex lockouts, (75224):12–13
      group lockouts, (75224):12, 13
      outside workers, (75224):12
      shift changes, (75224):13
      temporary removal, (75224):12
Log of Work-Related Injuries and Illnesses (Form 300) (OSHA), (75225):9, (75226):7–8
Lost Time (LTR) incidence rate, (75221):12
Lost work days, determining, (75226):5
LOTO. *See* Lockout/tagout (LOTO)
Lower explosive limit (LEL), (75224):23, 31, 50
Lower flammable limit (LFL), (75224):23, 31, 50
LTR. *See* Lost Time (LTR) incidence rate

M
"Making Zero Accidents a Reality" (CII), (75201):13
"Managing Your Hazardous Wastes" (EPA), (75219):40
Marijuana, (75201):10
Marlins Park Baseball Stadium, (75219):18
Material and tool storage, site safety plan, (75222):20
Media management, (75201):11, (75222):18
Medical records, (75226):13
Medical surveillance programs, (75219):42, 44–46
Medical treatment, defined, (75226):4
Mentoring programs, (75223):4–5
Mesothelioma, (75219):23, 24, 50

Message, (75205):1, 2, 17
Migratory Bird Treaty Act, (75219):33, (75222):27–28
Mine Act, (75201):19
Mines, surface, (75201):14, 18, 26
Mine Safety and Health Act, (75222):28
Mine Safety and Health Administration (MSHA)
    compliance, (75221):10
    incident investigations reporting requirements, (75225):8
    mission, (75222):28
    PPE standards, (75222):10
    regulatory requirements, (75201):18–19
    safety and health standards, (75222):30
Misconduct, (75220):4
Mold, (75219):35–37
Motivation
    ARCS Model of
        building relevance, (75205):8
        defined, (75205):1, 17
        fostering satisfaction, (75205):8–9
        gaining attention, (75205):7
        instilling confidence, (75205):8
    defined, (75205):6
    for at-risk behavior, (75220):4
MSHA. *See* Mine Safety and Health Administration (MSHA)

N

Nanometer, (75219):37
Nanotechnology, (75219):37–39
National Falls Campaign (OSHA), (75219):21
National Fire Protection Association (NFPA)
    common codes, (75222):30–31
    function, (75222):30–31
    Hazardous Materials Code, (75224):7
    regulatory requirements, (75201):20
    Standard for Electrical Safety in the Workplace, (75224):21
National Fire Protection Association (NFPA) label, (75219):40–41
National Historic Preservation Act (NHPA), (75219):33
National Institute for Occupational Safety and Health (NIOSH)
    ambient noise exposure recommendations, (75219):34
    Buy Quiet initiative, (75219):16, 34
    FACE program, (75220):1
    mandate, (75201):15, (75222):25
    medical surveillance programs, (75219):44
    PPE standards, (75222):10
    Prevention through Design (PtD) initiative, (75219):16
National Marine Fisheries Service (NMFS), (75219):33, (75222):27–28
National Occupational Research Agenda (NORA), (75201):15
National Pollutant Discharge Elimination System (NPDES) permits, (75219):31, (75222):21, 27
Naval Facilities Engineering Command (NAVFAC), (75201):21, (75222):32
NAVFAC. *See* Naval Facilities Engineering Command (NAVFAC)
Near miss, (75201):2, (75221):1, 28
Near miss management system, (75221):20
Near miss/near hit reporting, (75221):18–20
Net income, (75201):4, 6, 26
New employee orientation, (75223):1
NFPA. *See* National Fire Protection Association (NFPA); National Fire Protection Association (NFPA) label
NHPA. *See* National Historic Preservation Act (NHPA)
NIOSH. *See* National Institute for Occupational Safety and Health (NIOSH)
NMFS. *See* National Marine Fisheries Service (NMFS)

Noise
    ambient, (75219):12, 17, 34, 50
    Buy Quiet initiative (NIOSH), (75219):16, 34
    job-site concerns, (75219):34
    radiated, (75219):37
Non-permit required confined space, (75224):23, 24, 50
Non-recordable illnesses and injuries, (75226):2–6
Non-verbal communication, (75205):1, 2, 17
NORA. *See* National Occupational Research Agenda (NORA)
NPDES. *See* National Pollutant Discharge Elimination System (NPDES) permits

O

Observation
    defined, (75221):1, 28
    highlighted in text, (75221):2
    safety, (75221):8–10
Occupational Safety and Health Act, (75201):14
Occupational Safety and Health Administration (OSHA)
    asbestos standard, (75219):26
    competent person, defined, (75226):26
    confined space standard, (75224):41–42
    Days Away, Restricted, or Transferred (DART) case rate, (75226):24
    Employer Rights and Responsibilities Following a Federal OSHA Inspection, (75226):19
    excavation regulations, (75224):13–14
    fire protection requirements, (75222):17
    Focused Inspections Initiative, (75226):23
    forms
        Incident Report, (75225):11
        Injury and Illness Incident Report (Form 301), (75226):7, 9, 12
        Injury and Illness Recordkeeping Forms, (75225):33–34
        Log of Work-Related Injuries and Illnesses, (75225):9
        Log of Work-Related Injuries and Illnesses (Form 300), (75226):7–8
        mistakes in, common, (75225):33–34
        Optional Worksheet to Help You Fill Out the Summary (Form 300A optional), (75226):11
        privacy concerns, (75226):7
        records, keeping, (75226):9
        Report of Injury/Near Miss, (75220):24–28
        Report of Injury/Near Miss Form, (75220):24–28
        Summary of Work-Related Injuries and Illnesses, (75225):10
        Summary of Work-Related Injuries and Illnesses (Form 300A), (75226):7, 9, 10
    General Safety and Health Provisions, (75226):24
    Hazard Communication Standard (HCS), (75224):16, 18
    illnesses and injuries
        classifying illnesses, (75226):6
        classifying injuries, (75226):5
        criteria for classifying, (75226):2
        documentation, OSHA, (75226):7–12
        healthcare recommendations, (75226):3
        lost work days, determining, (75226):5
        medical treatment, defined, (75226):4
        recordable vs. not recordable, (75226):2–6
        recordkeeping requirements, (75226):4
        restricted work, (75226):4–5
        work-related, (75226):2–3
    illnesses and injury forms
        Injury and Illness Incident Report (Form 301), (75226):7, 9, 12
        Log of Work-Related Injuries and Illnesses (Form 300), (75226):7–8

Optional Worksheet to Help You Fill Out the Summary (Form 300A optional), (75226):11
privacy concerns, (75226):7
records, keeping, (75226):9
Summary of Work-Related Injuries and Illnesses (Form 300A), (75226):7, 9, 10
incidents
investigation recommendations, (75225):1
recordable incident rate collection, (75221):13–14
reporting requirements, (75225):8, (75226):2
inspections, (75221):10
"Keeping Workers Safe during Hurricane Cleanup and Recovery," (75219):35
lead Permissible Exposure Limit, (75219):25
medical surveillance programs, (75219):44
mold cleanup PPE recommendations, (75219):35–36
nanomaterials PELs, (75219):38
National Falls Campaign, (75219):21
Permissible Exposure Limit (PEL), (75224):16
permit requirements, (75224):2
PPE standards, (75222):10
problem-solving techniques
change analysis, (75225):17
job safety/job hazard analysis, (75225):17–19
Process Safety Management (PSM) regulations, (75222):20, 25–26, 36
The Process Safety Management of Highly Hazardous Chemicals, (75222):25
recordkeeping
for physical exams, (75219):46
problem areas, (75226):13
recordable incident rate collection, (75221):13–14
requirements, (75201):12, (75226):1–2, 13
violations, (75226):9
regulatory requirements, (75201):14–15
Safety and Health Management Guidelines, (75226):24
safety and health plan requirements, (75226):24
Safety and Health Regulations for Construction, (75222):25
$afety Pays program online calculator, (75201):6
safety training requirements, (75223):2
silica PEL, (75219):30
soil and rock, categories for, (75224):13
Special Government Employees (SGE) program, (75226):26
state programs, (75226):16, 23
trenching regulations, (75224):13–14
voluntary compliance assistance program, (75226):29
Voluntary Protection Programs (VPP), (75226):24, 26
Occupational Safety and Health Administration (OSHA)-approved health and safety programs, (75226):16, 23
Occupational Safety and Health Administration (OSHA) inspections
Checklist Provided as a General Guide, (75226):34–35
citations, (75226):20–21, 28–29
Fatal Four hazards focus, (75226):26
focused vs. wall-to-wall, (75226):23–27
limited, (75226):16
notifications in advance of, (75226):17
penalties, (75226):19, 21–23
post-inspection
abatement/abatement certification, (75226):28
affirmative defenses, (75226):28
appeals, (75226):28
corrective actions, (75226):27–28
hearings, (75226):28
informal conferences, (75226):28
reports to management and employees, (75226):28
preparing for, (75226):17
process
closing conference, (75226):19
opening conference, (75226):18
presentation of inspector credentials, (75226):18
selection of representatives, (75226):18
walk-around inspection, (75226):18–19
purpose of, (75226):16
reasons triggering, (75226):16–17
requirements, (75221):4
roles and responsibilities
CSHOs, (75226):16, 17–19, 20, 23–24, 27
safety technician, (75226):17–19
safety requirements, (75221):4
statistics, (75226):17
triage policy for prioritizing, (75226):16–17
violations
categories of, (75226):19–20
consequences of, (75226):16
correcting, (75226):19, 21, 27–28
defined, (75226):19
fines for, (75226):19, 21–23
good faith compliance, (75226):21
imprisonment resulting from, (75226):21, 23
issues, resolving, (75226):21
most frequent, (75226):20
penalties, (75226):19, 21–23
statistics, (75226):20, 23
wall-to-wall vs. focused, (75226):23–27
Oil Pollution Act, (75201):15, (75222):27
On-the-job training, structured (SOJT), (75220):10, 11, 21
Open-ended questions, (75223):16, 23, 26
Optional Worksheet to Help You Fill Out the Summary (Form 300A optional) (OSHA), (75226):11
OSHA. *See* Occupational Safety and Health Administration (OSHA)
Other employees, (75224):1, 9, 50
Oxygen concentration testing, (75224):34, 35
Oxygen-deficient atmosphere, (75224):23, 26, 50
Oxygen-enriched atmosphere, (75224):24, 26, 50

P

Paraphrase, (75205):1, 3, 17
Pareto diagram, (75225):13, 21, 25
Parts per million (ppm), (75224):23, 32, 50
Pattern, (75220):10, 12, 21
PCBs. *See* Polychlorinated biphenyls (PCBs)
Peer pressure, (75220):4
PEL. *See* Permissible exposure limit (PEL)
Penalties
defined, (75226):1, 9, 33
OSHA inspections, (75226):19, 21–23
People-based safety, (75205):1, 12, 17
Performance barriers, (75220):2–3
Permissible exposure limit (PEL)
defined, (75224):1, 50
OSHA standard, (75224):16
for toxic substances, (75224):32–33
Permit-required confined space (PRCS), (75224):1, 3, 23–24, 50
Permits
aerial lifts, (75224):21
chemical hazards, (75224):16–20
electrical work, (75224):21
excavations and trenching, (75224):13
facility citing, (75224):21
function, (75224):2–3

Permits (*continued*)
    hot work
        Burning - Welding example, (75224):5–6
        common types, (75224):3, 6
        inspections prior to issuing, (75224):7
        requirement for, (75224):3
        signage, (75224):8
    issuing body, (75224):2
    OSHA requirements, (75224):2
    policies, variation in, (75224):3
    pre-inspection requirements, (75224):3, 4
    roles and responsibilities
        controlling employer, (75224):2
        safety technicians, (75224):2
        subcontractors, (75224):2
    time restrictions, (75224):2
    types of, (75224):2
    vehicles, (75224):21
Personal fall arrest systems, (75219):19
Personal protective equipment (PPE)
    compliance, (75224):21
    for emergency situations, (75222):10–14
    for hazard control, (75219):17
    health risks, (75219):34–35
    for mold cleanup, (75219):35–36
    in the site safety plan, (75222):22
Pesticides, (75219):32
Pharyngitis, (75226):6
Physical barriers, (75220):2
Planning Safety Checklist, (75222):43–56
Pneumonitis, (75226):6
    hypersensitivity, (75226):6
Point of operation guards, (75219):19
Poisoning, (75226):6
Polychlorinated biphenyls (PCBs), (75219):23, 28–30, 50
Potentially responsible party (PRP), (75219):23, 50
Power transmission guards, (75219):18
PPE. *See* Personal protective equipment (PPE)
ppm. *See* Parts per million (ppm)
PQF. *See* Pre-qualification form (PQF)
Practicing Perfection®, (75205):12
PRCS. *See* Permit-required confined space (PRCS)
Pre-bid checklist, (75222):1, 36
Pre-job planning checklists, (75219):8
Pre-qualification form (PQF), (75222):37–42
Prerequisite training, (75223):1, 6, 26
Pre-task plan, (75219):9, (75221):12, 22, 28
Prevention through Design (PtD) initiative (NIOSH), (75219):16
Privacy concerns, (75226):7
Probability, (75219):12, 50
Probability of occurrence, (75222):1, 2, 36
Process Safety Management (PSM) regulations (OSHA), (75222):20, 25–26, 36
Profit, (75201):4, 6, 26
PRP. *See* Potentially responsible party (PRP)
PSM. *See* Process Safety Management (PSM) regulations (OSHA)
Psychology, applying to workplace safety, (75205):9
PtD. *See* Prevention through Design (PtD) initiative (NIOSH)
Punishment, (75205):9–10

Q
Questions
    in incident investigations, (75225):5
    open-ended, (75223):16, 23, 26

R
Radiation, electromagnetic, (75219):37
Radio frequency (RF) emissions, (75219):37
RCRA. *See* Resource Conservation and Recovery Act (RCRA)
Receiver, (75205):1, 2–3, 17
Recertification, (75223):2
Recognition, (75220):1, 7, 8, 21
Recordable, (75226):1, 2, 33
Recordable illnesses and injuries, (75226):2–6
Recordable incident rate, (75223):16, 20, 26
Records, safety and health
    audits, (75226):13, 15
    commonly kept, (75226):13
    confidentiality, ensuring, (75226):7, 13
    data gathering and analysis, four factors in, (75226):2
    employee access to, (75226):13
    employer requirements, (75226):1–2
    illnesses and injury
        functions for, (75226):15
        OSHA requirements, (75226):4
    OSHA required, (75226):13
    OSHA requirements, (75226):1–2, 13
    privacy concerns, (75226):7
    problem areas, (75226):13
    purpose of, (75226):1
    records location list, example, (75226):14
    safety technician responsibilities, (75226):13
    subcontractor requirements, (75226):1–2
    time requirements, (75226):9
    violations, (75226):9
Regulatory requirements, safety
    agencies of, (75201):14
    ANSI, (75201):19–20
    ASME, (75201):19–20
    Coast Guard, (75201):20–21
    DHS, (75201):20–21
    DOT CDL Program, (75201):16–18
    EPA, (75201):15–16
    MSHA, (75201):18–19
    NAVFAC, (75201):21
    NFPA, (75201):20
    OSHA, (75201):14–15
    USACE, (75201):21
Reinforcement, (75205):9, (75220):1, 2, 21
Release of energy, (75201):1, 2, 26
Relevance, ARCS Model of building, (75205):8
Remedial training, (75223):1, 26
Reportable quantity (RQ), (75219):23, 32, 50
Report of Injury/Near Miss Form (OSHA), (75220):24–28
Rescue procedures, confined space, (75224):44–45
Residential Lead-Based Paint Reduction Act, (75219):27
Resource Conservation and Recovery Act (RCRA), (75201):15, (75219):39, (75222):26
Respirators
    assigning and using, conditions for, (75222):11, 14
    categories of, (75222):10–11, 12
    for confined spaces, (75224):24–25, 34, 37
    protection factor, (75222):13
    selecting, (75222):11
Respiratory conditions, (75226):6
Respiratory protection
    lead removal, (75219):27, 28
    silica dust exposure, (75219):30, 31
Restricted work activity, (75226):1, 3, 4–5, 33
Reverse chronological order, (75225):1, 3, 25
Reward vs. punishment, (75221):10
RF. *See* Radio frequency (RF) emissions

Rhinitis, (75226):6
Risk
　acceptable level of, (75219):12, 50
　acceptance of, (75220):4–6
　assessing, (75222):1, 2, 36
　calculating, (75219):12–14
　defined, (75201):2, (75219):12
　formula, (75219):12–13
　matrix, (75219):13–14
Rock, categories for, (75224):13
Root cause, (75201):5, (75219):14–16, (75221):12, 19, 20, 21, 28, (75225):15–16
Root cause analysis, (75221):20, (75225):1, 2, 25
RQ. See Reportable quantity (RQ)
Run-off, storm water, (75219):23, 31, 50

S

Safe Drinking Water Act (SDWA), (75201):15, (75222):27
Safety
　audits, (75221):1, 8, 28
　Behavioral-based (BBS), (75205):1, 9, 17
　confined space, (75224):24
　defined, (75201):2
　hot work, (75224):7
　inspections, (75201):12, (75219):8, 10, (75221):1, 3, 4–7, 28
　people-based, (75205):1, 12, 17
　responsibility for, (75221):2, (75224):3
　terminology, (75201):2
　trenching, (75224):13–15
Safety, audits, (75201):12
Safety and Health Management Guidelines (OSHA), (75226):24
Safety coordination, site safety plan, (75222):22–23
Safety data, analysis of, (75221):22–24, 25
Safety data sheet (SDS), (75224):1, 16–19, 50
Safety devices for hazard control, (75219):17–19
Safety gate interlocks, (75219):19
Safety management programs, (75221):22, 24
Safety meetings
　formal
　　appropriate topics, (75223):17
　　attendance sheets, (75223):22
　　audience pet peeves, (75223):17
　　costs of, (75223):20
　　experts, using, (75223):17
　　group composition, (75223):17
　　guest speakers in, (75223):17
　　length and location, (75223):16, 17
　　materials, (75223):17
　　meeting site preparations, (75223):16–17
　　participant preparation, (75223):17, 27
　　presenter preparation, (75223):17
　　presenter self-evaluation, (75223):21
　　recordkeeping, (75223):20
　　sign-in form, (75223):22
　　speaker evaluation, (75223):18–19
　function, (75201):11
　length, (75223):16
　location, (75223):16
　toolbox/tailgate talks
　　Five Ps for Successful Safety Talks, (75223):22
　　function, (75223):20
　　length and location, (75223):20
　　participation, encouraging, (75223):27
　　preparations, (75223):22
　　required, (75223):20
　topics, (75223):16
Safety nets, (75219):18
Safety observation, (75221):8–10
$afety Pays program online calculator, (75201):6
Safety performance, improving
　administrative controls, (75219):17
　best practices, (75219):20
　engineering/design, (75219):16–17
　job site interventions, (75219):20–23
　personal protective equipment, (75219):17
　safety devices, (75219):17–19
Safety performance, measuring
　incidence rates, (75221):12–15
　injuries, analyzing, (75221):16, 22
　lagging indicators for, (75221):25
　predictive methods
　　audits, (75221):17–18
　　behavior observations, (75221):17
　　inspections, (75221):17–18
　　near miss/near hit reporting, (75221):18–20
　Workers' Compensation Experience Modification Rate (EMR), (75221):17
Safety plan, site-wide
　access considerations, (75222):20–21
　adjacent hazards, (75222):21
　administration, (75222):23
　benefits of, (75222):20
　emergency procedures, (75222):22
　environmental considerations, (75222):21
　generic, (75222):57–60
　material and tool storage, (75222):20
　personnel
　　communicating to, (75222):23
　　health and safety, (75222):22
　　pedestrian traffic, (75222):21
　　sanitary needs, (75222):21
　　training, (75222):23
　preparing a, (75222):2–3
　regulatory requirements
　　ANSI, (75222):30
　　ASME, (75222):30
　　Coast Guard, (75222):31
　　DHS, (75222):31
　　DOT, (75222):28
　　EPA, (75222):26–28
　　MSHA, (75222):28, 30
　　NAVFAC, (75222):31–32
　　NFPA, (75222):30
　　OSHA, (75222):25–26
　　for training, (75222):32
　　USACE, (75222):31–32
　required, (75222):20
　requirements, (75222):3
　safety coordination, (75222):22–23
　scope of work, (75222):21–22
　site location and layout in the, (75222):20
　standards, (75222):20
　traffic patterns, (75222):20–21
　training, (75222):23
　utilities, locating, (75222):21
Safety planning, pre-bid
　benefits of, (75222):1
　checklists for, (75222):1, 43–56
　general format, (75222):1
　hazard controls, (75222):2
　hazard identification, (75222):1
　risk assessment, (75222):2
　site safety plan
　　preparing a, (75222):2–3
　　requirements, (75222):3

Safety programs
- administration costs, (75201):8
- audits, (75201):11–12
- basic components, (75201):9
- best practices, (75201):13
- chemical-specific
  - asbestos, (75219):24–25, 26
  - lead, (75219):25, 27–28
  - PCBs, (75219):28–30
  - silica, (75219):30
- development resources, (75201):9
- emergency action plans, (75201):11
- employee involvement, (75201):11–12
- evaluation and follow-up, (75201):12
- incident investigation and analysis, (75201):12
- inspections, (75201):11–12
- media management, (75201):11
- for mine operators, (75201):19
- policies and procedures
  - alcohol abuse, (75201):9–10
  - drug abuse, (75201):9–10
  - policy statement, (75201):9
  - safety, (75201):10
- recordkeeping, (75201):12
- savings from, (75201):8

Safety pyramid, (75221):2, 3
Safety stand-downs, (75219):21
Safety technicians
- responsibilities
  - advising site management, (75224):2
  - audits, (75201):12, (75221):2
  - behavioral problems, identifying and handling, (75219):20, (75220):1
  - competent person, (75226):26
  - confined spaces, (75224):24
  - coordination, (75221):2
  - data analysis, (75225):13
  - examples of, (75224):3
  - excavations and trenching, (75224):13
  - hazard identification and control, (75219):1
  - hazard identification and correction, (75224):3
  - hazardous waste management, (75219):41
  - hot work permits, (75224):7
  - incident investigations, (75225):1, 13
  - inspections, (75201):12, (75221):2
  - observation, (75221):2
  - OSHA inspections, (75226):17–19
  - OSHA rule enforcement, (75226):16
  - possible, (75224):2
  - PPE compliance, (75224):21
  - recordkeeping, (75226):13
  - resource to site management and personnel, (75201):11–12, (75224):3
  - safety meetings, (75201):11
  - for training, (75224):2, 9
  - training requirements, (75224):2
  - work permits, (75224):2–3
- skills needed, (75221):2–3

Sanitation, (75222):21
Satisfaction, ARCS Model of fostering, (75205):8–9
SCBA. *See* Self-contained breathing apparatus (SCBA)
Scenario workshops, (75223):5
Scope of work, site safety plan, (75222):21–22
SDS. *See* Safety data sheet (SDS)
SDWA. *See* Safe Drinking Water Act (SDWA)
Secondary containment, (75219):23, 39, 50
Self-contained breathing apparatus (SCBA), (75222):11, 12, 14, (75224):23, 25, 34, 50

Sender, (75205):1, 2–3, 17
SGE. *See* Special Government Employees (SGE) program (OSHA)
Silica, (75219):23, 24, 30, 50
Silicosis, (75219):23, 30, 50, (75226):6
Simulators, (75223):1, 4, 26
Site safety plan
- access considerations, (75222):20–21
- adjacent hazards, (75222):21
- administration, (75222):23
- benefits of, (75222):20
- emergency procedures, (75222):22
- environmental considerations, (75222):21
- generic, (75222):57–60
- material and tool storage, (75222):20
- personnel
  - communicating to, (75222):23
  - health and safety, (75222):22
  - pedestrian traffic, (75222):21
  - sanitary needs, (75222):21
  - training, (75222):23
- preparing a, (75222):2–3
- regulatory requirements
  - ANSI, (75222):30
  - ASME, (75222):30
  - Coast Guard, (75222):31
  - DHS, (75222):31
  - DOT, (75222):28
  - EPA, (75222):26–28
  - MSHA, (75222):28, 30
  - NAVFAC, (75222):31–32
  - NFPA, (75222):30
  - OSHA, (75222):25–26
  - for training, (75222):32
  - USACE, (75222):31–32
- required, (75222):20
- requirements, (75222):3
- safety coordination, (75222):22–23
- scope of work, (75222):21–22
- site location and layout in the, (75222):20
- standards, (75222):20
- traffic patterns, (75222):20–21
- training, (75222):23
- utilities, locating, (75222):21

Site-specific orientation, (75223):1
Skin diseases/disorders, (75226):6
Small quantity generators (SQGs), (75219):39
Social conformity, (75220):4
Soil, categories for, (75224):13–14
SOJT. *See* Structured on-the-job training (SOJT)
Spark hazards, (75224):33
Special Government Employees (SGE) program (OSHA), (75226):26
Specific gravity, (75224):23, 33, 50
SQGs. *See* Small quantity generators (SQGs)
Standard for Electrical Safety in the Workplace (NFPA), (75224):21
Standards, consensus, (75201):14, 26
Stand-down, (75219):21
State programs, OSHA-approved, (75226):16, 23
Static electricity, (75224):27–28
STOP WORK policy, (75219):12
Storm water run-off, (75219):23, 31, 50
Stress, workplace, (75205):6
Structured on-the-job training (SOJT), (75220):10, 11, 21
Summary of Work-Related Injuries and Illnesses (Form 300A) (OSHA), (75225):10, (75226):7, 9, 10
Superfund, (75219):23, 32, 50, (75222):26

Supervisors
  confined space roles and duties, (75224):43
  safety training, (75223):2
Surface mines, (75201):14, 18, 26
Swales, (75219):23, 31, 50

T
Tagout, (75224):1, 8, 50
Tagout devices, (75224):9–10
Take 5 for Safety form, (75224):4
Target organ, (75219):23, 44, 50
Task, (75220):1, 2, 21
Task barrier, (75220):3
Task safety analysis (TSA), (75223):2
Taurine, (75220):7
Threats, planning for, (75222):16
Three Levels of Incident Causation Model, (75225):13–16, 25
Time line in incident investigations, (75225):3–4
Toolbox/tailgate talks
  Five Ps for Successful Safety Talks, (75223):22
  function, (75223):20
  length and location, (75223):20
  participation, encouraging, (75223):27
  preparations, (75223):22
  required, (75223):20
Tools, confined spaces, (75224):33–34
Total Recordable (TRIR) incidence rate, (75221):12
Toxic atmospheres, (75224):32–33
Toxic Substances Control Act (TSCA), (75201):15, (75219):28, (75222):27
Traffic patterns in site safety plan, (75222):20–21
Training
  cross-training, (75219):12, 17, 50
  emergency action plans, (75222):8, 10
  emergency training, confined space
    inadequate illumination, (75224):43
    loss of air line, (75224):43
    loss of communications, (75224):43
    rescue procedures, (75224):44–45
  in hazardous materials, (75219):46
  lockout/tagout (LOTO), (75224):9
  PSM contractors, (75222):25–26
  regulatory requirements, (75201):11
  safety technician responsibility for, (75224):2, 9
  for safety technicians, (75224):2
  site safety plan, (75222):23
Training, safety
  basic components, (75201):11
  benefits of, (75201):10–11
  classroom management
    adherence to the schedule, (75223):11–12
    handling difficult situations, (75223):12
    pace of instruction, (75223):11
  concluding
    course completion verification, (75223):12
    course objectives review, (75223):12
    participant course evaluation, (75223):12, 13–14
    participant goals review, (75223):12
    performance assessment, (75223):12
  course evaluations, (75223):13–14
  delivery
    administrative details, (75223):10
    course introduction, (75223):9–10
    greeting participants, (75223):9
    ground rules, (75223):10
    icebreakers, (75223):10
    instructor introduction, (75223):9–10
    objectives review, (75223):10–11
  engaging learners, methods for
    audiovisuals, (75223):5–6, 7
    demonstration, (75223):3–4
    hands-on practice, (75223):3–4
    know-show-do, (75223):3
    mentoring programs, (75223):4–5
    scenario workshops, (75223):5
    sequence logically, (75223):3
    what-why-how, (75223):3
  follow-up, (75223):2
  as a leading indicator, (75221):21–22
  OSHA requirements, (75223):2
  preparations
    audiovisuals, (75223):9
    basics, (75223):8
    coordinating arrangements, (75223):6, 8
    coordinating room and participant availability, (75223):8
    identifying and inviting the correct people, (75223):6, 8
    materials, (75223):9
    obtaining and securing a classroom, (75223):8
    space arrangements, (75223):8–9
  prerequisite, (75223):1, 6, 26
  remedial, (75223):1, 26
  retention statistics, (75223):3
  types of
    continuing, (75223):2
    craft-specific, (75223):2
    new employee orientation, (75223):1
    recertification, (75223):2
    site-specific orientation, (75223):1
    supervisory, (75223):2
Trapped workers, rescuing, (75222):15–16
Trench, (75224):1, 50
Trenching, (75224):13–15
Trend, (75220):10, 12, 21
Trend analysis, (75225):21
TRIR. *See* Total Recordable (TRIR) incidence rate
TSA. *See* Task safety analysis (TSA)
TSCA. *See* Toxic Substances Control Act (TSCA)
Tuberculosis, (75226):6
Turnover, (75205):13–14

U
UEL. *See* Upper explosive limit (UEL)
UFL. *See* Upper flammable limit (UFL)
Ulcers, (75226):1, 6, 33
Uncontrolled release of energy, (75219):1, 50
Underground Storage Tank Regulation (EPA), (75201):15, (75222):26
Uniform Hazardous Waste Manifest, (75219):41–43
Universal precautions, (75222):4, 15, 36
Unsafe acts. *See also* Indirect cause
  blame, assigning, (75221):10
  common, (75201):4–5, (75221):8
  defined, (75201):2
Unsafe behavior, (75221):1, 8, 28
Unsafe conditions
  common, (75201):5, (75221):9
  defined, (75201):2
  identifying and correcting, (75221):3–4
Unwarrantable failures, (75201):14, 19, 26
Upper explosive limit (UEL), (75224):23, 31, 50
Upper flammable limit (UFL), (75224):23, 31, 50
USACE. *See* US Army Corps of Engineers (USACE/CoE)
US Army Corps of Engineers (USACE/CoE), (75201):21, (75222):31–32
USCG. *See* US Coast Guard (USCG)

US Coast Guard (USCG)
   medical surveillance programs, (75219):44
   regulatory requirements, (75201):20–21, (75222):31
US Fish and Wildlife Service (USFWS), (75219):33, (75222):27–28
USFWS. *See* US Fish and Wildlife Service (USFWS)
Utilities, locating, (75222):21
Utility lines, locating, (75224):14

V

Vapor density, (75224):23, 26, 50
Vehicles permits, (75224):21
Vehicle weight, gross unloaded (GUVW), (75201):14, 17, 26
Vehicle weight rating, gross (GVWR), (75201):14, 17, 26
Ventilation
   confined space, (75224):37–39
   hot work, (75224):7
Verbal communication, (75205):1, 2–3, 17
Violations
   citations, (75226):9
   defined, (75226):1, 33
   OSHA inspections
      categories of, (75226):19–20
      consequences of, (75226):16
      correcting, (75226):19, 21, 27–28
      defined, (75226):19
      fines for, (75226):19, 21–23
      good faith compliance, (75226):21
      imprisonment resulting from, (75226):21, 23
      issues, resolving, (75226):21
      most frequent, (75226):20
      penalties, (75226):19, 21–23
      statistics, (75226):20, 23
   penalty, (75226):9
Visual communication, (75205):1, 3–5, 17
Vitamins, B12, (75220):7
VOCs. *See* Volatile organic compounds (VOCs)
Volatile organic compounds (VOCs), (75224):36
Voluntary compliance assistance program (OSHA), (75226):29
Voluntary Protection Programs (VPP) (OSHA), (75226):24, 26
VPP. *See* Voluntary Protection Programs (VPP) (OSHA)

W

Wall-to-wall vs. focused inspections, (75226):23–27
Warning devices for hazard control, (75219):17
Waste, hazardous, (75219):23, 50
Waste manifest, hazardous waste, (75219):23, 41, 50
Water pollution regulation, (75219):31
WC. *See* Workers' compensation (WC) insurance
Weather hazards, (75219):34–35
Weather related emergencies, (75222):16
Wetlands, (75219):23, 31, 50
What-why-how, (75223):1, 3, 26
Why Method, (75225):13, 16–17, 25
Worker rotation for hazard control, (75219):17
Workers' compensation (WC) insurance, (75201):6, 7
Workers' Compensation Experience Modification Rate (EMR), (75201):4, 7, 26, (75221):12, 17, 28
Workplace
   alcohol use, (75220):7
   camera restrictions in the, (75225):2
   injury statistics, (75201):1
Workplace statistics
   fatalities, (75201):1, (75219):35, (75221):1, 12
   incidence rates
      calculating, (75221):14
      Days Away, Restricted, or Transferred (DART), (75221):12
      defined, (75221):12, 28
      Lost Time (LTR), (75221):12
      recordable, (75221):13–15, (75223):16, 20, 26
      recording, OSHA requirements, (75221):13–14
      statistics, evaluating, (75221):14–15
      Total Recordable (TRIR), (75221):12
   incidents, (75201):1, 10, 16
   injuries, (75201):1
Work site, multi-employer, (75226):29
Written communication, (75205):1, 3–5, 17